ARCHITECTURE COMPETITION WORKS

建筑竞标方案

凤凰空间 · 上海　编

凤凰出版传媒集团
江苏人民出版社
凤凰空间
IFENGSPACE

Contents 目录

Cultural & Sports 文化与体育

Office & Commercial 办公与商业

Public Facilities 公共设施

Residence 住宅

City Complex 城市综合体

ARCHITECTURE COMPETITION WORKS

Cultural & Sports
文化与体育

Richard Wagner Museum Extension

理查德·瓦格纳博物馆扩建 OFIS

Pre-invited Competition Project Proposal 2010
Bayreuth, Germany
2010年德国拜罗伊特预邀方案赛参赛方案

Credits
Project Team: Rok Oman, Spela Videcnik, Ana Kosi, Janez Martinčič, Katja Aljaz, Janja del Linz, Andrej Gregorič, Ieva Cicenaite, st.arch, Grzegorz , strowski, st.arch

这个美丽的历史建筑将迎来它的“二次革命”。我们计划在被三座迷人历史建筑环绕的主广场下设置入口和展示厅。

新介入空间只有四个天窗可以窥其内部，这赋予了原来的广场一种新的意义，他们让原来的建筑依然可见，同时也标志着地下发生了一些不同寻常的事。他们从视觉上将旧的部分和新的部分连接在了一起。天窗会给室内提供自然弥散的光线，对于不同的展示布局和各种不同的项目来说再适合不过了。四个天窗形成一个碗形，给外部创造了一个多功能的集会空间。在这里游客可以休息和冥想，同时还能举行小型活动和音乐会。通过斜坡的主入口强调了原来的轴线，这个斜坡使得原有的树木得以保存。游客进入有商店和售票处的主厅，接着能看见厅内有通向多功能厅和展厅以及咖啡馆和博物馆教学部分的入口。多功能厅是新大楼的心脏，天窗就在它上面，透过它可以在广场上看见室内的空间。展厅是一个清晰的环形轨道，围绕着多功能厅。

另一个嵌入空间是“园丁之家”新的扩建区域，新的学术中心就坐落在里面。

新的方案恢复了新的通讯室，连接着中心，广场和地下博物馆。扩建内容与现存的外观分离，并且全部由玻璃制成。这样一来，原有的建筑可以保持原来的情形，所有新的展览空间和大厅都建在地下。学术中心在原“园丁之家”中。

理查德·瓦格纳博物馆新建筑的主要材料是混凝土和玻璃。强化混凝土能够支撑起大型的，开放型地下广场空间和四个大型天窗，同时在地上的玻璃建筑和原有的建筑巧妙地融合在一起。

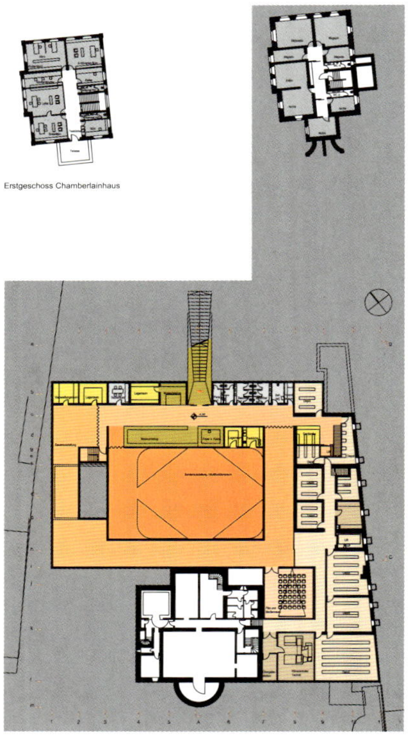
Erstgeschoss Chamberlainhaus
Untergeschoss

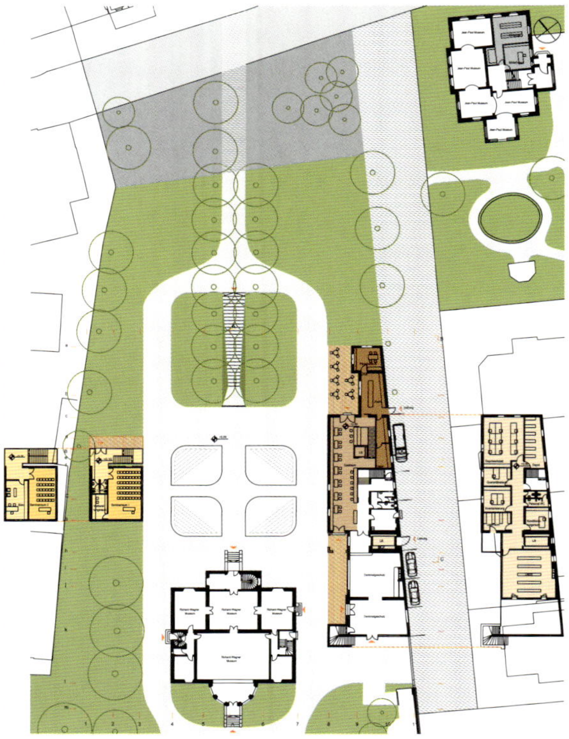
Erdgeschoss

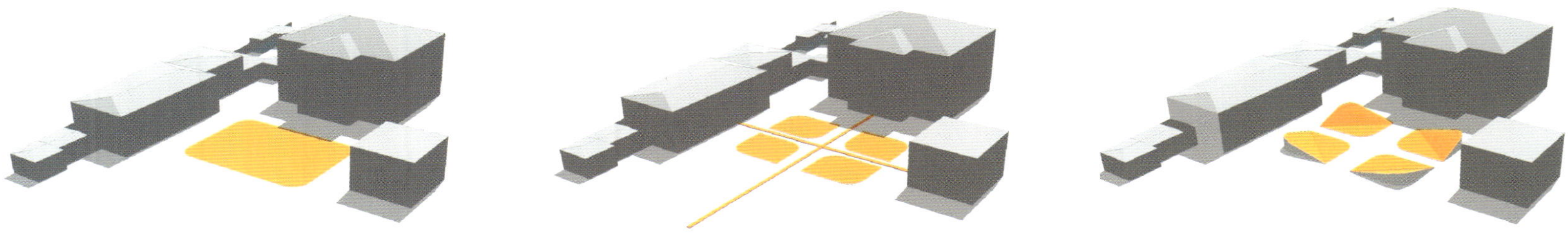

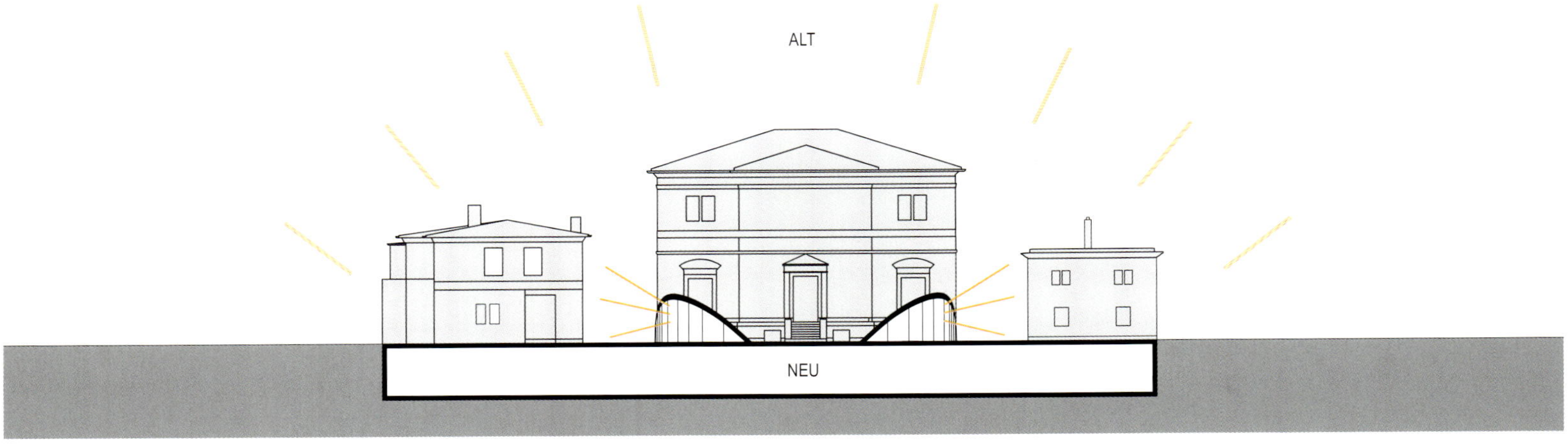

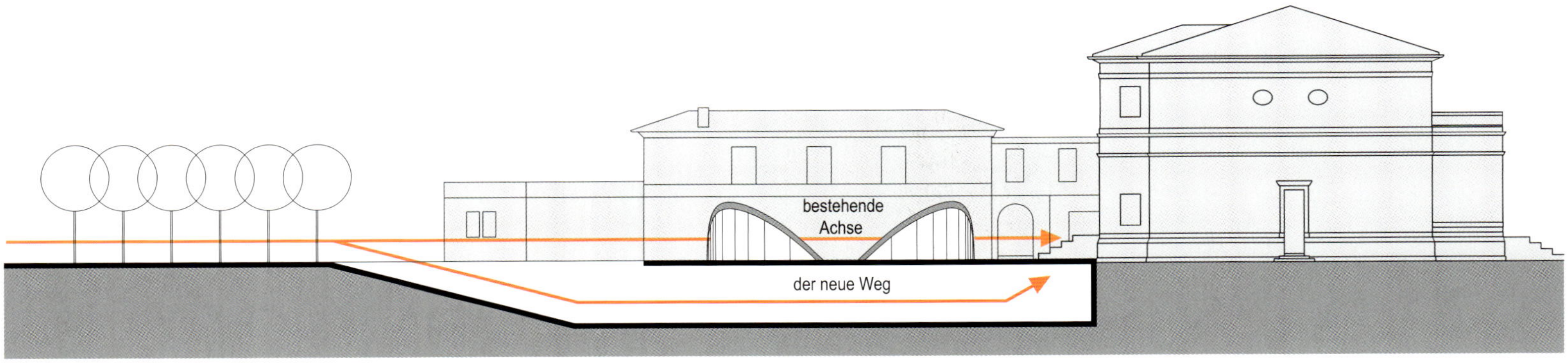

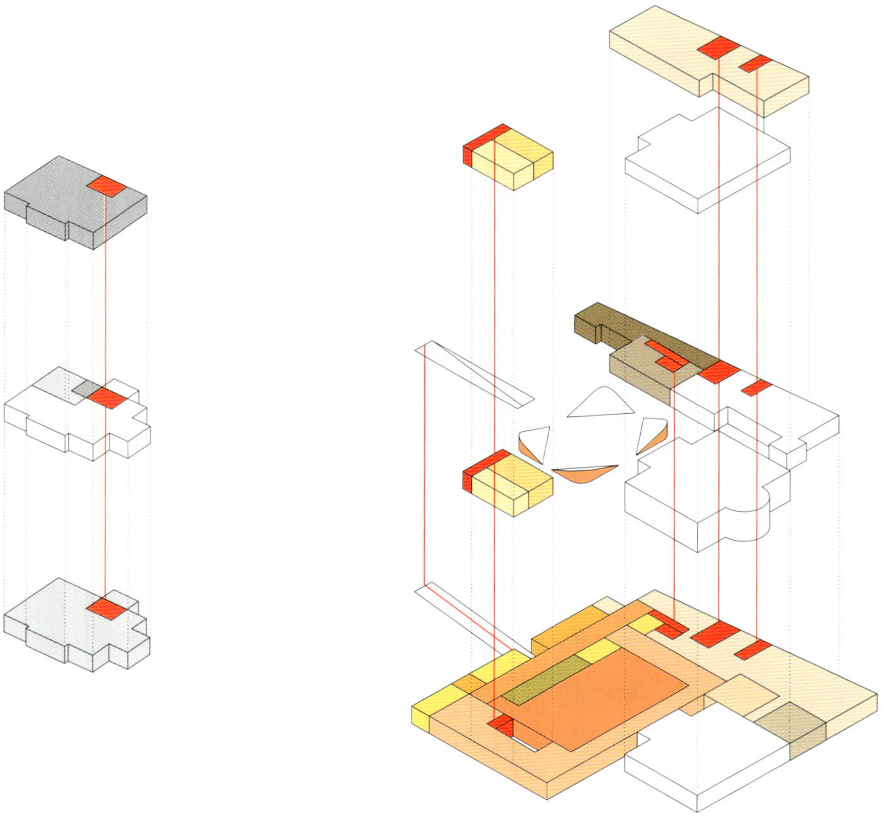

OBENGESCHOSS

Sondernausstellung
Museumspädagogik_Seminarraum
Museumspedagogik_Service
Administration

ERDGESCHOSS

Sondernausstellung
Museumsgastronomie_Gäste
Museumsgastronomie_Service
Museumspädagogik_Seminarraum
Museumspedagogik_Service
Bibliothek
Administration

UNTERGESCHOSS

Dauerausstellung
Sonderausstellung
Medienraum
Foyer
Kasse, Museumsshop
Sanitärräume
Service
Depot
Haustechnik
Magazin, Archiv

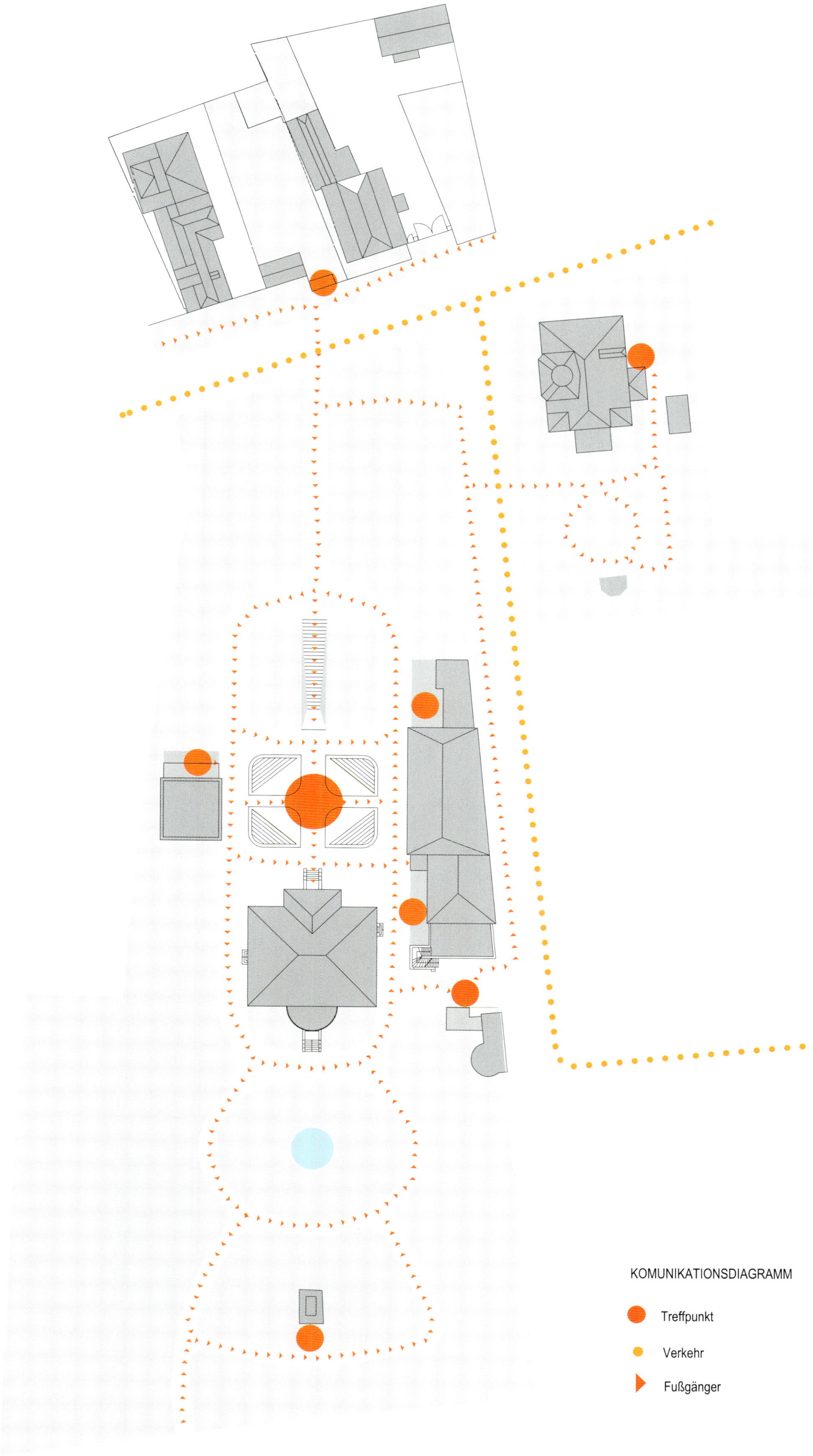
KOMUNIKATIONSDIAGRAMM
Treffpunkt
Verkehr
Fußgänger

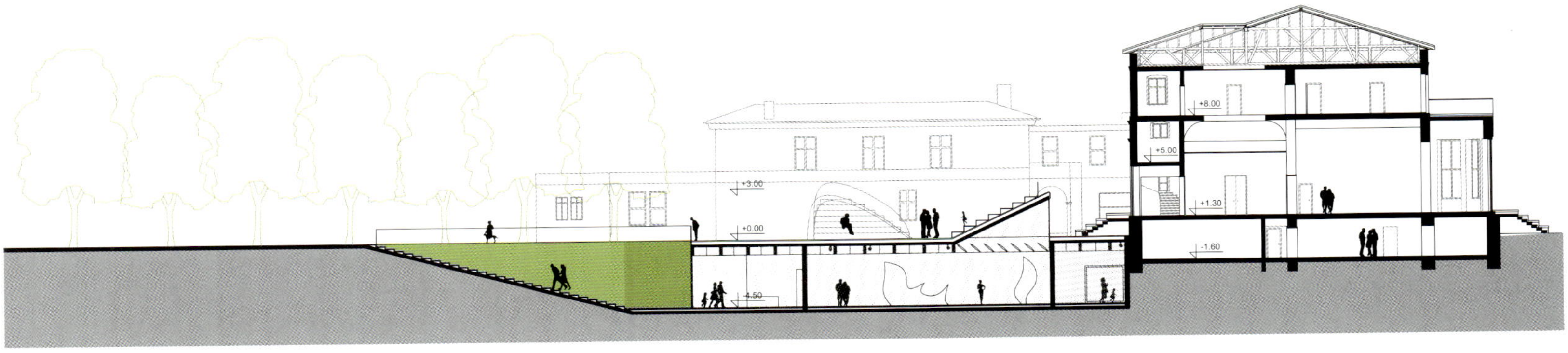
+8.00
+5.00
+3.00
+1.30
+0.00
-1.60
-4.50

A-A

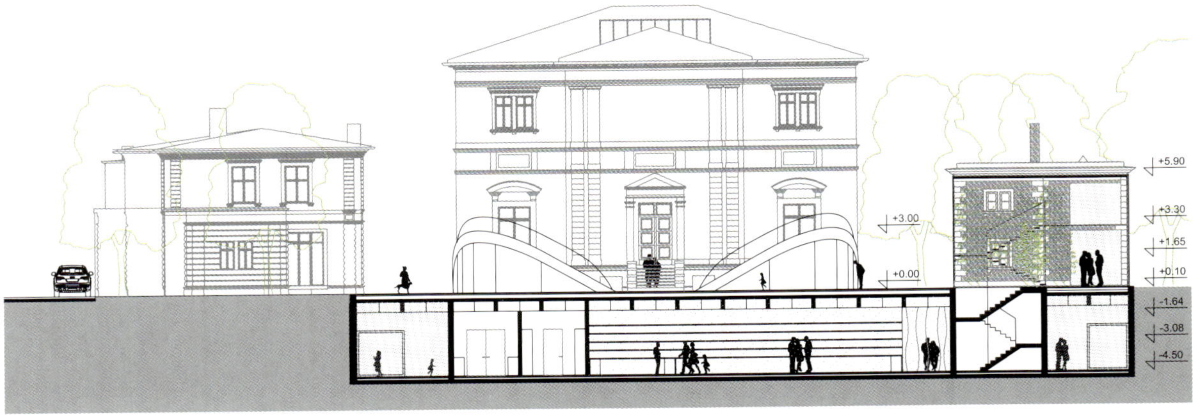
+5.90
+3.30
+3.00
+1.65
+0.10
+0.00
-1.64
-3.08
-4.50

B-B

+3.30
+3.00
+3.90
+0.00
+0.00
+0.00
-4.50
-4.50
-4.50

C-C

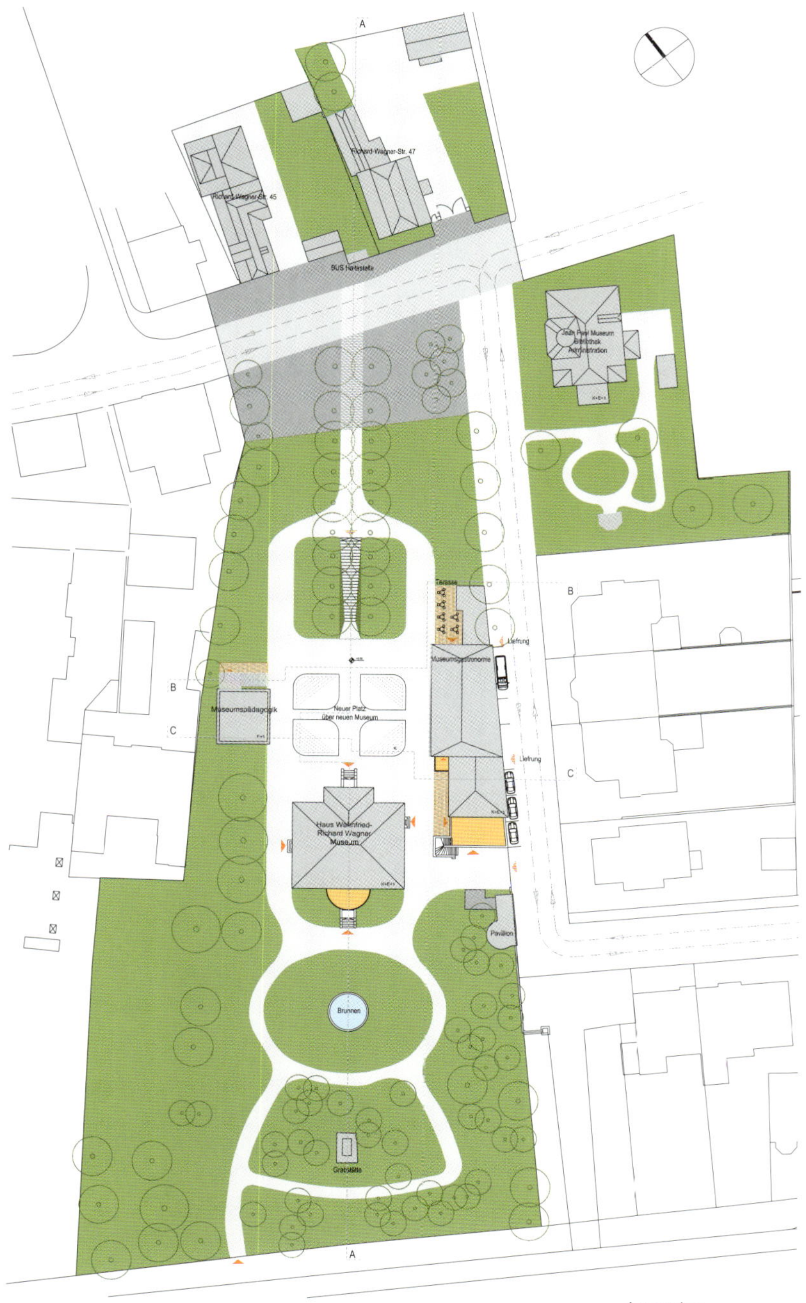

Lageplan

The beautiful historical site represents a chance for subtle unique rearrangement with its own identity. We propose to arrange entrance and exhibition hall under the main square, which is surrounded by three charming, historical buildings.

The new space intervention is seen only by four gentle skylights. which give a new meaning to the existing square. they let the existing architecture to still be seen, as well as they alert that there is something else going on underground. They visually connect old part of the complex with the new one. Skylight would provide natural diffuse light inside the interior that is perfect for different exhibition layouts and variable events. The shape of the skylights creates a form of a bowl that creates exterior multifunctional gathering space. This is a space where visitors can rest and contemplate and it also offers possibility for small outdoor events and concerts. Main entrance is through the ramp that emphasizes existing axis. The ramp allows existing trees to be kept. One enters the main lobby where shop and ticket sales are organised. From the lobby there are entrances to Sonderausstellung / Multifunktionsraum and Dauerausstellung, as well as to the cafe and the pedagogical part of the museum. The Sonderausstellung is in the heart of the new building. The skylights are above it, so one can see inside this space from the square. The Dauerasustelung is arranged as a clear, circular path around the Sonderausstellung space.

Another intervention in the space is the new extension to the Gärtnershaus in which the new Seminar center is arranged.

The new proposal reinstates new communication box, which connects the Center, the square and the underground museum. The extension is very subtile, detached from the existing facade and made completely out of glass. It lets the existing house to remain in original state. All the new exhibition spaces as well as the lobby is arranged underground. From the lobby there is the direct connection to the cafe and the Pedagogikscentrum on the groundfloor. The Pedagogical center is positioned in the existing Gärtnerhaus.

The main materials used for the new building of the Richard Wagner Museum are concrete and glass. Preinforced concrete enables to organise the big, open underground space with four big skylights, while on the ground level, glass creates a subtile, gentle constructions, which do not oppose the existing architecture.

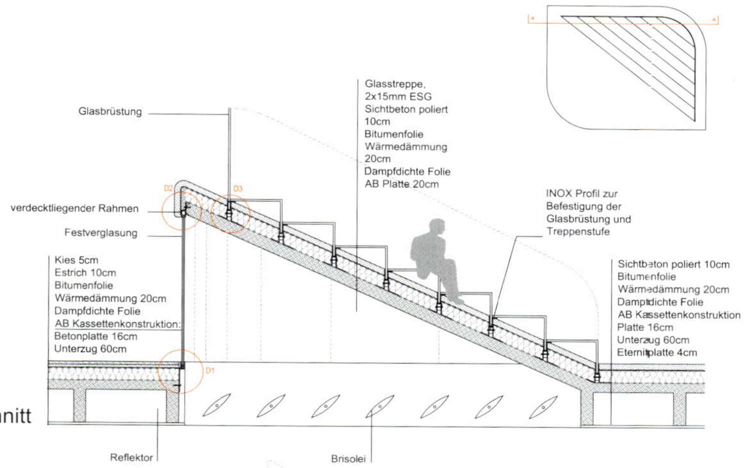

Fasadenschnitt

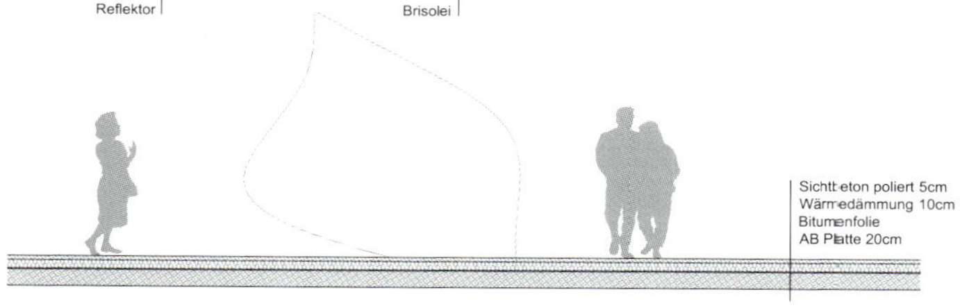

Fasadenschnitt _ A-A

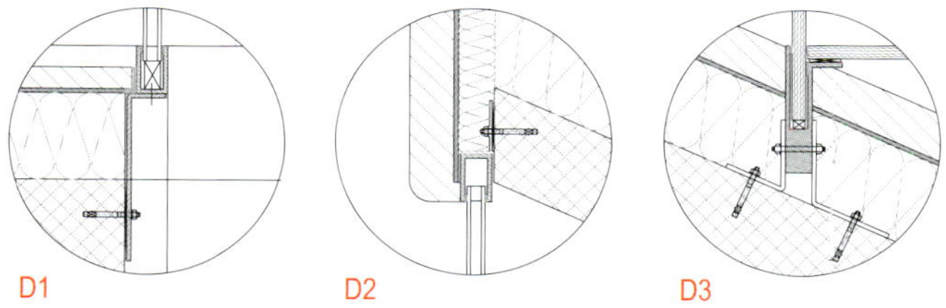

D1 D2 D3

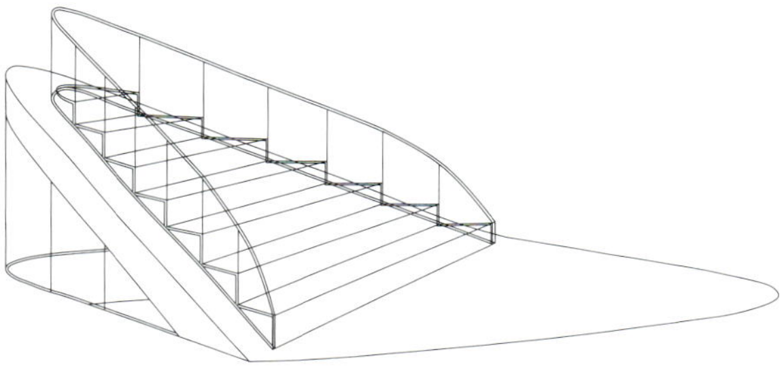

Zubau Ansicht M

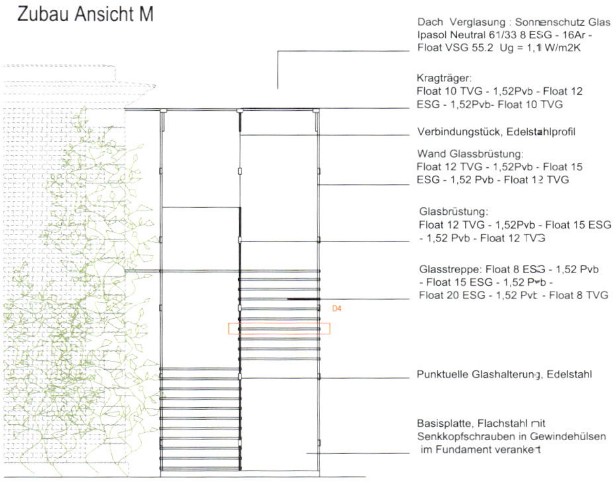

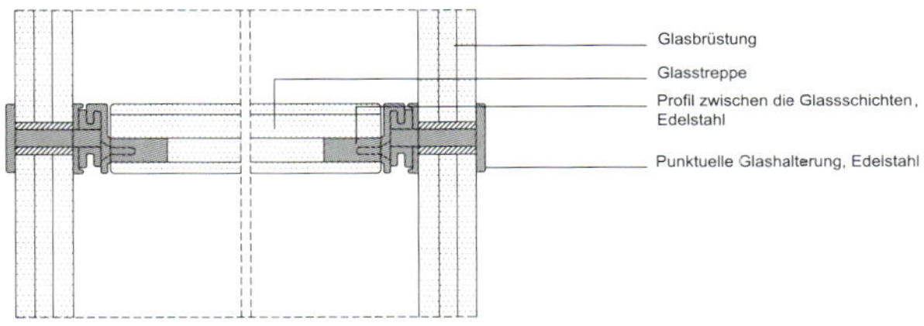

Verfastung der Glasstreppe zur Glasbrüstung

Hunan Provincial Museum

湖南省博物馆 n!studio

International Short-listed Competition for Reconstruction and Expansion of Hunan Provincial Museum
湖南省博物馆改扩建工程概念国际竞赛参赛方案

Credits
Design: n!studio Susanna Ferrini, Antonello Stella
Team: Marco Biuzzi, Simone Bove, Valerio Rompietti, Angelo Venturi, Alessandro Macaluso, Anna Laura Pinto, Lucio Zappalorti, Celine Bosal, Maria Carla Lini, Francesca Viglione

Collaborators: Fabio Bonfà - Engram Studios (Rendering); 3DSIGN (Animation, Rendering)
Model: Marco Galofaro
Year: 2011

湖南省博物馆改扩建项目为创造一座杰出的当代博物馆提供了一个独特机会，该博物馆将在中国和国际舞台上扮演重要角色。

在城市构造与市内最大的综合性公园中间嵌入这样一座新建筑，使湖南省博物馆成为一个象征符号、长沙市和湖南省的图标、湖南文化的地标建筑和基本组成，以及诠释湖南历史和文化的重要窗口。

设计方案中，项目区域内将没有任何边界将其与周围环境隔开，博物馆本身是一个开放的综合性建筑，欢迎人们进入其公共空间，公共空间内有花园、池塘和临时的露天展览和活动区域。博物馆区域将成为城市生活的直接延伸。博物馆将成为城市和公园的交叉点，同时，建筑物本身也成为一个城市地标，以及烈士陵园的一个新大门。

因此，新博物馆建筑围绕现有陈列大楼发展起来，如同一座堡垒围绕其财富建造。现有陈列大楼完工于1999年，围绕这座建筑设计了一座矩形环状建筑，这样的设计是受到楚国文化阶梯构造建筑特点的启发，并尽最大可能利用建筑语言阐释湖南文化。

新博物馆是一个几何学的规则形状建筑，与无规则的丘陵景观相并列，从而强调自然与人造、建筑与景观的关系。

现有陈列大楼的新远景以一种惊人的方式向参观者展开。从东风路方向看上去，丘陵景观和新博物馆悬浮拐角就如同现有大楼的一层外框，而从烈士陵园方向看，大楼成为入口散步长廊的背景。

"新"与"旧"的关系是以"正确的距离"这一概念为基础的。对现有陈列大楼保持极高的敬意的同时，湖南省博物馆的扩建为长沙市和湖南省呈现出一个新标志。

新博物馆围绕现有陈列大楼建起，通过一系列绿色露台为参观者提供一个欣赏建筑物和公园的新奇视角。建筑物平滑而连续的线条将游人与年嘉湖、浏阳河和烈士陵园的自然环境融为一体。

朝向现有陈列大楼的新建筑的内侧大部分被玻璃外墙所覆盖，不仅丰富了视觉关系，而且在夜晚如同一盏古灯照耀着现有陈列大楼。

建筑物的内侧完全由乳白色玻璃面板覆盖，这种乳白色玻璃面板就是为人熟知的U形玻璃。这种设计可让自然光在白天过滤到博物馆内部，晚上照耀现有陈列大楼。自然光被广泛地应用在公共空间和临时展览区域。根据日光照明情况和内部空间需求，自然光的强度可通过自动遮阳板的移动适当地进行调节。特制透明玻璃板的巧妙位置，以一种出人意料的方式向参观者提供了在现有陈列大楼上的新视角，同时，在顶层参观者可以欣赏到周围的迷人景观。

SECTION 5-5
剖面图 5-5

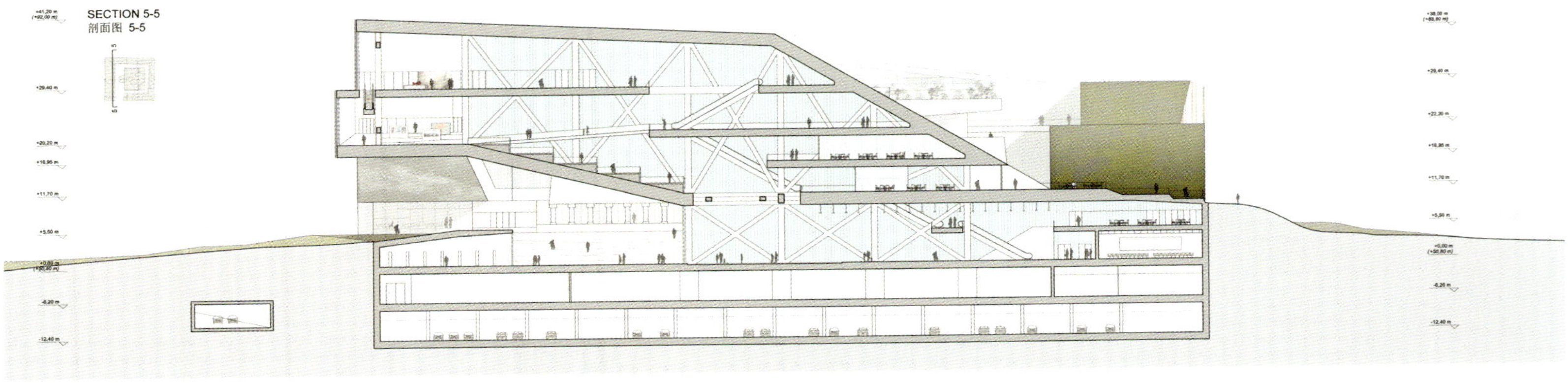

SECTION 3-3
剖面图 3-3

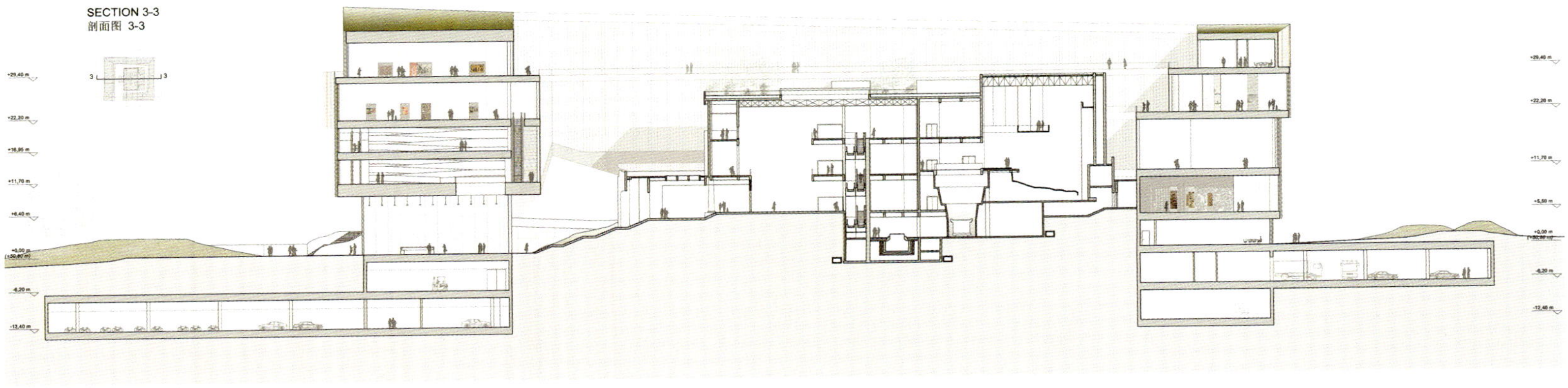

ELEVATION SOUTH
正面图 南

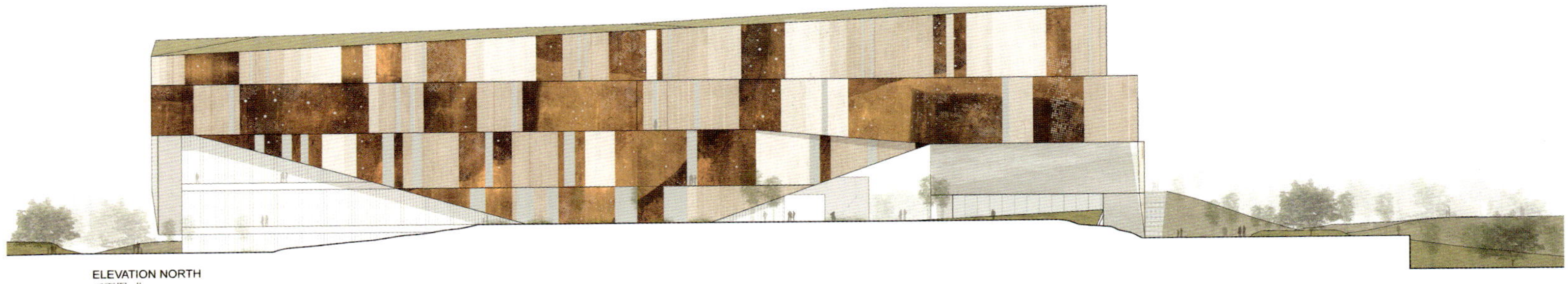

ELEVATION NORTH
正面图 北

PLAN LEVEL 2
2楼

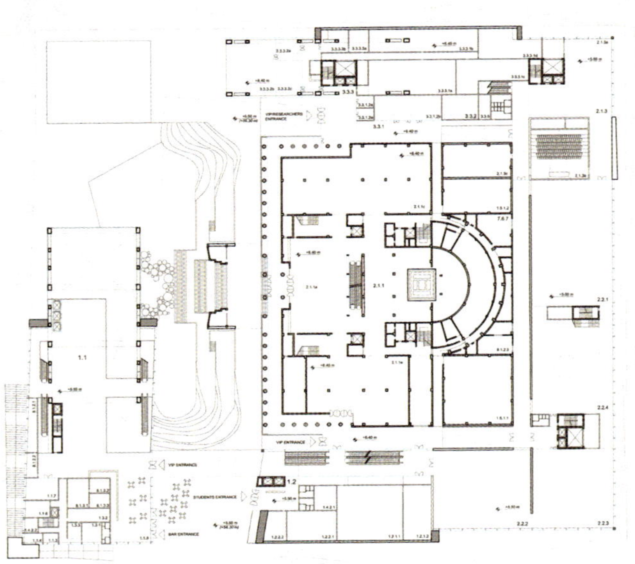

SITEPLAN
位置图

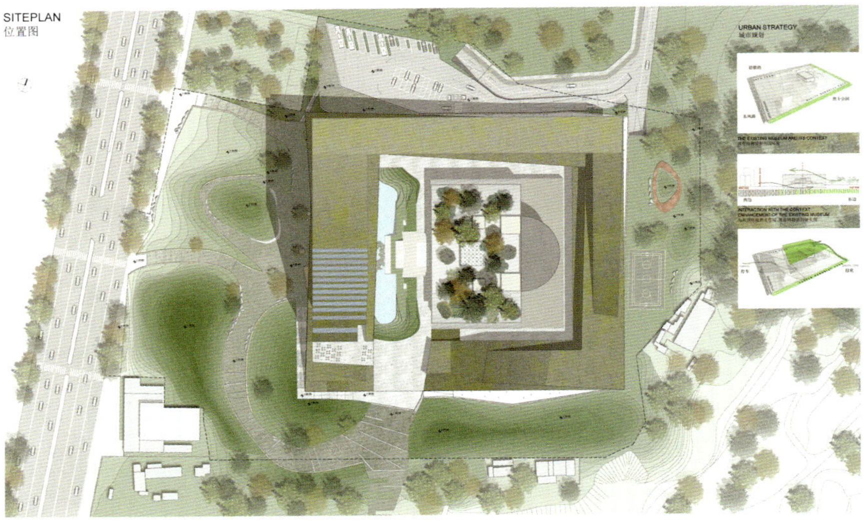

The project for the Reconstruction and Expansion of Hunan Provincial Museum represents the unique opportunity to create an outstanding contemporary museum with a prominent role in the Chinese and International scene.
The insertion of the new building between the urban fabric of the city and the city's largest comprehensive park makes Hunan Provincial Museum a symbol and an icon of the city and the province itself, a landmark and fundamental component of Hunan culture as well as an important window through which to interpret the history and culture of Hunan.
The project area has no borders to separate it from its surroundings. The museum presents itself as an open complex that welcomes people inside its public areas with gardens, ponds and temporary open-air exhibitions and events. The museum area becomes a direct extension of city life. It plays the role of urban junction between the city and the park, and the building itself becomes at the same time an urban landmark and a new gate to the Martyrs' Park.
The new building thus grows around the existing museum like a fortress grows around its treasure. A square ring-shape construction that entirely surrounds the exhibition building is completed in 1999, which was inspired by the architectural features of terraced structures of Chu-state culture and made large use of architectural language to interpret Hunan Culture.
The new museum building grows with a geometrical and regular shape juxtaposed to the irregular hilly image of the landscape thus emphasizing the relationship between the natural and artificial, between architecture and landscape.
New perspectives of the existing buildings surprisingly open up to the visitor.

From Dongfeng Road the existing building appears framed between the hilly landscape and the suspended corner of the new museum whereas from the Martyr's Park it becomes the background of the entrance promenade.

The relationship between "new" and "old" is based on the concept of "the right distance". Though extremely respectful of the existing building, the Extension of Hunan Provincial Museum presents itself as a new symbol for the city of Changsha and the Hunan Province.

The new museum grows around the existing one and through a series of green terraces provides visitors with unexpected points of view both towards the building and the park. It emerges from its site in a smooth, continuous gesture that connects people to the natural atmosphere of Nianjia Lake, Liuyang River and the Martyr's park.

The inner side of the new building facing the existing one is mostly covered with glass facades that enrich visual relationship and, at night, enlighten the existing building like an ancient lantern.

The inner side of the building is totally covered with opalescent glass panels, better known as U-Glass. This solution allows natural light to filter inside the museum spaces during the day and illuminate the existing building at night. Natural light is widely employed in the public areas and temporary exhibition areas. The intensity of natural light is properly regulated by automatic shading panels that move according to the conditions of solar radiations and the requirements of interior spaces. Extra clear glass panels will be carefully located to provide unexpected views on the existing building and, at upper floors, charming views on the surrounding landscape.

Oslo Museum

奥斯陆博物馆 PATTERNS

International Competition, 2009
2009年国际竞赛参赛方案

Credits
Project Type: Museum
Location: Oslo, Norway
Area: 33,600 m^2
Client: City of Oslo

Principals in charge : Marcelo Spina and Georgina Huljich
Project Designer : Geoffrey Elander
Project Assistants : Bobby Frederick, Robie Eleazer, Dan Manino, Giuliana Haro and Joshua Moratto

设计理念——博物馆中庭

考虑到项目的规模及其重要的文化意义，PATTERNS建筑事务所将在挪威国家艺术、建筑和设计博物馆（NMAAD）的设计方案中，提出奥斯陆这座城市新的文化特性。除了其丰富的收藏品展示、现代展览与其他公共的活动，博物馆还将通过参与活跃的公共活动深入到奥斯陆城市其他各式各样的城市活动中。

该地公共生活空间与民用建筑形成一个密集的大型网络，分布于周围。设计的目的是要把博物馆、馆内画廊及其公共活动空间融入这个网络并成为城市的标志性建筑。这个城市网络将现今保留的车站建筑连接在一起，同时将奥斯陆市政厅、Radhuspladsen广场、Pipervika广场与阿肯修弗斯城堡统一搁置在河岸另一边。由于休息区的空隙与表面华丽的装饰，博物馆中庭和邻近博物馆不同，没有了那种纪念性与严肃感，反而像是中世纪的建筑。外表的修饰使其形式多样化，具有明显的现代特征，极富动感，显得活跃、狂热。

如此一来，新的NMAAD在功能上，超越了博物馆本身的性质，为城市提供了更加丰富的活动空间；在外观上，与现存的历史建筑及西部铁路城市空间交错，形成了形式上的相近与视觉上的交叉。新的博物馆在推广一种“集聚地”与“共同记忆”概念的同时，还增强了其丰富背景的视觉效果。

整体结构

项目规划有利于公共功能区域沿泳滩的布置，加强了滨水区的公共环路的使用。所有的画廊呈堆叠式分布于不同的楼层，其中，现代艺术画廊邻近前厅，以双层展示空间呈现，适合大型雕塑展览，但也可根据需要灵活改变其功能。

从接待室可乘一个大型的货运电梯直接到达储藏室。储藏室位于接待室上面的二层，呈线性分布。储藏室上方是双层展示空间的工作室，这样可使艺术品收藏与运输效率最大化。大楼管理处与图书馆外形如同两个相对的翅膀，面向阿肯桥与另一个接待室，垂直矗立在Radhuspladsen广场上。

▼ 全景效果图 Aerial rendering view

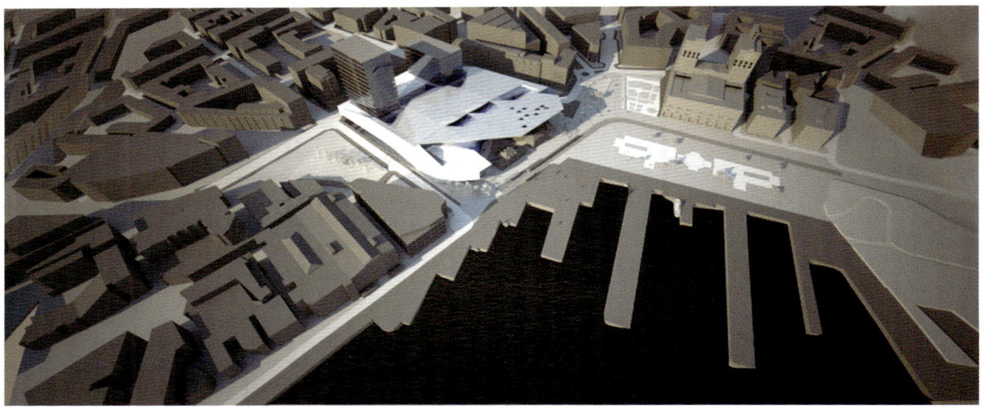

▼ 纵剖面透视图：每一层都运用了混凝土与结节密肋楼盖体系
Perspective longitudinal section: A concrete and nodular waffle slab system is used along every floor

向中庭望去：中庭极具感染力与空间诱惑力。它可作为一个户外的活动空间，在这里，可以举办各种公共活动与重大事项。其外表面向内凹，在空间上造成一种围绕的感觉，这使中庭形成一种文化沉浸的氛围

View towards atrium space: The museum atrium creates a warmly affective and alluring sense of spatial inclusion. It works as an outdoor performance space whereas a number of public activities and events could take place. Spatially conceived as a surrounding experience by means of its near envelope concave shape, the atrium behaves as a deeply immersive environment

从户外公共区域向博物馆广场望去：进入博物馆大堂的入口带领参观者穿越博物馆广场。这个广场在博物馆的步道上设计了大型雕塑，从这里可以清晰地看到凌驾于正前方古老车站建筑的滨水区

View from public open area towards museum plaza. The museum entry sequence leading to the main lobby takes visitors through the museum plaza, which houses large public sculptures to the museum promenade at a level that can overlook the waterfront over the old station buildings right in front

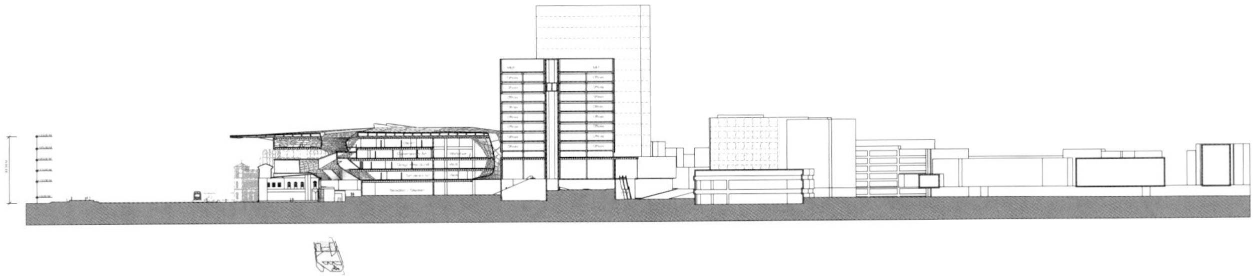

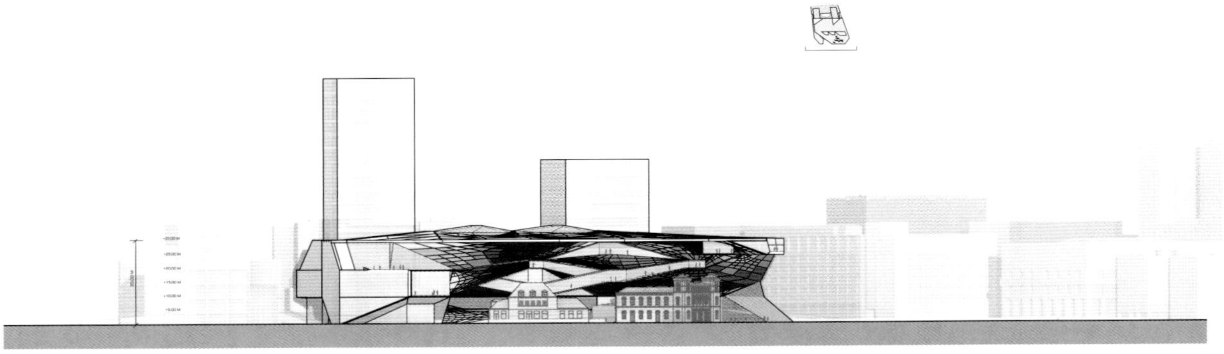

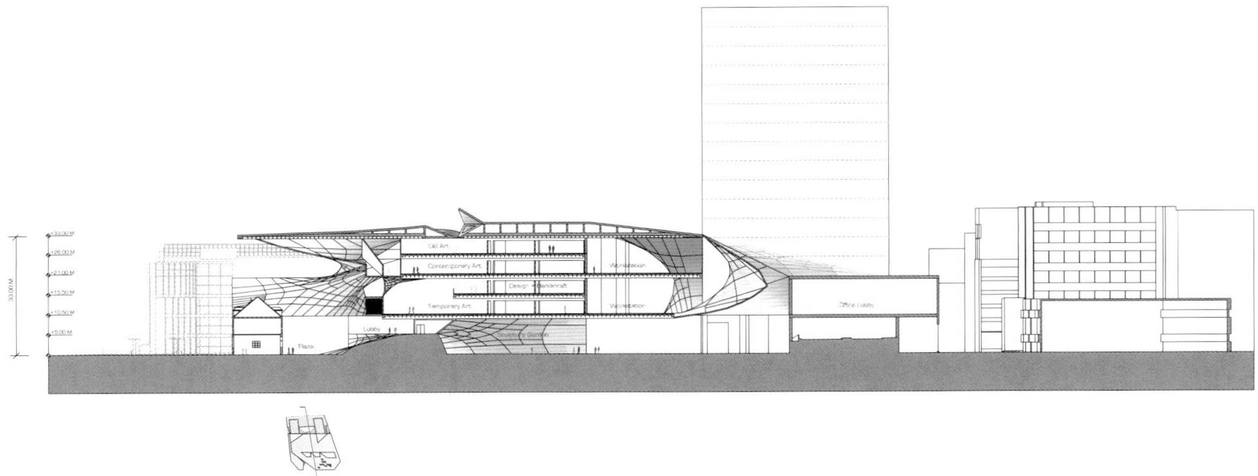
+33.00 M
+26.00 M
+21.00 M
+15.50 M
+10.50 M
+5.00 M
33.00 M
Old Art
Contemporary Art
Workstation
Design + Handicraft
Temporary Art
Workstation
Office Lobby
Lobby
Plaza

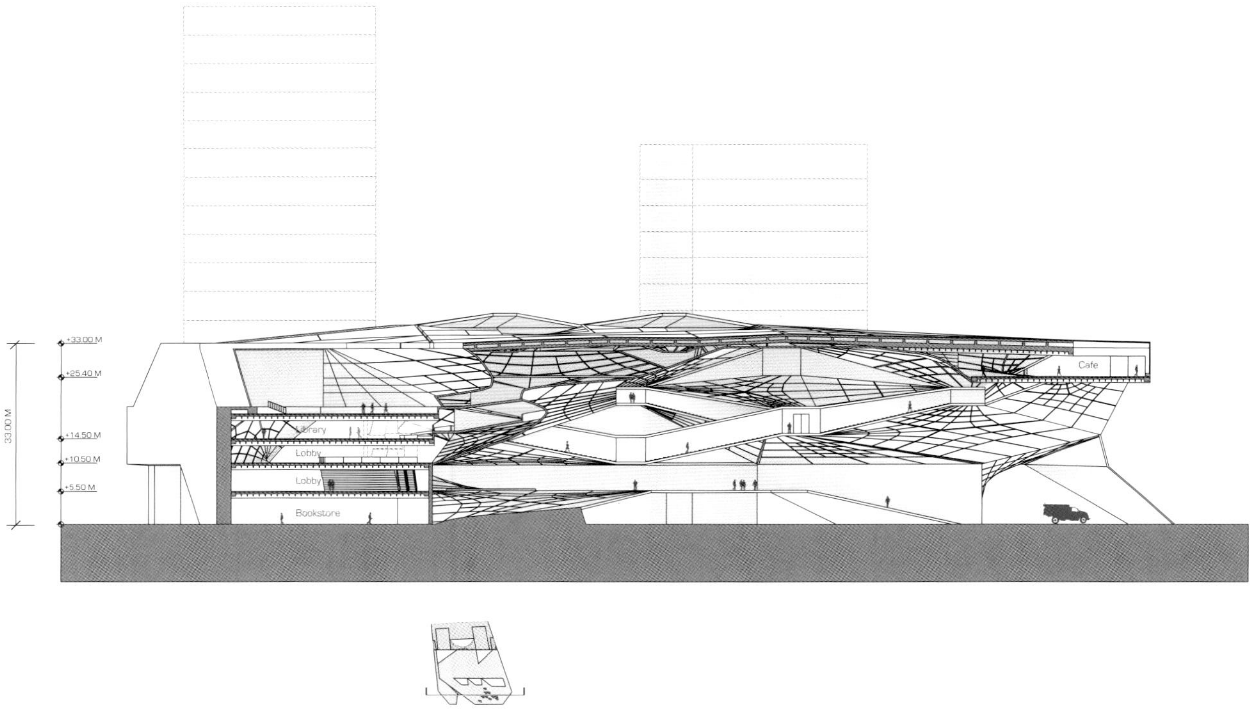
+33.00 M
+25.40 M
+14.50 M
+10.50 M
+5.50 M
33.00 M
Cafe
Library
Lobby
Lobby
Bookstore

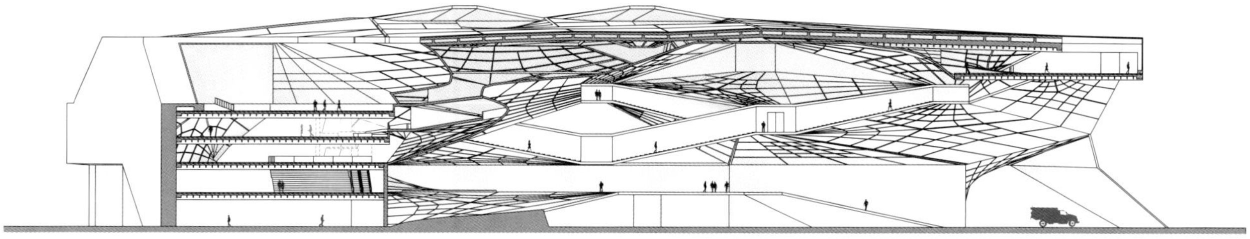

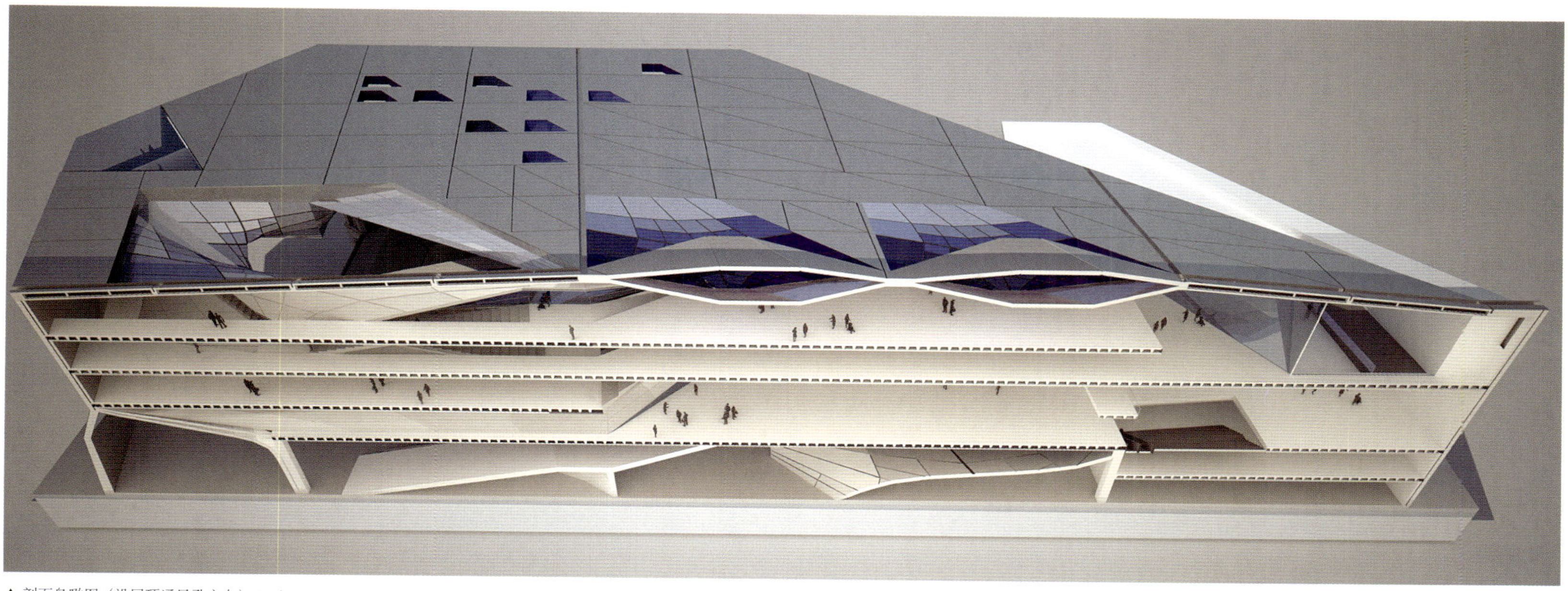

▲ 剖面鸟瞰图（沿屋顶通风孔方向） Birds eye sectional view along roof apertures

▲ 纵剖面透视图：每一层都运用了混凝土与结节密肋楼盖体系 Perspective longitudinal section: A concrete and nodular waffle slab system is used along every floor

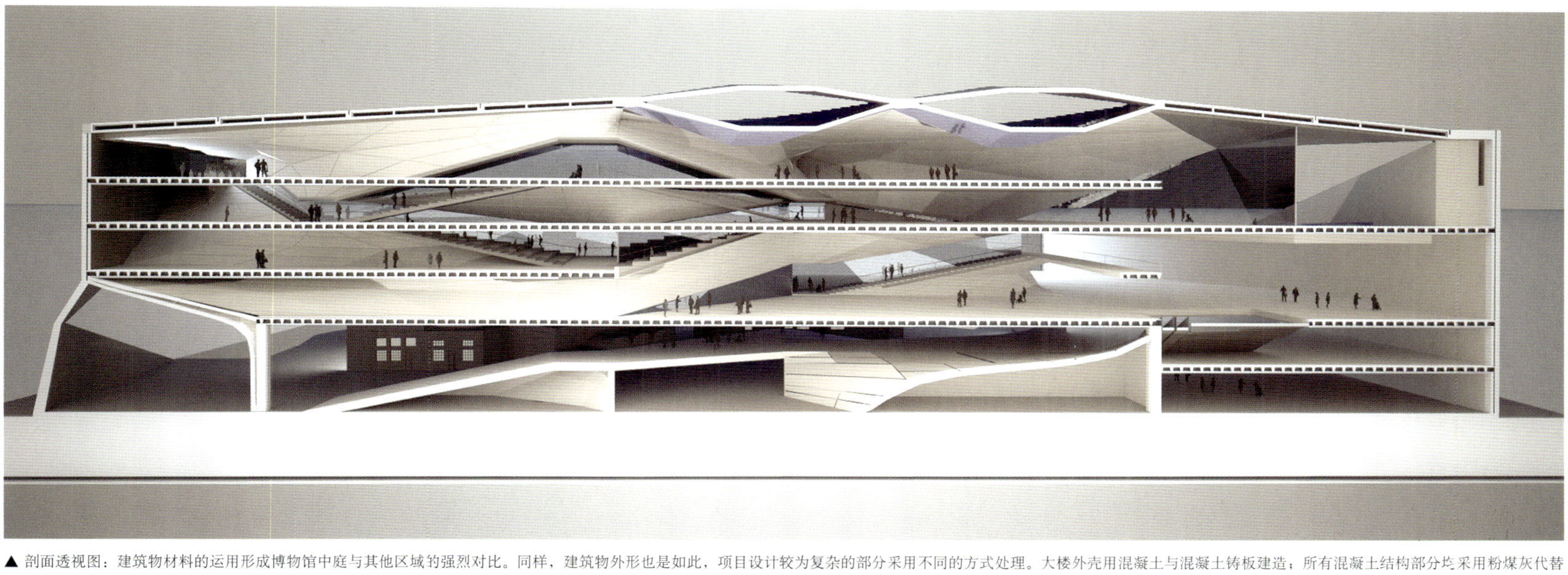

▲ 剖面透视图：建筑物材料的运用形成博物馆中庭与其他区域的强烈对比。同样，建筑物外形也是如此，项目设计较为复杂的部分采用不同的方式处理。大楼外壳用混凝土与混凝土铸板建造；所有混凝土结构部分均采用粉煤灰代替水泥，大大减少了环境污染

Perspective longitudinal section: The materiality of the building responds to the need of creating a stark contrast between the public open atrium and the rest of the project. Similarly as in the form and geometry of the building, the inflected and convoluted portions of the project are treated in a distinct manner. The exterior shell of the building will be finished in concrete and concrete cast panels. Fly ash will be utilized as a replacement for cement in all concrete portions of the project. This diminishes the environmental impact of the project

Concept I Urban Atrium

Our proposal for the National Museum of Art, Architecture and Design [NMAAD] at Vestbanen aims to produce a new urbanity within Oslo—one that engages and understands both the scale of the site and the significant cultural importance of the program. The new museum will be a vital institution, not only by promoting high culture through its considerable collection, temporary exhibitions and other public programs, but also by advancing other kinds of urban activities through its vigorous engagement with public life.

A network of public spaces and civic buildings is housed under one large civic condenser. Our intention is to make the museum, its galleries, and other public spaces part of a vibrant physical and visual network of civic icons and urban spaces. This network ties together the existing historical Station buildings, while framing spectacular views of Oslo City Hall, Radhuspladsen Square, Pipervika, and Akershus Fortress on the other side of the water. The resulting inner square, or Museum Atrium, is not a typical monumental and oppressive space next to many national museums, but one that is medieval in nature due to the interstitial quality of its meandering space, baroque in its appearance given its superficial manifold composition of its circulation, and remarkably contemporary in its dynamic spirit as a result of its frenetic, voyeuristic and extroverted character.

By producing a new civic space that transcends the museum function and by inducing a physical proximity and visual interface with existing historic buildings and urban spaces within the Vestbanen site, the new NMAAD enhances the visibility of its rich context while promoting a sense of both "Meeting Place" and "Collective Memory" as described in the project brief.

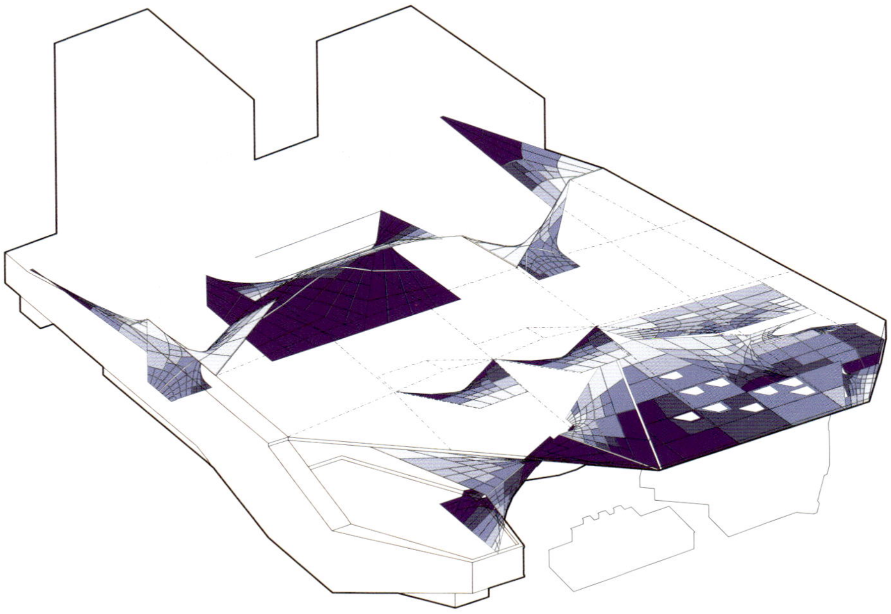

Programmatic Organization

The program distribution privileges the deployment of public functions along the Stranden, consolidating the public circuit within the waterfront. Galleries are stacked in four successive levels, with the contemporary art gallery located adjacent to the lobby, and featuring a double-height space appropriate for exhibiting large sculpture while allowing for functional flexibility to be subdivided if necessary.

With direct access from the reception area and served by a large freight elevator, vaults are located on two subsequent levels and organized in a linear fashion. Double-height space workshops are placed on top of the vaults, allowing for maximum efficiency in the handling and transport of art. Two opposed wings, which are perpendicular to Radhuspladsen Square, house administration and library towards Akerbrigge and reception on the other.

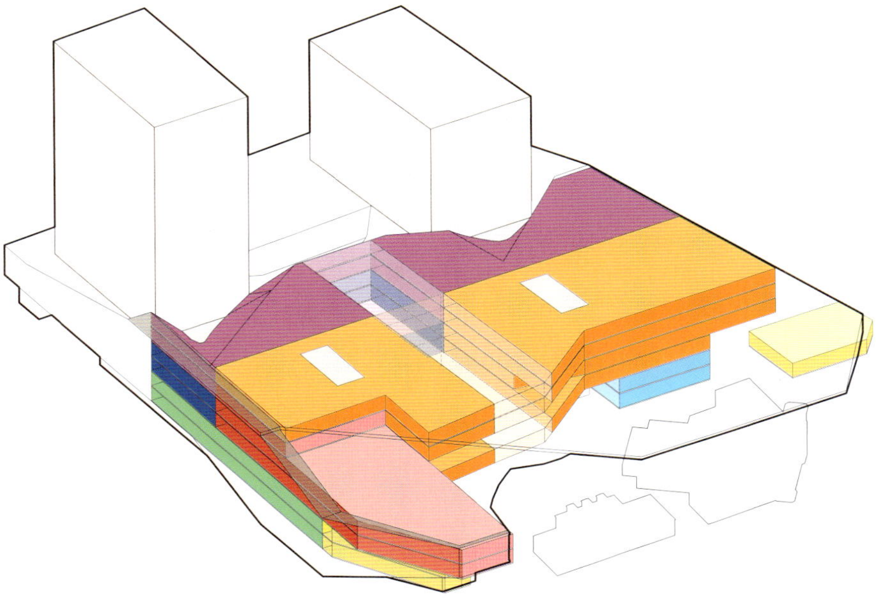

▼ 室内效果图（由展览画廊向现代艺术画廊望去）Interior rendering view from exhibition gallery towards contemporary art space

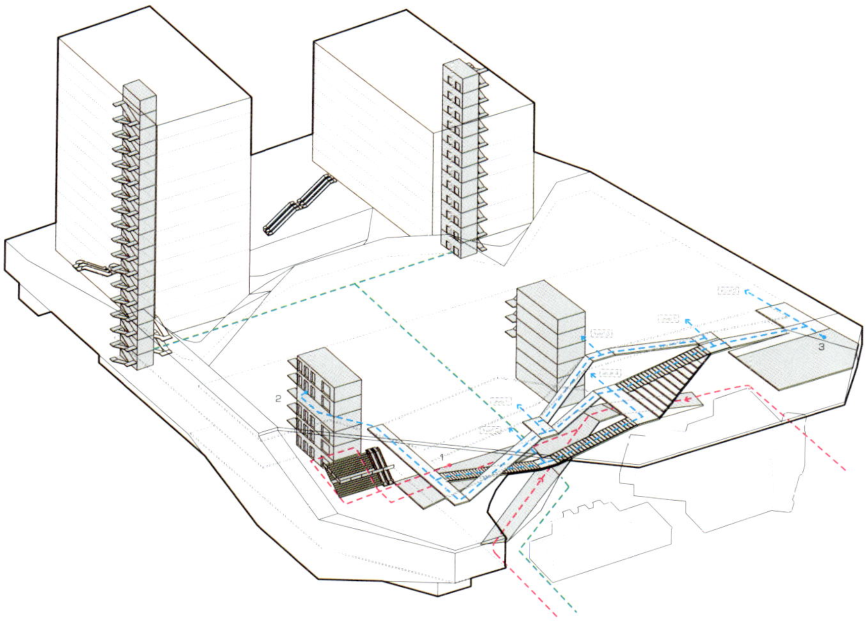

1
2
3

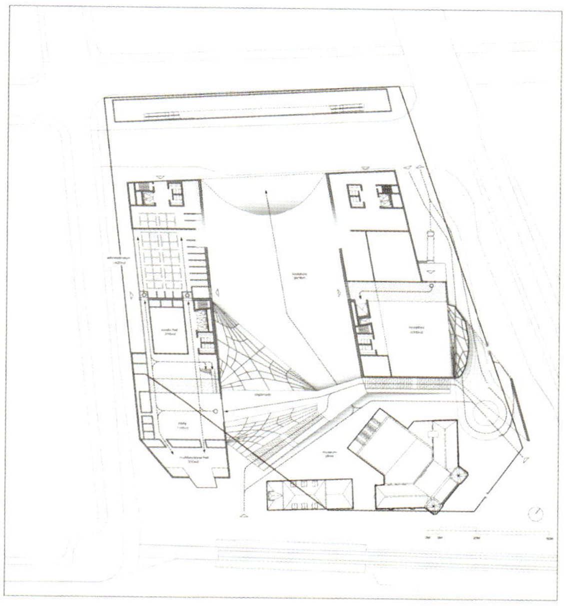

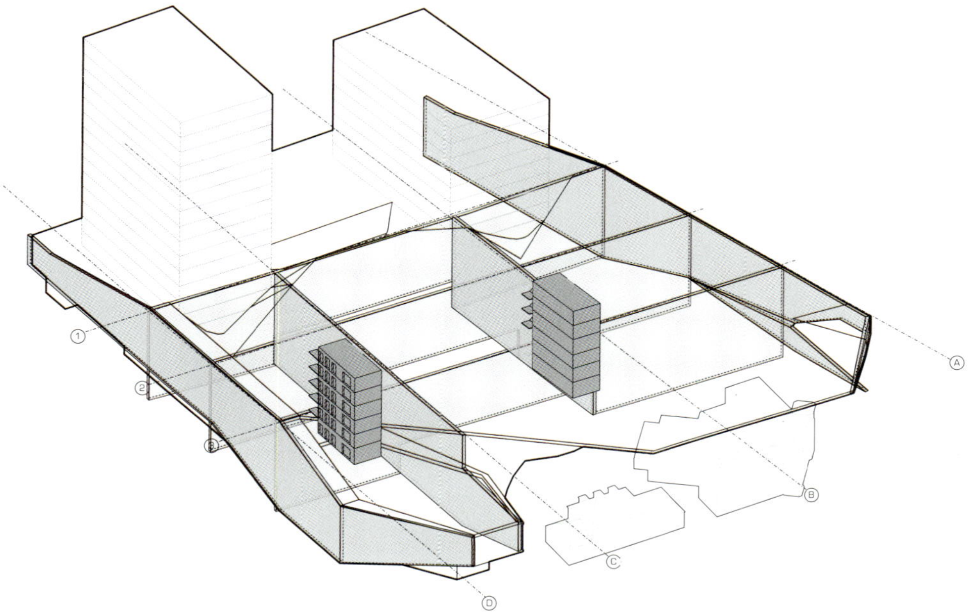

1
2
3
A
B
C
D

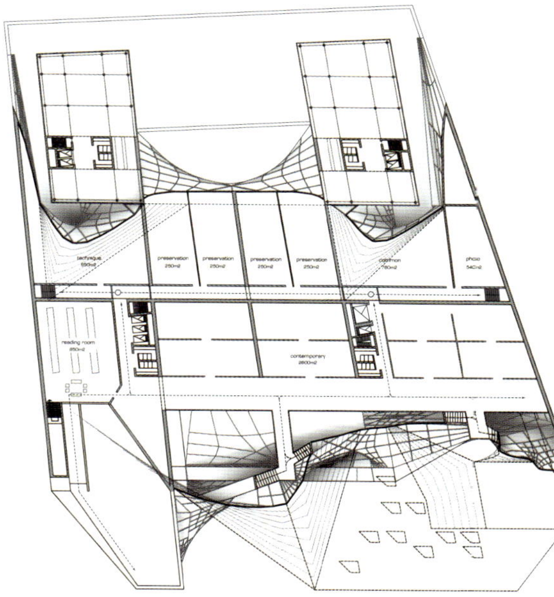

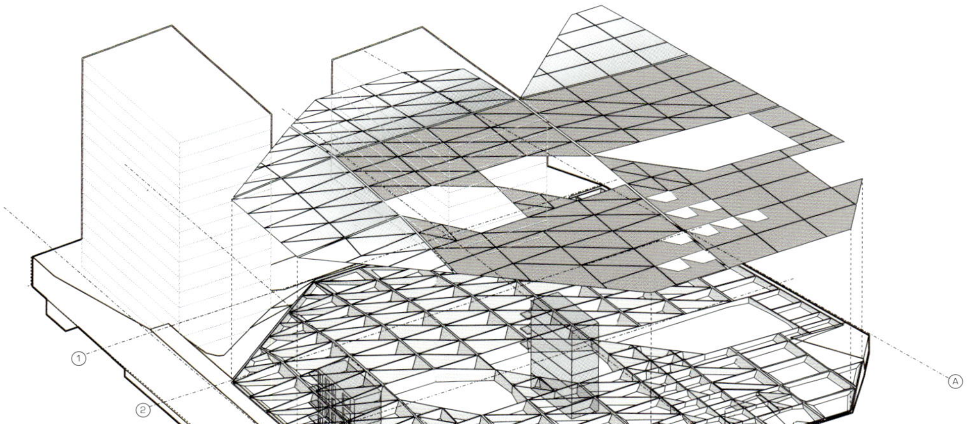

1
2
3
A
B
C
D

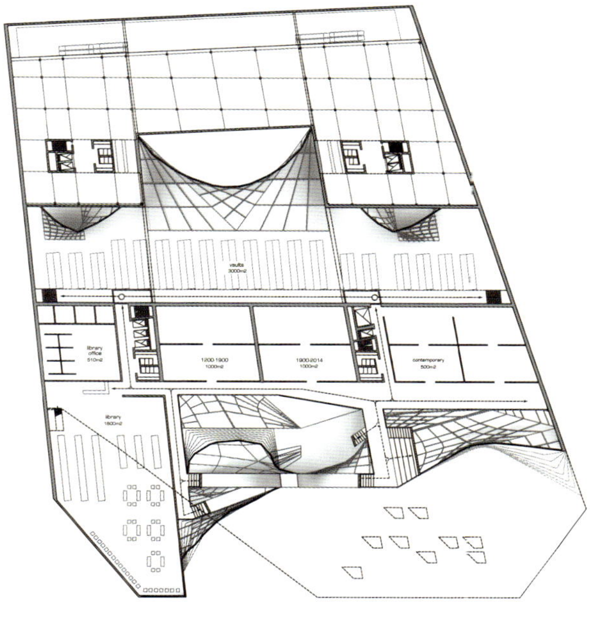

Water and Sculpture Hills Ichihara

市原市水与雕塑之丘博物馆

Katsuhiro Miyamoto & Associates

Renovation Proposal for Ichihara City Art Museum "Water and Sculpture Hills"
The 2nd Prize
市原市水与雕塑之丘博物馆整修方案竞赛 二等奖

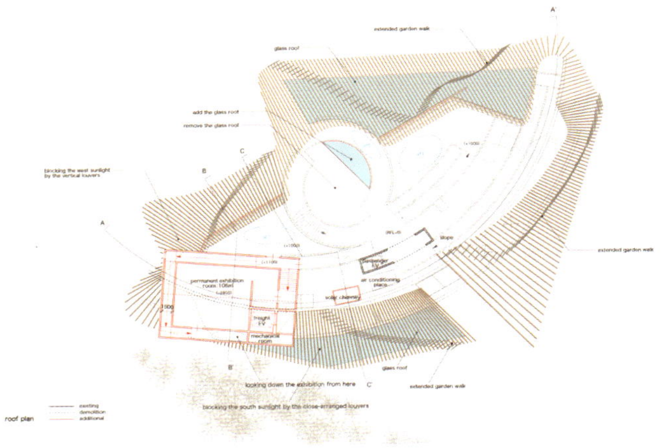

Credits
Prize: Excellent Prize (The 2nd Prize)
Location: Ichihara-city, Chiba
Principal Use: museum
Competition Year: 2010
Total Floor Area: 1,740 m²
Structure: Wood, 3 stories
Design: Katsuhiro Miyamoto & Associates

这是宫本佳明建筑设计事务所为千叶县市原市的水与雕塑之丘博物馆整修方案竞赛提交的设计，该竞赛是日本最大型的竞赛活动之一。该事务所的设计方案在231项方案中位居第二。

该设计的创意之处是在原有的钢筋混凝土博物馆外面加一层木结构，完全改变了博物馆的立面与功能。木制的栅格表层形成温和的圆锥形表面，将内部空间遮掩起来，并与建筑原有的风格奇异的外层相结合，在两层立面之间形成一个新的活动空间。

建筑师并没有对博物馆进行大规模的改造，而是充分利用了其原有结构，通过简单地添加一个新的框架减少了整修成本。同时，博物馆的现有外墙还可以用来展示作品。

竖向栅格采用的材料是防风化的工程格木（150毫米×450毫米），斜靠在现有的矮护墙和地面之间。这些栅格不仅构成了博物馆的新外层，同时也可以调节日光的射入，实现通风性与私密性。跟日本传统的葭箦一样，外面的人看不到里面，但是里面的人却可以看见外面。

栅格下面的一系列通往博物馆的空间也可以为参观者提供一些乐趣，建筑师称之为“艺术环”，它不仅是一种艺术作品，同时也是一种功能性设施（如咖啡厅），以及一种建筑形式。

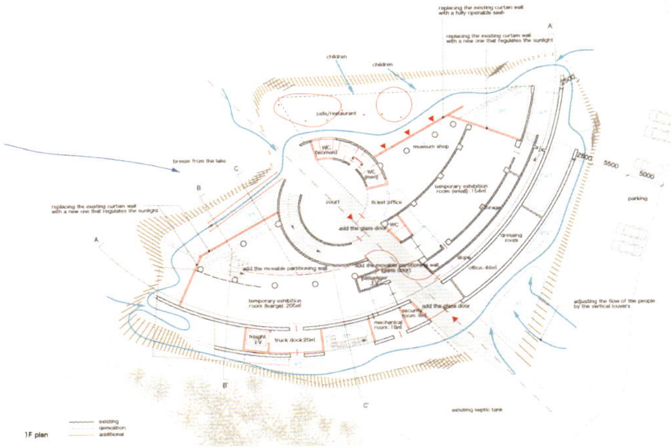

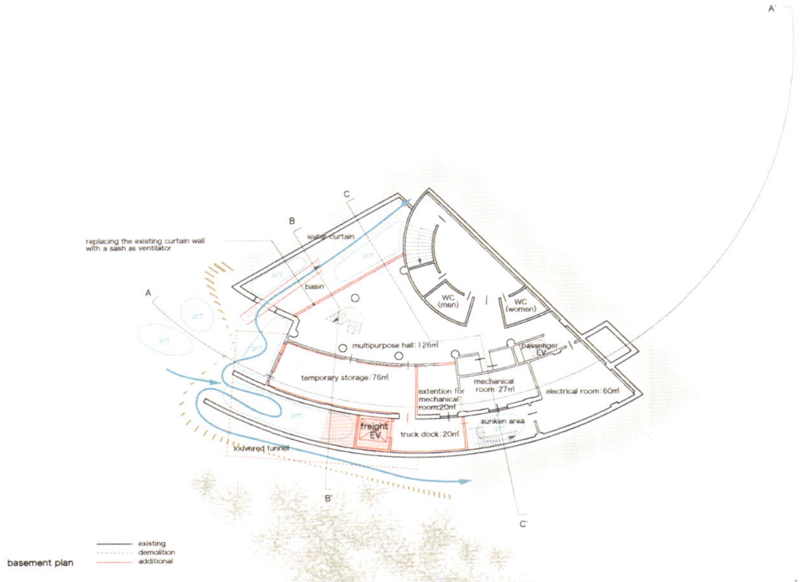

This is our proposal for the renovation of a museum at Ichihara-city in Chiba, one of the biggest competitions in Japan. Our proposal was selected as the 2nd prize from among 231 proposals.

What is innovative about our concept is our idea of enveloping the existing museum made of RC with a wooden structure and changing totally the façade and function. The wooden louvers create a gentle conoid-shape surface as a new layer. The new layer conceals the existing volume and integrates the old layer, which has a complex and grotesque façade. A new activity space then appears in between these two layers.

We take full advantage of the existing building structure, so we wouldn't add any large-scale structural improvement to it. We can reduce the renovation cost with the simple additional new frame, and we can use all of the existing structure as the current façade would become an exhibition's wall.

The vertical wooden louvers are made of weatherproof, 'Tali', engineered wood (150mm×450mm). They are simply leaned between the existing parapet and the ground, However, they would work not only as a new layer but also as a regulator of the sunlight, ventilation and privacy, that is, they can prevent people on the outside from looking inside, while permitting those inside to look out, similar to 'Yoshizu', which is one of Japan's traditional screening devices.

The sequential spaces beneath the louvers would be enjoyable approaches to the museum. We would like to call it 'Art Loop' as a space combined with not only art works but also something like cafe facilities, and additionally, architecture itself.

Froissy Archaeological Museum

弗鲁瓦西考古博物馆

n!studio

Winning Project of International Shortlisted Competition for the New Gallic-Roman Museum, Froissy, France

法国弗鲁瓦西新高卢罗马博物馆国际设计赛获胜项目

Credits

Design: n!studio Susanna Ferrini, Antonello Stella with David Serero

Team: Simone Bove, Stefano Battaglia, Mariangela Corsi, Tiziana Monti, Paolo Rossi, Giulia Scaglietta, Teresa Zanaria

Consultants: Carolina De Camillis, Riccardo Fibbi (Lighting Design)

Engineering: Betom, 3ti

Client: Communaute des Communes des Vallees de la Breche et de la Noyre

Model: Marco Galofaro

该项目位于地下，在高卢罗马竞技场中几乎看不到。其设计为地貌景观的延伸，被置入峡谷链以及皮卡第地貌沿地平线形成的凹陷中。此博物馆成为了一座标志性的建筑。其意义并非来源于其存在性或物质性，反而是因为凹入地下所造成的缺失感。该博物馆尝试着像其周围的考古区一样，通过空穴和开口的设置，将其隐匿的内涵展示给世人。该博物馆的设置使得游客可以沿着从公路层直到位于较低区域的博物馆入口的低合金高强钢坡道行走，在移动过程中不断发现其奥秘之处。该坡道沿着计划将建造的通往圆形露天剧场的通道轴线建造，几乎使游客滑入了地层的深处，沿途展示出了其不同的层次，经过天然存在的地貌风景层进入了博物馆的人工建筑区，最后抵达更深的考古发掘层。该博物馆同时也是地平面的提升，在地貌风景中设置了一个开口，形成了博物馆的入口，并使自然光线能够进入其内部。沿着入口坡道，博物馆的唯一正立面就是容纳了该坡道的低合金高强钢突出体，也是该博物馆的防护墙和展示区。低合金高强钢可以通过氧化作用自动形成防潮防腐蚀的保护层。铜质色调使其几乎像有机物一样，光泽和色彩会随时间而变化。

MUSEE-CENTRE DE RESSOURCES ARCHEOLOGIQUES

The project is set into the earth, almost invisible from the Gallic-Roman amphitheater. It is designed as an extension of the landscape, an insertion within the chain of valleys and depressions that the Picardie topography forms along the horizon. The museum is a sign, excavated into the earth, a building that is not defined by its presence or its materiality but, on the contrary, by its absence. It is through a play of excavations and openings that the museum attempts, like the archaeological site that surrounds it, to reveal the invisible. The museum allows itself to be discovered by visitors through movements, walking along a COR-TEN steel ramp from the level of the road towards the entrance to the museum, in the lower area. This ramp, positioned along the axis of the future path that will lead to the amphitheater, almost slides the visitor into the depth of the earth, displaying its various layers, passing from the natural, the existing landscape, into the artificial, that of the museum, arriving at the layers of archaeology, which sink even deeper. The museum is also simultaneously the rising of the ground plane, allowing for the opening of a cut in the landscape to create the entrance and bring natural light into its interior. Along the access ramp, the only façade of the museum is a blade of COR-TEN steel that contains the ramp, a retaining wall and the display space of the museum. Oxidation naturally gives COR-TEN steel its own protective layer against humidity and corrosion. Its copper hues render it almost organic as its patina and coloring change over time.

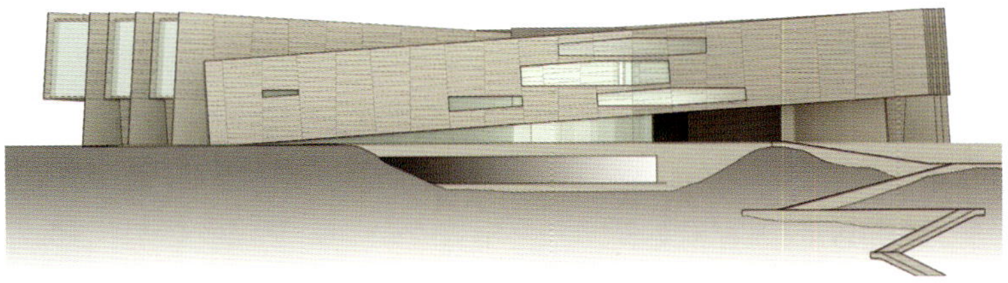

Embankment Elevation

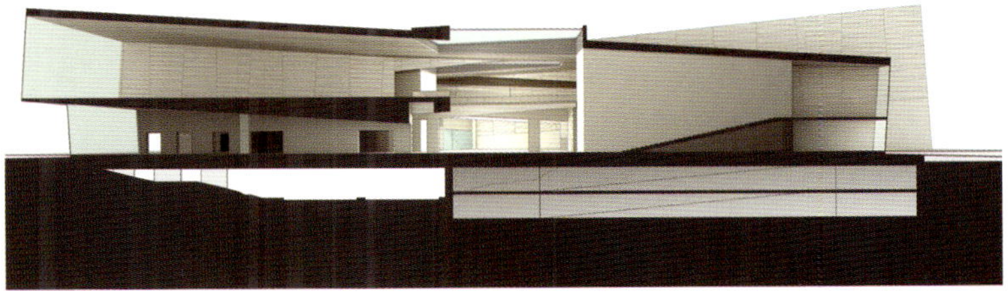

Section 1

Jazdow Elevation

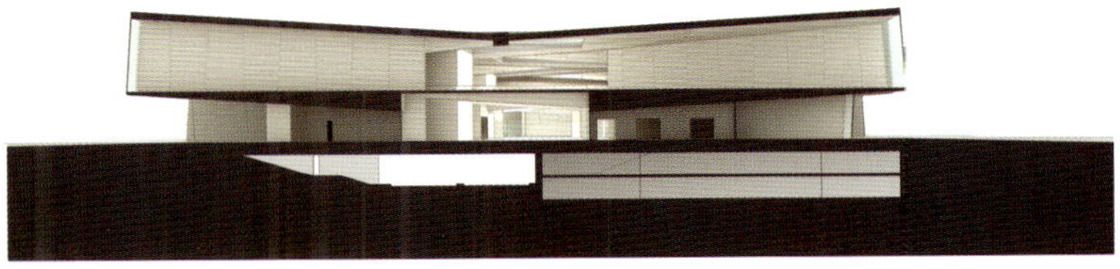

Section 2

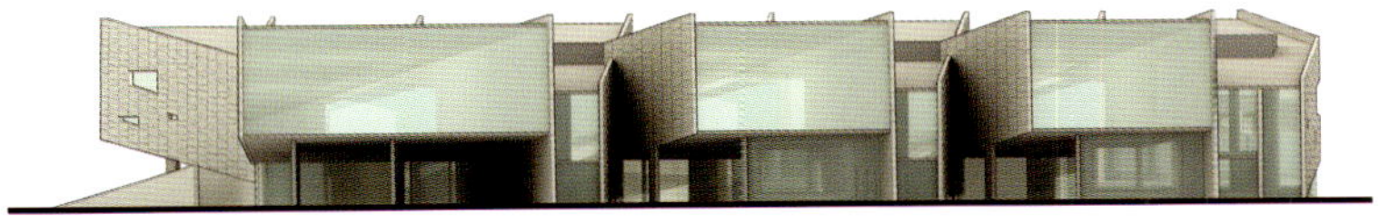

Castle Elevation

Section 3

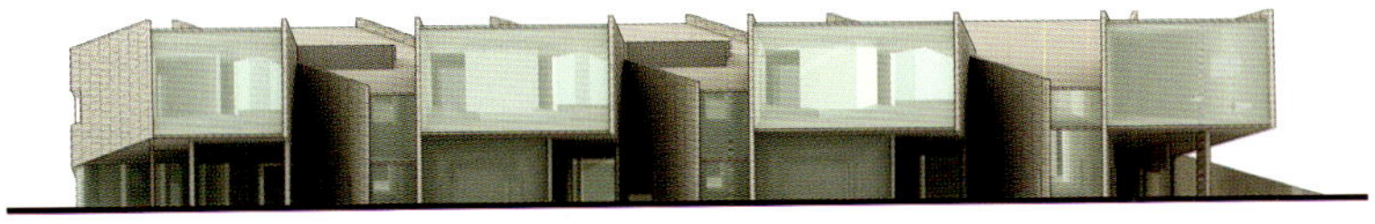

Cottage Elevation

Section 4

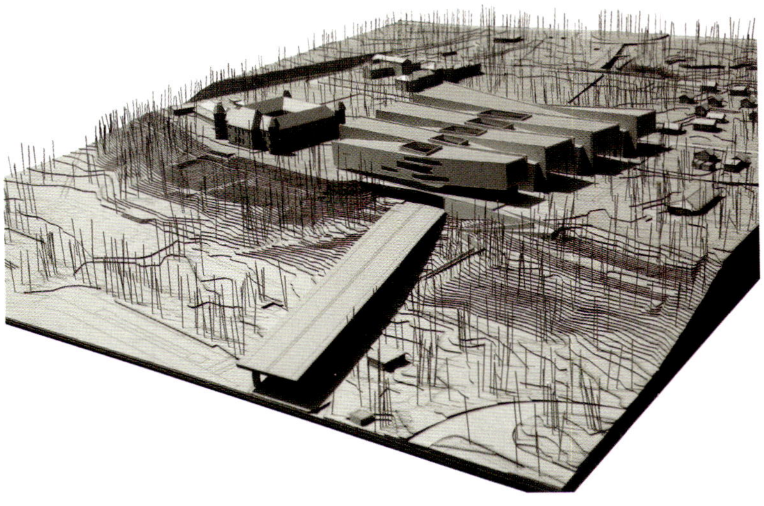

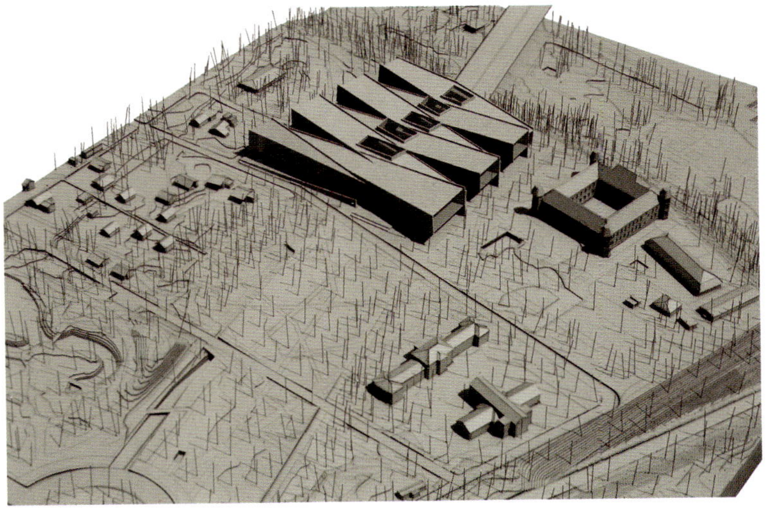

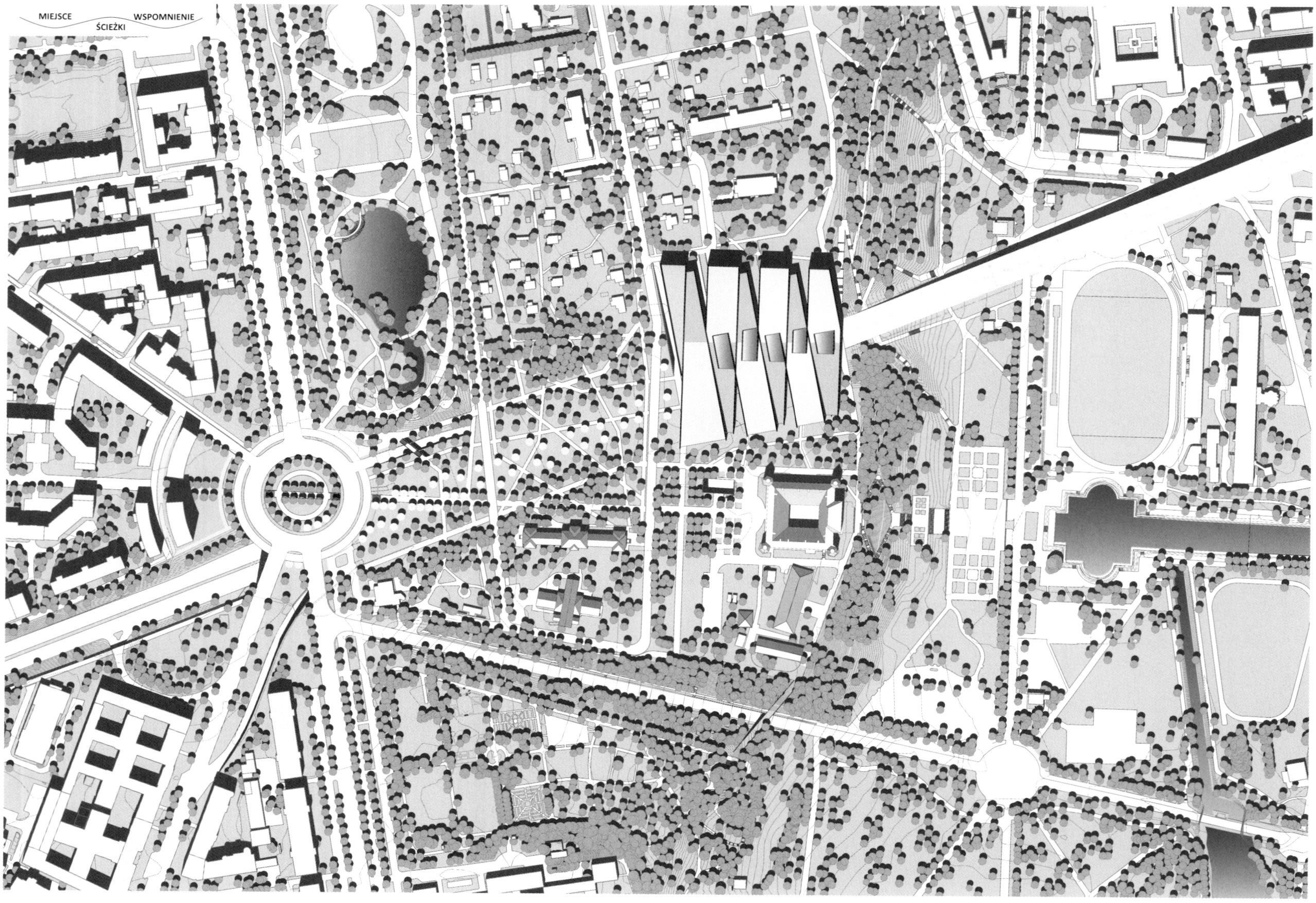

DEVELOPMENT STRATEGY

URBAN CORRECTION

1. ROUNDABOUT
   improve traffic flow and re-establish Baroque urban arrangement at Plac na Rozdrożu
2. COVERING OF TRASA ŁAZKIENKOWSKA
   integrate northern and southern sections of Park Sobieskiego, create public place for the MPH, and improve environmental air and sound quality
3. DEVELOPMENT PARCELS
   add buildings to strengthen new circular urban figure on Aleje Ujazdowskie
4. BUS STATION
   improve access to the MPH, Park Sobieskiego, and Zamek Ujazdowski

SITE MAINTENANCE/IMPROVEMENT

5. OŚ STANISŁAWOWSKA
   Baroque urban form is restored, lined with trees, and extended through Plac na Rozdrożu
6. UL. JAZDÓW
   reconnect north and south sections; widen and improve for one way traffic moving south; primary vehicular access to the MPH
7. UL. JOHNA LENNONA
   re-connect north and south sections; widen and improve for one-way traffic moving north
8. EXISTING TREES
   preserve and maintain existing trees and greenery
9. WALKING PATHS
   preserve and improve existing pedestrian paths
10. ESCARPMENT ROUTE
    improved walking and biking path extending from Aleja na Scarpie and continuing over ul. Agrykoli
11. FOOTBRIDGE
    new footbridge connecting Ogród Botaniczny with Zamek Ujazdowski along the escarpment route
12. ZAMEK UJAZDOWSKI
    improve landscape surrounding Zamek Ujazdowski by removing surface parking and introducing geometric planting arrangement

NEW ELEMENTS

13. MUSEUM OF POLISH HISTORY
    integration of museum within an attractive public place.
14. PEDESTRIAN PATHS
    network of public paths through Park Sobieskiego focused on the new public place of the MPH
15. PEDESTRIAN PATH
    connection between Plac Na Rozdrożu and the MPH
16. PEDESTRIAN PATH
    connection between the lower Skarpa Warszawska and the upper embankment
17. UNDERGROUND PARKING STRUCTURE
    250 space parking structure located beneathe the MPH and accessible from ul. Jazdów; the garage is constructed on the north slope of the Trasa Łazkienkowska trench stabilizing the previously unstable condition
18. COACH PARKING
    dedicated offsite parking for buses and cars under the elevated portions of Trasa Łazkienkowska

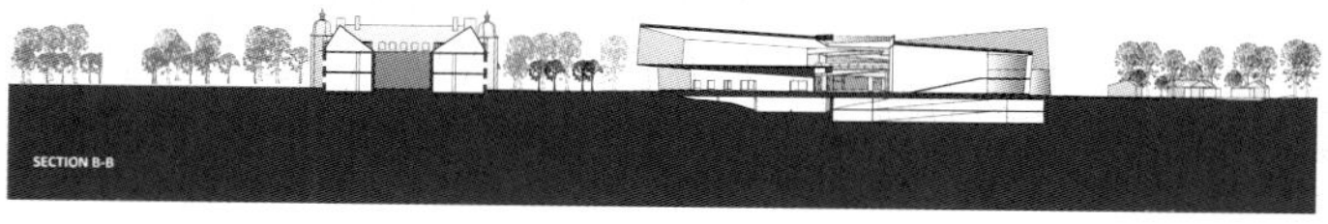

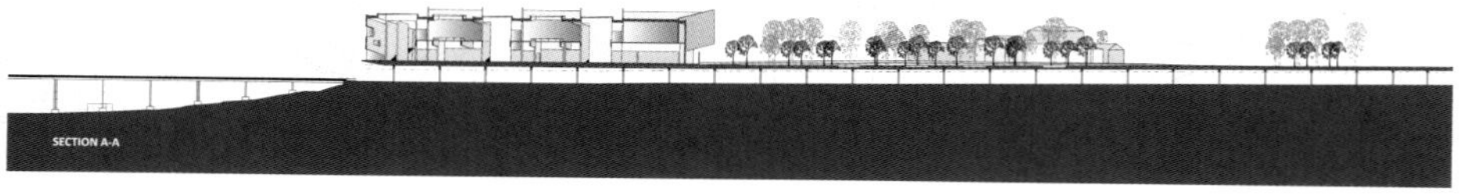

Museum of Polish History

波兰历史博物馆 3 Six o Architecture

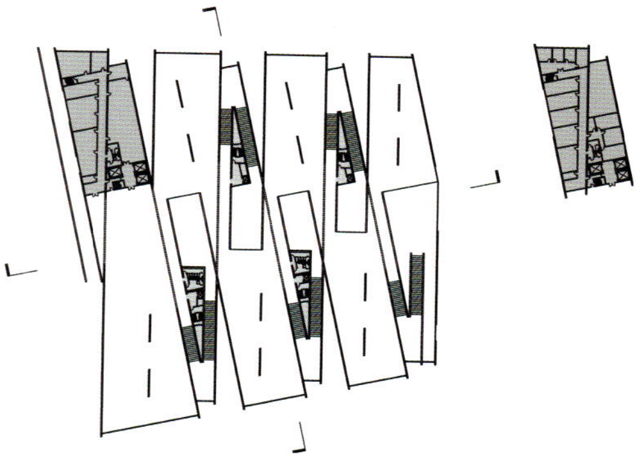

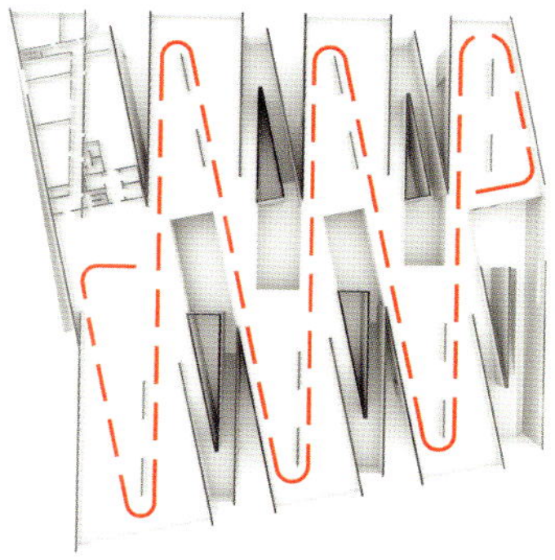

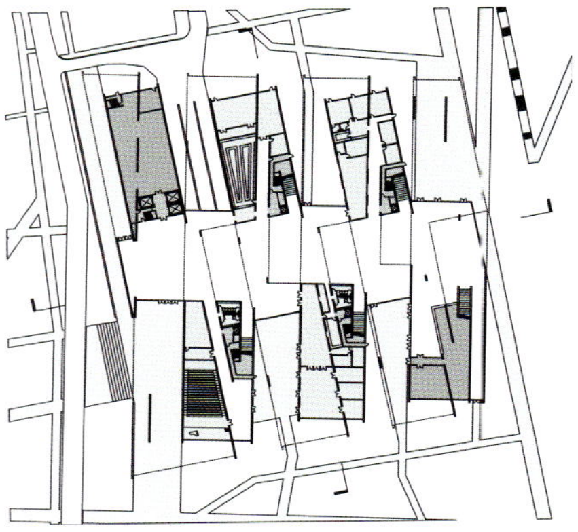

建筑理念

在这项投标方案中，路径把场地和往事串联起来：从ul. Jazdow穿针引线至Oś Stanisšawowska的途径；跨越Trasa łazienkowska的途径；令人激昂、沿着Skarpa Warszawska的途径；从pl. na Rozdrożu出发的途径；重新连接Zamek Ujczdowski与Sobieski花园的途径。所有的途径汇聚起来形成了博物馆的建筑体系。

该建筑由ul. Jazdow和Stanisšawowska Axis.之间运动的线条具体勾勒出来。这些线条排列着博物馆的主要内部流通通道。通过位于ul. Jazdow的酒吧建筑之下蜿蜒流淌的道路，连接了博物馆的入口。一个位于浅埋的楼梯网络底部，自西向东形成开放的公共中庭。这些循环的要素是建筑的主要式样要素。论坛与楼梯界定了展览场地，同时确定了辅助功能区——礼堂、会议室、多媒体图书馆、阅览室、博物馆商店以及咖啡厅的位置。浅埋的楼梯通过历史顺序分割区划，并且带领参观者来到位于上方的画廊。一条"全面的路径"起讫于ul. Jazdow入口以连续循环的方式来回穿越论坛中庭，连接画廊。意识到记忆变化的多样性，一条"快速路径"与按时间顺序排列的建筑式样相交，并为具体时期的展览或进行非正式访问提供了通道。画廊一起跨越建筑，还提供了多重视角和接回博物馆的庭院空间。

规划方法

该位置的总体规划方法包括三个相互依存的策略：pl. na Rozdrożu城市校正、总体现场维护与改良以及新要素的加入。

城市校正

- Plac Na Rozdrożu的城市形态与交通模式重新配置如下：
- 通过在Aleje Ujazdowskie引入一种循环的交通流量要素增加交通流量。
- 通过在Trasa łazienkowska和半圆形出口匝道覆盖新型城市建筑提高环境质量。
- 通过在Aleje Ujazdowskie加入新建筑进一步树立新型循环城市的形象。
- 新型的/改造过的汽车站。

现场维护/改良

- 根据需要保持和维持现有树木。
- 改进和升级现有人行道。
- 一条南北向的悬崖路线正在规划中。这将会是人行/自行车混合车道。
- 拓宽和修缮Ul. Jazdow。重新连接南北部分。
- 拓宽和修缮Ul. Johna Lennona。重新连接南北部分。
- 延伸Oś Stanisławowska至Plac Na Rozdrożu，并栽种新树。

新要素

- 波兰历史博物馆的公共场所。
- 聚焦新波兰历史博物馆的新公共场所的人行道网络。
- 连接Plac Na Rozdrożu与波兰历史博物馆（MPH）的人行道。
- 连接南部SkarpaWarszawska与北部筑河堤之间的人行道。
- 位于规划中的波兰历史博物馆地下，连通Ul. Jazdow，拥有250个车位的地下停车库。车库建造于Trasa łazienkowska沟渠北坡之上，稳定之前不稳定的环境状况。
- 位于Trasa łazienkowska抬升部分以下，SkarpaWarszawska以东，巴士与汽车专用的停车场。

MUSEE-CENTRE DE RESSOURCES ARCHEOLOGIQUES

MUSEE-RESERVE ARCHEOLOGIQUE FROISSY

Architectural Concept: Paths between Place & Memory

In this proposal, a gathering of paths links place with memory: paths that stitch ul. Jazdow to Oś Stanisławowska; paths that straddle Trasa Łazienkowska; paths that lead one up and along the Skarpa Warszawska; paths from pl. na Rozdrożu and paths relinking Zamek Ujazdowski with Sobieski gardens. All the paths gather to form the architecture of the museum. The architecture is shaped by the lines of movement between ul. Jazdow and the Stanisławowska Axis. These lines place the primary internal circulation of the museum. One arrives at the museum's entry through paths that bend and slide below a bar building, located on ul. Jazdow. One is crawn into the forum, an open public atrium which runs west to east at the base of a network of shallow stairs. These circulatory elements are the major ordering elements of the building. The forum and stairs place and shape the exhibition spaces, and locate the supporting functions—auditorium, seminar rooms, multimedia library, reading room, museum store and café. The shallow stairs cut sections through history and bring the visitor to the galleries above. A "comprehensive path" links the galleries back and forth across the forum atrium in a continuous loop that begins and ends at the ul. Jazdow entry. Recognizing the variability of memory, a "fast path" intersects the chronological ordering and provides access for a specific course of study or an informal visi-. The galleries stitch together across the building and give multiple views and courtyard spaces to connect back to the site. The forum culminates in a panoramic view eastward, towards the Wisła River.

Planning Approach

The general planning approach to the site consists of three interdependent tactics: urban correction at the pl. na Rozdrożu, general site maintenance and improvement and the addition of new elements.

Urban correction: The traffic pattern and urban morphology at Plac Na Rozdrożu are reconfigured to

- Improve traffic flow by introducing a circular traffic flow element on Aleje Ujazdowskie.
- Improve environmental quality by covering Trasa Łazkienkowska and the semi-circular exit ramp with new urban building.
- Add new building to strengthen the new circular urban figure on Aleje Ujazdowskie.
- New/improved bus station.

Site maintenance/improvement:

- Existing trees are kept and maintained where possible.
- Existing pedestrian paths are improved and upgraded.
- An improved north-south escarpment route is proposed. This would be a combined pedestrian/bike path.
- Ul. Jazdow is widened and improved. North and South portions are reconnected.
- Ul. Johna Lennona is widened and improved. North and South portions are reconnected.
- Oś Stanisławowska is lengthened to meet Plac Na Rozdrożu and lined with new trees.

New elements:

- A public place for the Museum of Polish History.
- A network of pedestrian paths (public condenser) focused on the new public place of the new Museum of Polish History.
- A pedestrian connection between Plac Na Rozdrożu and the MPH.
- A pedestrian connection between the lower Skarpa Warszawska and the upper embankment.
- A 250 space underground parking structure, underneath the proposed MPH, accessible from Ul. Jazdow. The garage is constructed on the north slope of the Trasa Łazkienkowska trench, stabilizing the previously unstable condition.
- Dedicated parking for buses and cars under the elevated po-tion of Trasa Łazkienkowska immediately east of the Skarpa Warszawska.

Taipei Performing Arts Center

台北艺术表演中心

Ivo Buda Architetto

Credits

Project Design: Ivo Buda
Project Team: Ivo Buda
Client: Department of Cultural Affairs, Taipei City
Location: Taipei, Taiwan
Date: August 2008

High: 29 m
Levels: 3 basements + Ground floor + 5 levels
Total Floor Area above Grade: 23,100m^2
Structure: Steel-concrete structure
Parking Spaces: 204

设计理念

这个圆形的艺术表演中心外观像一朵美丽的花，她用她那灵巧精美的姿态展现设计构造的多种理念。这朵花带领大家探索剧院的精髓与剧院内部表演。里面的表演散发着花的香气，并流露着晚霞的光彩。实际上，太阳与大自然每天都在上演着重复但却令人惊喜的场景；人类情愿接受这个重复的惊喜，这也揭示了我们对现实理解的关键。但是，建筑能像花朵那样在风和日丽的天空下跳舞吗？是的，这栋楼在图片上的确是静止不动的，但是在现实生活中，它随着光与人的移动而呈动态移动。这个铰链式建筑在光线下有多种状态：在晚霞的余晖里，它是那样风情万种；在正午的照射下，它又是那样严格谨慎。室内的设计延续外部构造，紧致、私密却又光线明亮。

整体布局

艺术表演中心的整体布局是三种不同类型剧院的结合：

1.大剧院；

2.舞台剧院；

3.多剧种剧院；

这三类剧院都有各自独立的操作控制系统，与一个休息室连接；在休息室内剧院入口处设有操作台。这个休息室除了特殊情况（区域管制）外，24小时免费开放。这里环境舒适，视觉上连通艺术表演中心所有楼层。

大剧院包括古典歌剧、大规模舞蹈表演与音乐喜剧。可容1,500人，其中管弦乐区域1,000座位，楼厅500座位。

舞台剧院与多剧种剧院主要是戏剧与舞蹈表演，可容800人。

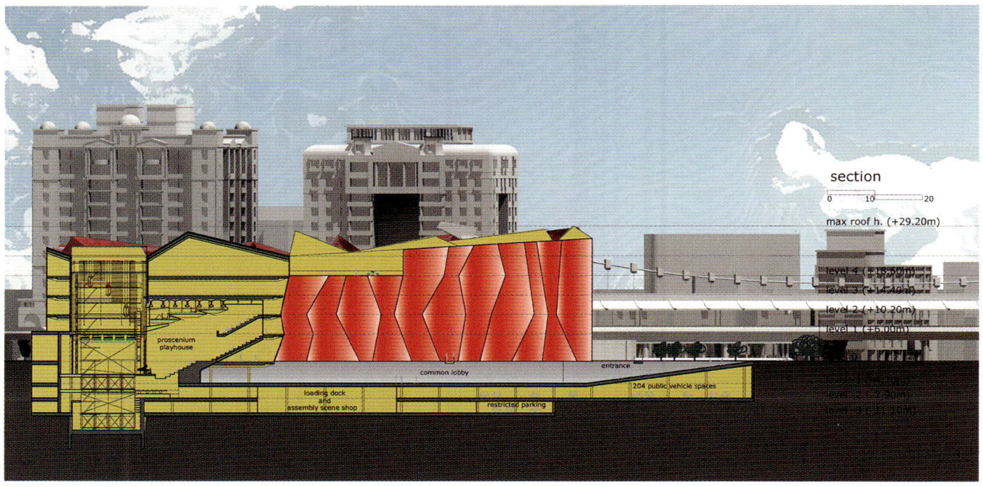

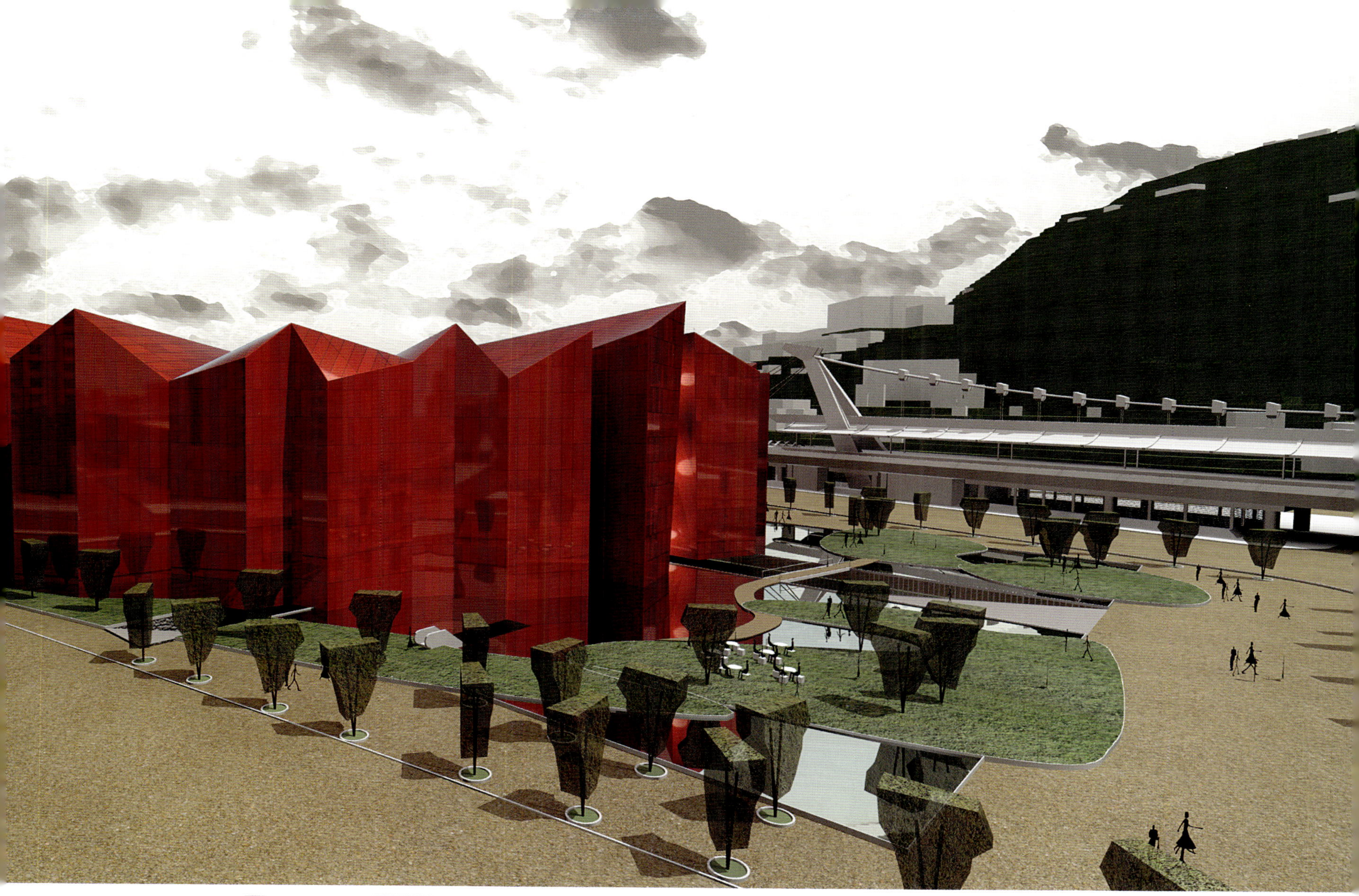

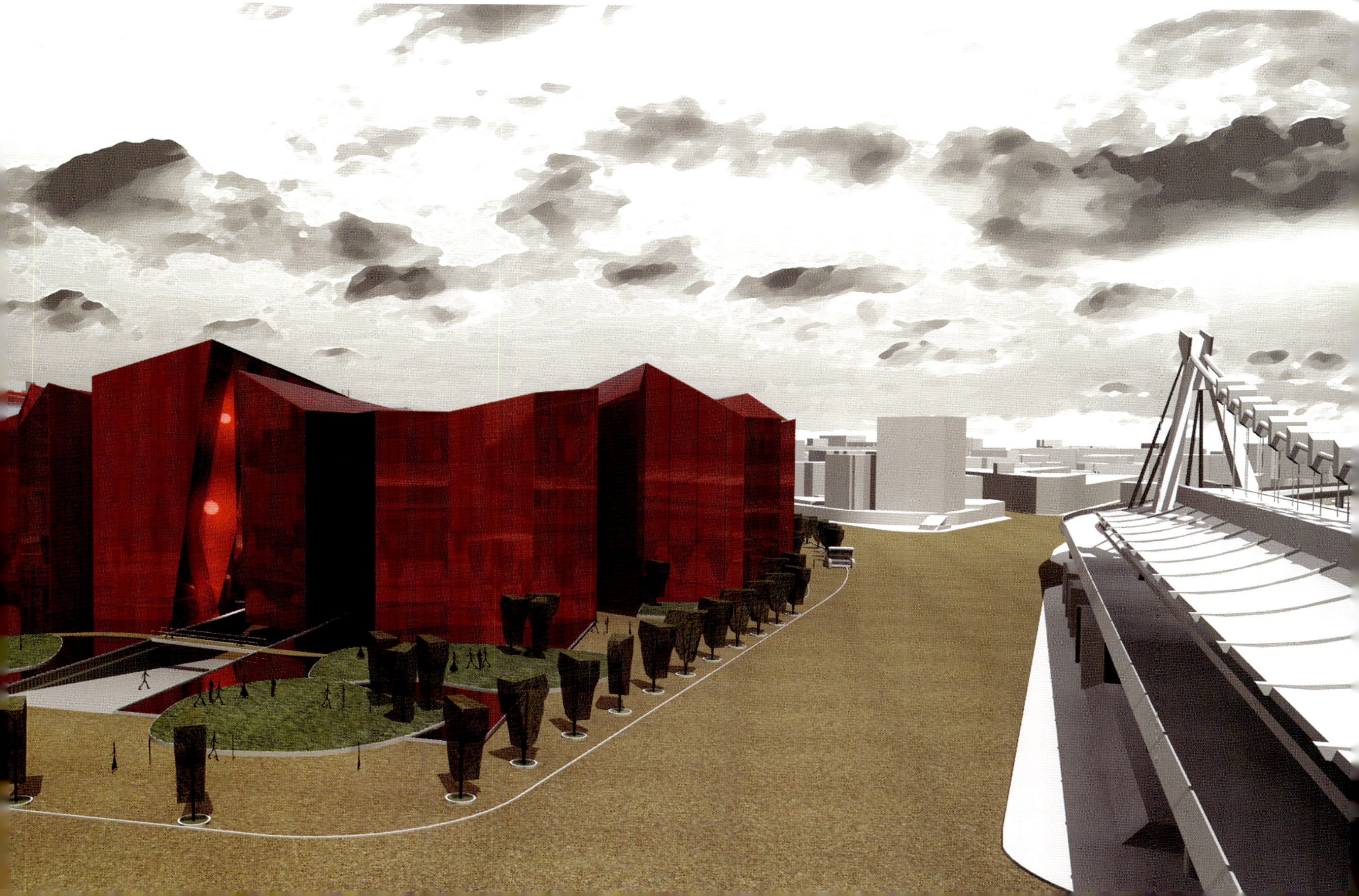

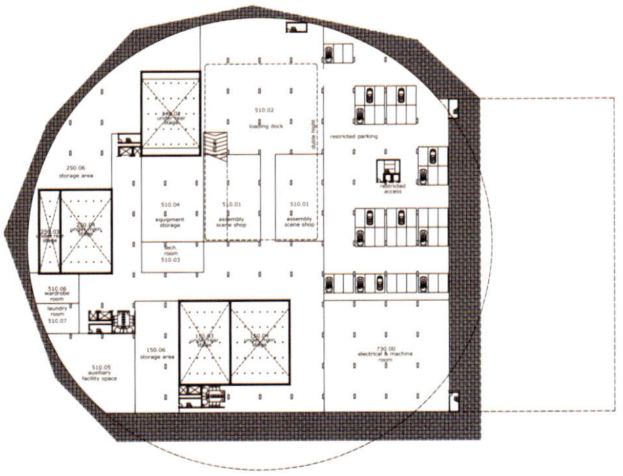

loading dock &
restricted parking
3100 m²

technical space &
storage
6500 m²

tot. area level -3
9600 m²

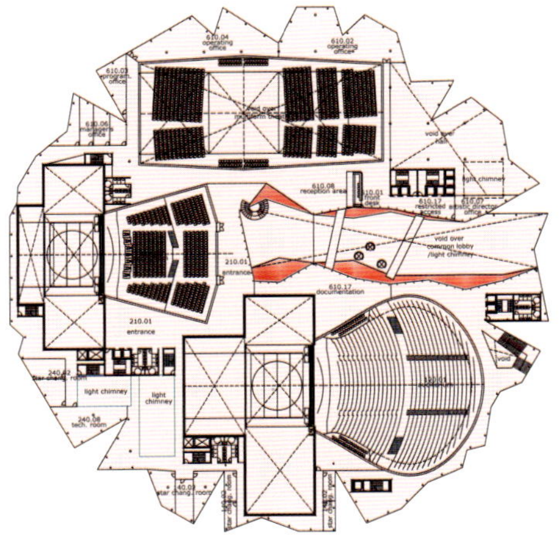

grand theatre &
back stage
870 m²

p. playhouse &
back stage
740 m²

administrative
space
1400 m²

tot. area level 0
3100 m²

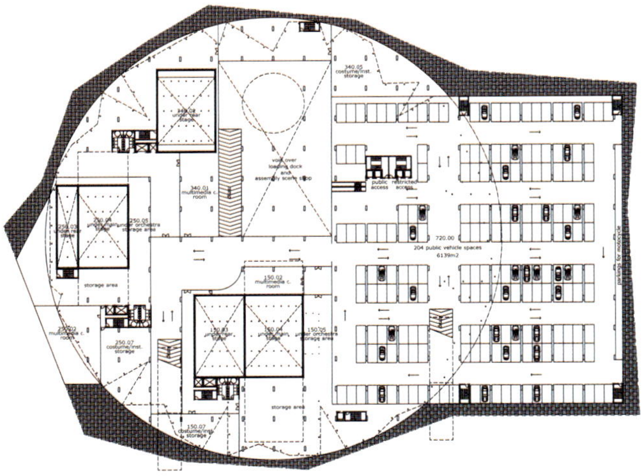

public parking
6100 m²

technical space &
storage
1800 m²

technical space &
storage
1050 m²

technical space &
storage
1260 m²

tot. area level -2
10210 m²

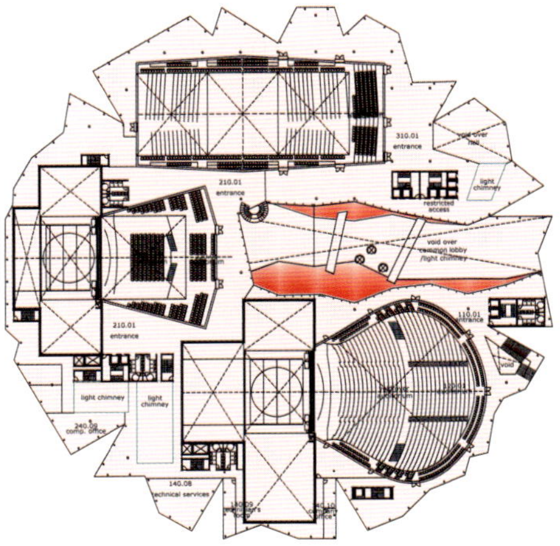

grand theatre &
back stage
1420 m²

p. playhouse &
back stage
950 m²

multiform theatre &
back stage
1060 m²

tot. area level 0
3430 m²

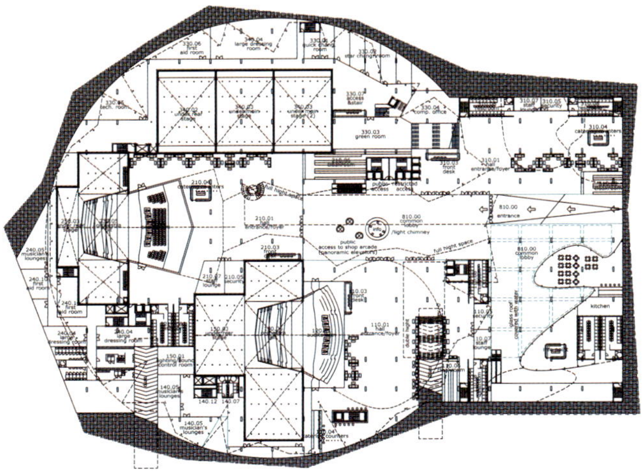

common lobby
2580 m²

grand theatre &
back stage
3520 m²

p. playhouse &
back stage
2860 m²

multiform theatre &
back stage
3610 m²

tot. area level -3
12570 m²

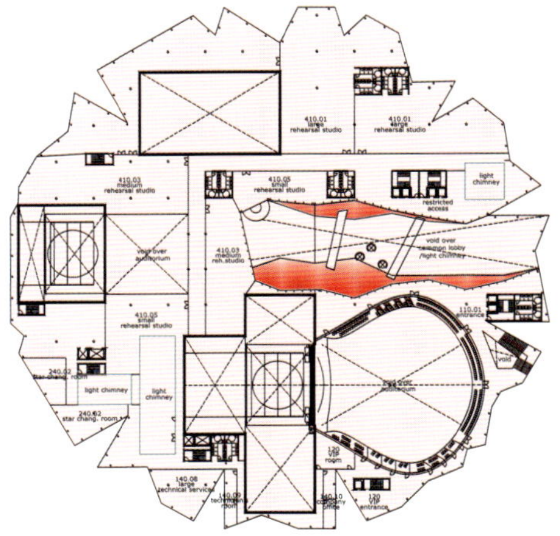

grand theatre &
back stage
840 m²

p. playhouse &
back stage
350 m²

rehearsal
space
3120 m²

tot. area level 0
4310 m²

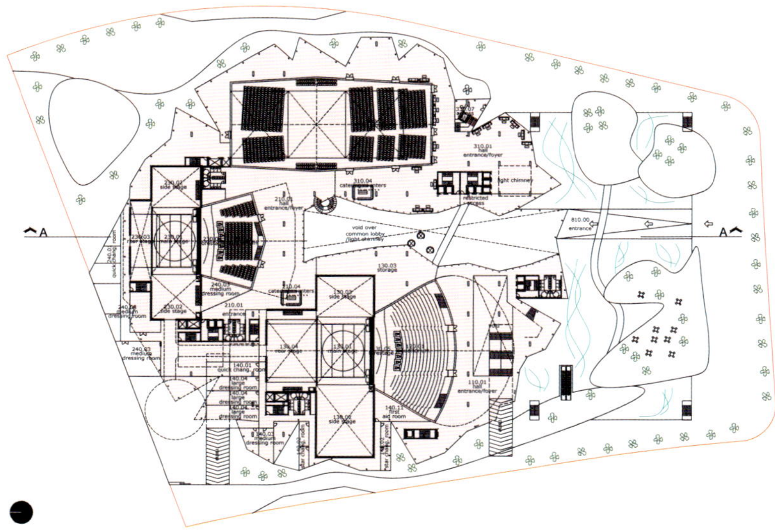

grand theatre &
back stage
1420 m²

p. playhouse &
back stage
950 m²

multiform theatre &
back stage
2560 m²

tot. area level 0
4930 m²

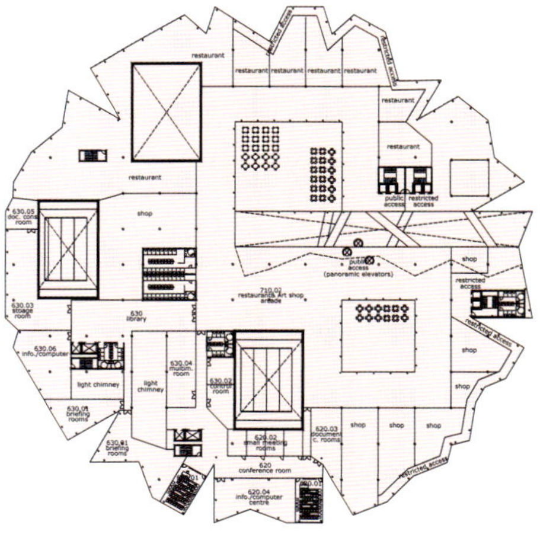

restaurant & art shop
arcade
5480 m²

library &
conference room
1850 m²

tot. area level 0
7330 m²

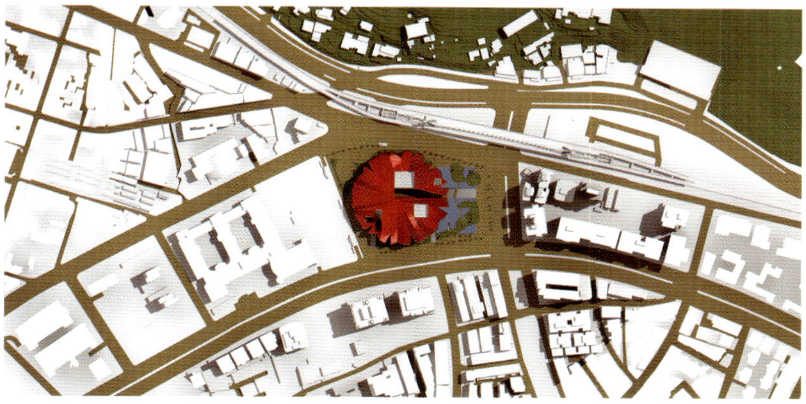

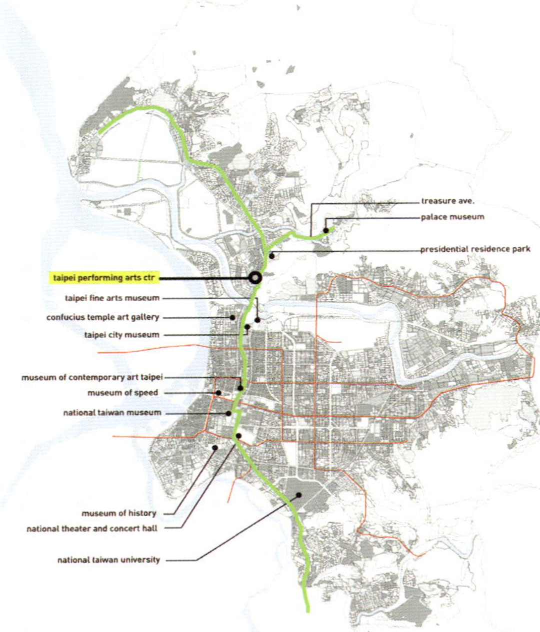

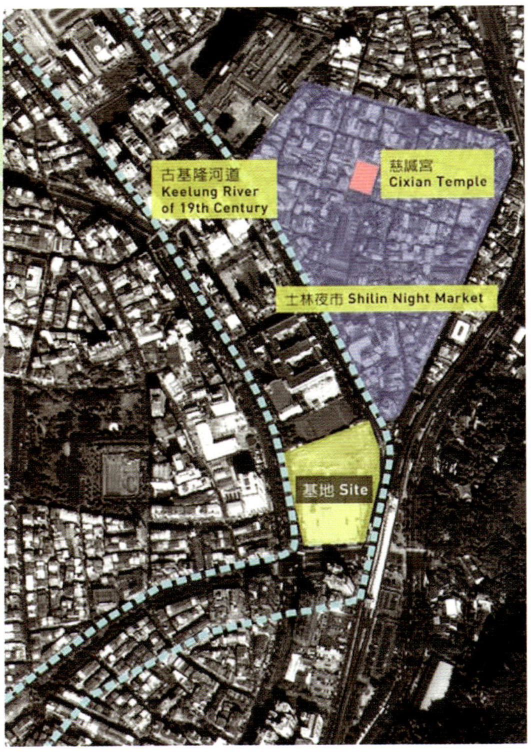

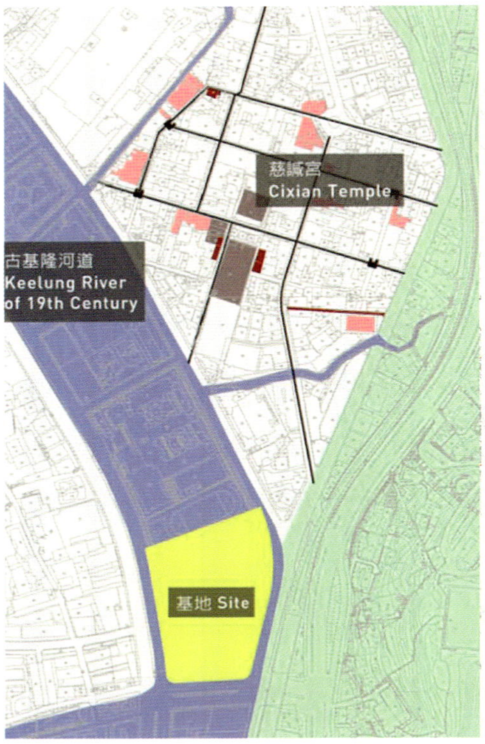

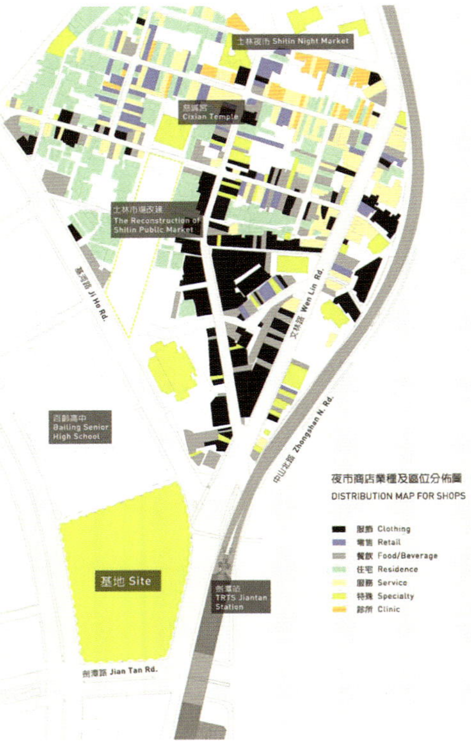

夜市商店業種及區位分佈圖

DISTRIBUTION MAP FOR SHOPS

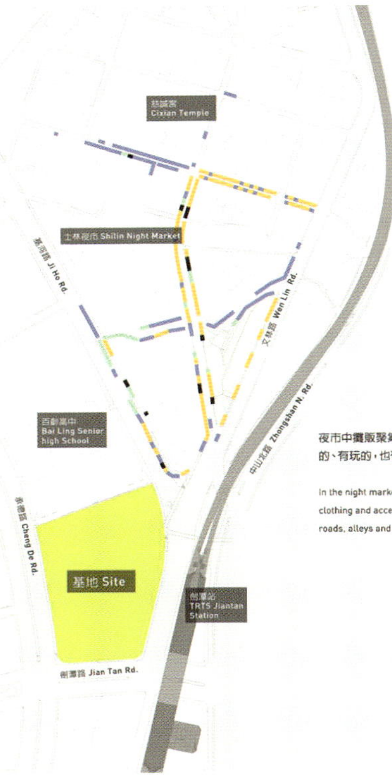

士林夜市商攤販分佈圖

DISTRIBUTION MAP FOR STALLS

夜市中攤販聚集在道路、巷道與騎樓之中，有吃的、有玩的，也有販賣服飾與生活用品的小店面。

In the night market, food stalls and shops selling toys, clothing and accessories can be found clustered along roads, alleys and arcades.

士林日市分佈圖

DISTRIBUTION MAP FOR DAY MARKET

士林夜市也是全台少見的早夜市混合使用的地方。早上，不見夜晚的小吃、服務攤販，而是聚集蔬果、海鮮及肉類等攤販，成為社區的市集。

士林地區的家庭主婦與居民，經步行、騎乘機車或是搭乘大衆運輸工具，包含公車、捷運到此購買三餐食材。

Shilin Night Market is one of the few marketplaces in Taiwan which provides the functions of both a day and a night market – during the daytime, the market is a community bazaar full of vegetables, seafood and meat vendors. Housewives and residents in the Shilin district arrive here by foot, motorcycle, bus or rapid transit to shop for a variety of food.

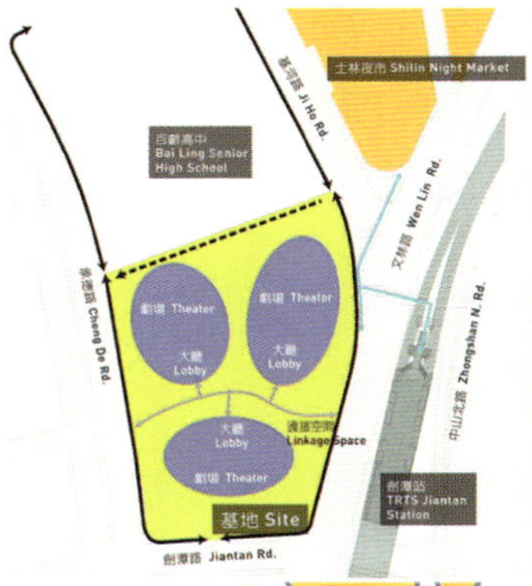

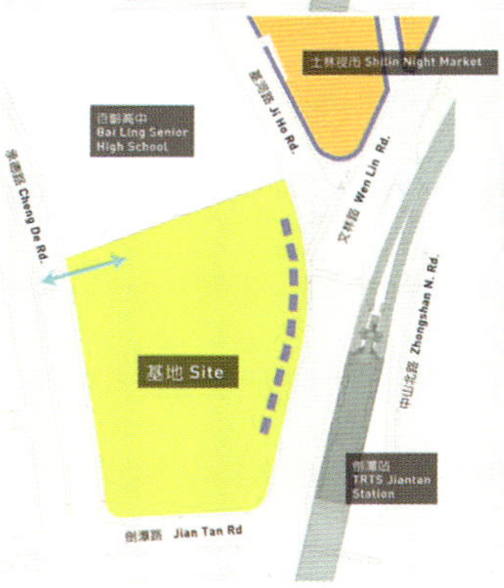

台北藝術中心附設商（店）業與停車場出入口規劃原則

SHOPS & PARKING ENTRANCE/EXIT OF TAIPEI PERFORMING ARTS CENTER

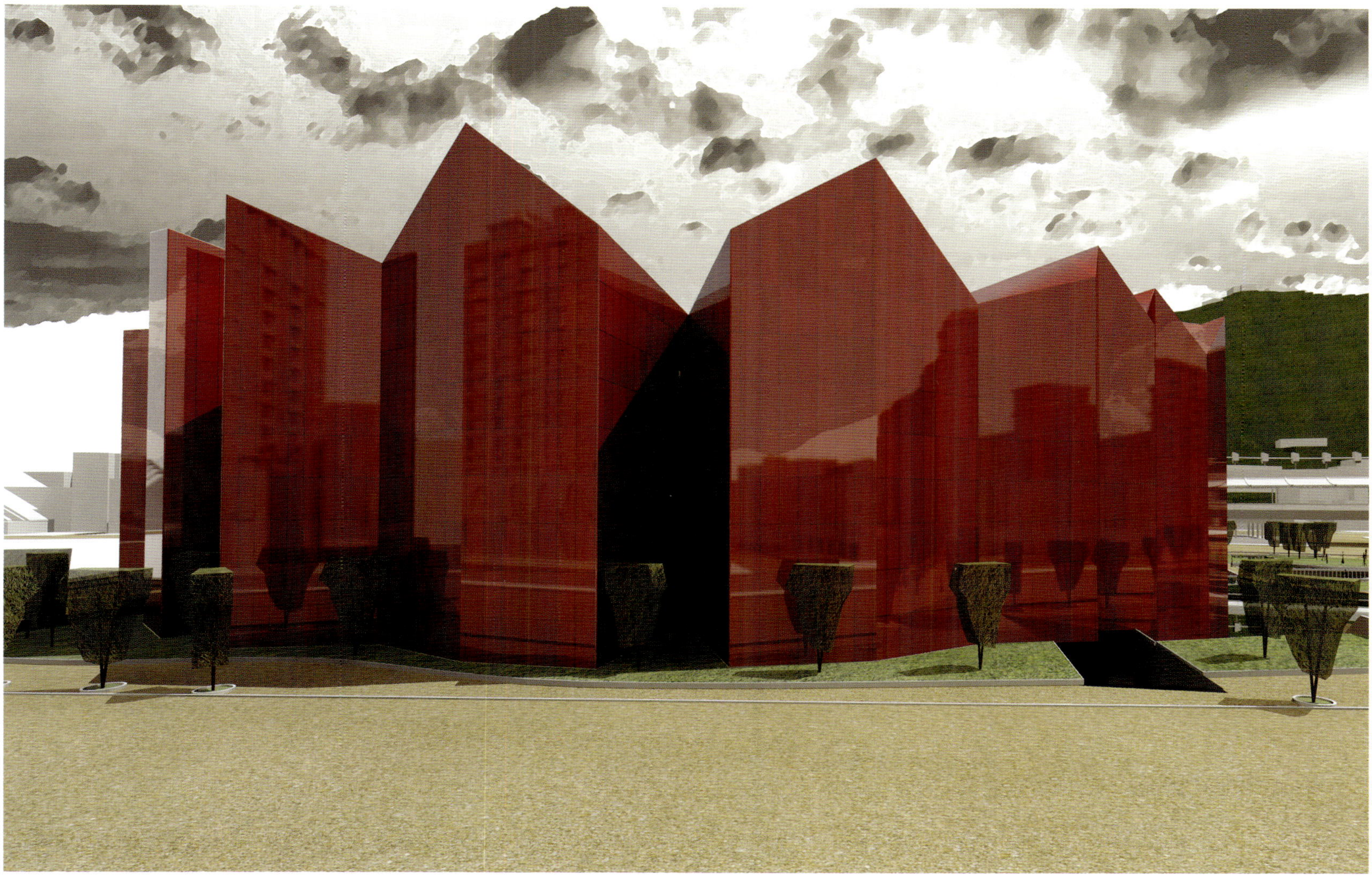

Architectural Design

Like a big flower, a round shape, corroded little by little through a precise gesture, embraces all the civerse elements of the composition. This reference is an opportunity to research the essence of the theater and the activity inside it. That activity has the fragrance of a flower and the intensity of a sunset. In fact the sun and the nature give us every day a repeated, but unexpected spectacle, and yet the willingness to be permeated by the unexpected often reveals new keys to the comprehension of reality. But like a flower dances in the sun and in the wind, can the architecture darce too? Yes, in the images the architecture is still, in its real life it gets cross by light and people. The articulated volume in the light undertakes multiples possibilities of definition: in the sunset light it is carnal and sensual, in the midday light acute and precise. Inside the space is a continuum with the outside, compressed, secret, but luminous.

General Organization

The overall organization of the building derives from the connection of three types of theaters:

1 The Grand Theater

2 The Proscenium Playhouse

3 The Multiform Theater

The three theaters all have independent control and operation systems; they are connected with a common lobby with control points at the theater entrance. The common lobby is a 24-hour open area with no requiring ticket, except only when area control is needed. The environment in the lobby is comfortable and visually is linking all the floors of the center.

The program for the Grand Theater includes classical operas, large-scale dance performances and musicals. The sitting capacity is of 1,500 seats with seating of 1,000 seats at the orchestra level and 500 seats at balcony.

The major programs for the Proscenium Playhouse and the Multiform Theater are drama and dance performances. The sitting capacity is of 800 seats.

Ulvi Cemal Erkin Concert Hall

"Ulvi Cemal Erkin"音乐厅

Yazgan Design Architecture

Ulvi Cemal Erkin Concert Hall National Architectural Project Competition
"Ulvi Cemal Erkin"音乐厅设计竞赛参赛方案

Credits
Location: Turan Gunes Boulevard, Cankaya, Ankara, Turkey
Employer: Cankaya Municipality
Project Year: 2010
Total Construction Area: 52,330 m²
Architectural Project: Yazgan Design Architecture
Landscape Project: Yazgan Design Architecture Co. Ltd.
Consultants: Erduman Consultancy Engineering Co. Ltd. (Structural), Bahri Türkmen Engineering Consultancy Co. Ltd. (Mechanical), Ovacık Engineering Consultancy and Trading Co. Ltd. (Electrical), Levent Topaktas (Acoustical)
Construction Type: Reinforced Concrete, Steel

"Ulvi Cemal Erkin"音乐厅的设计目标主要在于让建筑物对城市开放，并组建一系列多功能的城市空间。它为公众提供了自由的流通与活动区域，并将公共空间引入建筑物内。对公众来说，这座音乐厅的主要功能是创造、展览以及娱乐。建筑主要通过大道及城市规模来与城市建立联系。该设计将音乐厅内部分为多个区域，各个区域在空间布局、体积、技术要点及调配等方面各具特色。各区域相互独立但同时都与门厅相连。除了公共入口，厅内还设有为艺术家、大楼管理人员、服务人员、电台及市政府工作人员提供的私人入口，不会造成人员流通的冲突。

根据功能、体积以及空间需求，音乐厅可以从X、Y、Z三个方向进行区域划分。从X方向可以分为三片狭长的区域：第一片是面对大道的门厅；第二片是各个大厅；第三片是面对主街的各种服务区域。各个区域的大小是根据其功能来确定的。音乐厅从Y方向可以划分为4个区块：北边的排练厅、主厅、小厅以及市政当局服务区。同X轴上的各个区域一样，Y轴上的4个区块的大小也是根据其功能来决定。四个区块的纵向长度从北到南依次缩短，这种层次性淡化了整个建筑的厚重感，

X轴与Y轴相交区域，也就是各个大厅被抬高，这样就构成了建筑的两个楼层。4个区块如同悬臂一般挂起，影子投射在大道之上。区块的末端设有门厅及露台，可以欣赏到城市的美景。在悬起的区块下面有透明玻璃墙围起的空间，用作门厅以及对广场开放的展示区。这一透明区域还用来提供户外公共活动的延续。

人们可以通过观光电梯、自动扶梯以及斜坡道到达上部的4个区块。外面的斜坡道将庭院与艺术家工作室连接起来。斜坡道也作为了残疾人与老人上楼的通道。

In Ulvi Cemal Erkin Concert Hall Project, the main goal of the design is to open the building to the city and create multi-purpose urban spaces. It provides free circulation and activity areas for the public and takes the public spaces into the building. Production, exhibition and recreation are the main activities feeding the public face of the building. Panoramic relations with the city are created both at the level of the boulevard and the urban scale. Program of the building is classified into blocks and each block is differentiated according to its spatial, dimensional, technical necessities and regulations. The blocks are designed to work separately but connected with foyers. Besides public entrance, private entrance for artists, building management, service, radio-TV and the municipality is also provided. The circulations are not coincided but, the visual relation is created.

The building is classified programmatically on the directions of X, Y and Z according to the functional, dimensional and spatial needs. It is classified on the X direction into 3 strips: the first strip facing the Boulevard consists of foyers; the second strip of halls; the third strip facing the main street consists of services. The dimensions of these strips are defined according to their functional needs. The building is composed of 4 blocks on the Y direction. The first block at the north; 'rehearsal halls block', the second one; 'main hall block', the third one: "small hall block" and the fourth one is the 'municipality service unit's block'. Like the X axis directed strips, the dimensions of these blocks are defined according to their functional needs. The mass effect of the building is decreased by gradual leveling of the blocks, in which the blocks are pulled back gradually from north to south.

The blocks coincided with the strips are heightened on the Z direction and so, the floors of the building are composed. Each block works as cantilever and projects onto the Boulevard. At the tips of these cantilevers there are foyers and terraces with open view to the city. Under the cantilevered blocks, there are transparent volumes housing foyers and exhibition areas open to the plazas. This is a transparent volume providing the exterior public activities continue inside the building.

The accessibility to the heightened blocks is provided through panoramic elevators, escalators, and ramps. Ramps at outside connect the courtyards and the ateliers physically and visually. The accessibility of the disabled and the old people is also provided through these ramps.

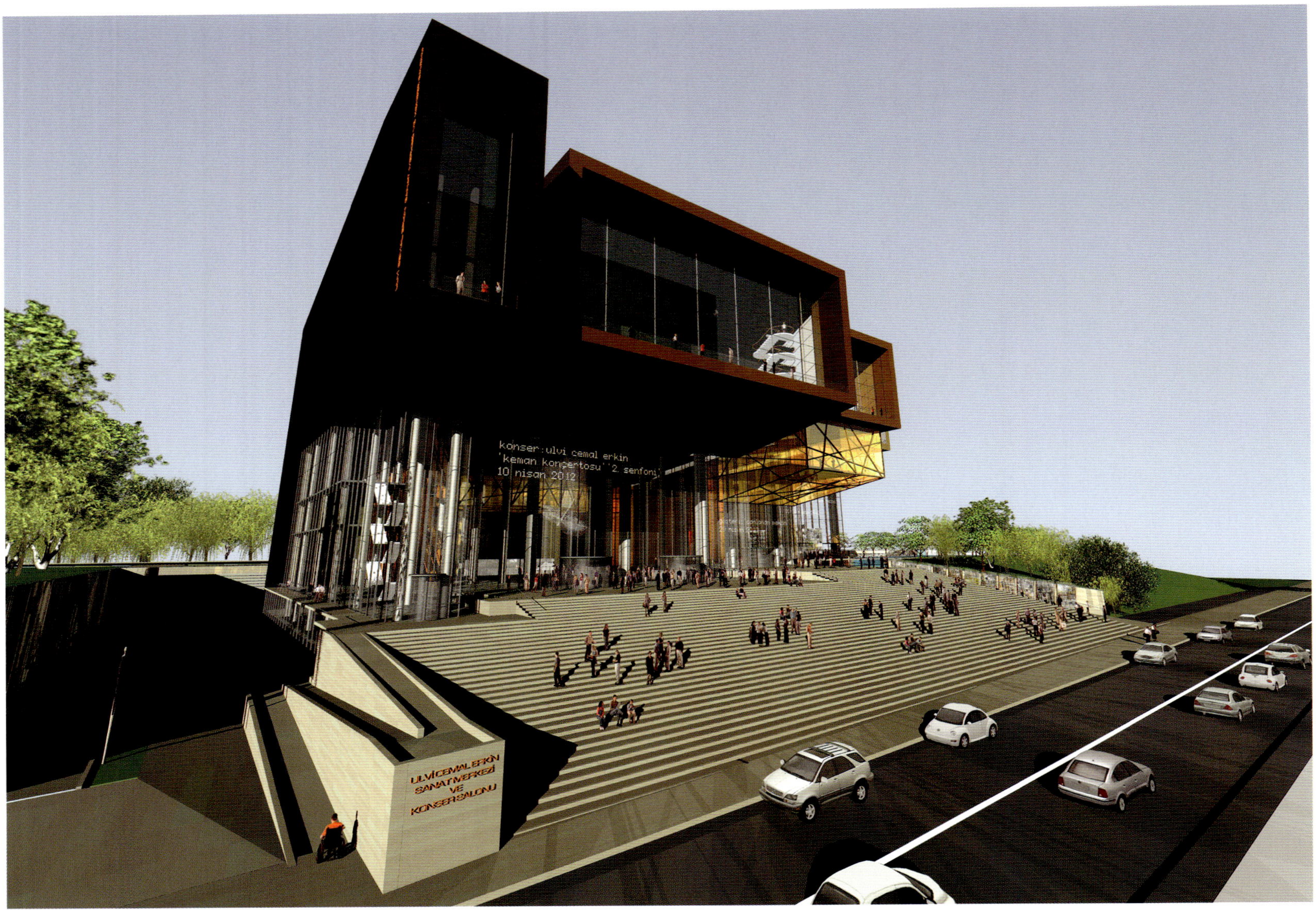
konser:ulvi cemal erkin
'keman konçertosu''2 senfoni'
10 nisan 2012
ULVİ CEMAL ERKİN
SANAT MERKEZİ
VE
KONSER SALONU

ÇANKAYA BELEDİYESİ BAŞKANLIK HİZMET BİNASI

konser: ulvi cemal erkin
keman konçertosu '2 senfoni
nisan 2012

Kristiansund Opera and Culture Center

克里斯蒂安松歌剧院和文化中心 C. F. Møller Architects

First Prize in International Competition
国际竞赛一等奖

Credits

Client: Kristiansund Municipality and FG Eiendom
Lacation: Kristiansund, Norway
Architect: C. F. Møller Architects
Landscape: C. F. Møller Architects
Engineers: COWI, Buro Happold
Stage Technics: Torsten Nobling - AIX Arkitekter AB-Arteno
Area: 15,400 m²
Competition Year: 2010
Construction Year: 2014

克里斯蒂安松歌剧院和文化中心是挪威历史最悠久的歌剧院，历史可以追溯至1805年，每一年都会举办持续数周的歌剧艺术节——挪威规模最大、剧目最齐全的歌剧和音乐剧演出，也是斯堪的纳维亚半岛规模最大的歌剧艺术节之一。

新歌剧院的愿景是为了要在克里斯蒂安松——诺德摩尔地区首府，特隆赫姆西部——设立一个文化力量中心，也是整个地区的地标性建筑。

除了歌剧院之外，新文化中心还会提供一个芭蕾舞中心、一个图书馆、一个大学中心、一个会议室、一个餐厅、一个咖啡馆以及一个天空酒吧。该项目还将整合目前已有的两幢建筑，其中一幢是从前的Folkets Hus，并在建筑物四周，通过连接人行步道、广场和附近的公园，建立生意盎然的城市空间。

Kulturkvartalet，如同项目制作方C. F. Møller建筑事务所称呼的那样，在动态创新的环境内，使不同的文化功能区域，于主音乐大厅的中心处，浑然一体。音乐大厅将提供卓越的音响效果和超灵活的剧场设施。

克里斯蒂安松歌剧院是近年来由C. F. Møller建筑事务所设计的若干具有重大意义的文化建筑之一。其他建筑还有，伦敦达尔文中心、伦敦自然历史博物馆蚕茧型扩建区，其中包括达尔文独一无二的收藏和世界领先的科学研究中心等研究设施。C. F. Møller建筑事务所还负责设计了奥尔胡斯音乐厅扩建部分，建筑面积翻了一番，带有一个新的交响音乐厅，为奥尔胡斯交响乐管弦乐队和皇家音乐学院提供卓越的音响效果和设施。而且，该事务所还在新奥登塞音乐剧场——一座拥有4个音乐厅，为卡尔·尼尔森音乐学院提供设施的文化中心设计大赛中获得一等奖。

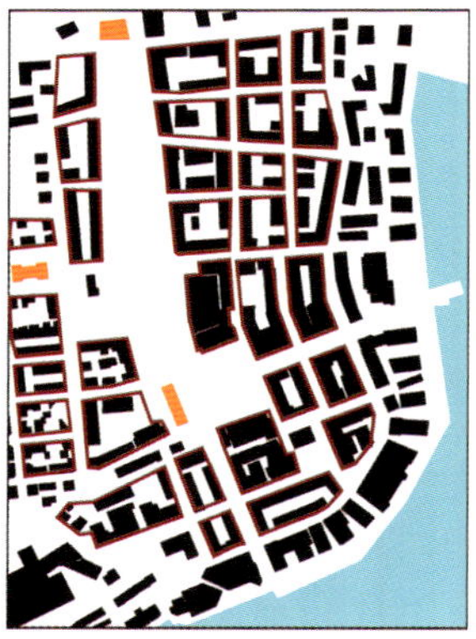
URBAN BLOCK STRUCTURE AND LANDMARKS

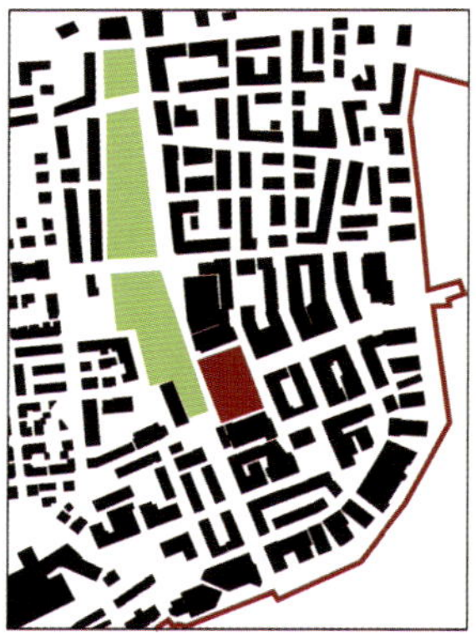
MAIN URBAN SPACES

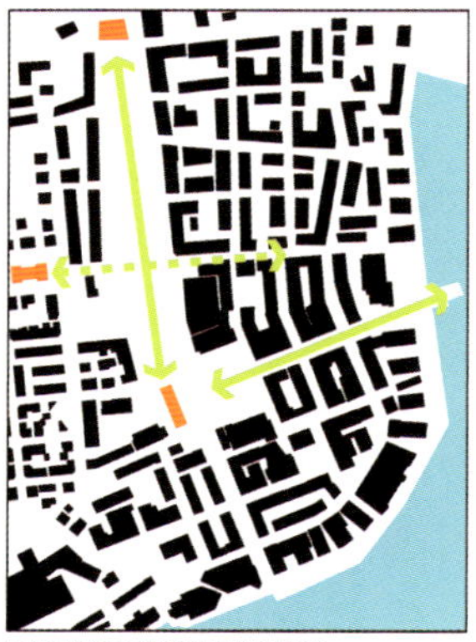
LANDMARK BUILDINGS AND MAIN VIEWS

Kristiansund Opera and Cultural Center will house Norway's oldest opera, dating back to 1805, which every year arranges the Opera Festival Weeks - Norway's largest and most comprehensive presentation of operc and musical theater and one of Scandinavia's largest opera festivals.

The vision of the new Opera House is to create a cultural power center in Kristiansund, the capital of the region of Nordmøre, west of Trondheim – a landmark to the entire region.

Apart from the opera the new cultural center will house a ballet center, library, college center, conference rooms, a restaurant, a café, and a skybar. The task has been to integrate two existing buildings, one of them the former Folkets Hus, in the new and create a vibrant urban space around the building by connecting the cultural center with the pedestrian streets, squares and a nearby park.

Kulturkvartalet, as the project made by C. F. Møller Architects is called, unites the different cultural functions in a dynamic and creative environment, centers round the main concert hall. The concert hall has been designed to offer sublime acoustics and super flexible theater settings.

The Kristiansund Opera is one of several significant cultural buildings designed by C. F. Møller Architects in recent years. These include the Darwin Center in London, the cocoon shaped extension of the Natural History Museum in London, which contains both Darwin's unique collection and a world-leading science research center research facilities. C. F. Møller Architects has also been responsible for extension of the Concert Hall Aarhus, which has been doubled in size with a new Symphonic Hall with sublime acoustics and facilities for the Aarhus Symphony Orchestra and the Royal Academy of Music. Furthermore the company has won first prize in the competition for the new Odense Music and Theater Hall, a cultural center with four halls and facilities for the Carl Nielsen Academy of Music.

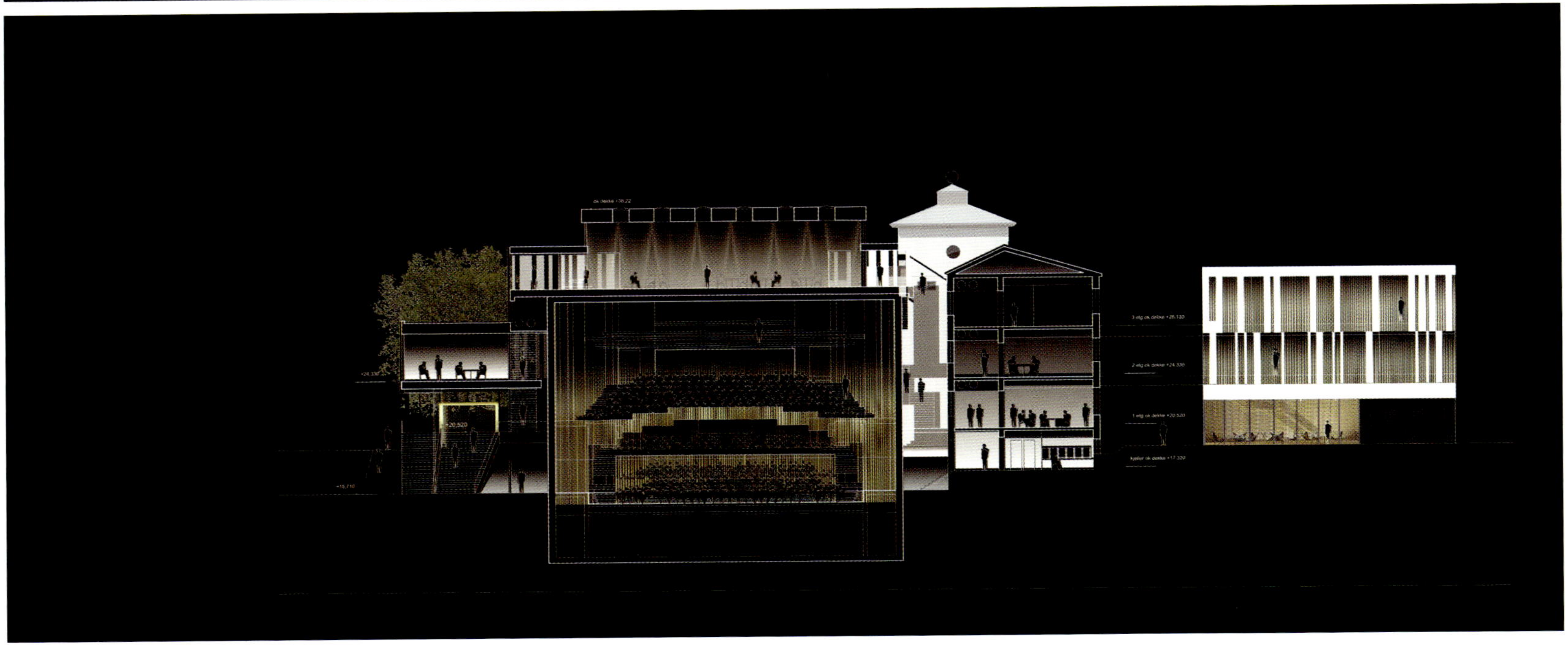

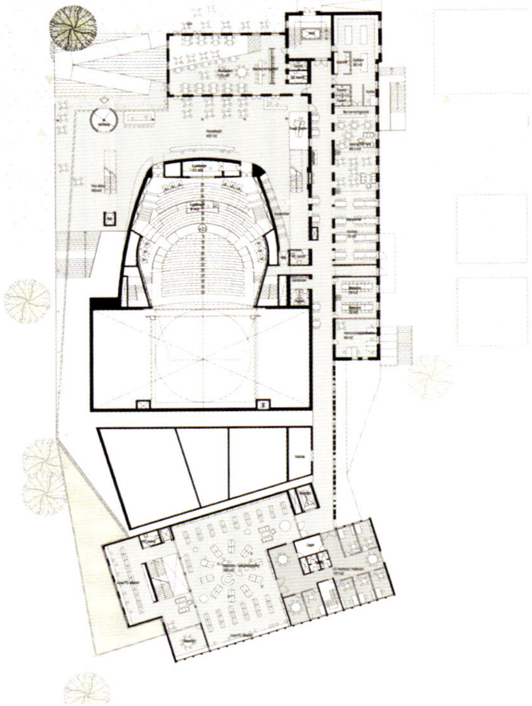
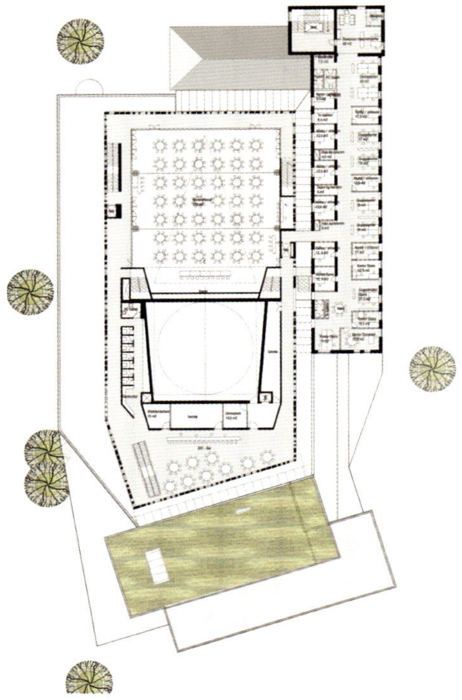
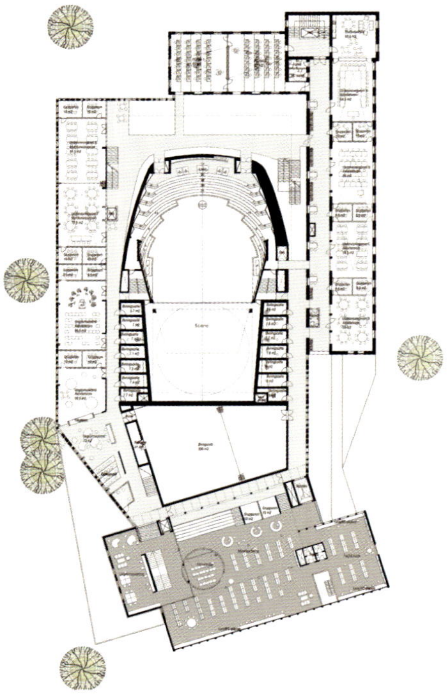

Odense Music and Theater Hall
奥登塞音乐戏剧大厅 C. F. Møller Architects

2008 First Prize in Architectural Competition
2008年建筑竞赛一等奖

Credits

Client/PFI Group: Odense Municipality (initiators and users)
The Carl Nielsen Academy of Music (users)
H. C. Andersen Congres Center A/S (owner, operator and user)
Lillebælt Gruppen A/S (contractor)
Architect: C. F. Møller Architects
Landscape Architect: C. F. Møller Architects

Engineers: COWI A/S and Anders Chr. Gade (acoustics)
Collaborators: Maths Nystrøm (theater technics)
Focus Advokater I/S (legal advice)
Nykredit (finances)
Area: 18,500 m²
Address: H. C. Andersen Quarter, Odense, Denmark
Competition Year: 2008

新音乐戏剧大厅的愿景是为了建立一个在同一屋檐下拥有4座舞台和卡尔·尼尔森音乐学院的文化中坚力量。大厅能容纳约1,800名观众，小厅能容纳约300名观众。剧场还拥有一个小型室内音乐厅以及一个韵律厅。文化中心同时还提供一间餐厅、通往屋顶露台的酒吧/休息室、办公室以及小卖部和一个地下停车场。

建筑地点位于汉斯·克里斯蒂安·安德森中心，邻近历史久远的小镇和多层居民楼。多变的小镇景观为建筑设计提供了灵感。开发项目由许多小镇住宅组成，楼房高度随邻里建筑变化而变化，两者和谐相处。

建筑的体积对城市结构的影响因此被降低了，细分之下的建筑帮助创造了各式各样的、小规模的城市空间，环绕着音乐剧场。

两幢建筑——盛典大厅和舞台之塔，以坚固的特性脱颖而出。灿烂的白色外景，与阴沉的建筑立面形成鲜明对比，给文化中心带来感官上的一致性。白色的表面可以用影像投射，告诉外界建筑内的活动。

文化中心将会带有城中之城的特性，正如同各式各样的建筑，将由玻璃顶的拱廊、广场和开放空间串联起来，同时又与毗邻社区相连接。

门厅整天扮演着“文化市场”的角色，另外高度灵活的音响效果以及盛典大厅的功能陈设，与其意大利巴洛克剧场启发的几何形状一道，保证了除音乐会和歌剧之外的多功能性，比如，跳舞表演、研讨会、时装秀等。

MUSIK- & TEATERHUSET OG
MUSIKKONSERVATORIUM ODENSE

MUSIK- OG TEATERHUS OG MUSIKKONSERVATORIUM

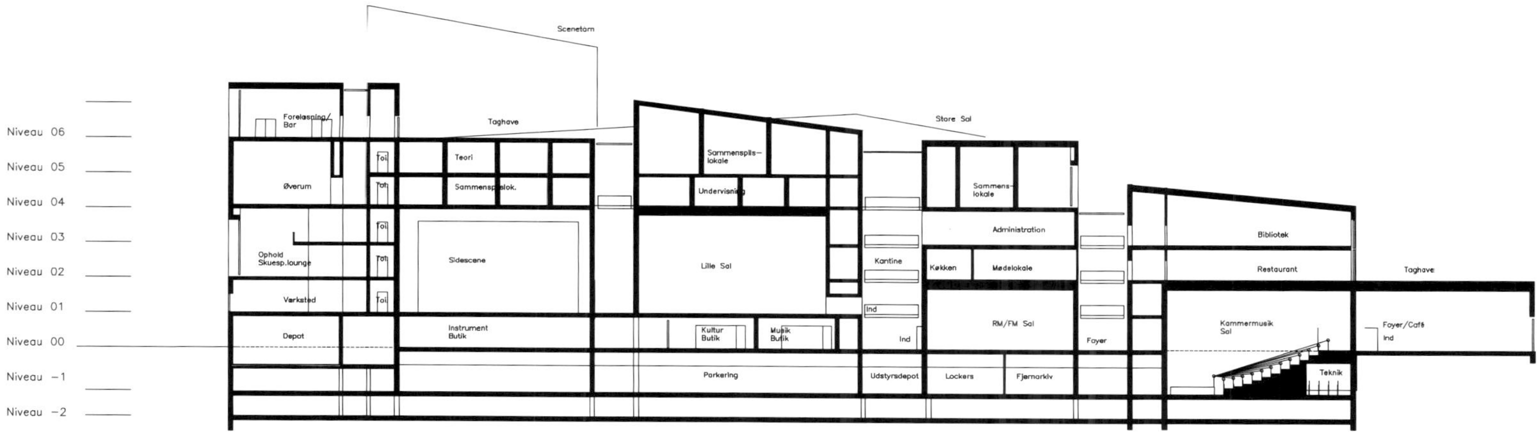

SNIT A–A

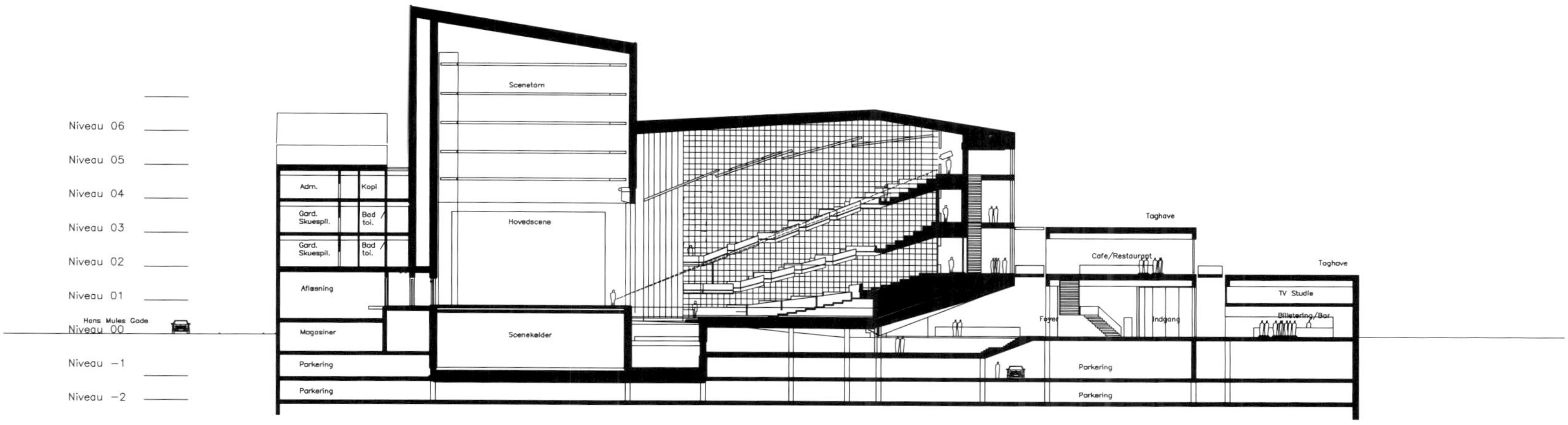

SNIT B–B

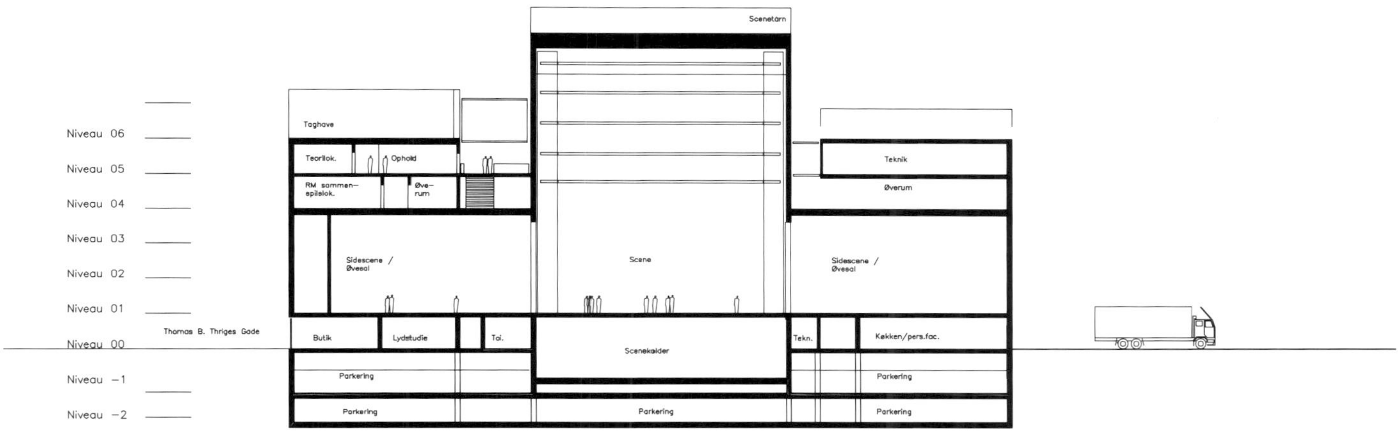

SNIT C–C

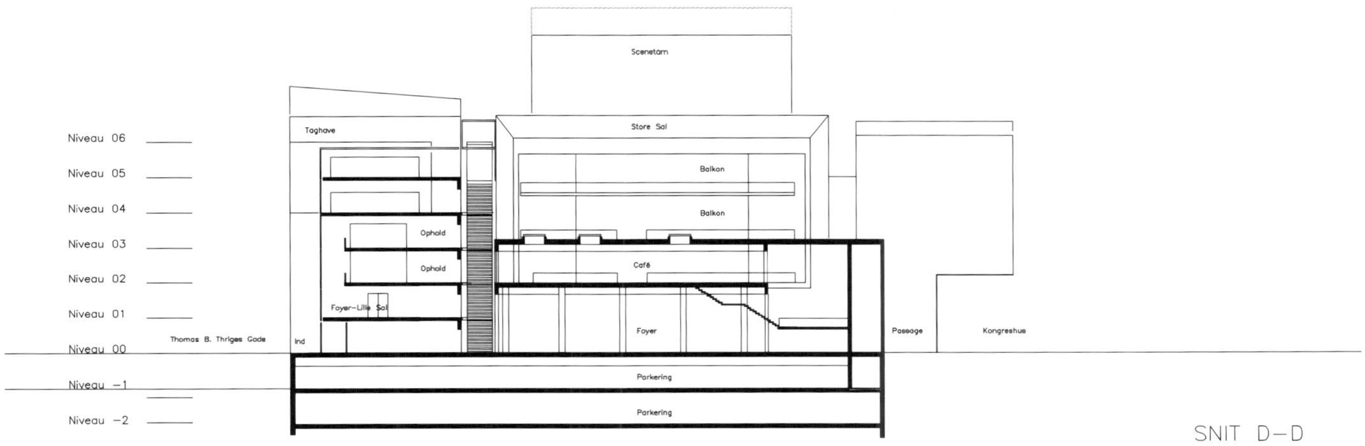

SNIT D–D

Nive

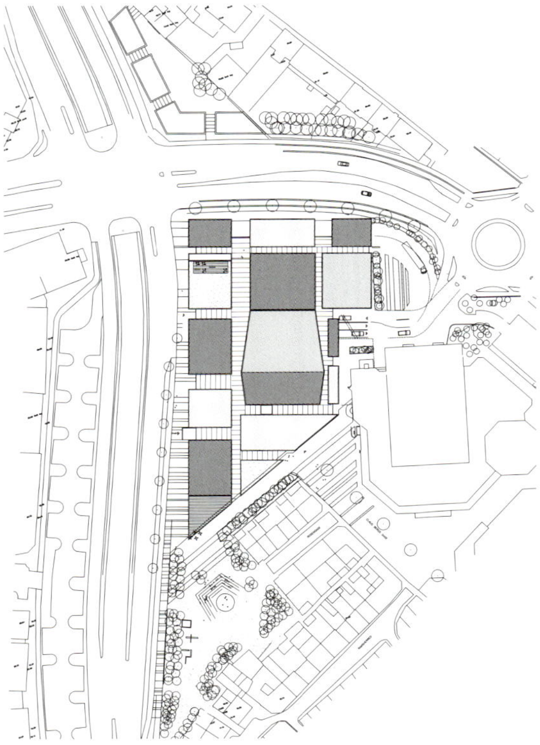

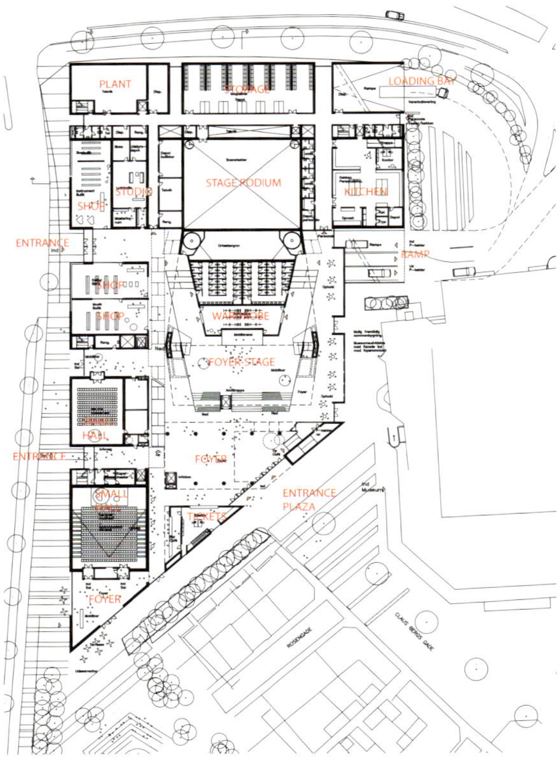
PLANT
STORAGE
LOADING BAY
STAGE PODIUM
STUDIO
SHOP
KITCHEN
ENTRANCE
RAMP
SHOP
SHOP
WARDROBE
FOYER STAGE
ENTRANCE
FOYER
ENTRANCE PLAZA
TICKETS
SMALL HALL
FOYER

STAGE PODIUM
PLANT
RAMP
155 p-pladser
PARKING 150 PLACES
5600 m² p-kælder
BICYCLE STORAGE
SMALL HALL

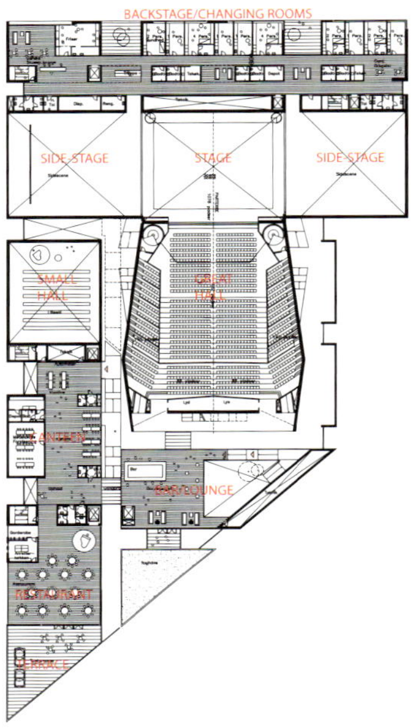
BACKSTAGE/CHANGING ROOMS
SIDE-STAGE
STAGE
SIDE-STAGE
SMALL HALL
GREAT HALL
LOUNGE
RESTAURANT

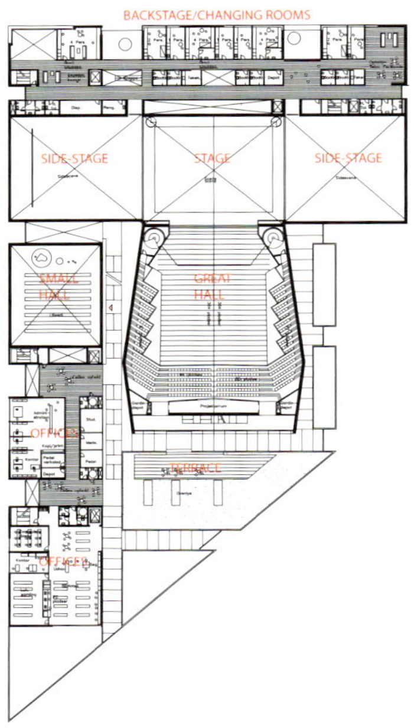
BACKSTAGE/CHANGING ROOMS
SIDE-STAGE
STAGE
SIDE-STAGE
SMALL HALL
GREAT HALL
OFFICES
TERRACE
OFFICES

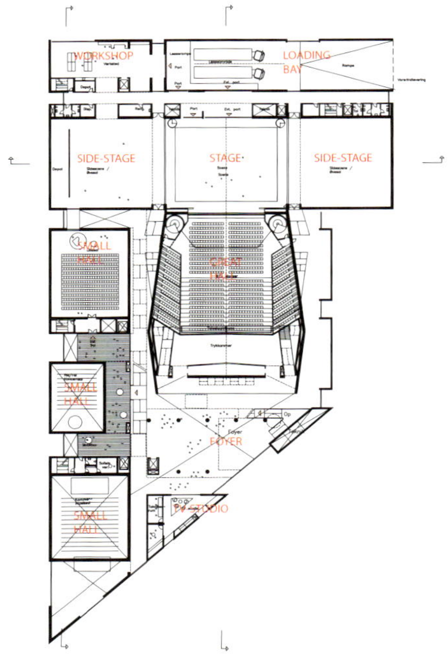
WORKSHOP
LOADING BAY
SIDE-STAGE
STAGE
SIDE-STAGE
SMALL HALL
GREAT HALL
SMALL HALL
FOYER
TV STUDIO
SMALL HALL

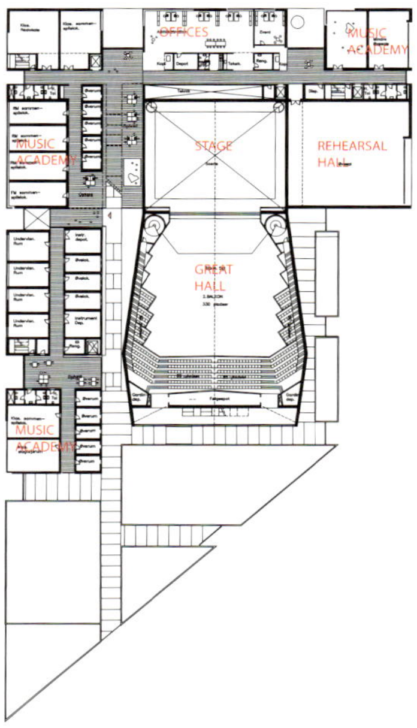
OFFICES
MUSIC ACADEMY
MUSIC ACADEMY
STAGE
REHEARSAL HALL
GREAT HALL
MUSIC ACADEMY

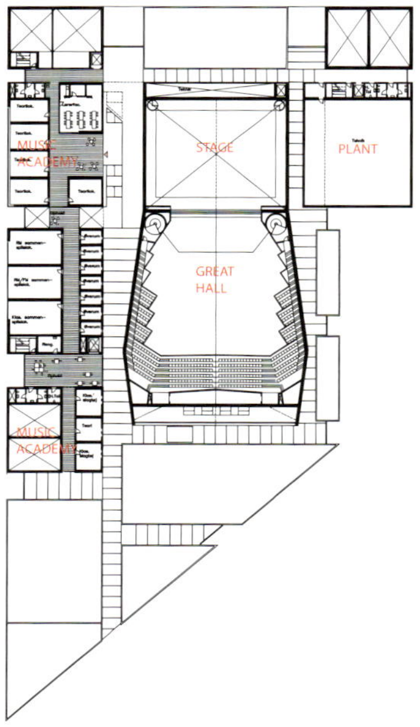
MUSIC ACADEMY
STAGE
PLANT
GREAT HALL
MUSIC ACADEMY

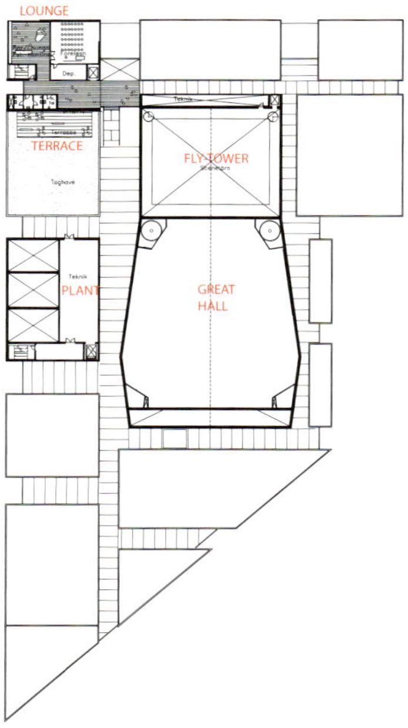
LOUNGE
TERRACE
FLY-TOWER
PLANT
GREAT HALL

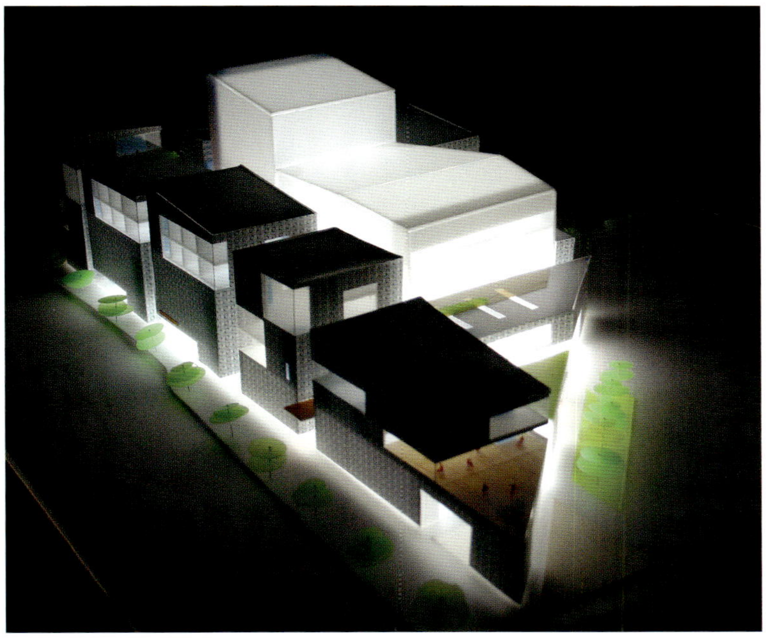

The vision of the new music and theater hall is to create a cultural power center with four stages and the Carl Nielsen Academy of Music under one roof. The great hall accommodates approximately 1,800 spectators, while the small hall accommodates approximately 300 spectators. There is also a chamber music hall and a rhythmic hall. The Culture center also has a restaurant, bar/lounge with access to a roof terrace, offices and canteen as well as a parking basement.

The site is located in the Hans Christian Andersen Quarter and borders both the historical town and contemporary multi-storey housing. The varied townscape has provided inspiration for the architecture. The development is composed of a number of town houses, the heights of which vary in harmony with the neighbouring buildings.

The volume of the building is thus reduced in its impact on the urban fabric, and the subdivision helps create varied and small-scaled urban spaces surrounding the Concert and Theater Hall.

Two of the buildings stand out in their solidity: the Great Hall and the Stage Tower. With a luminous white exterior, these form a contrast to the otherwise dark relief facades, and help to bring a sense of identity to the cultural center. The white surfaces can be used for image projections to tell the surrounding world about the activities in the building.

The cultural center will acquire the character of a city within a city, as the various buildings will be linked by glass-roofed arcades, squares and open spaces which connect with the surrounding neighborhoods.

The foyer acts as a 'cultural marketplace' all day long, and the highly flexible acoustic and functional layout of the great hall, with its basic geometry inspired by the Italian baroque theaters, guarantees a multitude of uses besides concerts and opera, such as dance performances, conferences, fashion shows etc.

BILLETSALG
RM / FM
KONCERTSAL

Theater Complex, Langfang, China

廊坊剧院综合体 Will Alsop at RMJM

Competition Entry, Third Place
竞赛第三名

Credits
Area: 150,000 m²
Location: Langfang, China

该项目为廊坊提供了一个地标性的文化平台来和邻近城市北京比拼。画廊、博物馆与酒店将核心剧院围绕在中间，设计理念基于"生活围绕戏剧，戏剧体现平淡生活，平淡生活就是戏剧"。

以剧院为中心，四周绿树成荫，是活动空间，停车库蜿蜒在地下，盘旋在一个封闭的表演舞台四周。另外，舞台由四根立柱支撑，每根立柱都光彩夺目。这使剧院成为整个区域的标志性建筑。

2,000人的核心剧院，800人的综合剧院和500人的儿童剧场，这三个剧院均设在同一栋楼里，构成主楼。核心剧院设计精细，使其能成为一个标准的270度舞台。同时，主楼还设有三个剧院公用的主休息室及后台技术员工休息区，它们结合在一起构造出中国传统景观建筑。

四大立柱之一作为剧院的入口。从地下入口，游客可以搭乘自动扶梯直接进入一层主剧场休息室。这里设有商店、售票处、垃圾回收及餐饮处。二楼休息室设有天桥，可直通剧院。剧院四周设有购物街，顶层是一个线性公园，公园内有统一喇叭状的餐厅，是举行婚礼的理想地点。

后侧两根立柱设有斜通道及大号铰链式手推车。该剧院将成为史上最大剧院，舞台、表演设备及主要楼层服务设施都可以直接运往后台。斜通道及大型货梯为紧急运输提供从地下进入楼层的直接通道。

剧院主楼立面采用模块化防水胶合板覆层系统。胶合板立面是一层钢格栅板结构，替换了部分上釉板，胶合板南面则装入光伏电池。尽管外表形式简单，但该系统采用了三组不同的模型。所有的设计都精心打造着剧院建筑完美的曲线。

5.0 THEATER BUILDING DESIGN 大剧院设计

5.7 BUILDING EXTERIOR- ELEVATIONS 建筑外观立面

NORTH ELEVATION 北向立面图

5.0 THEATER BUILDING DESIGN 大剧院设计

5.2 ORGANIZATION OF USES 使用者分析

1 Coach parking
2 Service Core
3 Theatre entrance, box office
4 Basement car parking, plant, servicing and waste storage
5 Entrance/exit to basement
6 Main theatre
7 External stage
8 Banked grass for outdoor performance seating
9 Theatre entrance, box office
10 Service core
11 Foyer
12 Classical performance theatre
13 Childrens theatre
14 Theatre servicing and back of house
15 Foyer
16 Proposed bus stop
17 Fly tower
18 Service Entrance
19 Proposed Taxi drop-off
20 Delivery bays at upper level

1- 大巴停车场
2- 服务核心筒
3- 剧院入口, 售票处
4- 地下停车库
5- 地下室出入口
6- 核心剧院
7- 户外舞台
8- 户外自然阶梯座位
9- 剧院入口, 售票处
10- 服务核心筒
11- 前厅
12- 古典剧院
13- 儿童剧场
14- 剧院服务空间与设备室
15- 前厅
16- 建议公交站点
17- 舞台塔及设备空间
18- 服务空间入口
19- 建议出租车下车点
20- 上层卸货空间

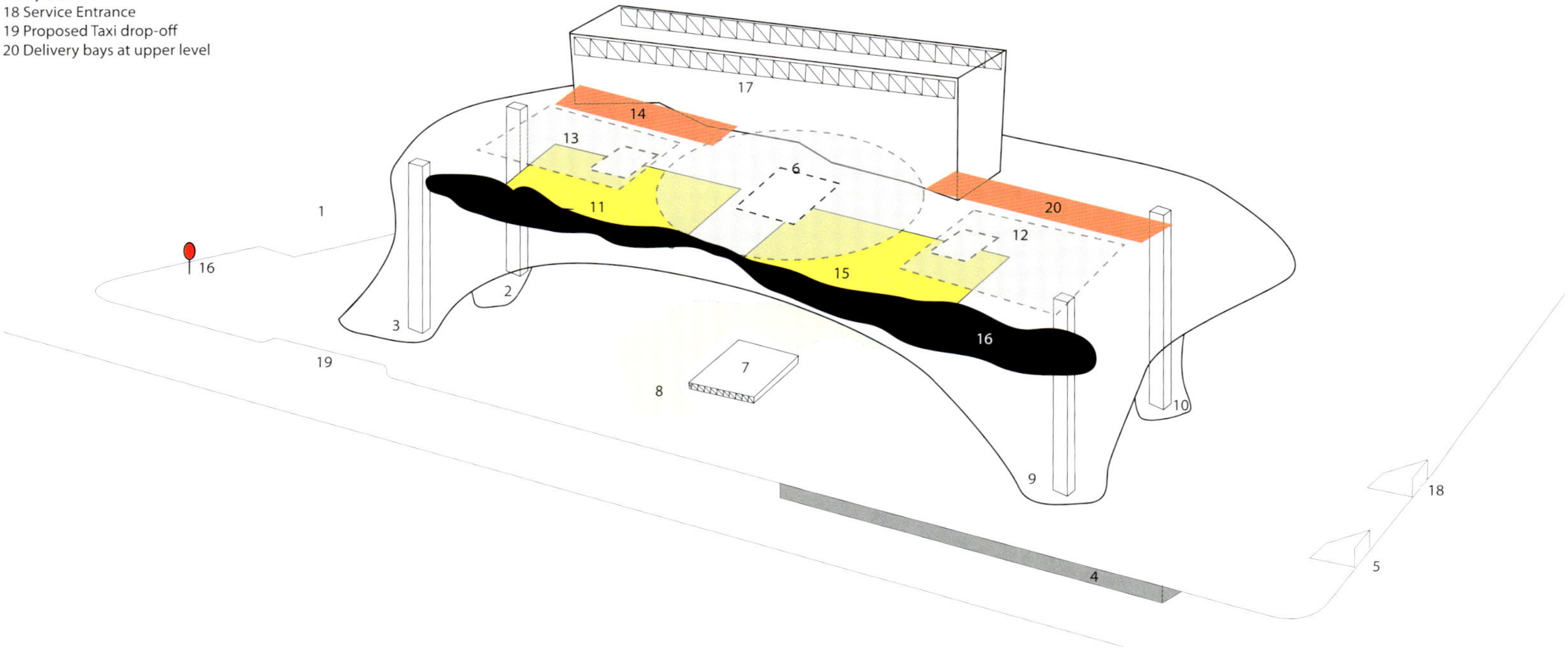

5.0 THEATER BUILDING DESIGN 大剧院设计

5.3 CIRCULATION 流线分析

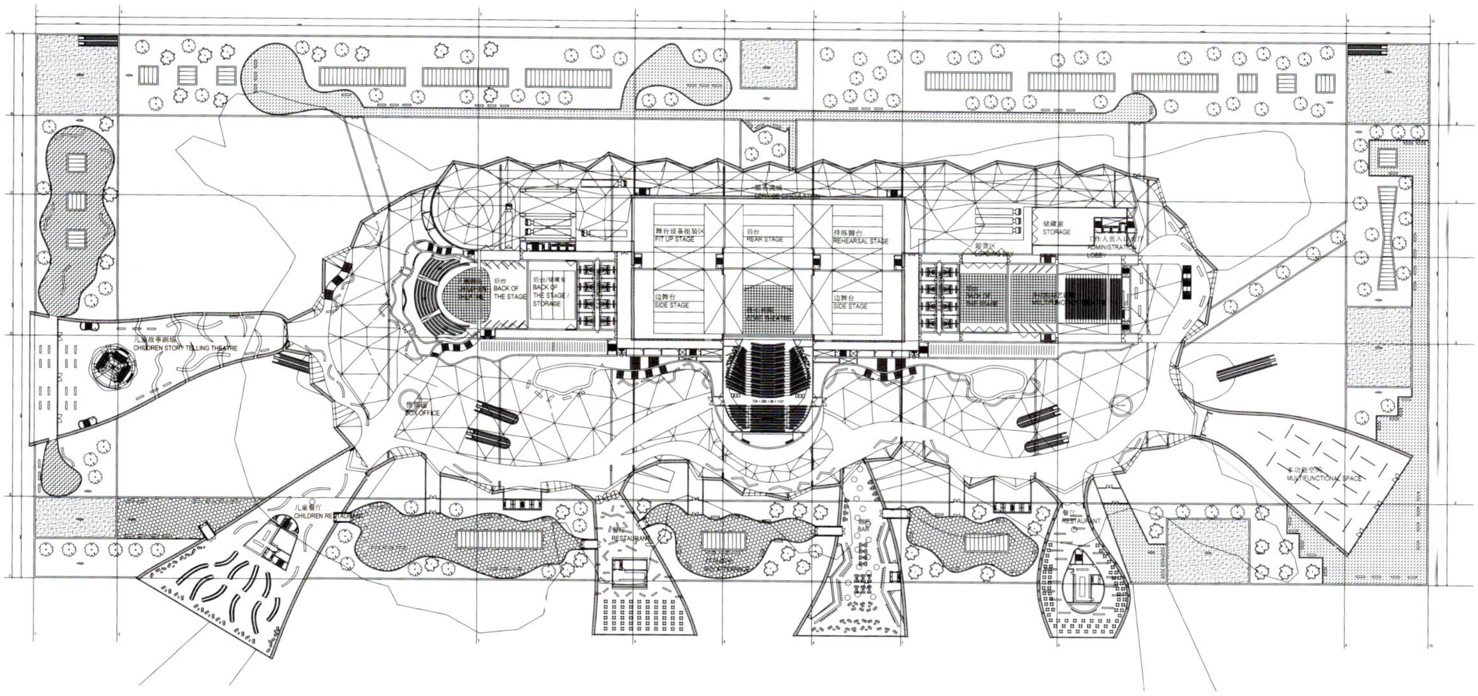

Theater & FOYER LEVEL PLAN 剧院层平面图

5.0 THEATER BUILDING DESIGN 大剧院设计

5.3 CIRCULATION 流线分析

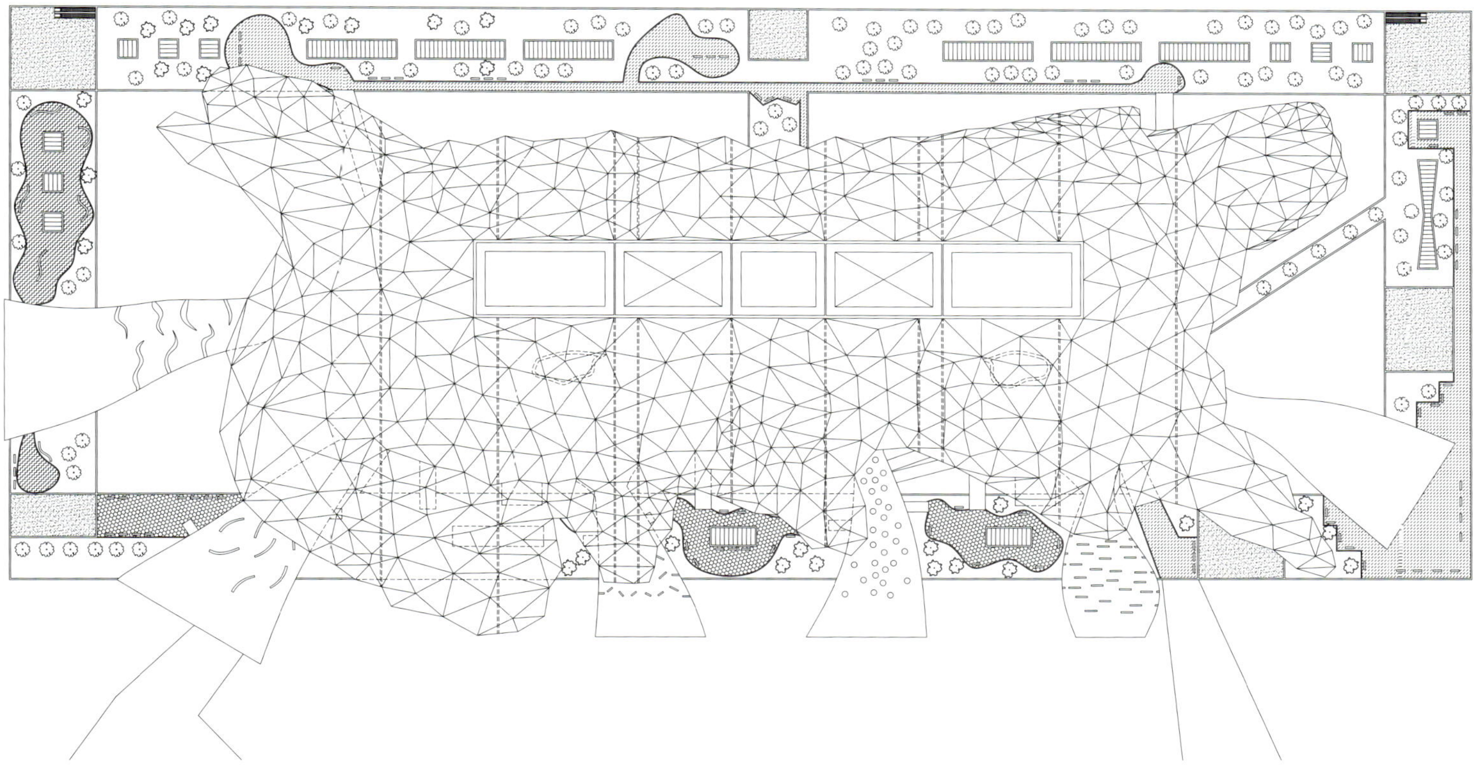

ROOF LEVEL PLAN 屋顶层平面图

5.3 CIRCULATION 流线分析

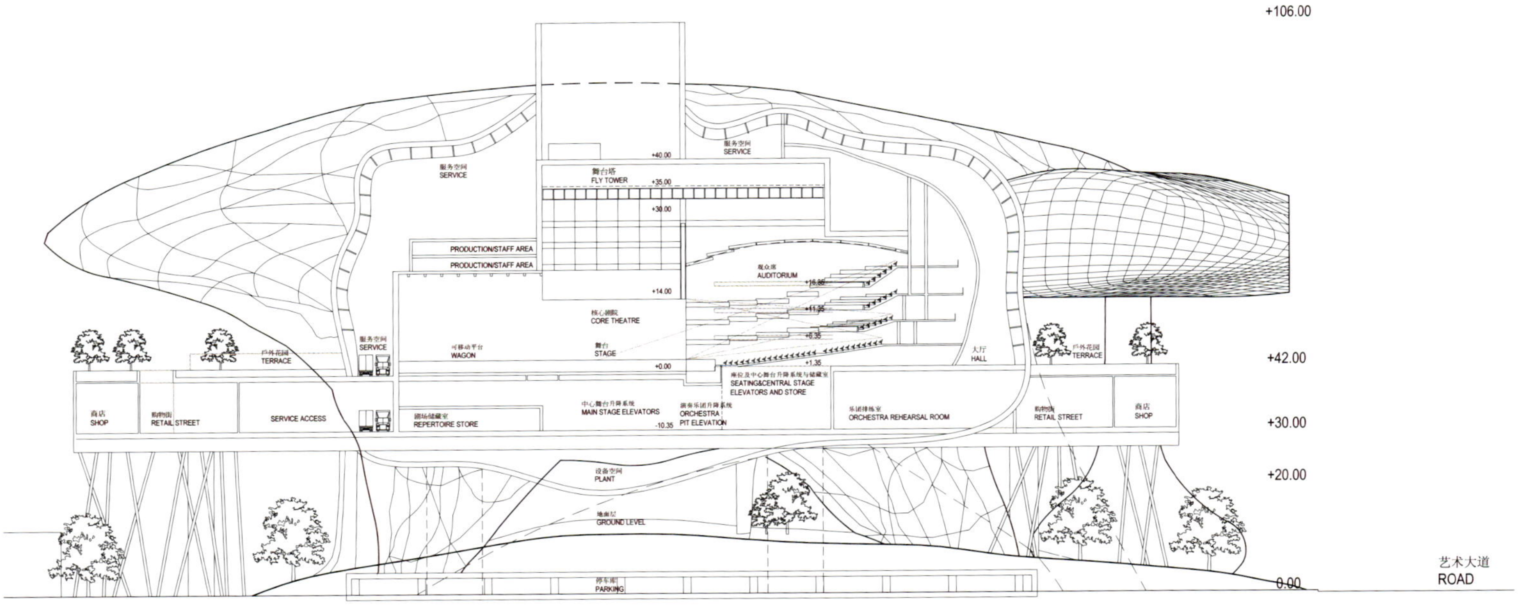

CROSS-SECTION 剖面图

The brief for this project called for an iconic cultural platform for the city to compete with nearby Beijing. The master plan is to include a cluster of galleries, museums and hotels around the central theater. The concept for this design is based upon the notion that "life revolves around the theater, the theater is placed in the center of ordinary things, life in ordinary things is theater." Centering the theater spatially in the site to allow life to revolve around it, involved raising it into the air so that the parking could flow underneath around a covered, shaded performance space. This also creates a sculpture within the park- a four-legged shiny element, unique to its surroundings.

This proposed main theater building contains three theaters - the main 2000-seat theater, 800-seat multifunction theater and 500-seat children theater under one roof. The main theater was carefully designed to be able to change layout to suit a standard, central or 270 degree stage. Meanwhile, the public main foyer shared by the 3 theaters and the back stage technical and personnel areas are also located in this main building. The 3 theaters and their back stage support area create a very sculptural integrated object as if the visitors are situated within traditional Chinese landscape.

The building is accessed via one of the four legs. From the ground level entrance, visitors take escalators directly up to the main public foyer of the theater building above where they can purchase and pick up tickets, drop off their belongings and have a drink or something to eat whilst enjoying the view. Visitors enter the theaters directly from the two-level foyer, with bridges across at upper level. A ring of retail surrounds the building with a linear park on its roof, ideal for weddings and functions with immediate access to the horn-shaped restaurants above.

An access ramp and a lift for full-size articulated lorries are provided within the rear two legs. This would be the largest theater in the world where stage sets, performance related equipments as well as general building services equipments can be delivered direct to back stage area. The ramp and the truck lifts also provide direct access from ground level into the building for emergency vehicles.

The facade of the main theater building is designed using a modular plywood rain screen cladding system. These panels sit onto a steel lattice secondary structure. Glazed panels can be substituted where necessary, and PV panels will be attached to the south side onto the plywood. Although freeform in appearance, this system uses a careful pattern of three different clusters. These are arranged strategically to suit the curve of the sculptural theater architecture.

Emeryville Center for the Arts

爱莫利维尔艺术中心

Aidlin Darling Design

Credits

Name: Emeryville Center for the Arts
Location: Emeryville, California
Area: 30,000 sf (2,787.09 m^2)
Cost: $10 Million
Status: Design Competition
Purpose: Visual & Performing Arts Center for Emerging Artists

爱莫利维尔艺术中心作为一座标志性的建筑，为艺术的创作者与欣赏者提供了一个聚会、交流的场所。建筑呈现出脚手架的形态，开放而灵活，参与者可以对其进行各种有创意的诠释。其多孔性将各种不同的活动联系起来。其中，美术展览馆和剧院作为建筑最基本的组成部分，成为城市居民活动领域的延伸。

设计师在建筑的西北角设置了一个全开放式的门厅，临街而建，将艺术中心与毗邻的城市大厅结合在一起。参观者进入室内，迎面可见各种丰富多彩的视觉与表演艺术作品。在建筑底层，门厅、剧院、咖啡厅和商店看上去如同要溢出可装卸的玻璃幕墙似的。从门厅处可以很明显地看到拱顶的开放式美术馆，美术馆的古木与砖砌结构淹没在从西边射来的光线之下。

在庭院中，沿着一系列的台阶与斜坡，参观者可以到达咖啡厅以及剧院的屋顶之上的露天区域。通过这样一个向上的阶梯，参观者不仅可以从不同角度欣赏美景，还能够直接进入到建筑内部，实现与艺术家多层次的交流。参观者在屋顶的空中庭院可以欣赏各种电影、雕塑、展示品以及互动性的表演节目，也可以随意聚会交流。晚上的时候，剧院上的高塔将成为闪亮的标志，表现建筑内部活动的同时，也体现出爱莫利维尔艺术中心的城市形象。

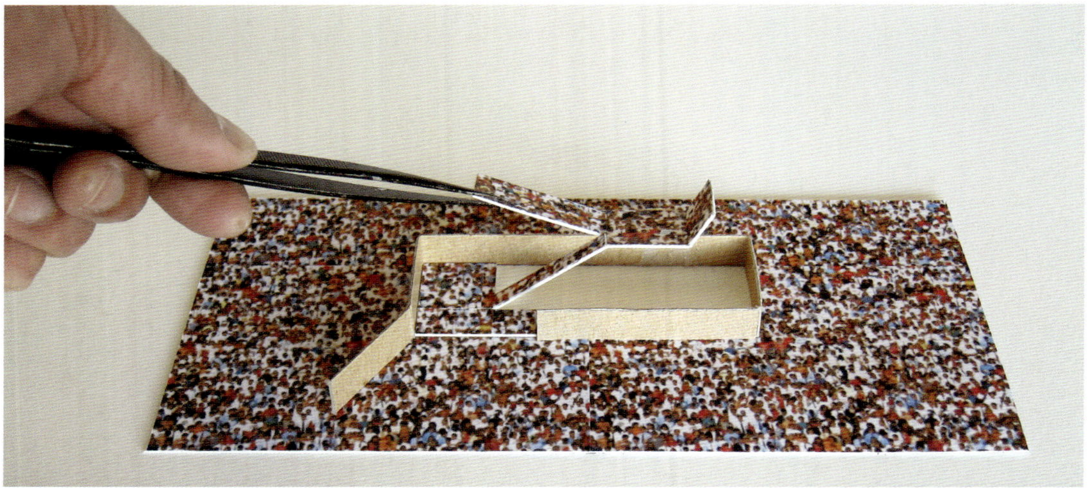

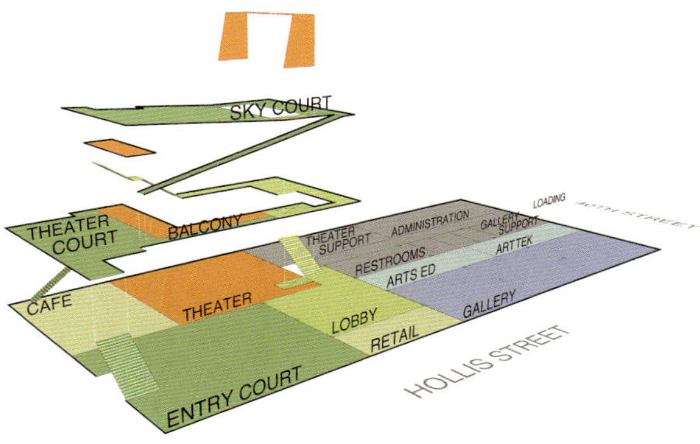

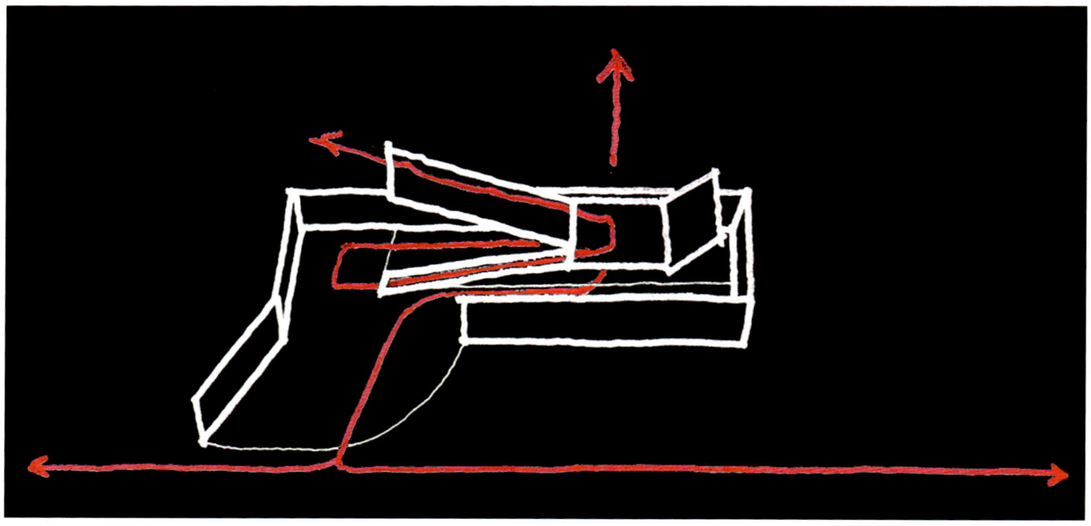

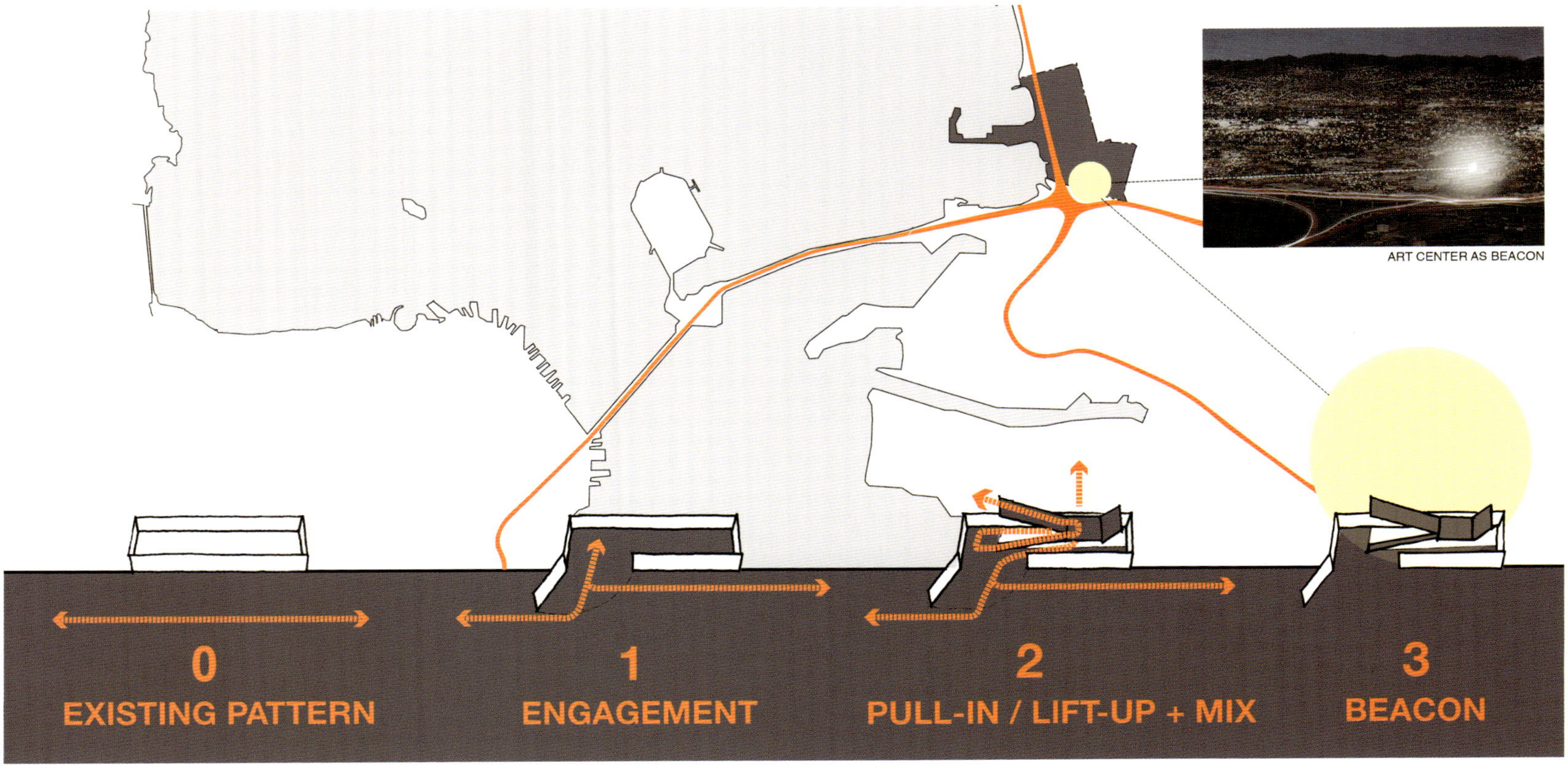

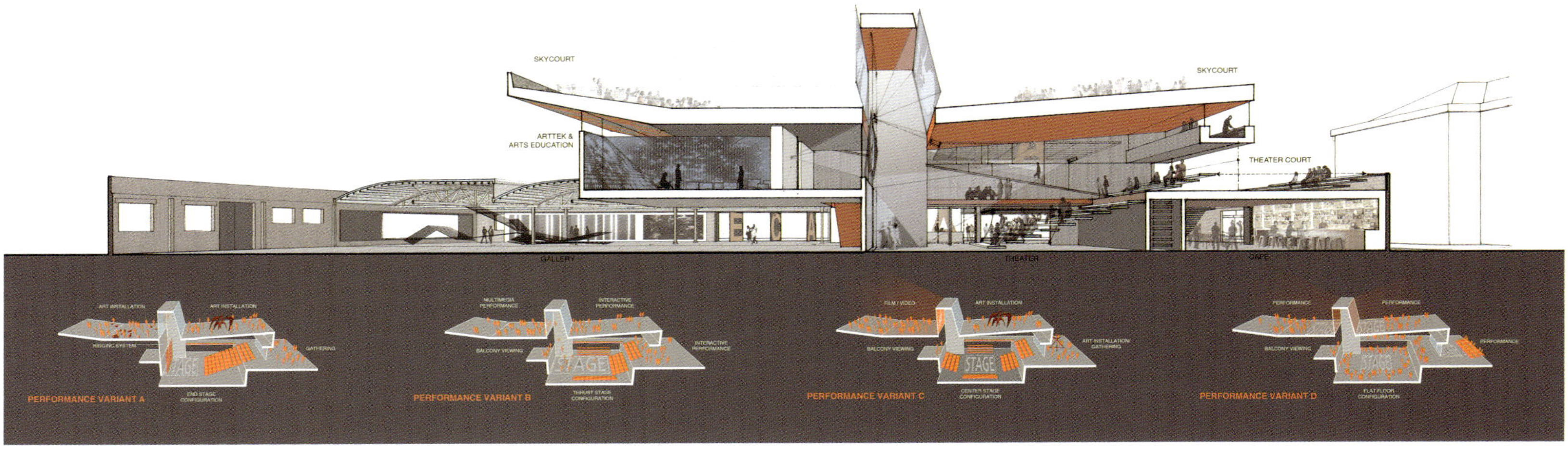

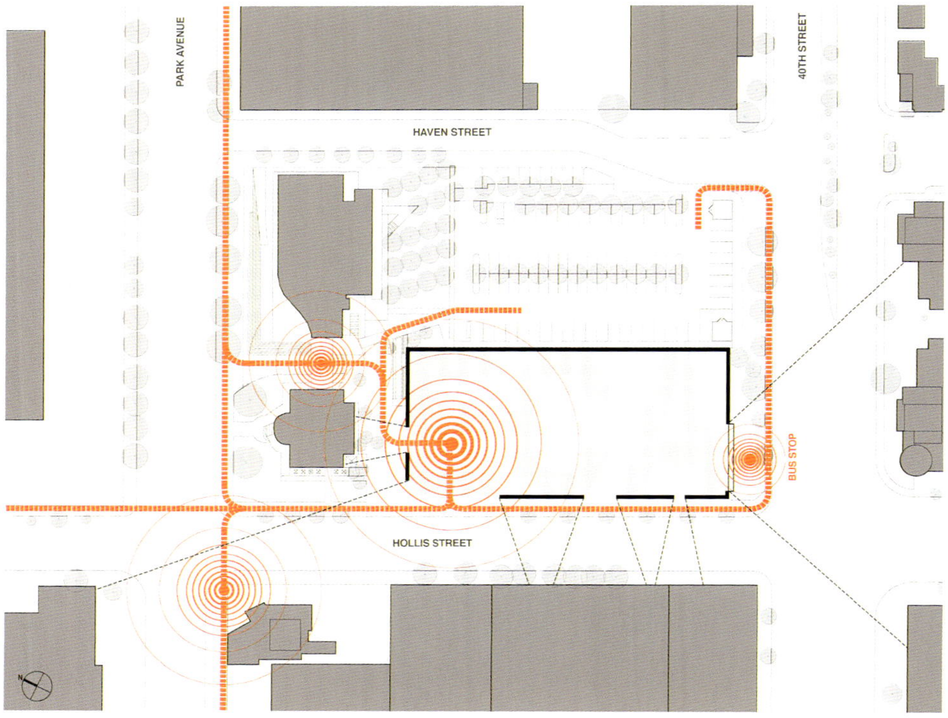

NODES OF PEDESTRIAN ACTIVITY

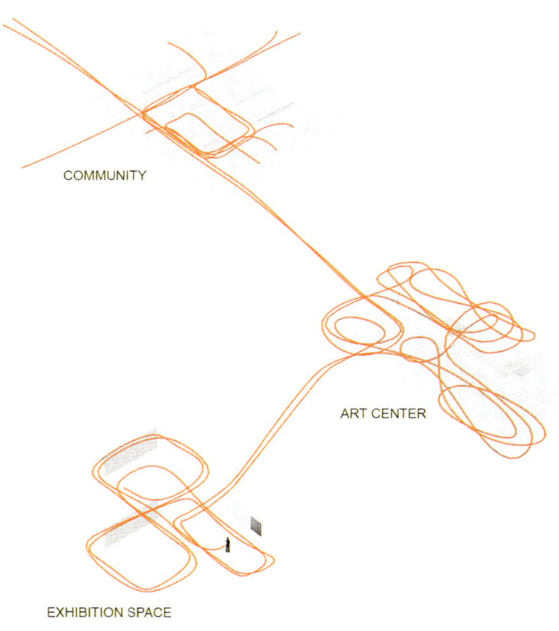

A CONTINUOUS SPLINE OF ENGAGEMENT AT MULTIPLE SCALES

The new Emeryville Center for the Arts is both a beacon and a gathering place, fostering dialogue among the makers and viewers of the arts. The building performs as a scaffold, open and flexible for creative interpretation by its participants; its porosity allows unexpected connections among diverse activities. Its principal components, a gallery and a theater, are conceived as extensions of the civic realm.

An entry court is carved into the northwest corner of the existing building, opening broadly to the street and uniting the Arts Center with the adjacent City Hall. Upon entry the visitor is offered numerous options to explore the visual and performing arts. At ground level, the lobby, theater, cafe, and retail spaces spill out into the courtyard through operable glass walls. Immediately visible through the lobby, the vaulted open gallery is set within the historic wood and brick structure and awash with northern light.

From the courtyard, visitors are invited up a series of stairs and ramps to outdoor spaces above the café and on the roof of the new theater structure. This climbing circulation spline provides views and direct access into spaces within the building, allowing for interaction between viewer and artist at multiple levels. The rooftop sky courts are elevated civic spaces accessible to the public for film, sculpture, exhibits, and interactive performances as well as for casual lunchtime gatherings. At night the theater's fly tower becomes a glowing icon on the skyline, reflecting and projecting both the activity within and the civic presence of the Emeryville Center for the Arts.

Mirage Dance Hall

梦幻舞厅 Kjellgren Kaminsky Architecture

Open International Competition First Prize

国际公开竞赛一等奖

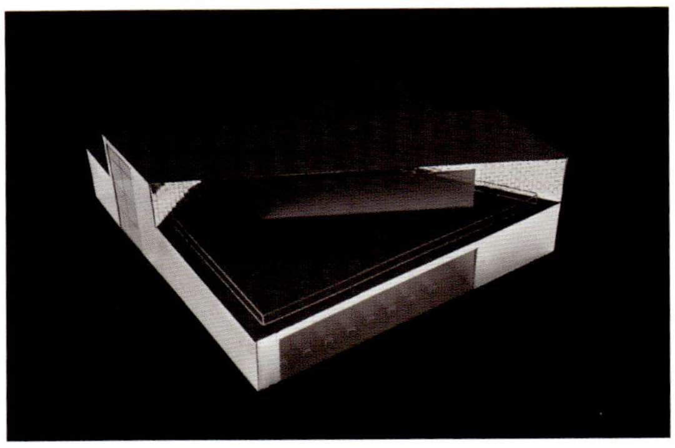

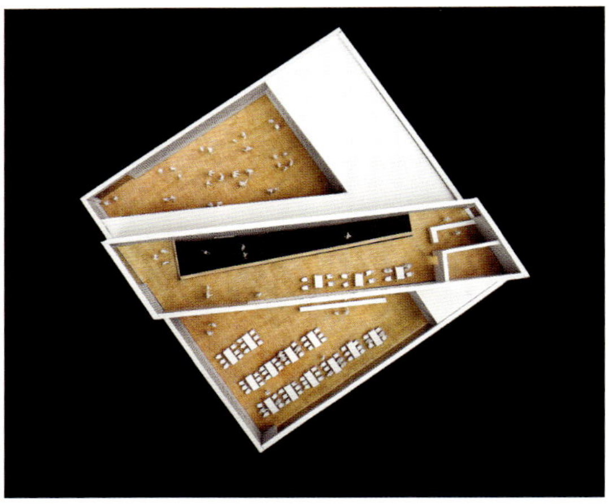

Credits

Name: Mirage Dance Hall

Type: Dance Hall and Function Center

Competition Completion Date: 2006

Building Completion Date: 2009

Location: Falsterbo, Sweden

Designer: Kjellgren Kaminsky Architecture AB

Photographer: building images: kalle sanner, model images: Malin Hedman

Competition Organizers: Vellinge Municipality and Sveriges Arkitekter

梦幻舞厅离瑞典最南端的一片松树林中的沙滩地约200米。该地原有一个深受人喜爱的舞厅，但是在2006年被一场大火摧毁。为在原地重新建造一个舞厅，市政当局举行了一场国际竞赛，而梦幻舞厅方案在竞赛中获得一等奖。该方案最终被采用而舞厅也于2009年落成。

新舞厅完全建立在原有舞厅所在场地之上，处在一片自然保护区之中。为了与周围环境相互衬托，建筑的立面采用的是镜面板材，可以反射出周围的景色，同时与原有建筑的立面保持一致。建筑内部抽象化的森林格局进一步将舞厅与周围的森林环境联系起来，淡化了室内和室外的界限。与建筑相连的木制露台和小径让人不禁联想到不远处沙滩的突堤与小径。

旧舞厅仅仅只是由舞池与吧台构成，而新舞厅增加了两间餐厅、一间厨房、一个乐队/表演场地以及带屋顶露台的二楼。新建筑有1,600平方米，可以容纳1,500人。

该国际竞赛更多地强调空间的音质状况，于是，建筑师与一位声学专家合作设计了一面定制的阻尼墙。阻尼墙主要是由黑色的吸声毡构成，吸声毡上覆盖有多种规格的白色木板，赋予了建筑强烈的画面感。

梦幻舞厅与周围环境建立起了关于历史与未来的对话，无论是从功能还是规模上看，均与场地环境与历史相协调。该设计成功地实现了尊崇历史与展望未来之间的平衡。

获奖情况

该设计在2006年的国际竞赛中获一等奖，并且在建成之后获得了斯科纳省建筑设计奖项。

评委会评语

“总而言之，该设计以其醒目的对角线入口成为一种极具魅力的艺术表现形态，其平面构思无论是从舒适度还是功能性方面来看，都是令人称赞的，值得一提的是，设计还考虑到了声学问题。”

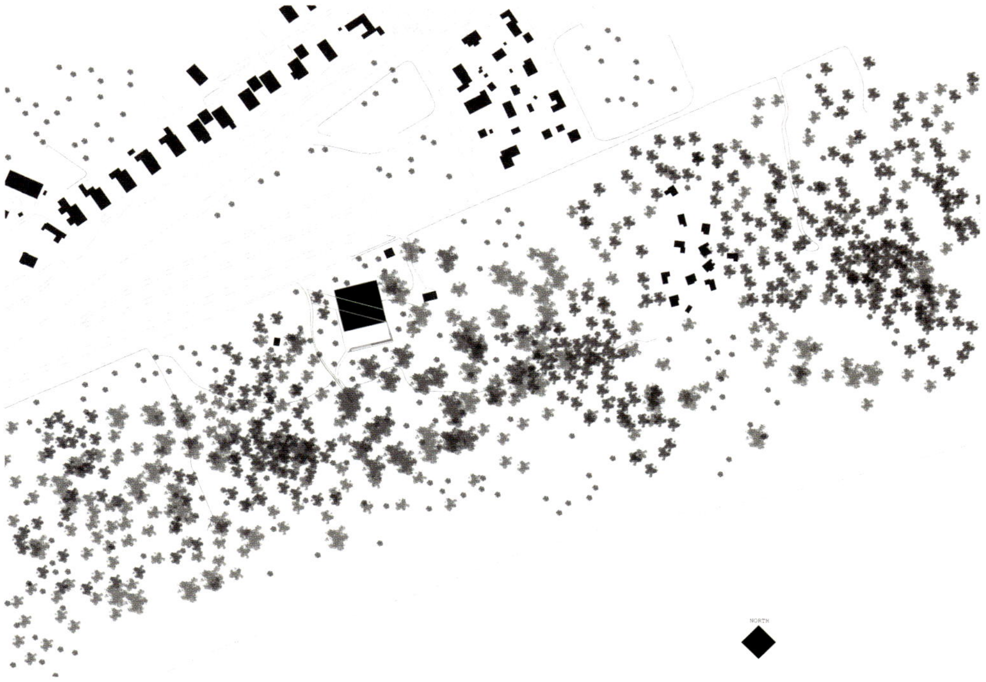

The Mirage Dance Hall is situated 200 meters from the beach in a small pine wood grove on the southernmost tip of Sweden. The project won first prize in an international competition to replace a much loved existing dance hall that was destroyed in a fire in 2006. The competition design was built and the building was inaugurated in 2009.

Set in a nature reserve, the new building took the sitting and footprint of its predecessor. Sympathetic to its surroundings, the buildings facades comprise mirrored panels that reflect the surroundings and echo the facade pattern of the old dance hall. The abstracted forest pattern used on the interior walls further acts to connect the building with the forest setting, blurring the boundaries between indoor and outdoor. The timber terraces and paths which connect to the building recall the paths and jetties of the nearby beach.

The old dance hall solely consisted of a dance floor and a bar, but the new building brief was extended to include two restaurants, a kitchen, a band/performance space and a second floor with roof terrace. The building is 1,600 m^2 and is able to accommodate 1,500 people.

The competition placed much emphasis on designing spaces with extraordinary acoustic qualities. Subsequently a custom made damping wall was designed in collaboration with an acoustic expert. The wall consists of a black acoustic felt covered by white wooden boards of various dimensions which acts as a strong graphic throughout the building.

The Mirage dance hall establishes a dialogue with the site, its history and future. It is functionally appropriate yet sympathetic in scale to the site context and history. It is a design that strikes a unique balance between respecting the past and embracing the future.

Award Information

The design was selected in an international architecture competition held in 2006. After inaugurated the building has also won the Skåne region architecture prize.

Jury Verdict from Competition

"In summary, the proposal is an inviting interpretation with its bright diagonal entrance. The plan solution is well-conceived in terms of both comfort and function. Acoustics are taken into account."

fasad mot öst

fasad mot syd

fasad mot nord

fasad mot väst

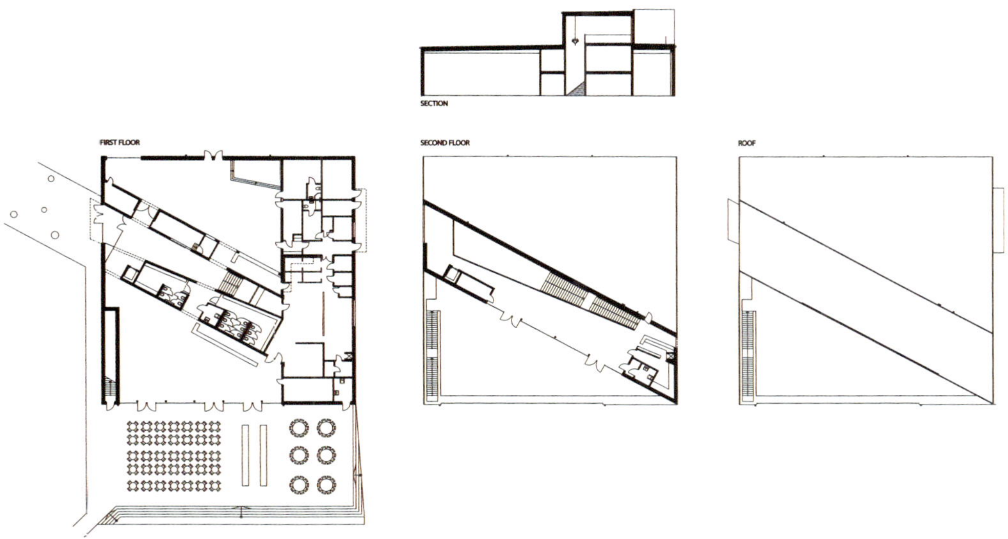

International Music and Dance Center – the Hague

海牙国际音乐舞蹈中心

Neutelings Riedijk Architects

International Competition 2010
2010年国际设计方案赛参赛方案

Credits
Surface Area: 45,000 m² floor surface
Location: Spuiplein, the Hague
Principal: City Council the Hague
Architectural Design: Neutelings Riedijk Architects, Rotterdam, the Netherlands
Structural Design: Aronsohn consulting engineers | Rotterdam, the Netherlands
Building Physics: DGMR Arnhem | Nederland
Installation Design: HE ingenieurs | Rotterdam, the Netherlands
Acoustic Advice: Larry Kirkegaard Associates | Chicago | USA
Theater Technical Advice: dUCKS scéno | Villeurbaine | France
Copyrights Photos: Neutelings Riedijk Architects

千人剧院

海牙国际音乐舞蹈中心（IDMC）为整个海牙市市民开放，其造型独特，而且拥有大都市的富丽堂皇。IDMC可容几千人，将在这个美丽的地方为观众呈现让人惊叹的表演。

剧院：真实性与协同作用

IDMC是荷兰舞蹈剧院、民间管弦乐队、皇家音乐学院与节目主持学院的集聚地。这里有学生也有艺术家，可以进行彩排与表演，既保持着透明化又保护着个人隐秘。但最重要的是它激发了不同人群之间的协作与配合。

中庭：台阶引领汇集人群

中庭是大楼的中心，引领着所有的来访者、艺术家、学生与合作人。如一个露天的林荫大道，中庭由一层层的台阶组成，中间穿越休息室、餐厅与图书馆，直通顶楼。

剧场：世界级层叠表演大厅

表演大厅是整栋楼的核心，由四层构成，每一层都有其独特之处。钢琴诺贝尔舞蹈与歌剧院位于一层，这里主要是通俗音乐会举行的地方。二层演奏厅是主要表演场所。表演大厅穹顶处可以俯瞰海牙城市的全貌——这里主要是戏剧院与合唱团演播厅。

学习区：防干扰

围绕表演大厅的通道分布着教室、彩排室与办公室。每一个部门，如舞蹈、音乐、电子音乐与青少年培训班，都有各自的楼层。这样，每个部门都有自己的空间与学习气氛，同时又保证了他们之间的协作。

练习区：专业创作的地方

练习区为海牙市民提供了灵活彩排与创作的空间。舞蹈大厅与皇家音乐学院彩排室位于音乐厅周围，旁边是荷兰国际舞蹈剧院工作室。更衣室、艺术家与技师用具设施则毗邻舞台。

建筑外形：密实紧凑

IDMC设计紧凑而密实。这样形成了从市政大楼到Spui文化广场的自然过渡。它比旁边的市政厅要矮很多，其锥形外立面则使大楼整体看上去更加纤细。

穹顶皇冠：城市标志

大楼壮丽的穹顶像皇冠一样，使其成为海牙大地的标志。穹顶传统而不失现代元素，像是为海牙皇家般的风格专门定做的。这座大型的圆形穹顶将深深地吸引海牙市每一位游客。

外立面：优雅、精致、时尚

精心设计的外立面使整个建筑物看起来优雅而稳重。颜色、材料与图案热情而深沉。波浪斜纹图案的设计来自舞蹈中优美的舞姿与和谐的韵律。建筑以一种愉悦的形式，穿越白天与夜晚，创造出光与影的美丽交织。

整体分布

设有1,800人音乐厅、1,100人歌剧院/舞厅、500人戏剧院与700人合唱厅，另有教室、工作室、休息室、后勤与停车场。

Philipszaal
Conservatorium
Lucent Danstheater

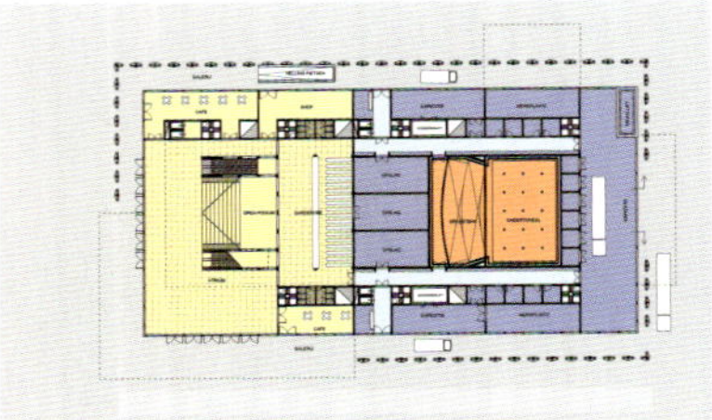

LAAG 0: Atrium en entreegebied / Logisitiek centrum

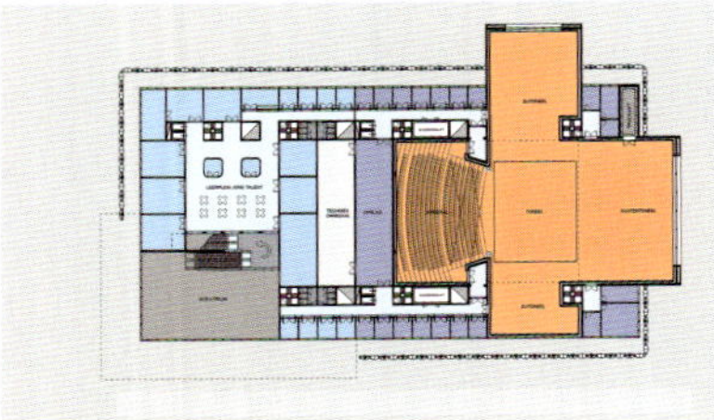

LAAG 1: School jong talent / Dans-Operazaal

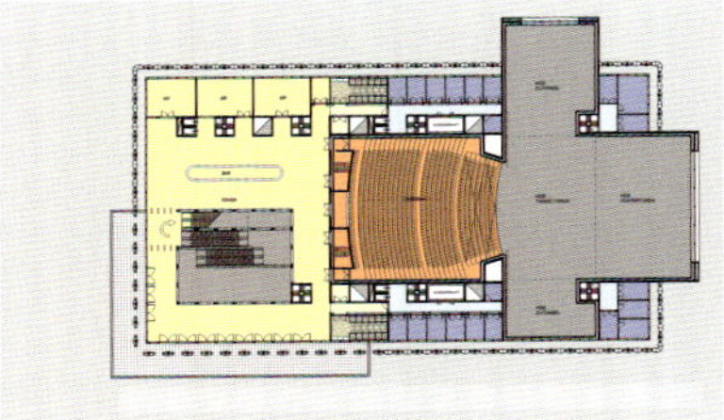

LAAG 2: Foyer / Dans-Operazaal parterre

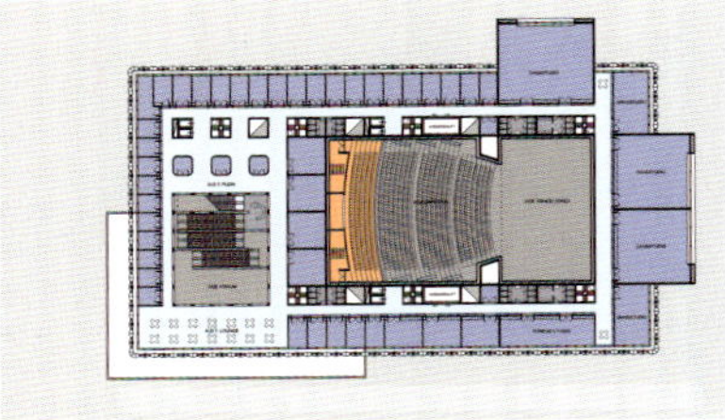

LAAG 3: INDT - Productiehuis / Dans / Operazaal balkon

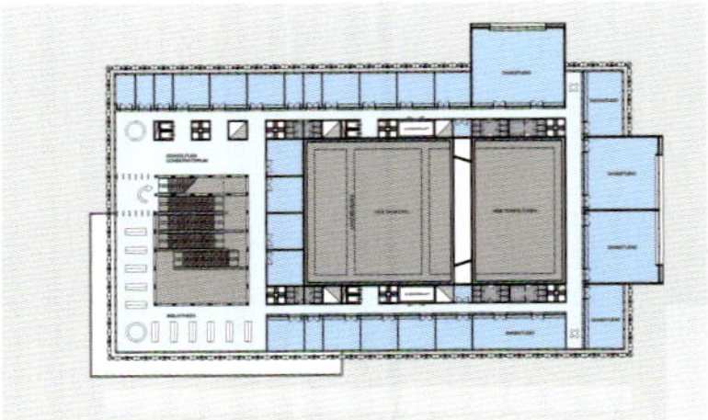

LAAG 4: Bibliotheek / Dansvakopleiding

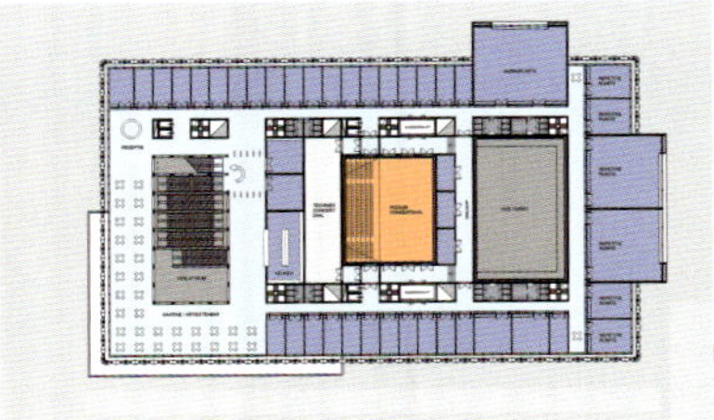

LAAG 5: Kantine / Concertzaal back of house / Residentie orkest

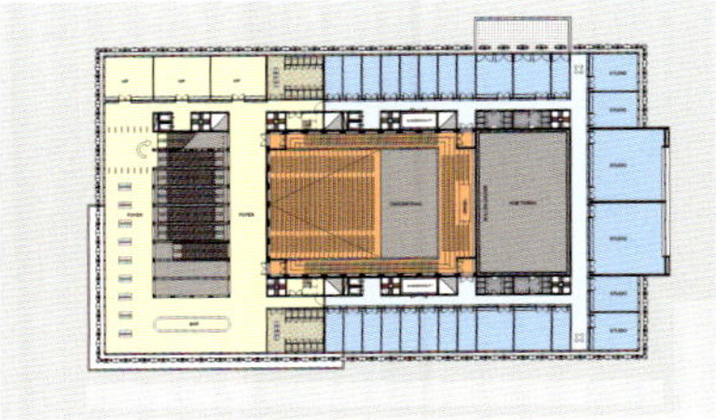

LAAG 6: Foyer / Concertzaal parterre / Instituut Sonologie

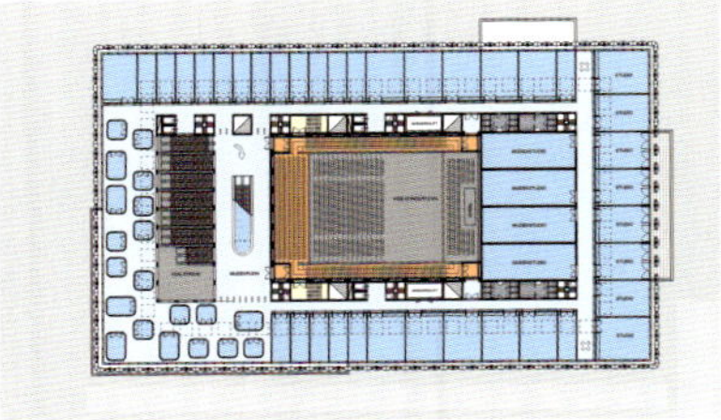

LAAG 7: Concertzaal balkon / Muziekvakonderwijs

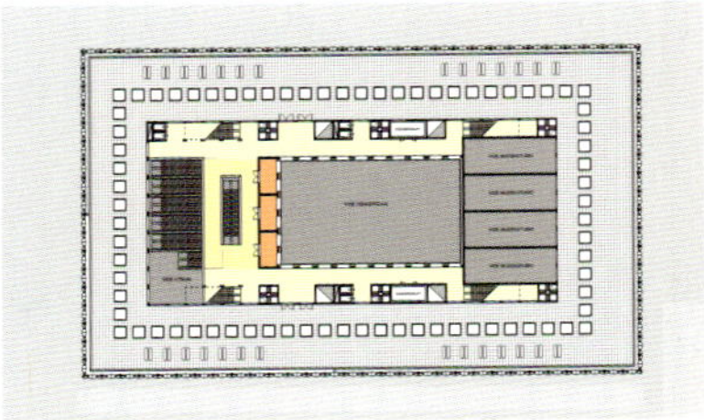

LAAG 8: Dakterras

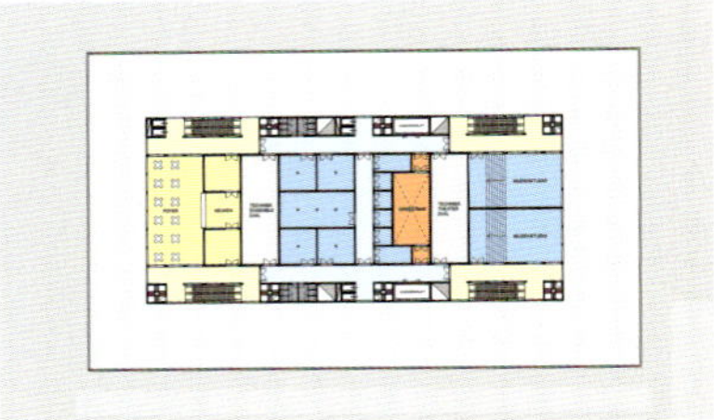

LAAG 9: Restaurant / Bakstage ensemble / theaterzaal / Studio's

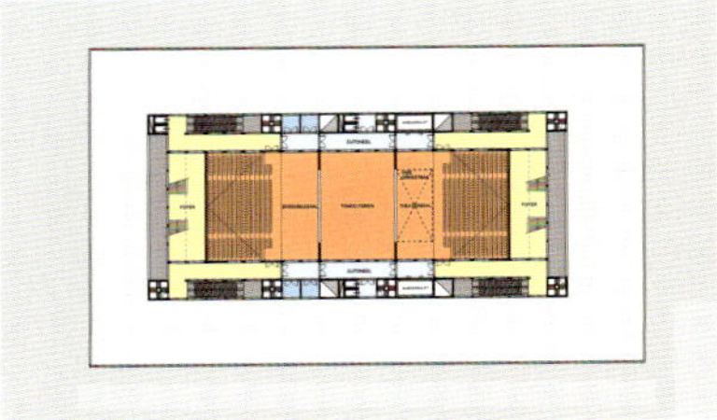

LAAG 10: Ensemblezaal / Toneeltoren / Theaterzaal

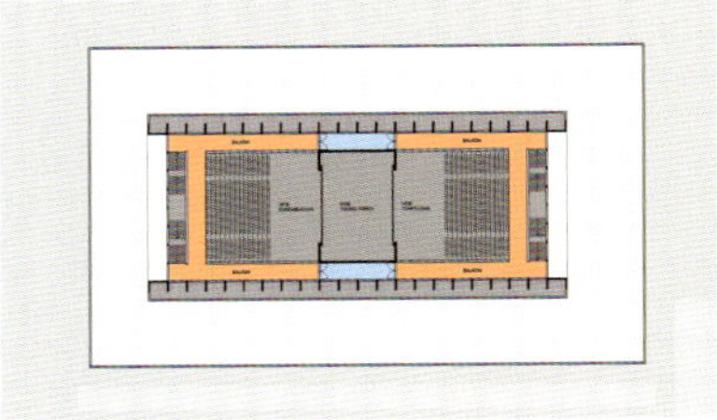

LAAG 11: Balkon ensemblezaal en theaterzaal

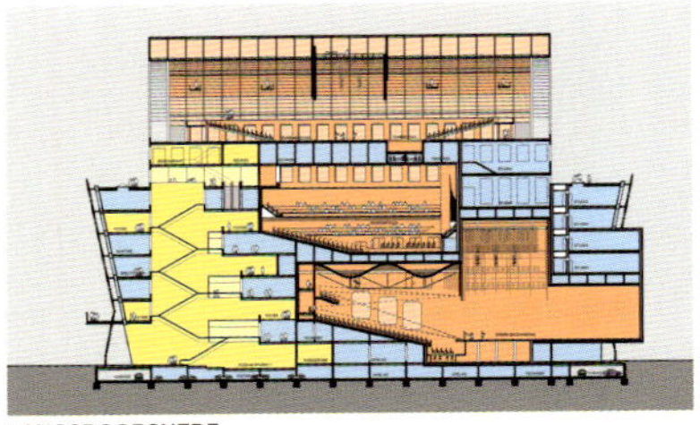

LANGSDOORSNEDE

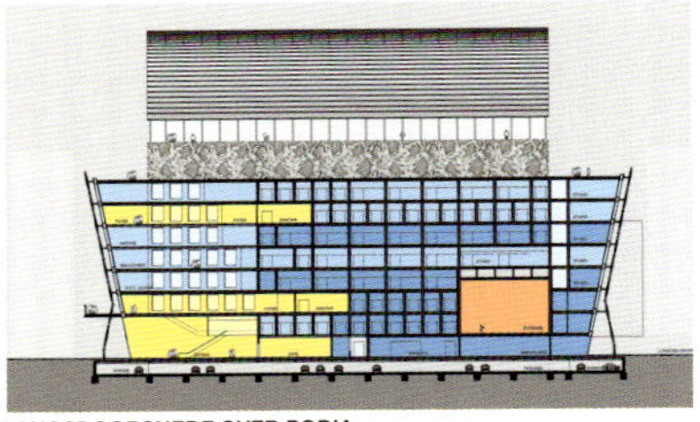

LANGSDOORSNEDE OVER PODIA

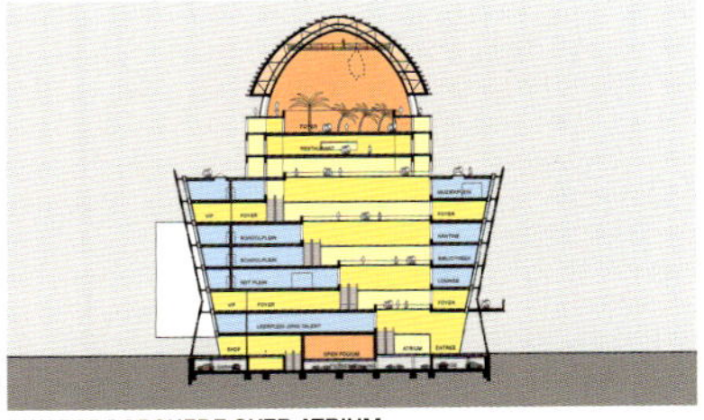

DWARSDOORSNEDE OVER ATRIUM

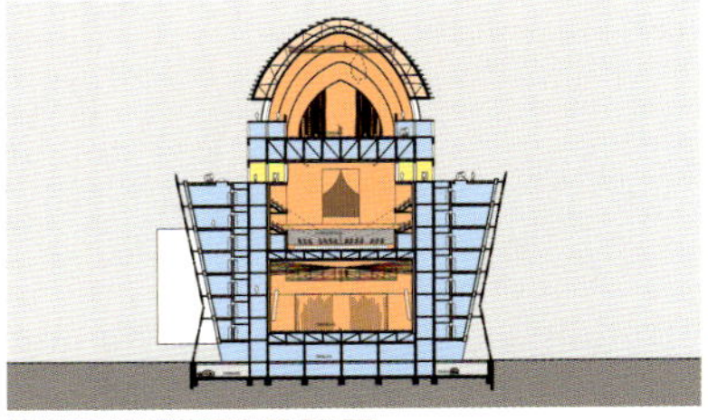

DWARSDOORSNEDE OVER ZALEN

IMPRESSIE SCHEDELDOEKSHAVEN

IMPRESSIE SPUIPLEIN VANAF DE NIEUWE KERK

IMPRESSIE TURFMARKT

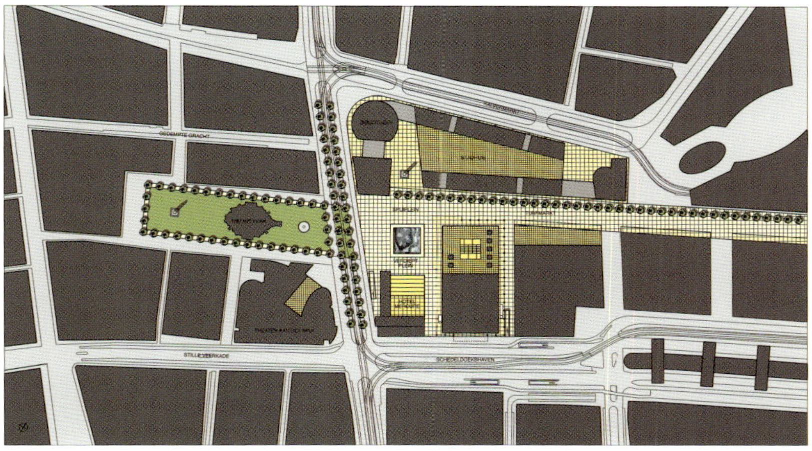

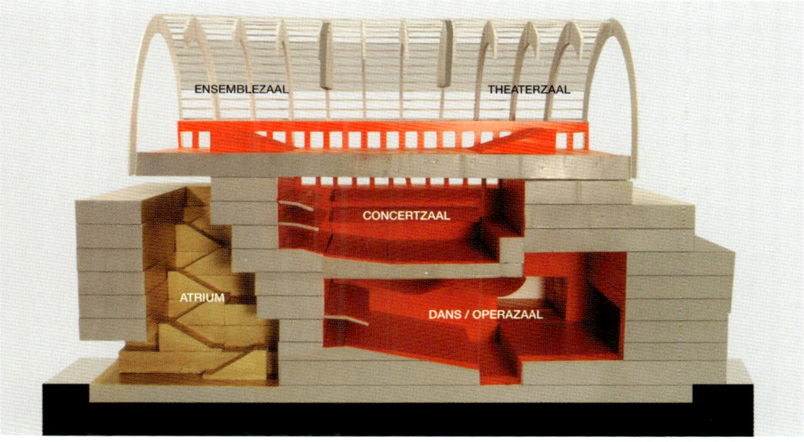

Reaching out to the City

The IDMC is a public building for all The Hague's citizens. The IDMC is a well-structured building with cosmopolitan grandeur where many thousands of visitors will be warmly welcomed and will enjoy spectacular and sparkling performances in festive surroundings.

Authenticity and Synergy

The IDMC is the communal residence for the Dutch Dance Theater, the Residential Orchestra, the Royal Conservatory and the Foundation for Host Programming. It accommodates students and artists, rehearsals and performances, transparency and privacy, but foremost wishes to stimulate synergy and cooperation between the different user groups.

The Atrium: Mountain of Stairs and Meeting Place

The Atrium is the heart of the building thct welcomes all visitors, artists, students and co-operators. It brings visitors all the way to the roof via a monumental mountain of stairs, as an open boulevard in a vertical city, along the public areas like foyers, restaurants and libraries.

The Play House: World Class Stacked Performance Halls

The four stacked performance halls form the core of the building, each having its own character and atmosphere. An open stage for informal concerts is located at street level. At the Piano Nobile the Dance and Opera Hall is situated. The Concert Hall, beating heart of the building, floats above it. The Dome on roof level hosts the Drama Theater Hall and the Ensemble Hall that have magnificent views of the city.

The Study House: Concentration and Security

Class rooms, rehearsal rooms and offices are located in the flexible aisles around the big performance halls. Every department, like Dance, Music, Sonology and Young Talent has its own floor. In this way every department has a clear structure and secure atmosphere, but also creates possibilities for cooperation and synergy.

The Work House: Professional Foundation for Creation

The work house is a flexible structure of rehearsal rooms and class rooms for the residents of the IDMC. Here they can work and create productions. The studios of the Dutch National Dance Theater are situated around the Dance Hall and the Royal Conservatory's rehearsal rooms around the Concert Hall. Dressing rooms and other facilities for artists and technicians are in close proximity to the stages.

The Building Volume:Low and Compact

The IDMC has a low and compact building volume. In this was a natural transition is made from the skyline of the ministry towers to an intimate Culture square at the Spui. The building's shoulder is significant lower than the neighboring Town Hall, while the tapered façade makes the building volume look slimmer.

The City Crown: An Icon for The Hague's Skyline

A majestic dome crowns the building and forms a monumental beacon in the city's skyline. The dome is traditional and contemporary, tailor-made to the character of royal The Hague. This city roof with a large cupola hall for all kinds of public events will mesmerize every visitor with a spectacular city view.

The Façade: Elegant, Filigree and Contemporary

The filigree exo-skeleton gives the building elegance and depth. The colors, materials and textures are warm and deep. The diagonally wave pattern is inspired by the agile movements of dancers and the melody and harmony in music. Day and night the festive expression of the building will create an elegant interplay of light and shadow.

General

With a 1,800 seat Concert Hall, a 1,100 seat Opera / Dance Hall, a 500 seat Theater and a 700 seat Ensemble Hall, educational spaces , studios, foyers, logistical spaces, parking.

Borusan Music and Art House

博鲁桑艺术音乐馆 GAD

Credits
Location: Beyoglu /Istanbul/ Turkey
Project Architects: Gokhan Avcioglu & GAD
Project Team: Ozan Ertug, Baris Ucar, Ayhan Urguplu, Arzu Meyvaci, Yeliz Ozsoy, Bora Soykut, Gozde Nur Demir, Serkan Cedetas, Ertugrul Morcol
Interior Design: GAD
Photographs: Ozlem Avcioglu
Project Date: 2007
Construction Date: 2007-2009
Construction Type: Steel
Site Area: 230 m²
Total Construction Area: 1,900 m²

博鲁桑艺术音乐馆位于历史文明的伊斯坦布尔贝尤鲁地区著名的伊斯提克拉大街，建筑以外观为特色，具有历史意义。新的设计将其改造为一个集展览、会议、彩排与各种文化培训于一体的多功能文化馆。它在保留伊斯坦布尔的设计文化遗产的基础上，代表着整座城市当前文化与艺术的复兴。

GAD建筑事务所将文化中心核心部分完全移除，形成中空的状态；同时在中央灵巧地置入一个很现代的“盒子”。在这个“盒子”周围采用了轻钢龙骨框架系统，使楼层呈露天式。这样一来，不同楼层之间也出现灵活有序的交叉。这个砖石结构的嵌入式“盒子”看上去能抵抗地心引力。该设计两种不同的建造方法与材料的共生激发出过去与现在之间一种微妙而紧张的关系。

这栋楼共六层，主要入口设在伊斯提克拉大街，次要入口则面向皮拉米斯大街。地下一层由大型玻璃窗连通大街，让人回忆起以前商店的玻璃橱窗。透明外立面让观众可以预先欣赏到里面的表演。大楼在视觉上稍稍高出一些，古老的大理石台阶在两根立柱的支撑下格外醒目。地下一层与一层是展览区，另外也可以灵活运用，比如举办重大会议。

音乐厅呈中庭式，连接二层与三层，可容纳100人。二层主要用来作彩排与音乐会的场所。观众在二、三层都能欣赏到现场表演。

4楼设有6个独立的彩排室。五楼功能性多样化，可举行会议。6楼可欣赏到博斯普鲁斯海峡的美景，可以作为休闲咖啡馆或者小酒吧之类的地方。六楼上面还有一层，可作为观景平台；由于没有电梯通往，游客可以走楼梯。

外立面

馆楼外立面的设计延续了建筑原有的古朴风格。窗户原本是单扇玻璃，但应市政当局的要求，在正面添加了竖框，以更多地保留建筑外立面的历史特色。侧面则保持原有设计。窗户是固定的，不能打开。建筑物主要通过空调通风换气。

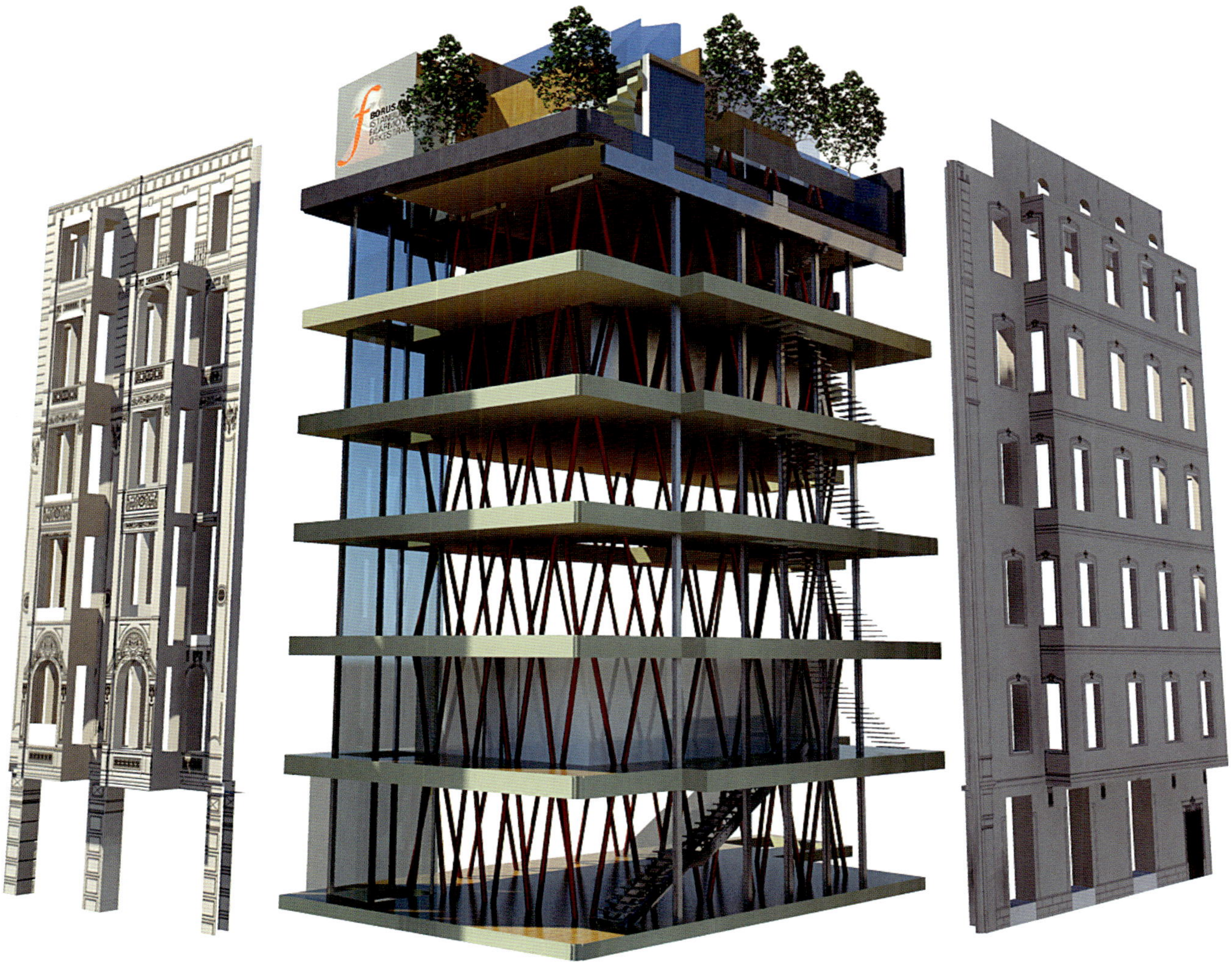

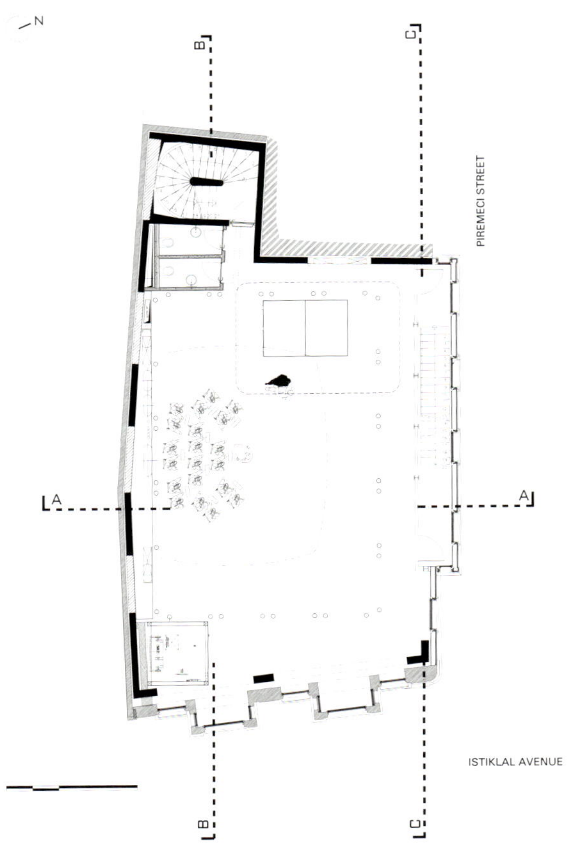

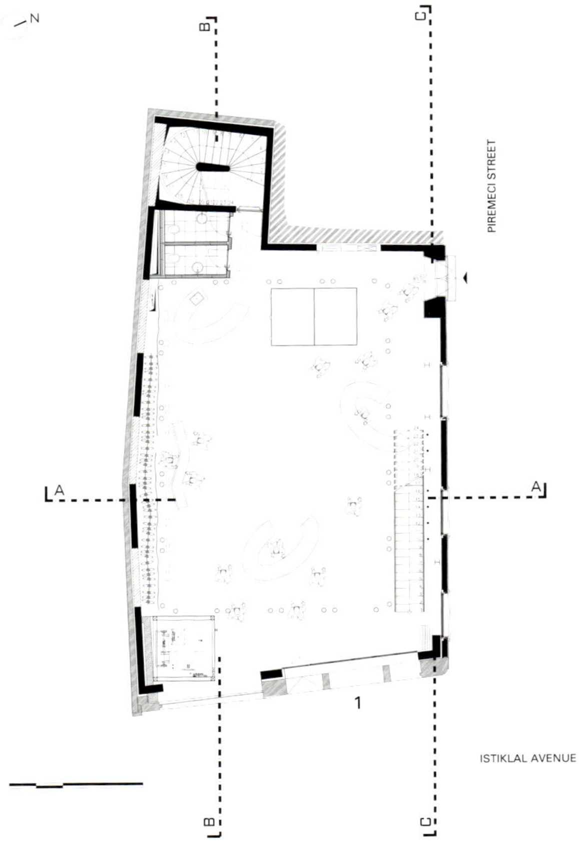

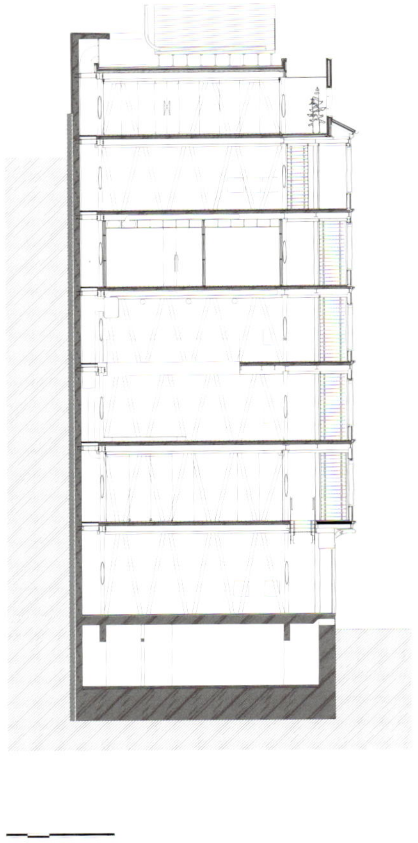

The new Borusan Music and Art Center in the historic neighborhood of Beyoglu, Istanbul is a multi-purpose space for exhibitions, events, rehearsals and various cultural training programs. This new cultural center symbolizes the cultural and artistic renaissance presently occurring throughout the city, while also preserving the architectural legacy of Istanbul.

GAD's design preserves and restores the historic shell of the existing building, which is typical of the facades along the famous Istiklal Caddesi. The building core has been completely removed, expressing an empty void. In a clever move, GAD inserted a contrasting contemporary 'box' within the core. The 'box' is designed using a lightweight steel diagrid frame system along its perimeter. This strategy allows for an open floor plan, and offers programmatic flexibility between all the floors. The inserted 'box' seems to defy gravity as it responds to, and is framed by, the historic masonry shell. The architectural symbiosis of the two contrasting construction methods and materials activates a playful tension between the past and present.

The main entrance to the six story building is from Istiklal Street. There is a secondary entrance from the intersecting Piremeci Street. The ground floor is linked to the street visually by large glass windows, reminiscent of the shop's previous glass display. The transparency on the façade allows the visitor to preview the action inside. The visitors enter the building through an oversized glass door. The building is elevated slightly, and old marble stairs are punctuated by two steel columns that support the facade. The ground and first floors serve as exhibition spaces, and also have maximum flexibility for events.

The concert halls on the second and third floors are two-storey atriums, housing up to 100 people. The second floor is used primarily for rehearsals and concerts. The audience is able to listen and observe live performance from the second and third floors.

There are six individual rehearsal rooms on the fourth floor. The fifth floor is used for multi-purpose events. On the sixth floor, the building opens up to views of the Bosphorus. This space will be used as a café and bar. There is another level above the café & bar used as terrace. As there is no elevator access to the upper level, visitors use a staircase to reach this level.

Façade

The formal logic of the facade was to keep the historic façade in its simplest form. The window openings were originally covered with a single-panel glass system, and the municipality requested a more historical reference for the façade. Therefore, mullions were added to the front façade. The side façade remains without mullions. The windows are fixed and may not be opened. Ventilation is maintained through artificial air conditioning.

LISTENTOYOUREYES

2012 Korean Thematic Pavilion Competition Entry

2012韩国主题馆竞标方案

3 Six o Architecture

阴阳理论：阴意味着黑暗和寒冷，阳意味着光明和炎热。上层太阳城的天篷让游客沐浴在潮起潮落的影子中，瞬间逃离了中午的烈日。下层的月亮城是双夹层结构，夹层中间充满着冷却水，每天运作两次通过潮汐景象补充用水。

蓝色理念：伴随着四季循环而吞吐和谐的生活。

远观主题馆像一个标志性的形象——晶莹透亮的水和海浪。游客沿着防波堤漫步，这种景象瞬间变化为起伏的金属冠层，漂浮在巨大的横木平台之上。在天篷的遮蔽之下，玻璃主题馆覆盖在广阔的平台之上，是下面平台的扩展。主题馆的平台平行于防波堤，玻璃的主题馆和天篷的取向转移13°对应着其更大的天体顺序——真正的东方和西方。这个天篷是用来挡住高空中的阳光，同时在冬季、早上或傍晚时分让更多的阳光射入。天篷在每天正午阻挡直射的阳光，暗示了每日不同的时间段。

细长的圆柱支撑着天篷的一切；圆柱的形状可以折射弯曲阳光。每个和天篷相连的圆柱间隔一致，保证每个圆柱负荷均匀。天篷的起伏波动影响着圆柱的长度和角度。许多圆柱捆绑在一起增加抗弯强度，形成了轻微起伏的网格景象。

从防波堤走到平台被认为是一种"沉没"：当游客慢慢走过缓缓的小坡时，就像海岸线侵蚀了整个平台。一旦站在平台之上天篷之下，游客便可以从天篷的斑点灯光之下漫步到海景之中，或者漫步到玻璃主题馆的东端展厅入口。展厅的入口用细小的玻璃天篷作为标志，此处为展会组织人员的控制点，采用楼梯、电梯、自动扶梯以保证循环畅通无阻。

展览区域分为上下两层：上层为装有玻璃天花板的主题馆，下层为一个更大的最佳实践区，并完全包含地下的平台。自助餐厅、最佳实践区展厅和海洋研究图书馆等游乐设施位于可俯瞰最佳实践区的夹层之上。当游客沿着阳光四溢的上层走到最佳实践区，他们立即会感受到主题馆的天然纯粹、与海洋之间的交流和冷却技术。下层展厅按照热水瓶双层结构建造而成，其中夹层可储蓄冷却水，并每日两次通过潮汐景象补充用水。冷却水来自夹层，并可以通过像毛细管状流通管道，带走展厅中多余的热量。

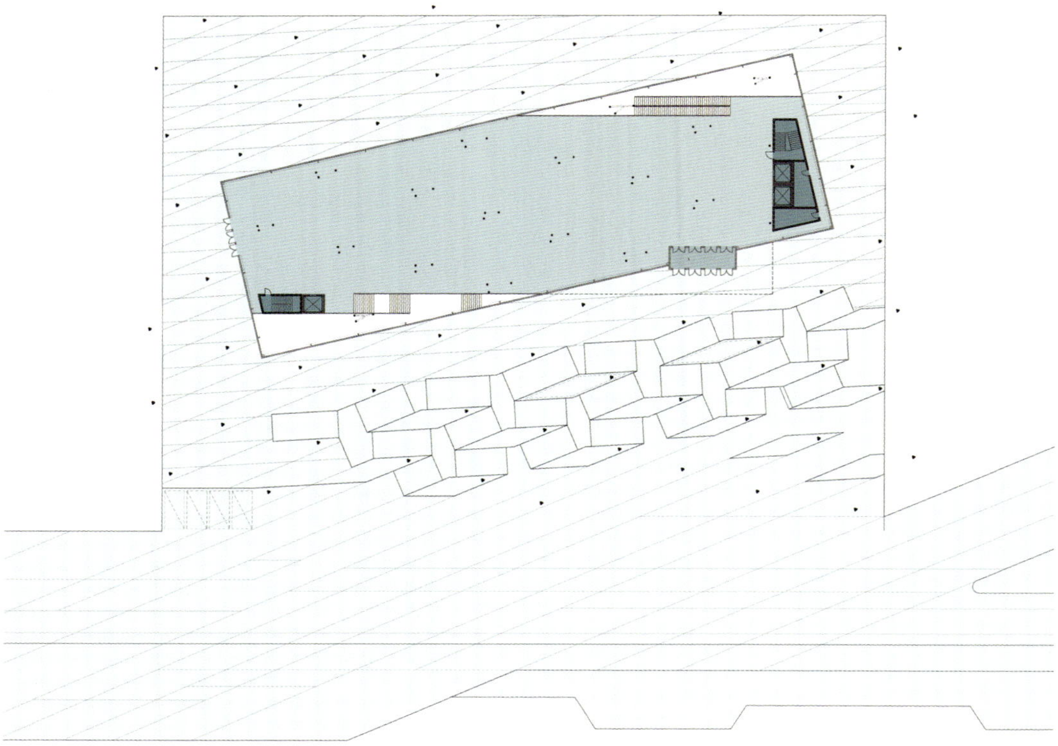

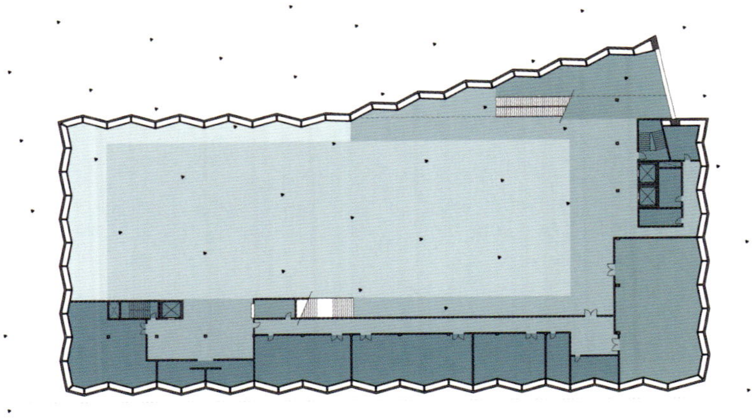

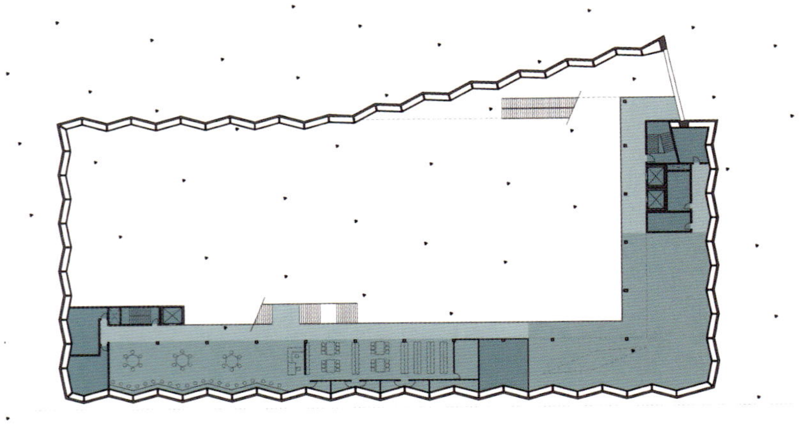

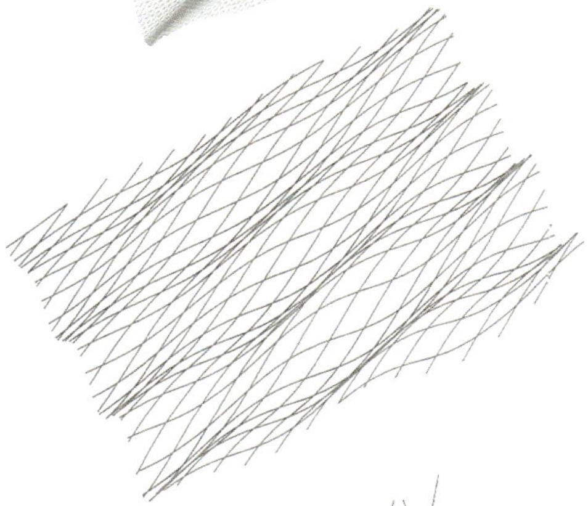

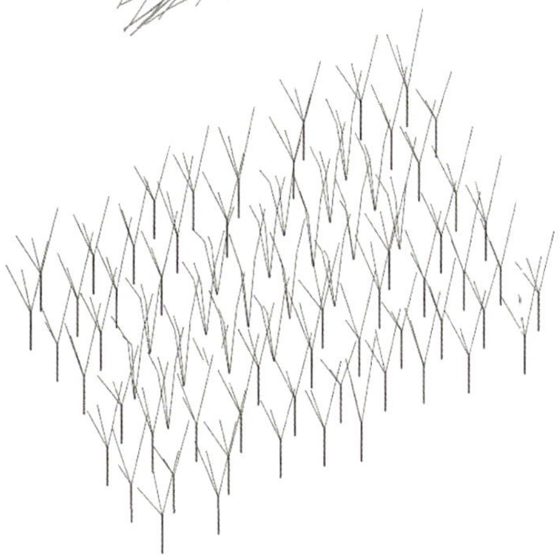

Um-Yang: The movement of dark and cold and bright and hot. The upper Sun Pavilion canopy bathes visitors in an ebb and flow of shadows, and momentarily eclipses the sun each day at solar noon. The lower Moon Pavilion is a double walled thermos that holds a reservoir of cooling water, replenished by tidal action twice daily.

The Promise of Blue: A life giving resource that inhales and exhales with the cycle of the day and seasons.

From a distance the pavilion appears as an iconic image-suggestive of water, waves, liquid transparency. As the visitor approaches along the breakwater, the image resolves into an undulating metal canopy, floating above an immense, open horizontal wood platform. Under the shade of the canopy, a glass pavilion covers the extensive exhibition spaces emergent from beneath the platform. The pavilion's public platform is parallel to the breakwater, while the glass pavilion and the tectonic orientation of the canopy above are shifted 13 degrees to align to the larger celestial order, true east/west. The canopy is designed to occlude high overhead sunlight allowing more light during the winter and morning and evening hours. Each day at solar noon the canopy momentarily eclipses all direct sunlight, eventfully marking the time of day.

Slender columns support the canopy above; their form alluding to the refraction and bending of light. The columns connect to the canopy equidistantly (in plan) from each other, allowing each column to carry an equivalent load. The undulations of the canopy effect column length and angle. As the many columns bundle together increasing buckling resistance, they in turn form a slightly undulating grid.

The ascent from the breakwater to platform is conceived as a kind of "ubmergence": as the visitor walks up and through a redundant field of gentle ramps which like a shoreline appear to erode the body of the platform. Once on the deck and under the canopy, the visitor can wander through the canopy-dappled light to the water view or to the exhibition entry at the eastern end of the glass pavilion. The entry, marked by a smaller glass canopy provides a control point for the exhibition organizers; stairs, elevators, escalators are organized to ensure ease of circulation from this point.

The exhibition area is divided between an upper floor in the glass roofed pavilion and a larger best practice area (BPA) wholly contained under the platform level. Visitor amenities such as cafeteria, seminar room and ocean research resource library are located on a mezzanine level overlooking the BPA exhibition. As the visitor descends from the light filled upper level to the BPA they directly experience the pavilion's natural, ocean interactive, cooling technology. The lower level, like a thermos is constructed of a double wall which acts as a reservoir of cooling water, replenished by tidal action twice daily. The cooler water is brought through the inner wall and allowed to circulate through a network of capillary-like pipes, drawing excess heat from the space.

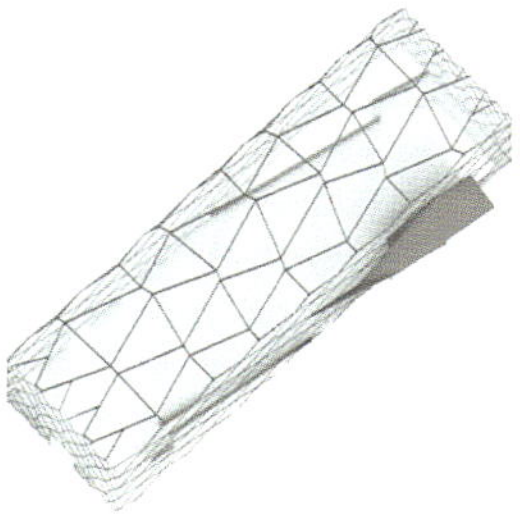

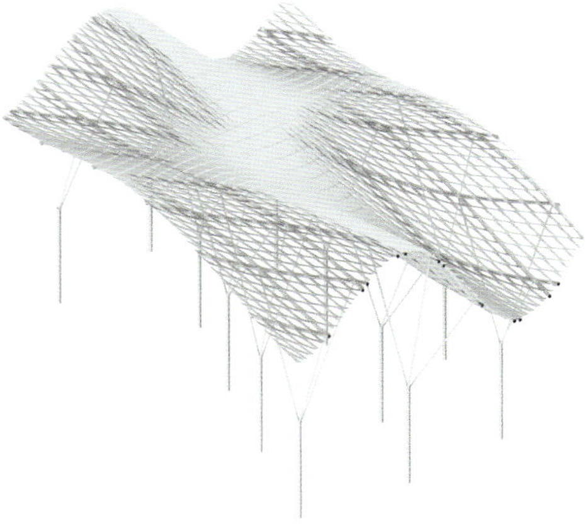

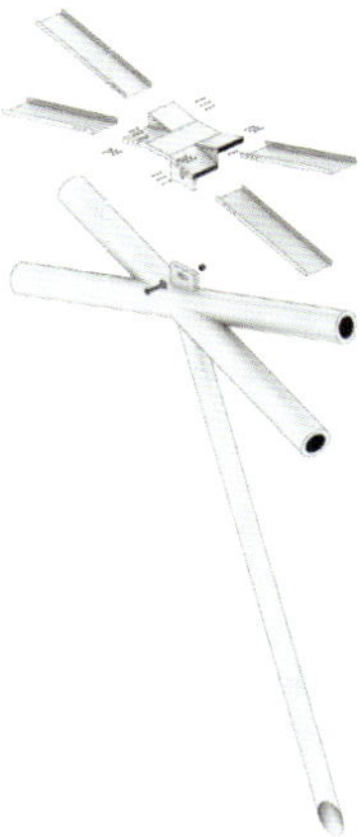

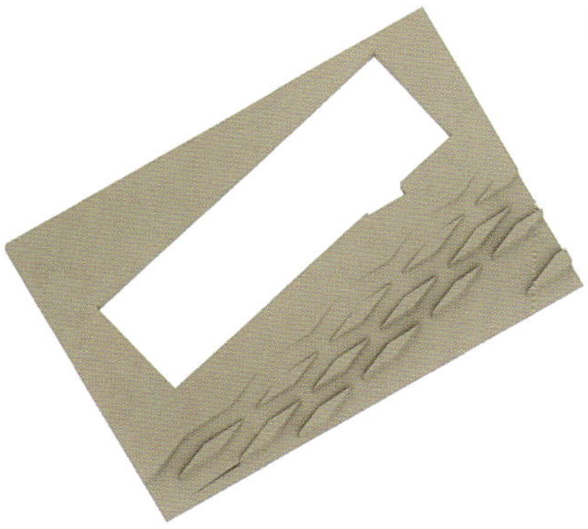

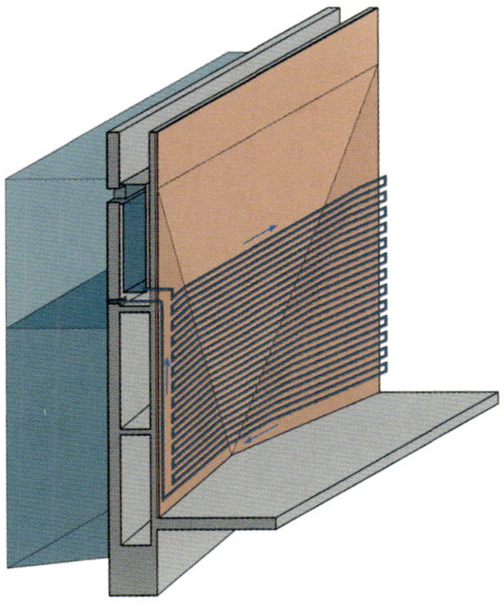

Edificio de ferias de Caceres

卡塞雷斯博览中心

Modostudio

International Architectural Competition, Honourable Mention
国际建筑设计竞赛荣誉提名奖

Credits
Location: Caceres - Spain
Client: INSTITUCION FERIAL DE CACERES - IFECA
Area: 7,687 m²
Year: 2010
Design Team: Modostudio (Fabio Cibinel, Roberto Laurenti, Giorgio Martocchia, Cristina Carmona Botana) with Luis Carmona Pla + Garcia Ruiz & Carmona Botana Arquitectos

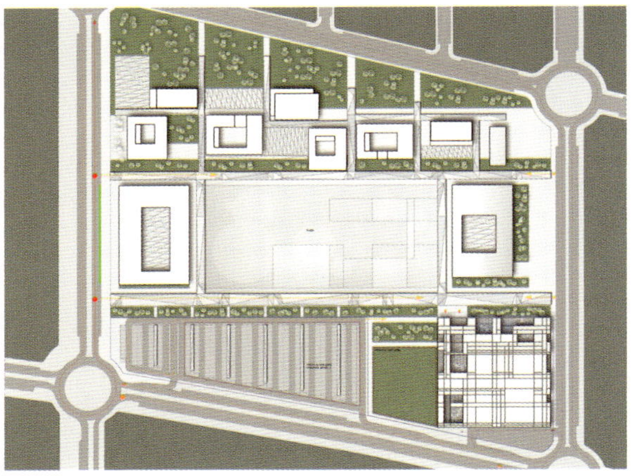

博览中心大部分空间用来作为商品交易会、展览、演出及各种重大会议的举办地点。中央大厅设计灵活，模块区分明确，实现了功能性多样化。入口处作为园艺展览区的延伸区域，毗邻中央大厅。整个设计简易地延伸出各种不同的功能区域。

太平梯都设在一端——为了在中央区域建造不同的夹层，以扩大中央大厅的面积。

另外，我们以简单的形式设计了一个大型独立的平台，在材料选用上，避免反差。这个平台具有它独特的极具代表性的功能，其中设有所有必要的办公部门。

项目以实用性、功能性、持久性与灵活性为基准，激发灵感，并以其成功而极富吸引力的外形打造完美的博览中心。

简而言之，这个博览中心：

——灵活多变，可依实际需要改变功能；

——设计方案、功能、质量与设施都与时俱进；

——最大化地节约能源。

A1

Pabellón A1

NIVEL +3.30m (p.1)

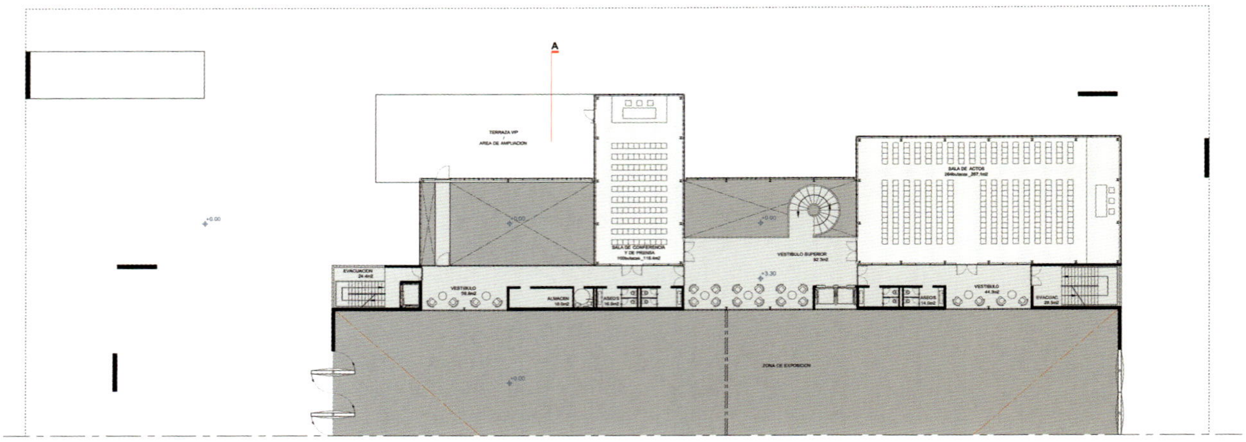

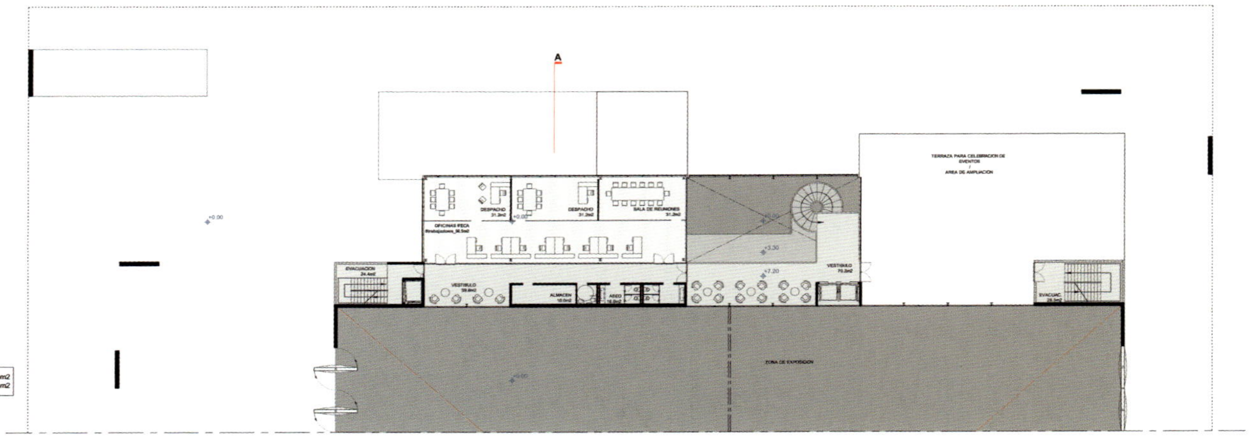

NIVEL +7.20m (p.2)

SUPERFICIE UTIL 6010.3m2 SUPERFICIE CONSTRUIDA 6202.6m2	p.0		
SUPERFICIE UTIL 839.0m2 SUPERFICIE CONSTRUIDA 967.8m2	p.1	SUPERFICIE UTIL TOTAL 7300.3m2 SUPERFICIE CONST. TOTAL 7690.4m2	
SUPERFICIE UTIL 451.0m2 SUPERFICIE CONSTRUIDA 516.0m2	p.2		

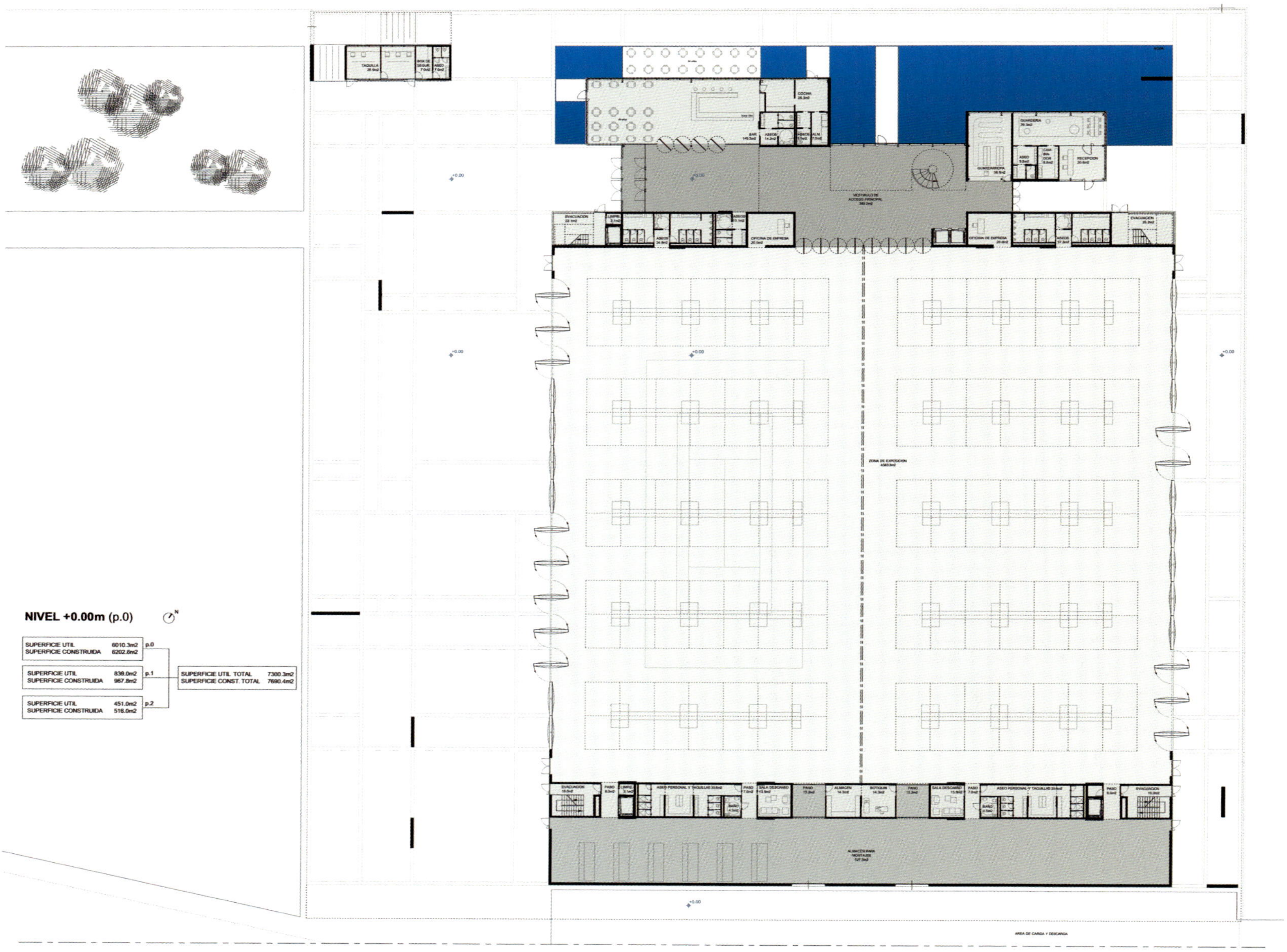

NIVEL +0.00m (p.0)

SUPERFICIE UTIL 6010.3m2 SUPERFICIE CONSTRUIDA 6202.6m2	p.0		
SUPERFICIE UTIL 839.0m2 SUPERFICIE CONSTRUIDA 967.8m2	p.1	SUPERFICIE UTIL TOTAL 7300.3m2 SUPERFICIE CONST. TOTAL 7690.4m2	
SUPERFICIE UTIL 451.0m2 SUPERFICIE CONSTRUIDA 516.0m2	p.2		

SECCION A-A

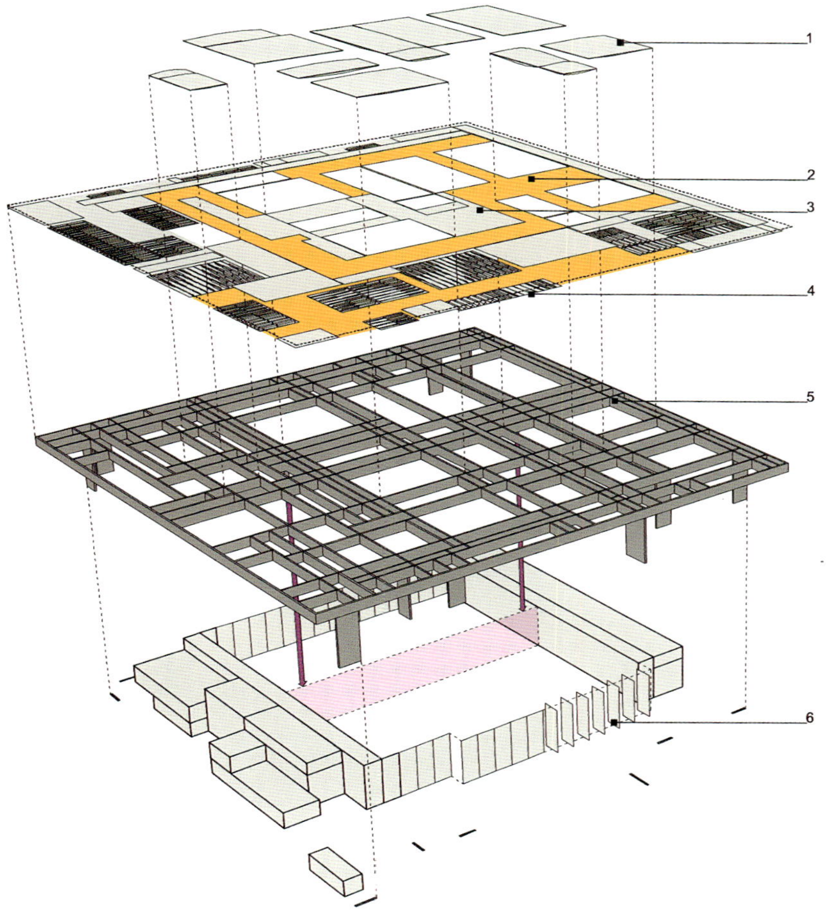

The building consists of a large main area where holding fairs and exhibitions, and performances and various events. The large central hall allows great flexibility and many possibilities partitioning, enabling work for lot of uses. Next to it stands the large porch as an extension of the garden exhibition. The building is designed to be easily extended to several forms.

By placing the escape stairways at the ends, it is possible to create different mezzanines in the central area which could expand certain areas of the main hall.

The architectural proposal is solved with a single volume made up of a large deck that hosts the various departments requested, with a large and representative character of its own function, without using complex shapes and contrasting materials.

The practicality, functionality, maintenance and flexibility are the basis of the project, along with a presence in the solar successful and attractive exterior image and identify the genus a package that perfectly meets the objective of the project.

In short, this is a building:

- Which responds to program needs raised with great flexibility of use and changes.
- In design, solution architecture, functionality, quality and facilities are in keeping with the times.
- Whose design features ensure maximum savings energy.

ESQUEMA ESTRUCTURAL

1-PANELES NEUMATICOS REGULADORES DE LUZ
2-PANELES FOTOVOLTAICOS
3-ELEMENTOS DE CUBRICION
4-BRISE-SOLEIL
5-ESTRUCTURA METALICA
6-PUERTAS NEUMATICAS MOVILES

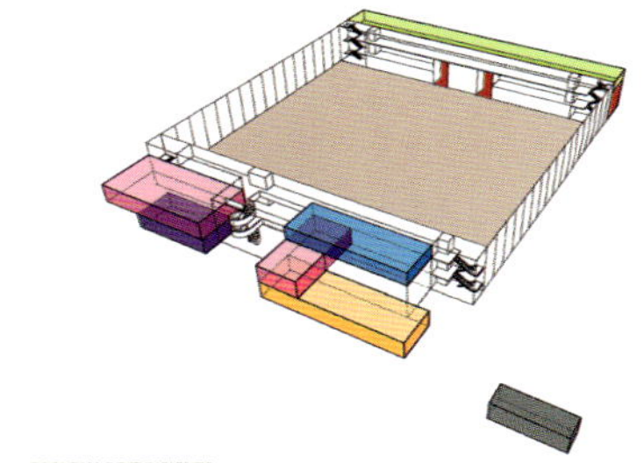

ESQUEMA DE FUNCIONES

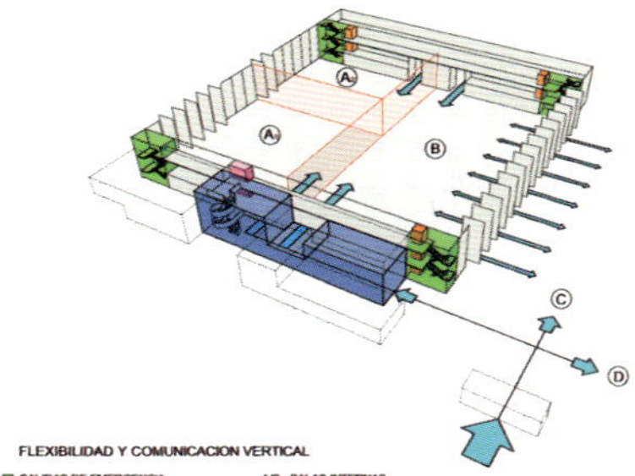

FLEXIBILIDAD Y COMUNICACION VERTICAL

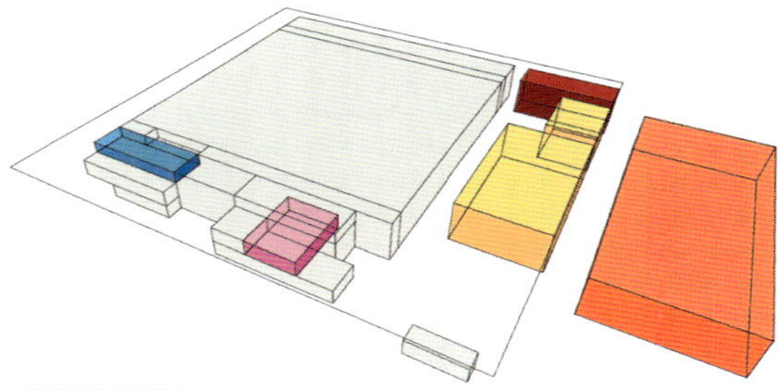

ESQUEMA DE AMPLIACION

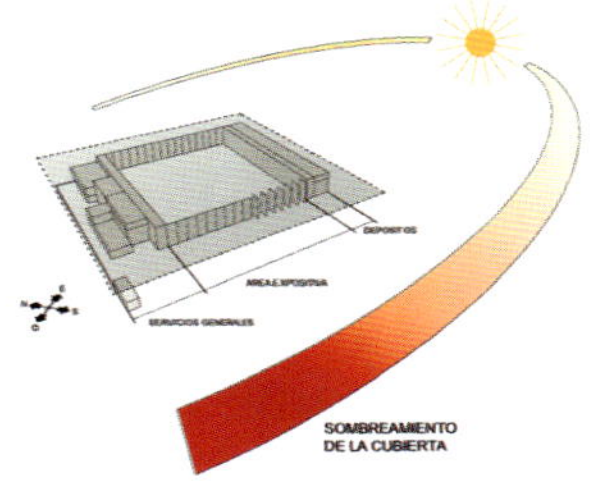

SISTEMA SOLAR PASIVO

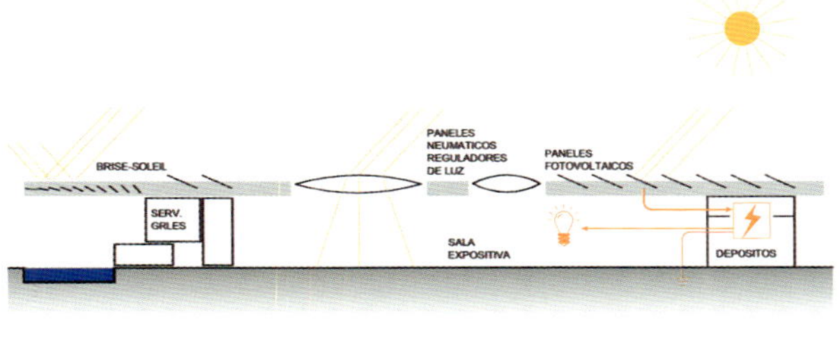

SISTEMA DE RECOGIDA DE AGUA DE LLUVIA

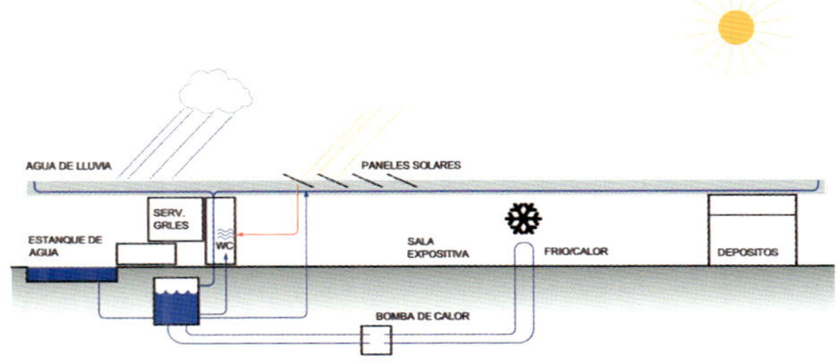

The Art Fund Pavilion

艺术基金馆

3 six o Architecture

场馆的设计从概念性的方法和建设性的逻辑之中产生，以三个相互交织的纽带的形式呈现。三根纽带凌空绕成三个线圈状的螺旋形。这些螺旋形宽度被限定在300毫米，并分割成不长于2,400毫米的数段，以满足制造和处理要求。三根纽带组装成相互交织的配置，构建了场馆的体积，置身其中的各条纽带从空间上和结构上加固彼此。

各纽带之间的接缝处也是该建筑的一大特色。纽带单独或全部伸出触及场馆空间内部，创造出釉质般光滑的开口、天窗以及水平的显示曲面。板式房的纽带之间的小间隙使照明条、轨道固定装置以及电气接线板排成线形。胶合板面板施工留于场馆内部展开，并使用透明涂料完工。

场馆的外墙是复合金属板材，与胶合板面板的接合处一致。所有的金属板立于后一块面板之上，这种叠法使得场馆仍然能够分解。纽带之间的开口涂上了釉质透明色料，场馆的东西两端向着庭院敞开，并由伸出屋顶的悬垂面板遮蔽。

The pavilion design is created from both a conceptual approach and a constructional logic that share the generative order of three intertwined bands. The bands coil in space to create three helical formations. These formations are limited in width to 300mm and are segmented into lengths no longer than 2,400mm to meet manufacturing and handling requirements. The three bands are assembled into an intertwined configuration to create the pavilion volume in which individual bands spatially and structurally strengthen each other.

The seams between the bands are celebrated for their architectural potential. Bands, individually or collectively, reach into the interior of the volume creating glazed openings, skylights and horizontal display surfaces. Small gaps between the bands of panels house linear strip lighting, track fixtures and electrical power strips. The plywood panel construction is left exposed on the interior of the pavilion and finished with a clear coating.

The exterior of the pavilion is clad in metal sheets that match the seaming of the plywood panels. All metal panels lap subsequent panels in such a way that the pavilion is still able to be disassembled. Openings between the bands are glazed while the west and east ends of the pavilion are left open to the courtyard and protected by the overhanging roof panels above.

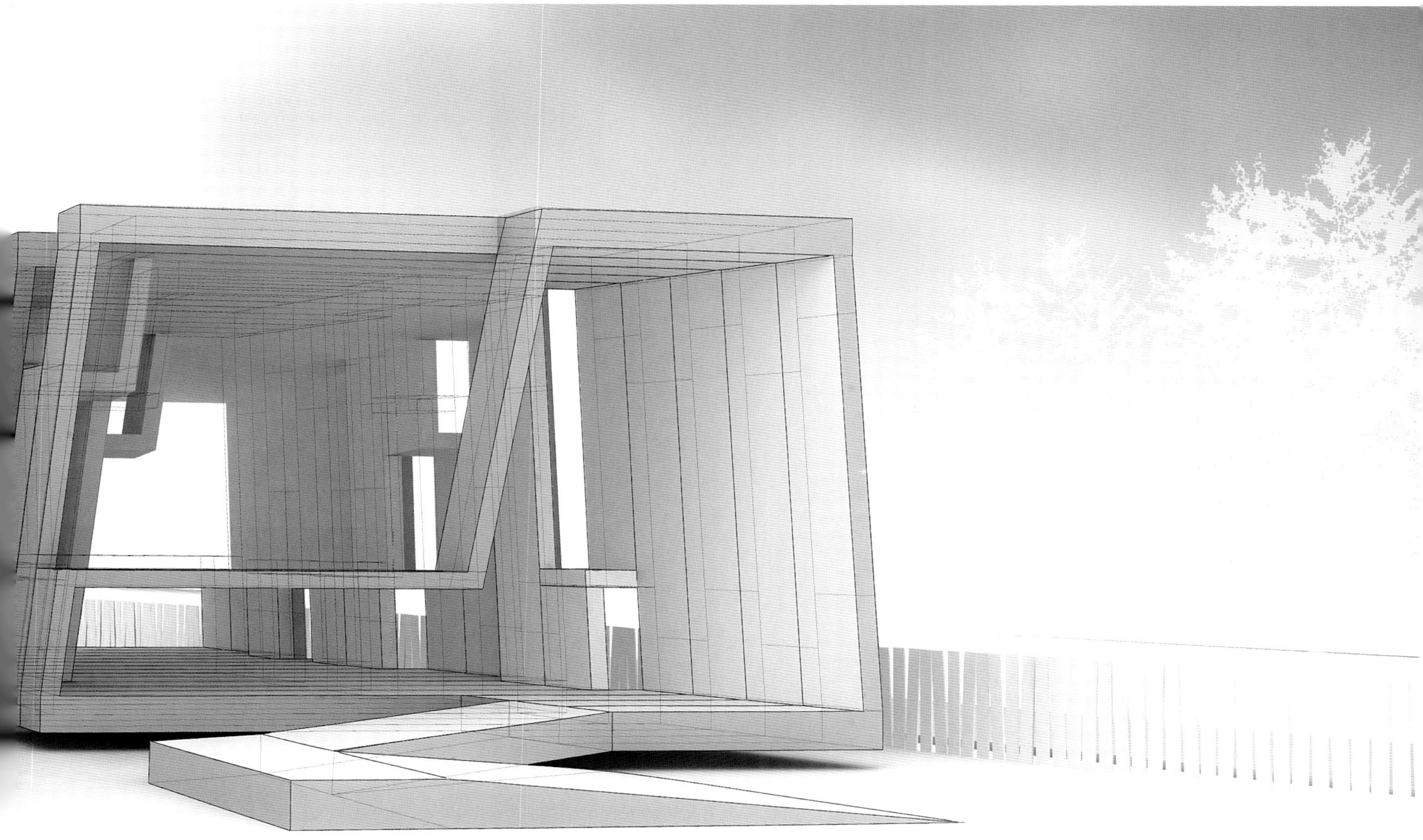

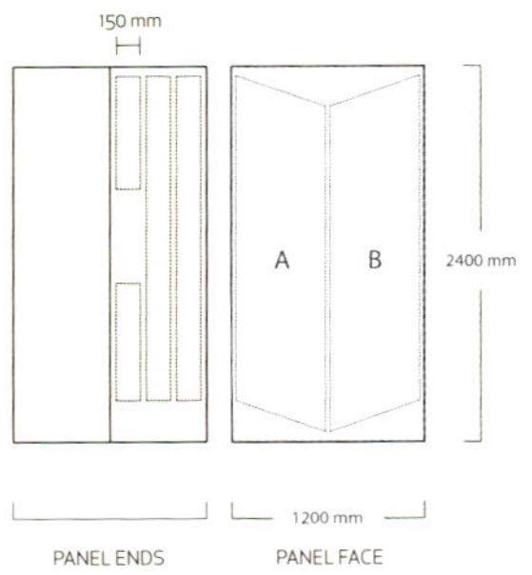

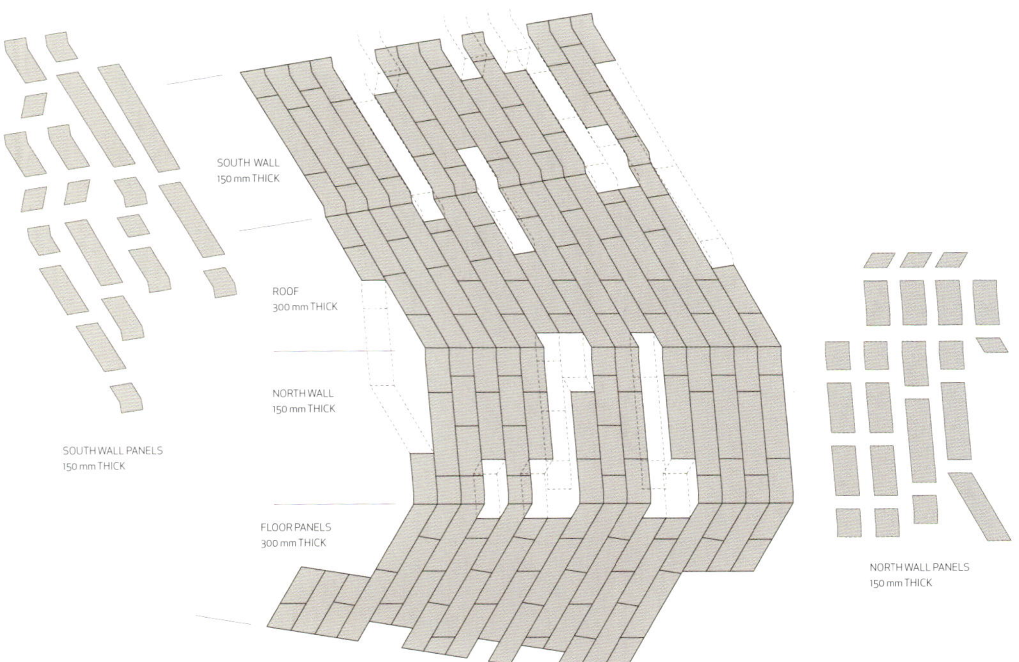

PANEL DIAGRAM

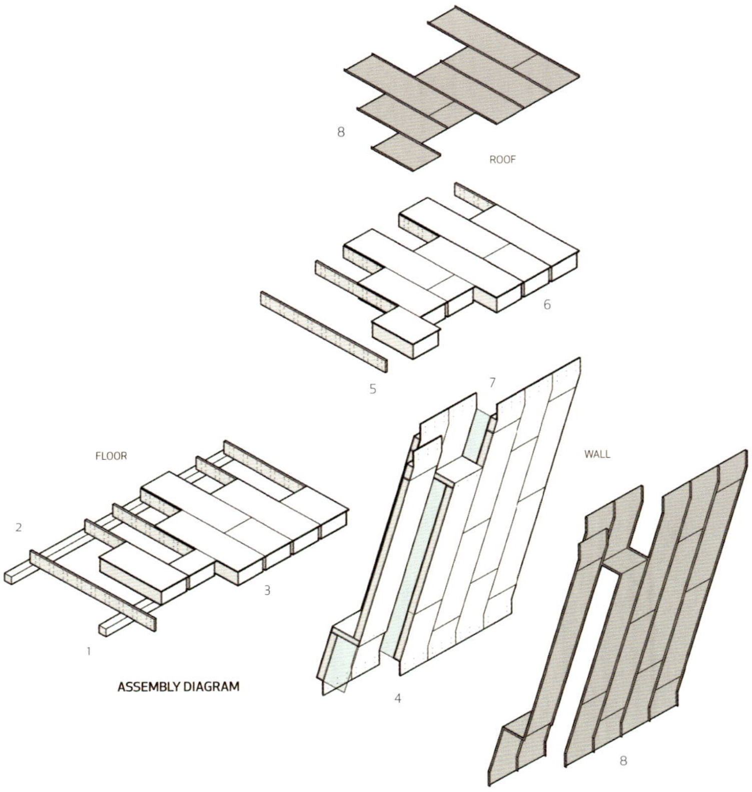

ASSEMBLY DIAGRAM

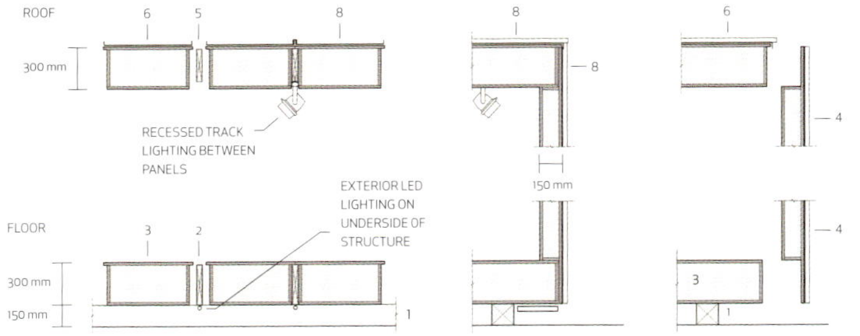

LONGITUDINAL SECTION DETAIL

TRANSVERSE SECTION DETAILS

POWER AND LIGHTING DIAGRAM

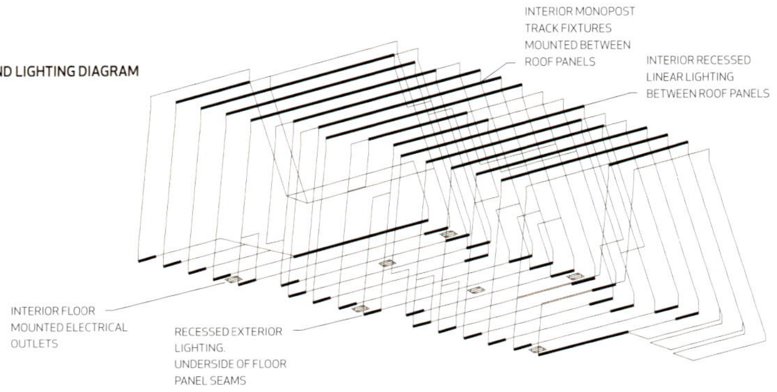

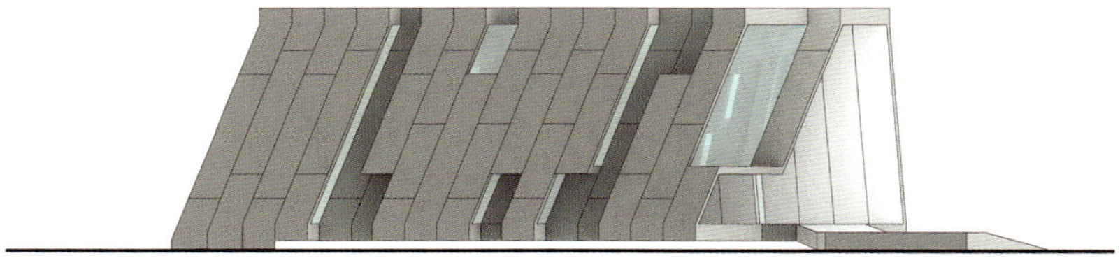

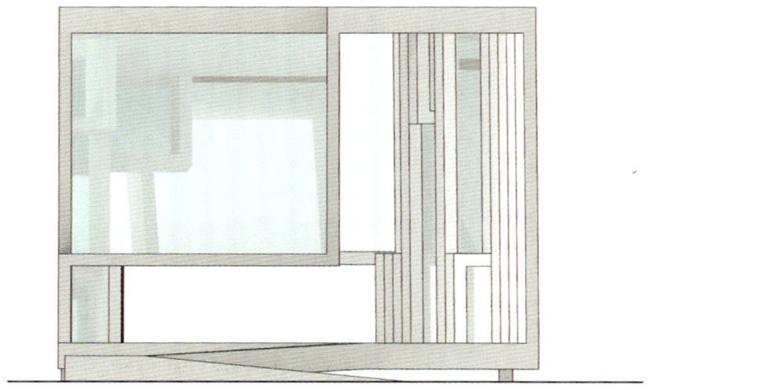

1

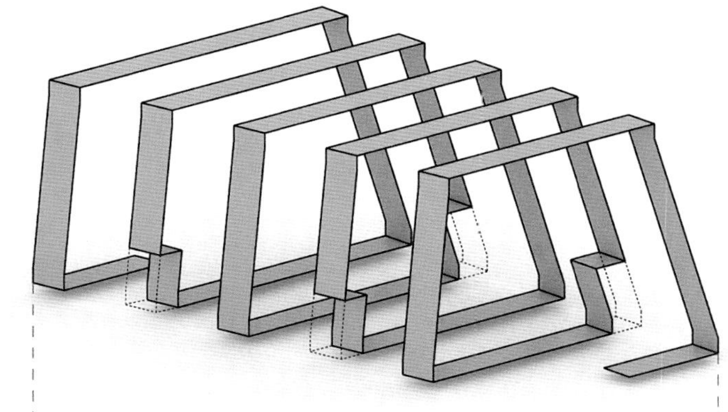

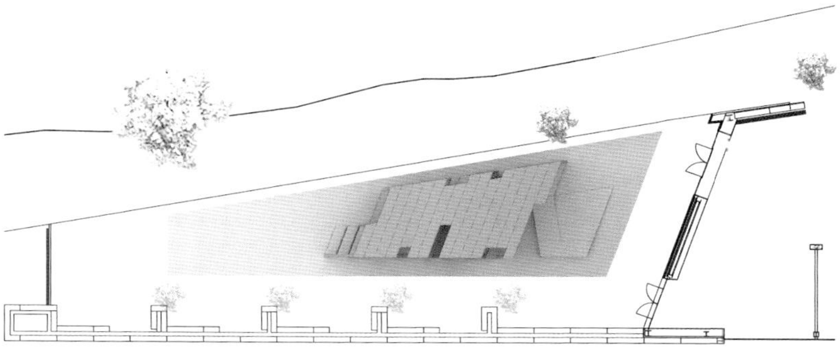

2

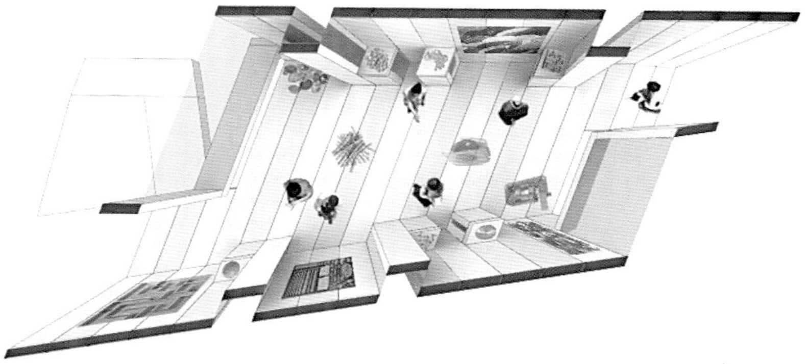

3

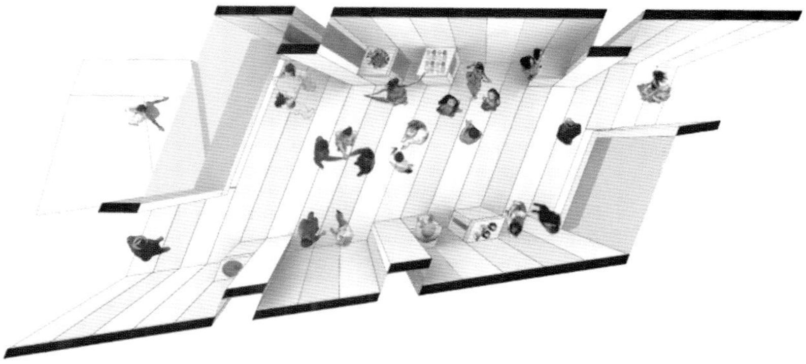

Ankara Arena Multi Purpose Sports Hall

安卡拉阿瑞娜多功能体育馆

Yazgan Design Architecture

2011 Nominee for European Union Prize for Contemporary Architecture Mies van der Rohe Award
2011年当代建筑欧洲联盟奖——密斯·凡·德·罗奖提名奖

Credits

Location: Cumhuriyet Caddesi, Altindağ, Ankara, Turkey
Employer: Turkish Basketball Federation
Construction Year: 2010
Total Construction Area: 55,000 m²
Design Consultants: Ersin Ersoy, Sport Concepts London
Architectural Project: Yazgan Design Architecture
Structural Project: Probi Construction Project Computing Inc. – Fikret Berker
Mechanical Project: GMD Engineering Co. Ltd.
Electrical Project: GMD Engineering Co. Ltd.
Acoustical Project: Mezzo Studio Architectural Acoustics – Prof. Dr. Mehmet Çalışkan
Interior Design Project: Yazgan Design Architecture Co. Ltd.
Landscape Project: Yazgan Design Architecture Co. Ltd.
Contractor: Turkerler Construction Inc.
Construction Type: Reinforced Concrete, Steel
Photography Credits with Dates: Yunus Özkazanç 2010, Cemal Emden 2010, E. Polat Demirtaş 2010

安卡拉阿瑞娜多功能体育馆占地46,000平方米，能容纳10,000人。体育馆所处场地在安卡拉总体规划中被划为体育活动区。体育馆周围是一些历史性的建筑，如第一议会大厦、跳伞塔和土耳其火车站。考虑到其特殊的地理位置——坐落在安卡拉这座老城的中心，体育馆的设计充分考虑了这种独特的环境与城市生活因素。

首先，我们设计了一个城市广场用来连接体育馆与周围的地方，广场同时也可以用来作为当地居民举行活动的地方，通过一条小道可以由广场到达火车站。另外，一条绿树成荫的人行道将国父文化中心与青少年娱乐公园联系起来。体育馆没有围墙或门等分界线，并且不会像典型的体育馆一样在没有体育活动的时候成为一片空地，周围的广场和道路使其成为城市生活的一部分。通过地铁、公交、火车等公共交通工具可以到达体育馆，为了鼓励人们选择步行或公共交通，体育馆的停车位被减到最少。

体育馆的设计更倾向于外放而非内敛。部分楼层上透明的立面及台阶造成了体育馆与城市在各个角度的视觉联系。各种各样的建筑设计元素，如拉杆、白色水平遮阳装置、橙色竖直遮阳装置、幕墙、粗制的混凝土柱、白色的门厅墙壁、绿色的入口以及橙色的公共厕所墙壁，由外至内，层叠累加，增加了从外向内看的景色深度。透明的立面以及15米的轻薄外檐淡化了建筑的立体效果，外檐如机翼一样悬挂建筑上。遮阳装置不仅增加了景色深度，同时赋予立面色彩及节奏感。

虽然该体育馆主要是用来举办篮球、排球、网球、冰上曲棍球、拳击等体育活动的，但音乐会和大型会议等文化活动也可以在这里进行。体育馆不仅达到了国际篮球联合会的所有设计标准，同时也遵循了普遍的设计原则。残疾人、老人和小孩可以很方便地穿行于体育馆中，馆中还专门设有为残疾人准备的私人休息室。抛物线形断面的观台使得每一个观众都能够从同等的角度观看比赛。建筑所处场地以及通往该地的道路上都种满了绿树，打造出一派春意盎然的绿色景象。

SPOR SALONU

219
120
119

Ankara Arena Sports Hall - with 10,000 people capacity - is built on a 46,000 m² land, which is located at the zone reserved for sports activities through the master planning of Ankara. The 53,000 m² building is in between historical buildings, such as, first. Parliament, Parachute Tower and Turkish State Railway Station. Being at the center of the old city of Ankara, it is designed considering the particularities of a special context and its urban life.

Firstly, a relationship was sought between the surrounding regions by implementing an urban plaza as a meeting point and an event space for inhabitants, with its way leading towards the Railway Station. In addition, a pedestrian axis - surrounded by trees in both sides - is designed to start a relationship in between the zone which harbors Ataturk Cultural Center and the one which harbors the youth entertainment park. The building has no boundaries, like fences or gates. Unlike typical sports halls that become a vacant place in times when there is not any sport activity taking place, Ankara Arena is part of a living city with its plazas and walkways. It is possible to access the place through public transport, like metro, bus or train. The number of car parking is put to minimum in order to enhance arrival through walking and public transport.

Secondly, rather than accomplishing an introverted typology, an extroverted design was considered. A transparent façade and terraces on certain levels were designed in order to render inhabitants develop visual relationships with the city in every direction. Diverse architectural elements, which are put layer by layer from the outside to the inside, such as tension rods, white horizontal sunshades, orange vertical sunshades, curtain wall, brut concrete columns, white foyer walls, green colored entrances, orange public toilet walls, increase the depth of view from the outside to the foyer walls behind. The building's volumetric effect is decreased not only through the transparent façade, but also through rendering 15 meters cantilevered eave structure thinner, like the wing of an air craft. Sunshades not only bring depth of view; but also give color and rhythm to the façade as well.

Although the building is designed mainly for sports activities, such as, basketball, volleyball, tennis, ice hockey, boxing, etc., it can harbor such cultural activities like concerts and congresses. It meets all the design criteria put out by FIBA (International Basketball Federation). Furthermore, it follows universal design principles; disabled and elderly people, and families with children can easily access and circulate throughout the building. There are private lodges for the disabled. The tribunes are designed via the development of parabolic sections in order to develop an equal angle of vision for every audience. A significant landscape planning was developed, with a considerable amount of tree, for not only the area itself, but also the surrounding approach routes.

south elevation

north elevation

east elevation

west elevation

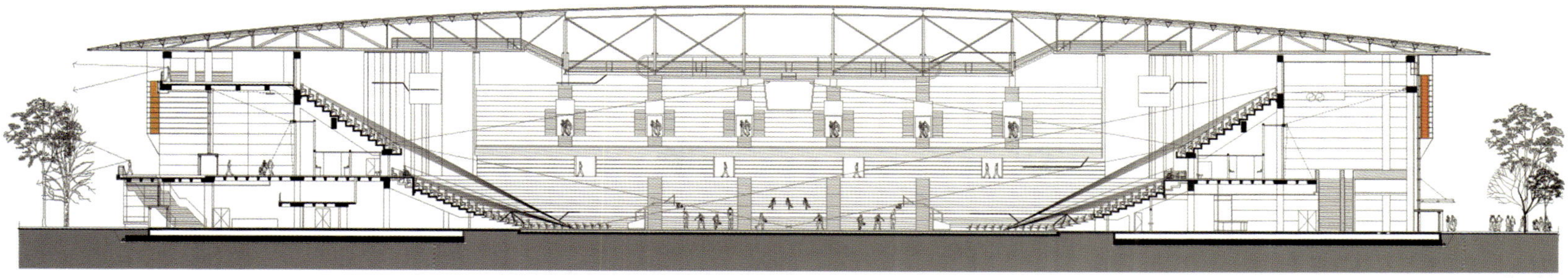

section bb

section cc

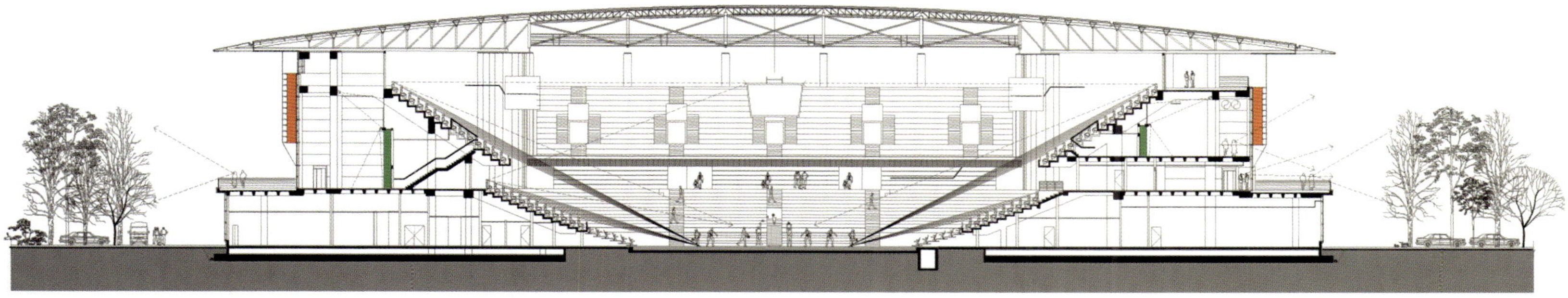

section dd

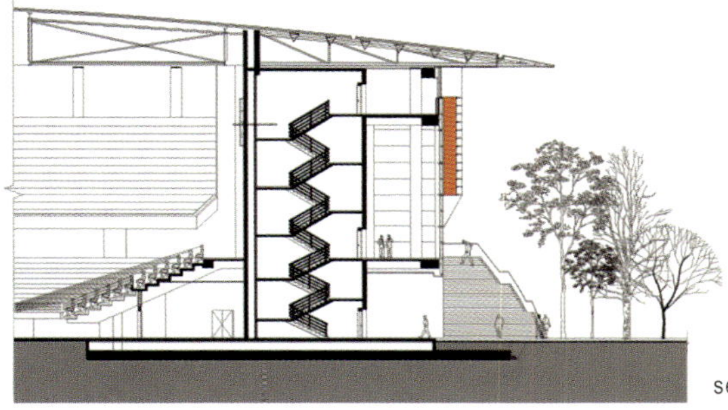

section aa

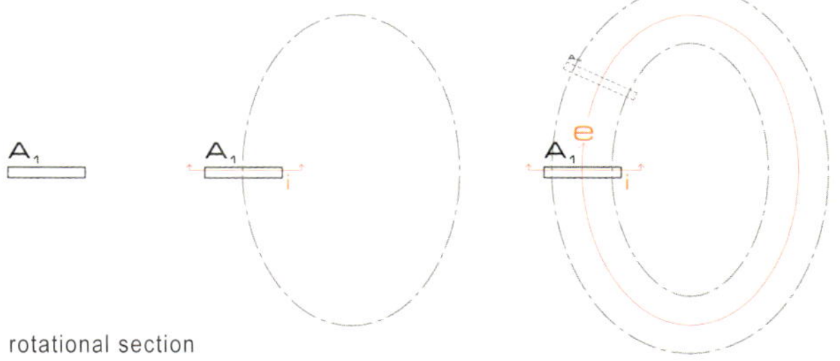

rotational section

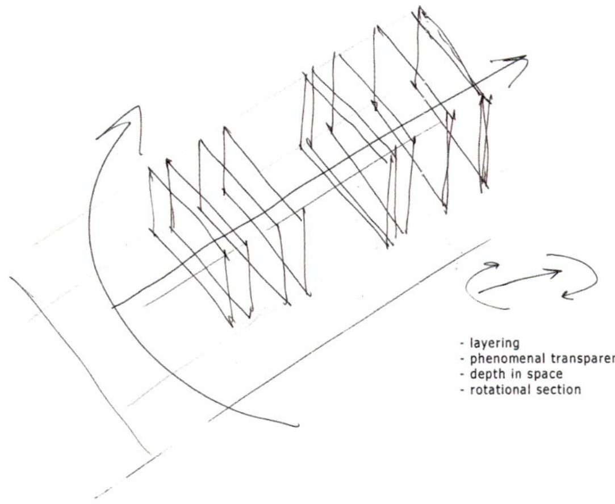

- layering
- phenomenal transparency
- depth in space
- rotational section

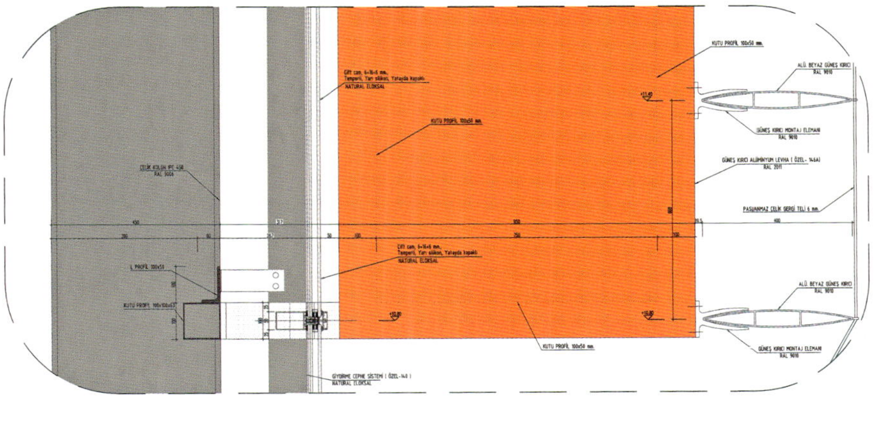

facade detail

Kajzerica City Stadium

Kajzerica城市体育场

Studio 3LHD

Public Competition-invited, Honorable Mention, 2008
2008年公共邀请赛荣誉提名奖

Credits

Program: public, business, sport
Time: 2008
Location: Kajzerica/ Hipodrom
City: Zagreb
Country: Croatia
Client: Grad Zagreb
Author: 3LHD
Team: Sasa Begovic, Marko Dabrovic, Tatjana Grozdanic Begovic, Silvije Novak, Tin Kavuric, Irena Mazer, Krunoslav Szorsen, Dragana Simic, Matija Crnogorac, UEFA and FIFA standards advisor
Team Collaborators: Ljiljana Dordevic, Vibor Granic, Jelena Mance, Ana Safar
Collaborators: Josip Dvornik, Faculty of civil engineering, Zagreb / Structural Analysis
Damir Lazarević, Faculty of civil engineering, Zagreb / Structural Analysis
Josip Atalić, Faculty of civil engineering, Zagreb / Structural Analysis
Sanja Hak, Faculty of civil engineering, Zagreb / Structural Analysis
Marta Šavor, Faculty of civil engineering, Zagreb / Structural Analysis
Mario Uroš, Faculty of civil engineering, Zagreb / Structural Analysis
Damir Pološki, Faculty of civil engineering, Zagreb / Traffic Planning
Željko Stepan, Faculty of civil engineering, Zagreb / Traffic Planning
Dražen Pejković, IGH PC Split / Urban Planning
Tomo Planinić, Termoinženjering-projektiranje d.o.o. / Mechanical Installation
Srđan Grujić, Projektni biro - grijanje d.o.o. / Plumbing and Drainage
Branimir Cindori, Apin d.o.o. / Sprinkler installation
Branko Čorko ing. el., IPZ Elektroinženjering-22 d.o.o. / Electrical Engineering
Hrvoje Domitrović, Faculty of Electrical Engineering and Computing, Zagreb / Acoustics
Zlatko Galić, Novalux d.o.o. / Lighting
Boris Goreta / 3D
Jelenko Hercog, Numen / Graphic Design
Studio HRG d.o.o. / Aerial Photography
Sandro Lendler, Photography
Ines Hrdalo, Landscape Design

萨格勒布的Kajzerica城市体育场是一个新地标。

该体育场坐落于与赛马场体育休闲区相邻的一处高原地带，可容纳至少50,000名观众，在周围Sava和Budek绿化带中占据着重要的地位。在Sava河流两岸的任何区域都可以欣赏到体育场。其所在区域分为三个部分，按照项目的规划，施工也分为三个阶段，与财务架构相一致。

北区是第一期工程，体育馆就独立于周围的其他住宅建筑巍然矗立在该区域；中区是第二期工程，第二期是整体工程中非常重要的环节，因为二期所在区域将建立不同的服务设施；南区是三期工程，包含了Velesajam整个区域以及其建筑，用于建设新商务、商业及住宅建筑，同时还有一个商业中心。在这个新兴的区域内，将会建设酒店、议会中心、商务中心和必要的住宅区。多种功能区的组合解决了该中心居民的生计问题，使人们期待已久的商业区愿望得以实现，同时为该区域的社会化进程提供了机会。

圆形竞技场的设计灵感来自于几种形式，其一是所谓的马格纳斯效应。在足球运动中，马格纳斯效应发生在足球运动轨迹改变的时候，是一个非常困难但又非常具有吸引力的动作。临近街区的名字kajzerica原是一种圆形点心的名字，体育场形状的设计就参照了这种点心的形状。在当代竞技体育场的背景之下，该设计唤起了〞面包与马戏〞的古语。

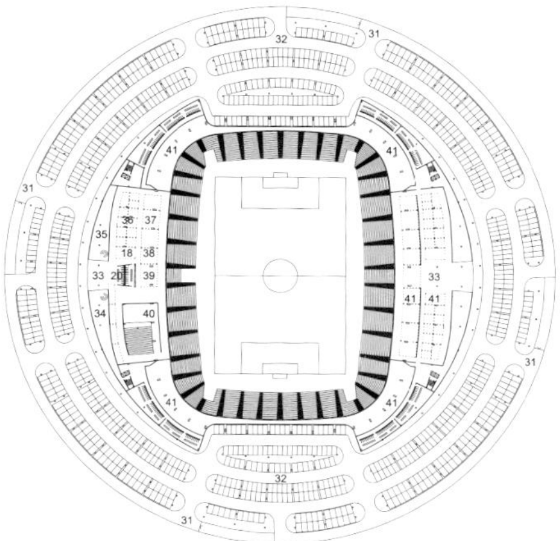

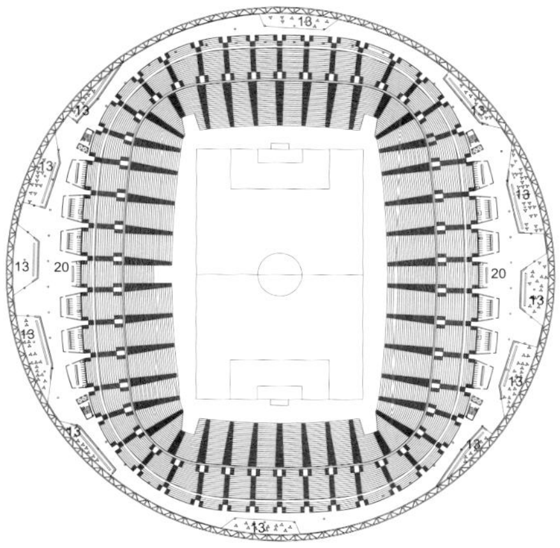

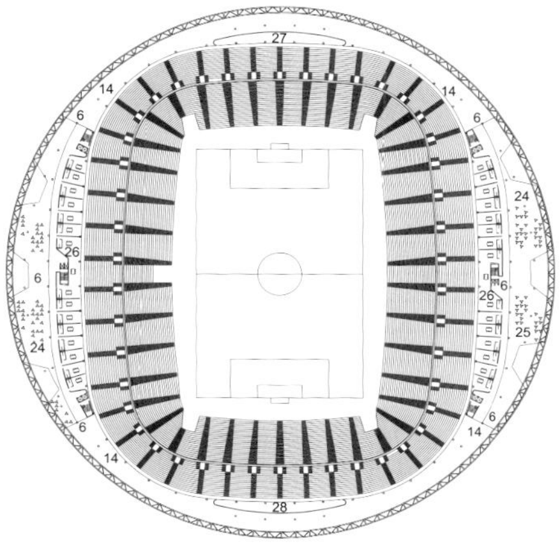

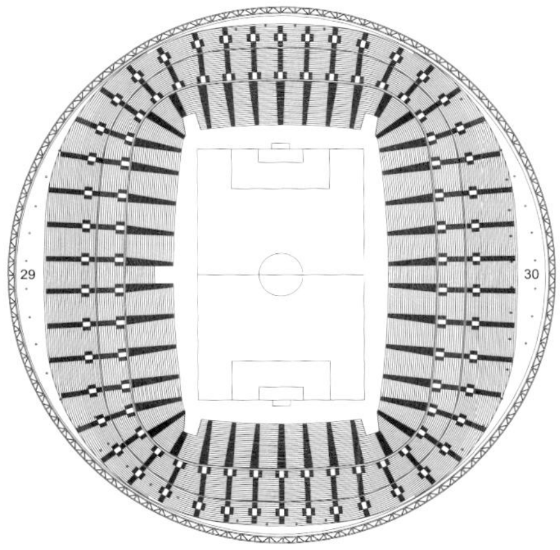

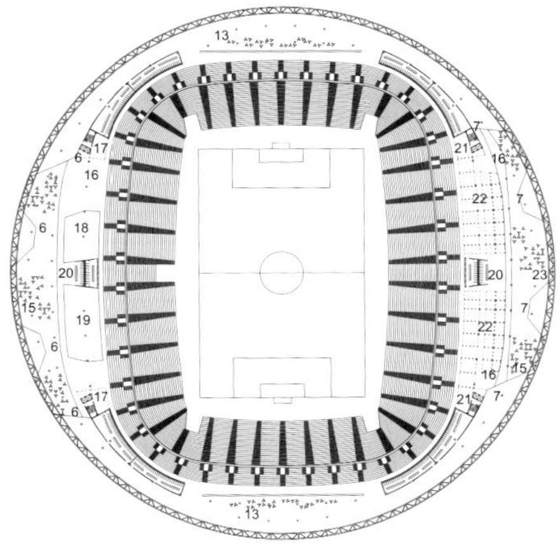

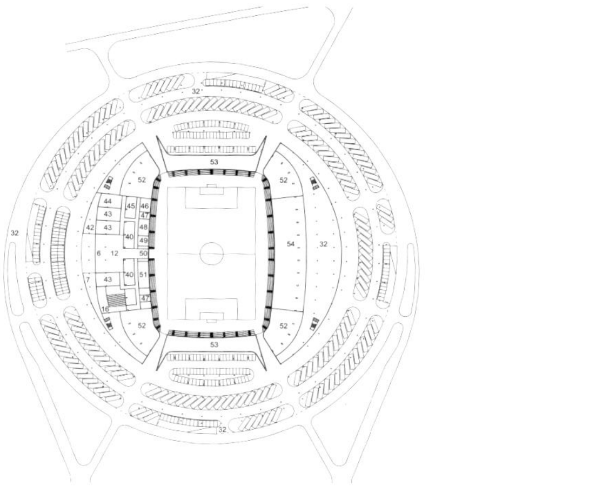

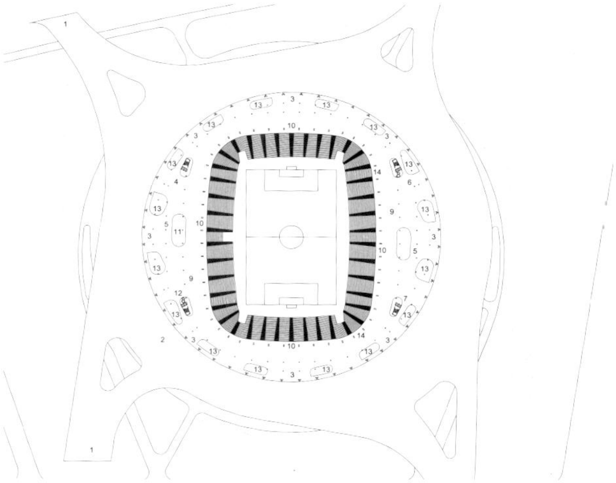

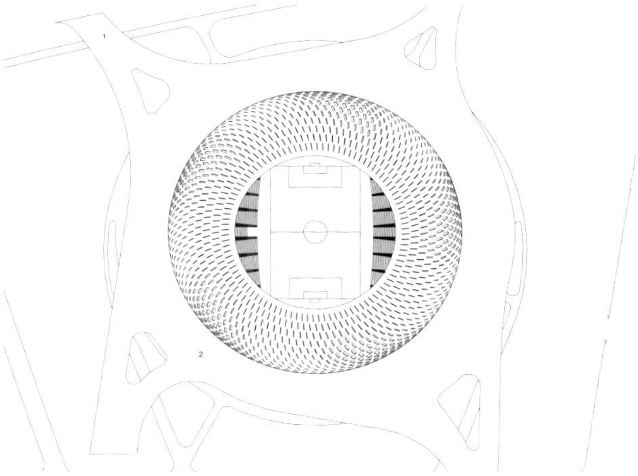

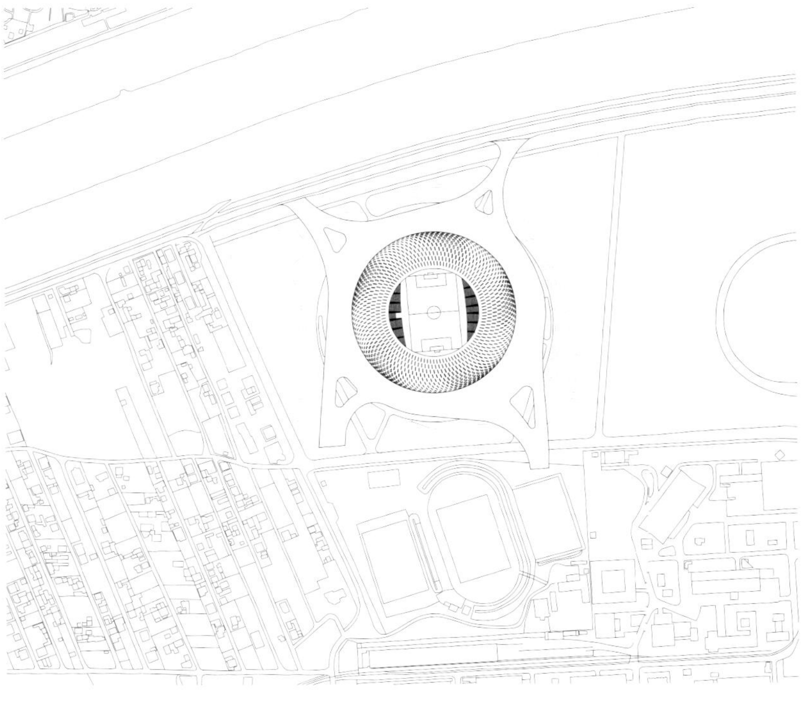

The city stadium Kajzerica in Novi Zagreb represents a new landmark in the town.

The stadium, which seats more than 50,000 spectators, is located on an elevated plateau next to the sports and recreational zone of the racetracks and it dominates the surrounding area, the green zone of Sava and Bundek. The stadium is visible from all surrounding areas on both sides of the Sava River.

The coverage area is divided into three zones, consistent with the project's program, and into three phases, consistent with the financial framework of construction.

The northern zone, the first building zone, enables the undisturbed building of the stadium, independent of the surrounding settlements, parcels and Zagrebački Velesajam. The middle zone, the second building phase is crucial for the development of the stadium because this is where different services will be situated. The southern zone, i.e. the third building phase, encompasses the whole area of Velesajam and all of its buildings. It is for the construction of new businesses, commercial and residential buildings as well as a new commercial center. Within the newly formed cassettes – "blocks" there will be hotels, a congress center, business centers and obligatory residential areas. Integration of these elements ensures livelihood of the new center and brings a long-awaited downtown feeling and an opportunity for socialization to Novi Zagreb.

The inspiration for the circular arena came from several suitable forms; the so-called Magnus' effect, one of the most difficult and attractive moves in soccer which occurs when a ball changes its course. The name of the neighborhood "kajzerica", which is a name for a certain kind of a bun whose shape is reflected in the stadium. In the context of a contemporary sports arena, this design evokes the saying from ancient times about "bread and circuses".

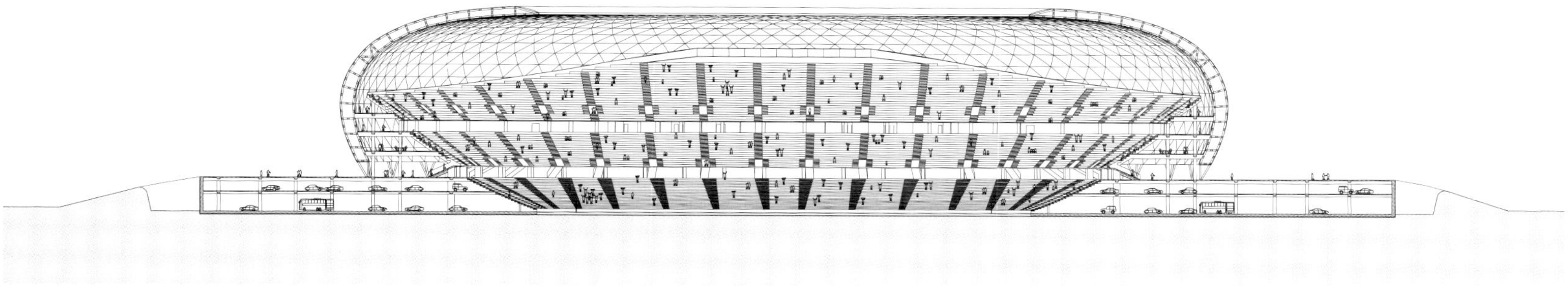

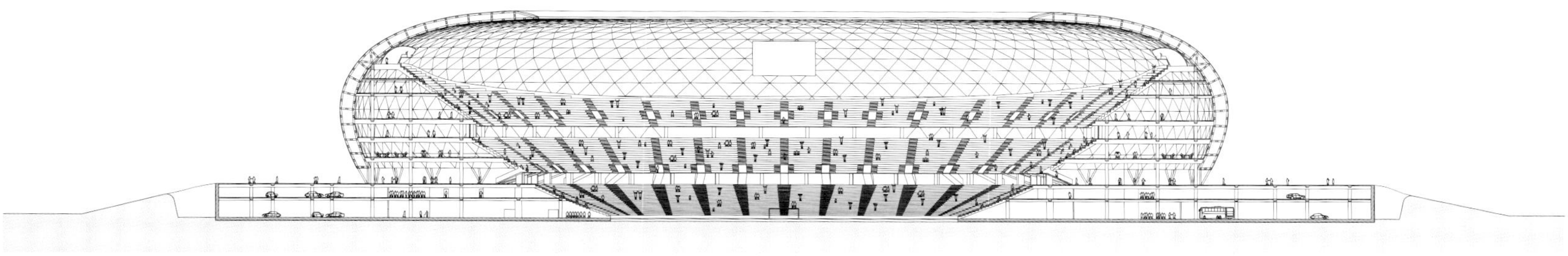

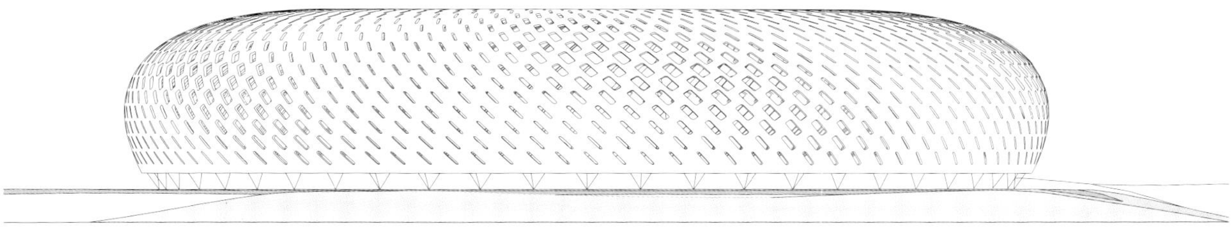

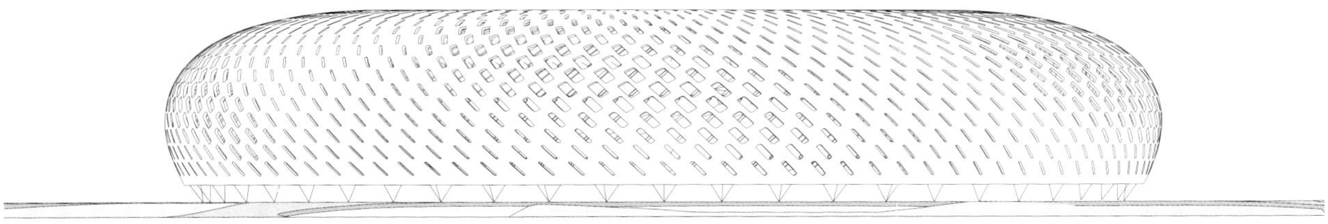

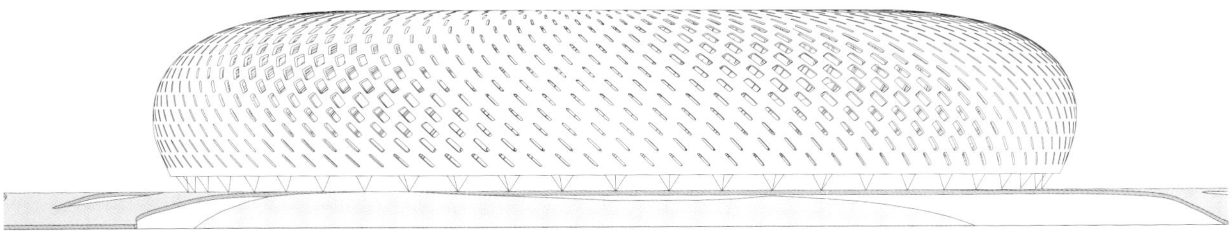

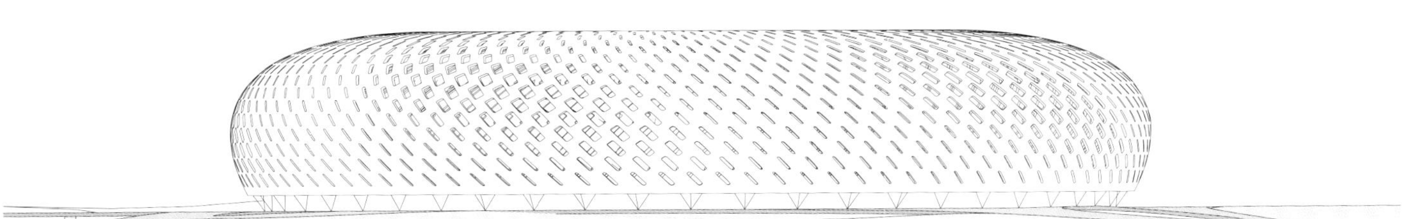

The KAIST Memorial Building, Deajeon, Republic of Korea

韩国大田科学技术院纪念馆

J. Mayer H. Architects

International Competition for the K.A.I.S.T. Memorial Building, Deajeon, Republic of Korea

韩国大田科学技术院建筑国际竞赛参赛方案

Credits

Competition Team: Juergen Mayer H., Mehrdad Mashaie, Jesko Malkolm Johnsson-Zahn, Paul Angelier

Invited competition 2009

Client: Korea Advanced Institute of Science and Technology

Model: cirp GmbH

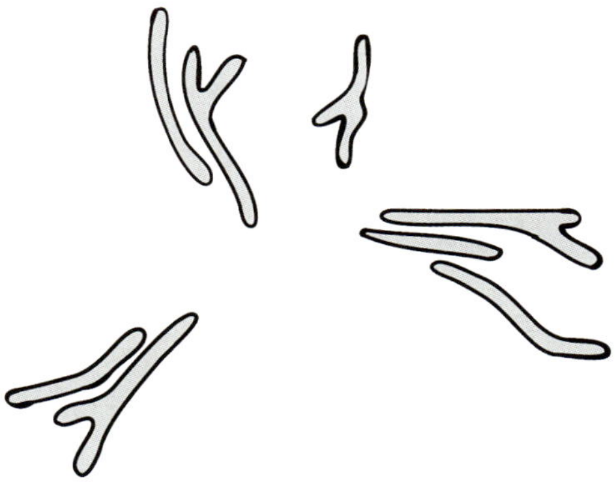

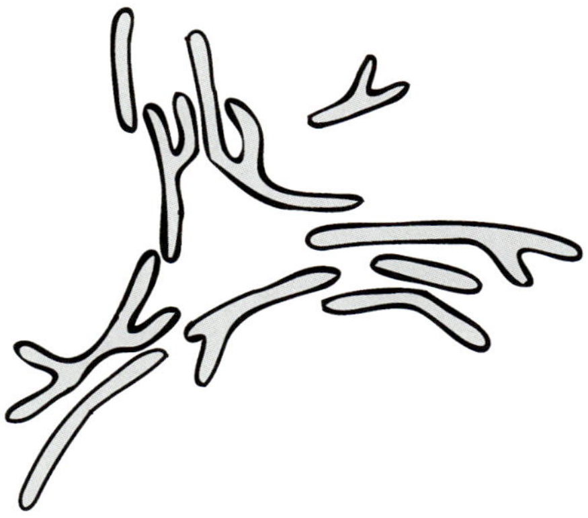

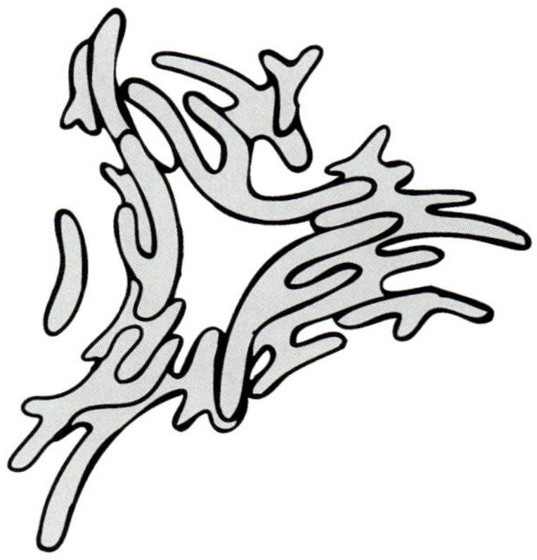

背景

韩国因其追求科学和技术领域的高精水平而久负盛名。在当今的国际环境下，韩国大田科学技术院（KAIST University）代表了这一非凡情况，并为未来承担可持续的责任。大田科学技术院纪念馆成为此成就和声誉的图标。

地点

韩国大田科学技术院校区是一个组织良好的、与户外范围和自然维持适当平衡的建筑系统。韩国大田科学技术院纪念馆坐落于校区的中央位置，迎接游客和许多学者的到来。自然与建筑之间密切的互动关系得益于周围的湖泊和森林生态。

理念

校区的景观设计是整个项目的出发点。纪念馆蹿出地面直入一座开放的雕塑建筑，其外形可以读作抽象的韩语字母。以科学、组合数学以及流线型作为语言的参考依据，为多样解读这幢建筑提供了丰富的内涵。该建筑代表着思想和知识强烈的推动变化的力量。

纪念馆的流体造型如同水寓意着变动和进展。这种建筑体系将有助于韩国大田科学技术院在科学、技术以及知识产生的领域内取得独一无二的全球地位。

建筑

韩国大田科学技术院纪念馆四面通达，且集成景观设计于底层楼面内部。内部的中央中庭径直通往屋顶露台，在那里可以欣赏整个校区内外的全景。中央螺旋形循环通道连接整个建筑内部的所有公共区域。这宽敞的中庭是一个交互空间，允许访客乘坐电梯往上抵达露台和展览区，然后往下直到他们再次抵达大堂。底楼和顶楼是开放式的公共区域。之间楼层的平面布局包括实验室、会议室以及办公室。

该建筑不断更改其外观。白天，悬臂梁下面的水帘使得纪念馆朝气蓬勃；夜晚，多媒体屏幕和投影同样使得纪念馆在日常办公时间之外栩栩如生。对于一幢学术建筑来说，纪念馆是一种原型，从白天到黑夜都是整个校区脉动的心脏。

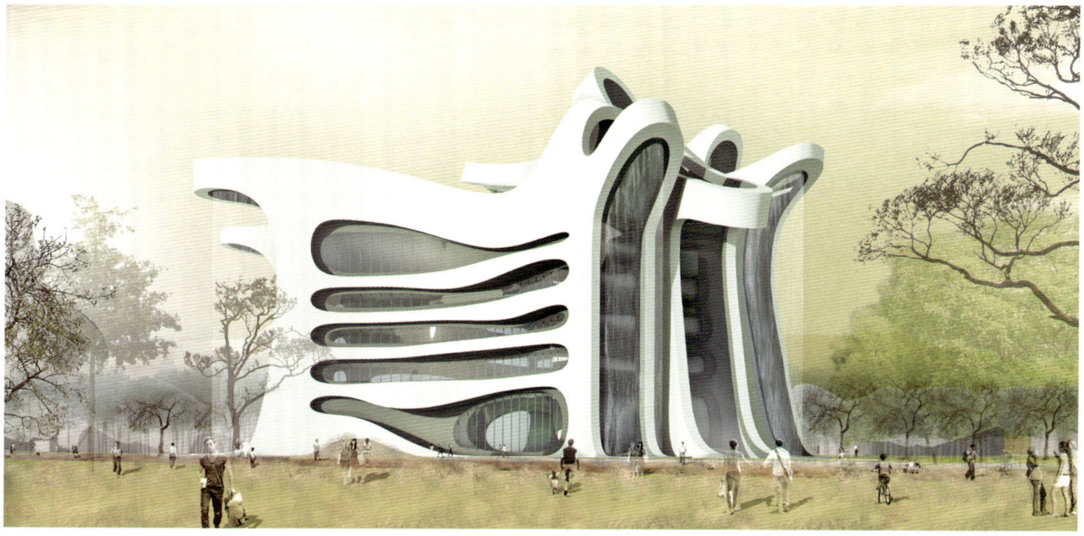

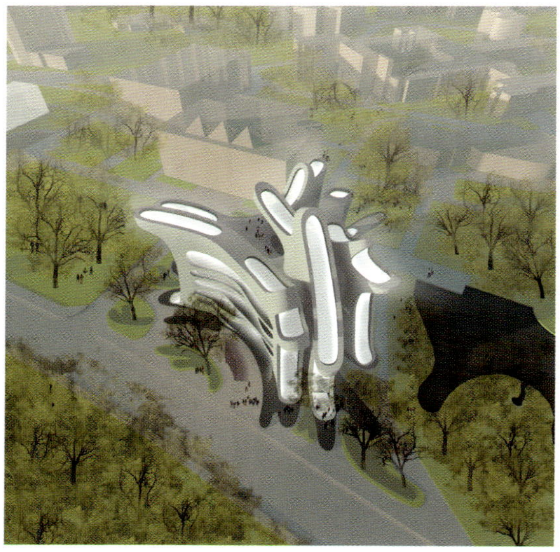

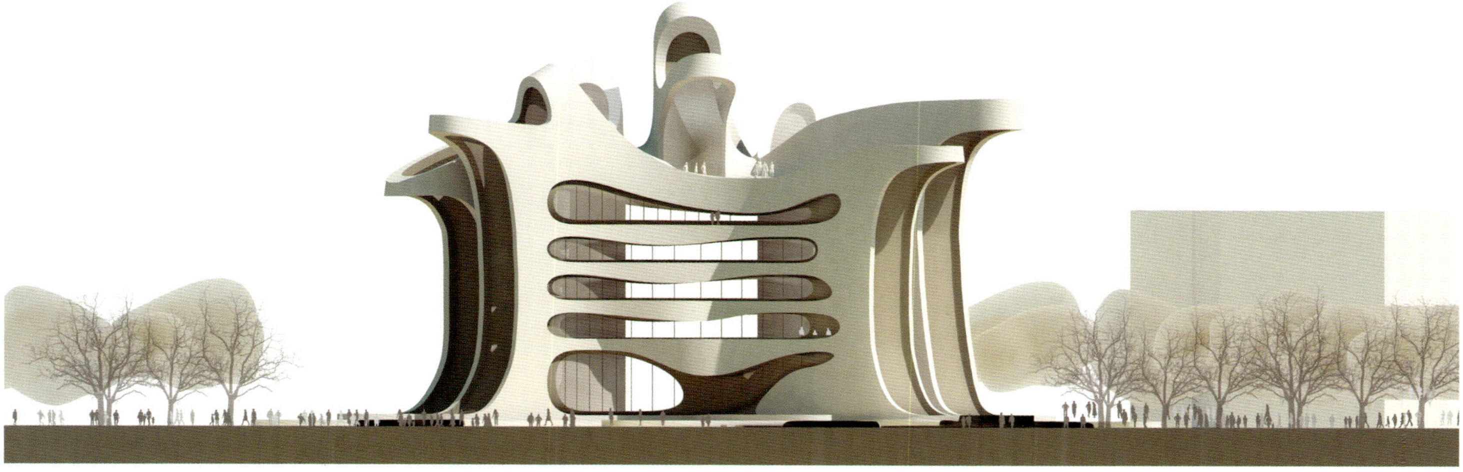

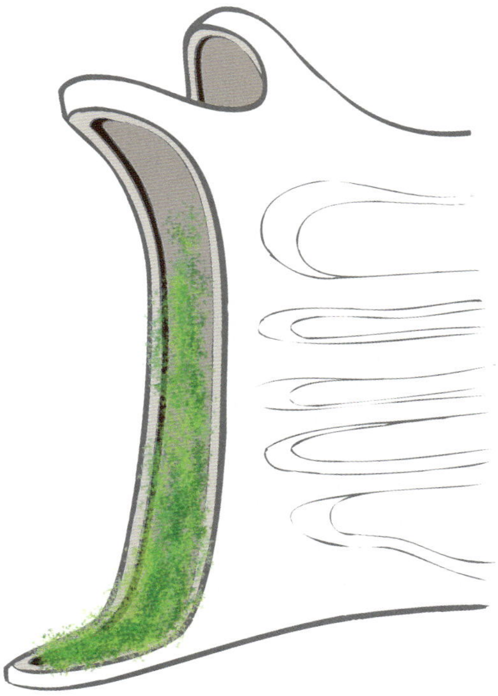

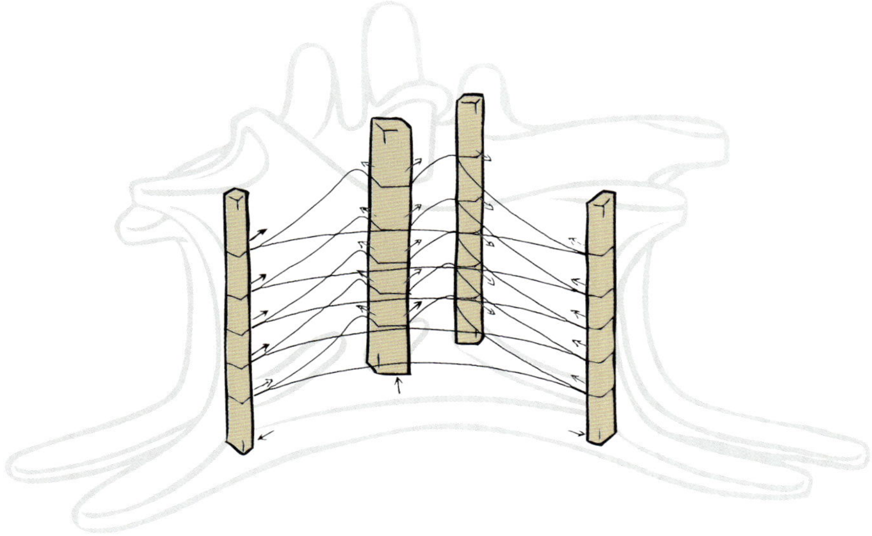

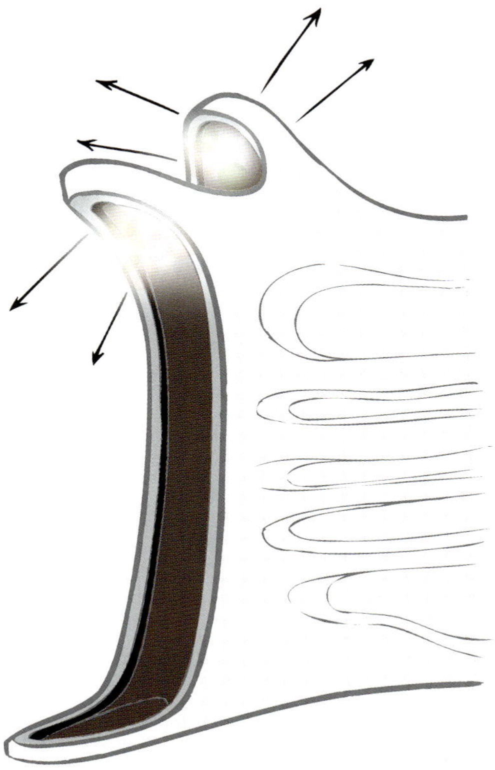

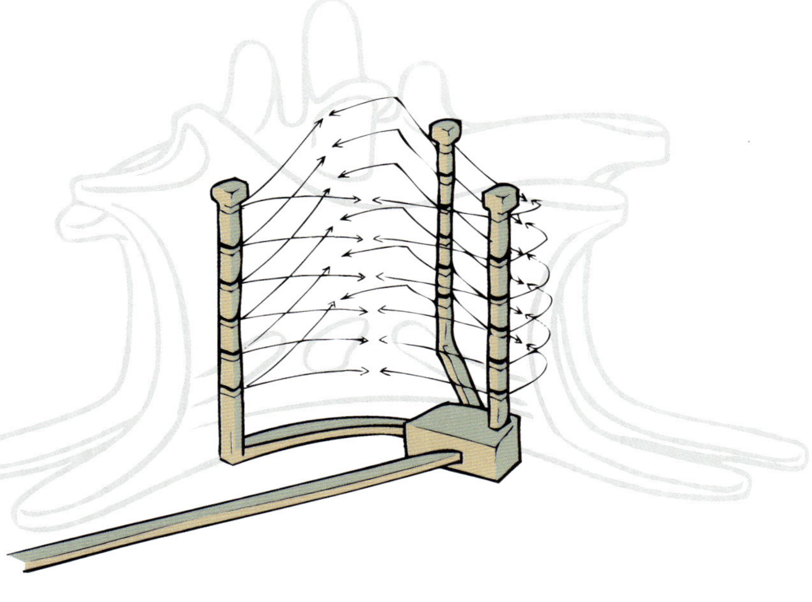

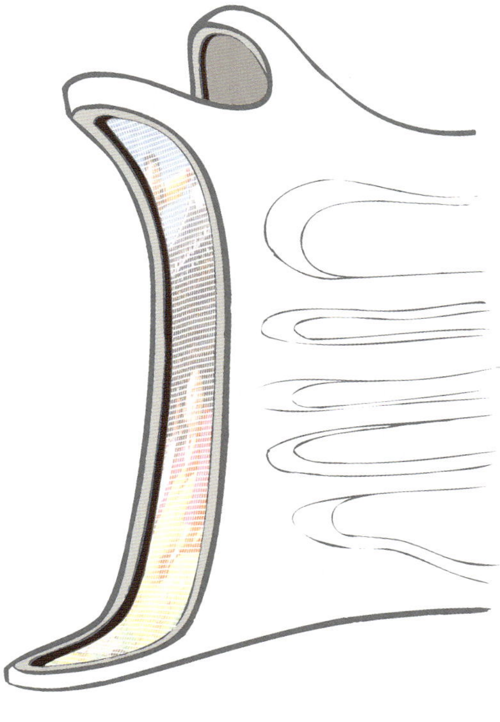

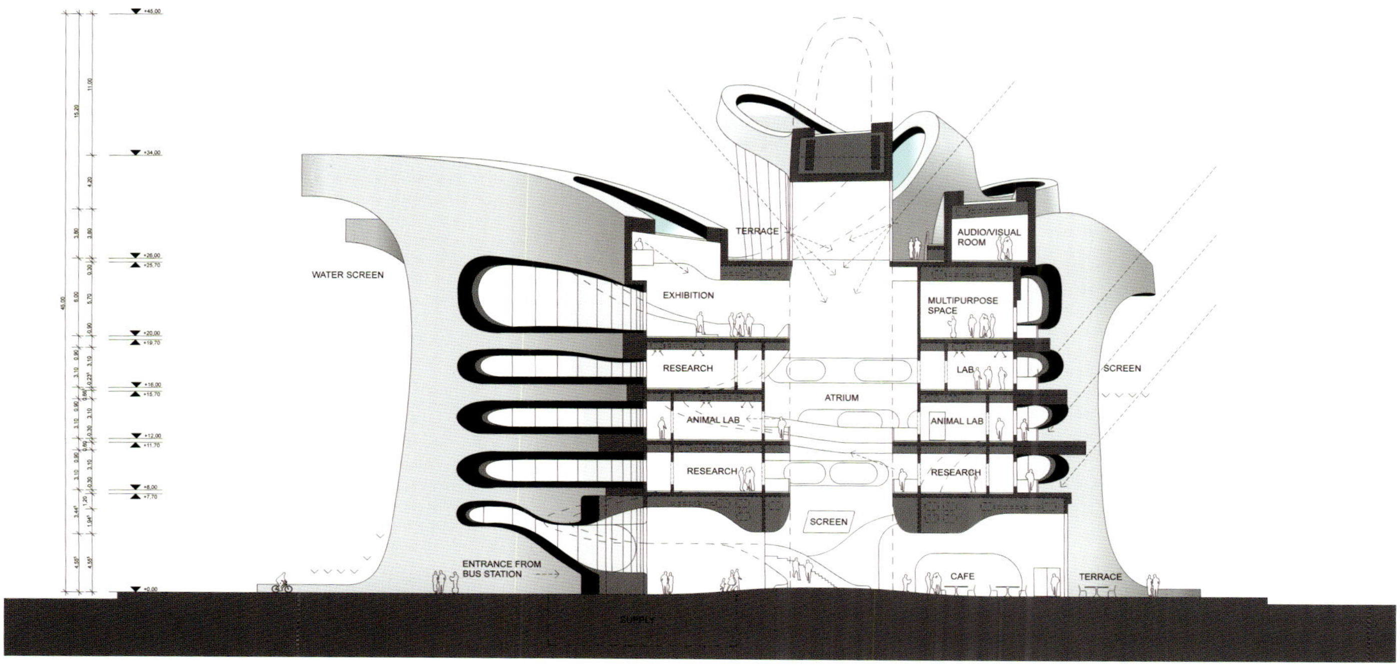
TERRACE
AUDIO/VISUAL ROOM
WATER SCREEN
EXHIBITION
MULTIPURPOSE SPACE
RESEARCH
LAB
SCREEN
ATRIUM
ANIMAL LAB
ANIMAL LAB
RESEARCH
RESEARCH
SCREEN
ENTRANCE FROM BUS STATION
CAFE
TERRACE
SUPPLY

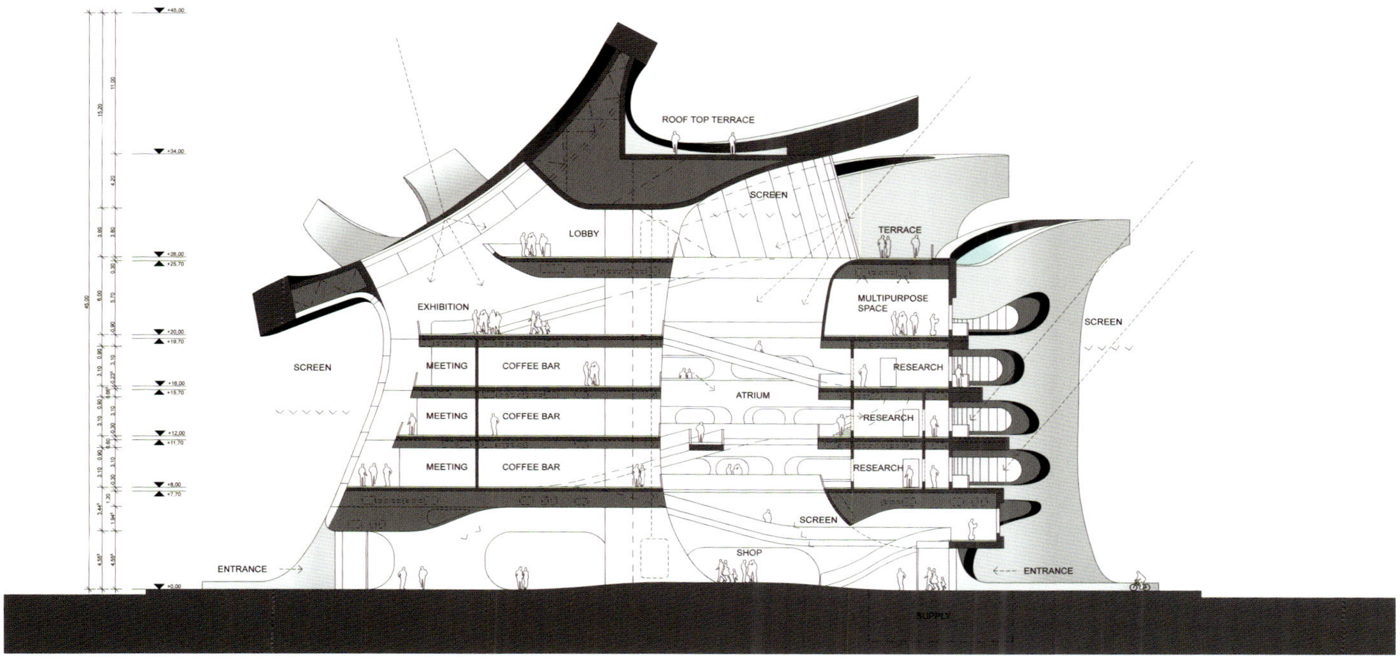
ROOF TOP TERRACE
SCREEN
LOBBY
TERRACE
EXHIBITION
MULTIPURPOSE SPACE
SCREEN
SCREEN
MEETING
COFFEE BAR
RESEARCH
ATRIUM
MEETING
COFFEE BAR
RESEARCH
MEETING
COFFEE BAR
RESEARCH
SCREEN
SHOP
ENTRANCE
ENTRANCE
SUPPLY

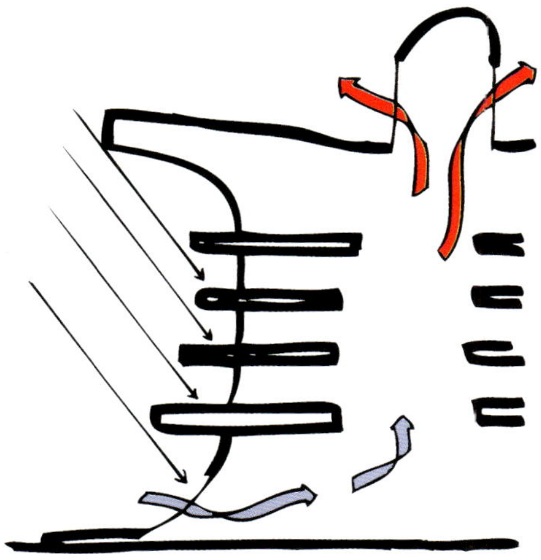

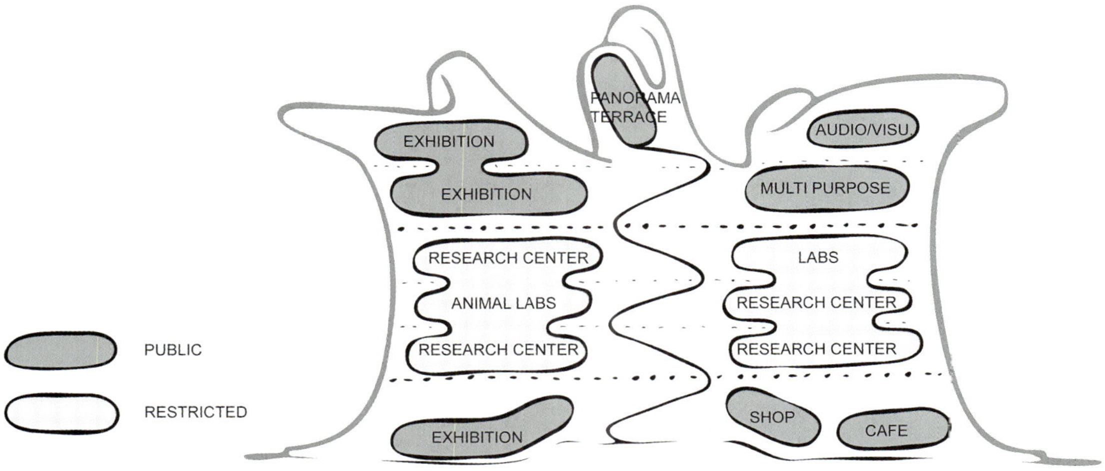
PUBLIC
RESTRICTED
PANORAMA TERRACE
EXHIBITION
AUDIO/VISU
EXHIBITION
MULTI PURPOSE
RESEARCH CENTER
LABS
ANIMAL LABS
RESEARCH CENTER
RESEARCH CENTER
RESEARCH CENTER
SHOP
CAFE
EXHIBITION

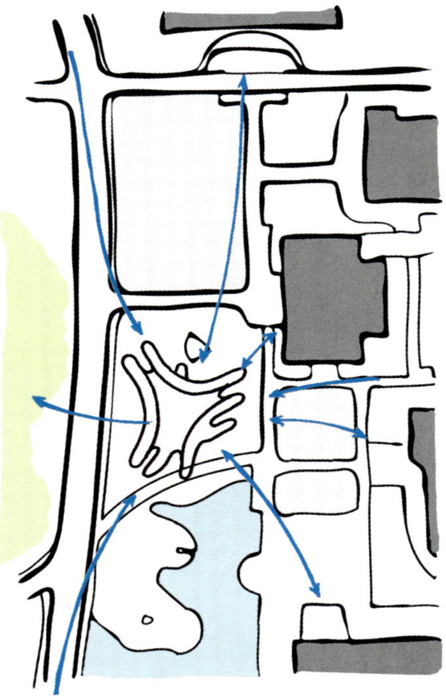

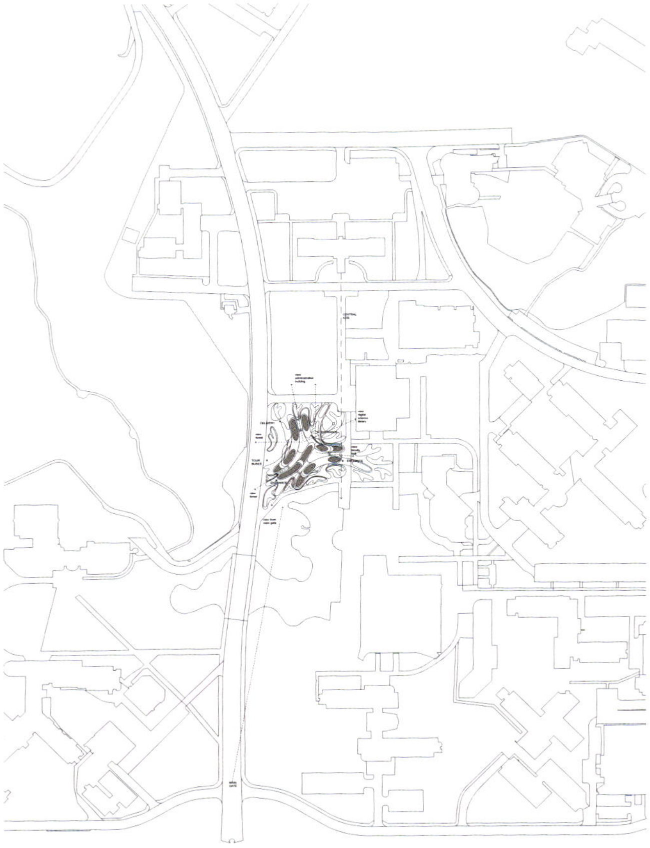

Context

Korea is known for its long tradition of highest standards in science and technology. KAIST University represents this exceptional situation within an international context of today and with its sustainable responsibility for the future. The KAIST Memorial Building becomes an icon for this achievement and reputation.

Site

The KAIST Campus is a well organized system of buildings in good balance with outdoor areas and nature. The KAIST Memorial Building is centrally situated on the Campus and welcomes visitors and academics alike. The close interaction is between nature and building profits from the natural surrounding of a lake and a forest. Outdoor and indoor spaces allow for a fluid circulation and open atmosphere.

Concept

The landscape of the campus is the starting point of the project. The building rises out of the ground into an open sculptural building which can be read as abstracted Korean letters of the alphabet. This reference of language as science, combinatorics and flow, gives this building a rich concept for multiple readings. This building represents the intense dynamic of thought and knowledge.

Its fluidity like water stands for movement and progress. This architecture will contribute to the unique global position that KAIST University takes in the field of science, technology and the production of knowledge.

Architecture

The KAIST Memorial building is open to all directions and integrates the landscape inside the ground floor. The central atrium is open all the way to the roof terrace with a panoramic view onto the entire campus and beyond. A central spiral circulation connects all public areas inside the entire building. This spacious atrium is an interactive space that allows visitors to take an elevator up to the terrace and exhibition spaces, and then move down until they arrive again in the lobby. The ground floor and top floors are open public areas. The floors in between are flexible floorplans that house laboratories, meeting rooms and office spaces.

The building constantly changes its appearance. Water curtains underneath the cantilevers animate the building during the day while at night, media screens and projections activate the building beyond official opening hours. It is a new prototype for an academic building that is a pulsating heart of the campus all day and night.

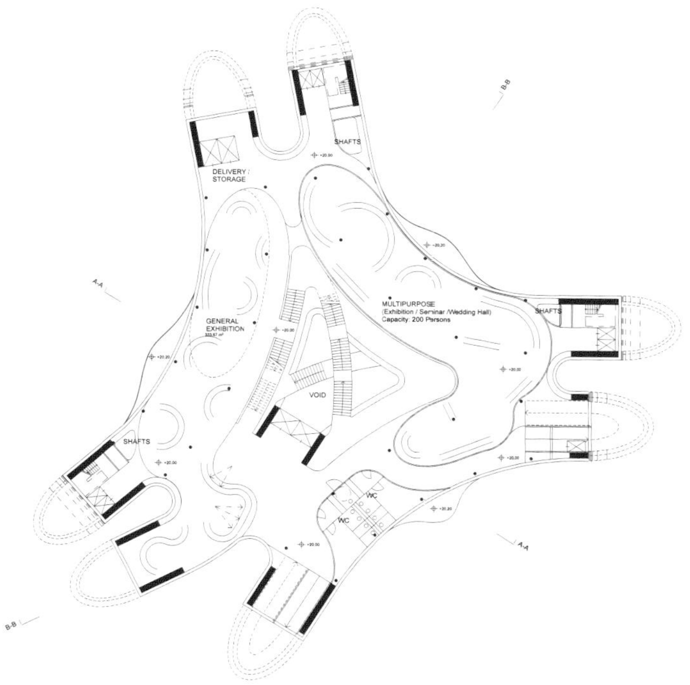
B-B
SHAFTS
DELIVERY / STORAGE
A-A
MULTIPURPOSE
(Exhibition / Seminar /Wedding Hall)
Capacity: 200 Persons
SHAFTS
GENERAL EXHIBITION
VOID
SHAFTS
WC
WC
A-A
B-B

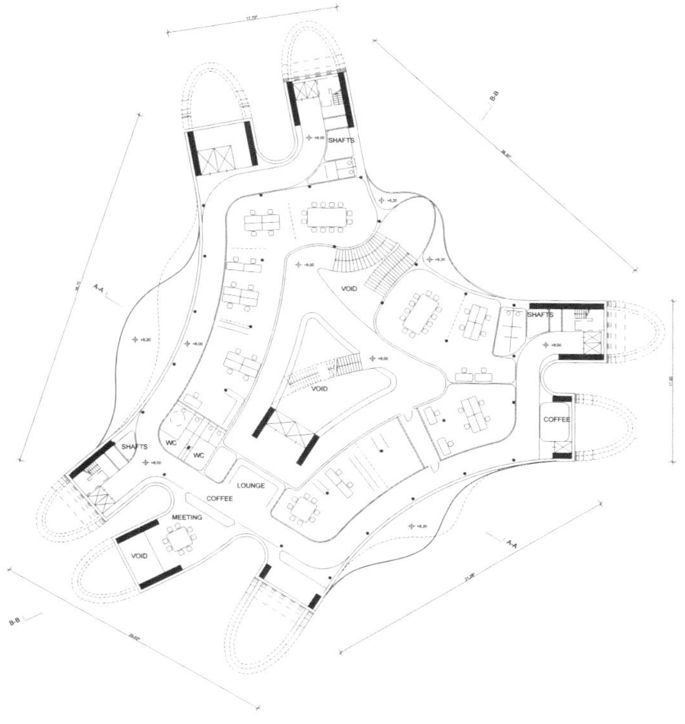
B-B
SHAFTS
A-A
VOID
SHAFTS
VOID
COFFEE
SHAFTS
WC
LOUNGE
COFFEE
MEETING
VOID
A-A
B-B

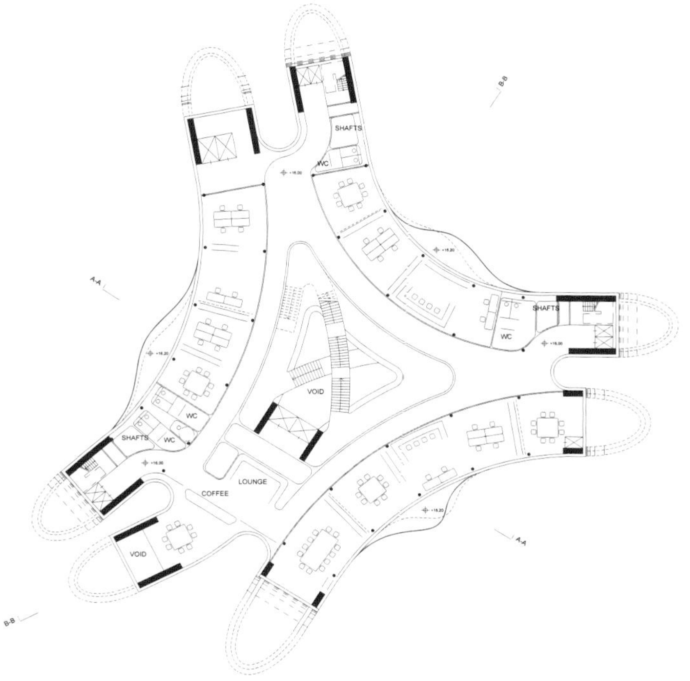
B-B
SHAFTS
WC
A-A
SHAFTS
WC
VOID
WC
SHAFTS
WC
LOUNGE
COFFEE
VOID
A-A
B-B

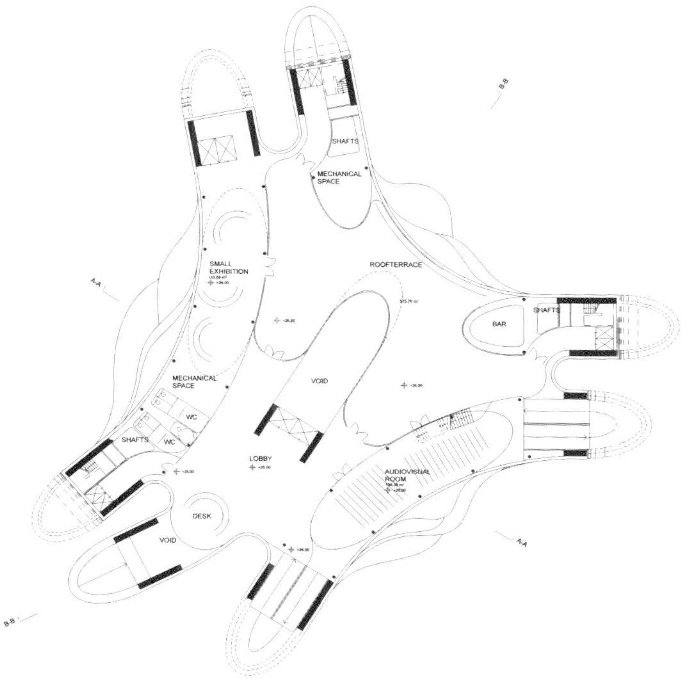
B-B
SHAFTS
MECHANICAL SPACE
A-A
SMALL EXHIBITION
ROOFTERRACE
BAR
SHAFTS
MECHANICAL SPACE
VOID
WC
SHAFTS
WC
LOBBY
AUDIOVISUAL ROOM
DESK
VOID
A-A
B-B

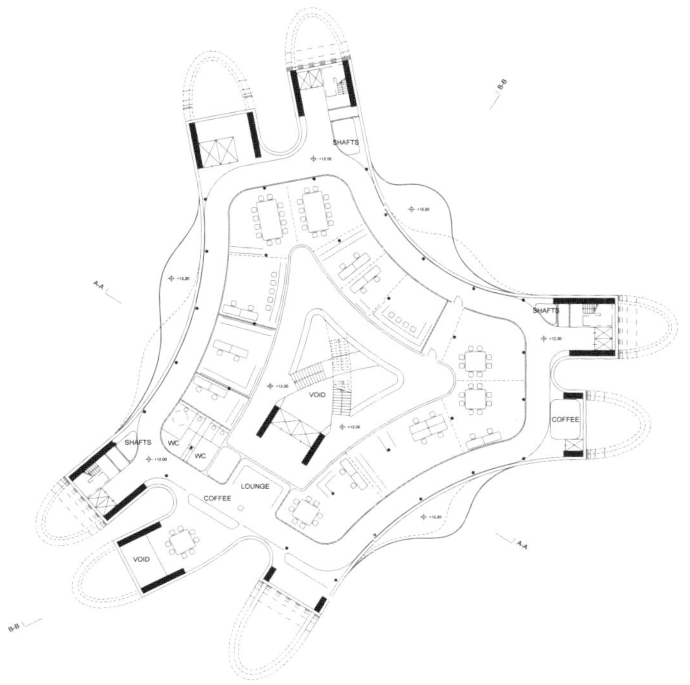
B-B
SHAFTS
A-A
SHAFTS
VOID
COFFEE
SHAFTS
WC
LOUNGE
COFFEE
VOID
A-A
B-B

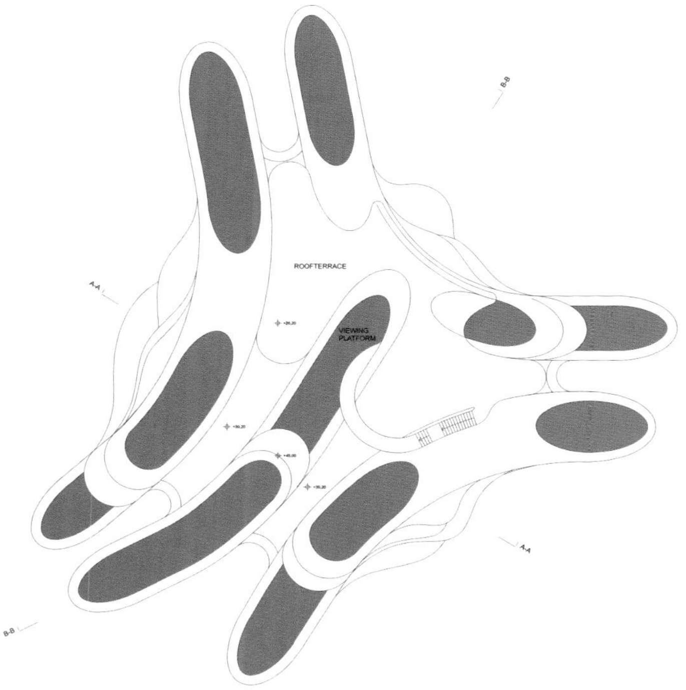
B-B
A-A
ROOFTERRACE
VIEWING PLATFORM
A-A
B-B

Dubai Camel

迪拜骆驼塔 Ivo Buda Architetto

XI ThssenKrupp Elevator Architecture Award
德国蒂森克虏伯集团第十一届建筑奖

Credits
Design: Ivo Buda Architetto
Team: Ivo Buda , Manfredo Bianchi, Tomaz Kristof
Status: Competition (unbuilt)
Client: ThyssenKrupp Elevators
Location: Dubai, UAE
Date: 2009
Program: Viewing tower with congress center, kid's library, gallery and restaurant.

High: 165m
Levels: Basement+Ground l.+3 levels in the legs, 5 levels in the head
Total Floor Area: 2,030m²
Structure: Steel-concrete structure
Parking Spaces: 30

设计背景与理念

迪拜这个世界性城市的大胆，正如穿越沙漠旅行的勇气。出于对这份大胆的敬意，ivo buda建筑事务所设计了这座骆驼塔——外形如一头走向绿洲饮水的骆驼——为整个城市带来一种沙漠旅行的形象。塔楼意欲揭示沙漠旅行的精神，那就是青年在大自然中探索未知领域的志向与冒险的快乐。迪拜在这个沙漠旅行的过程中，既是一个探险地，同时也充当着供给饮用水的绿洲。塔楼以优雅的外形活灵活现地呈现在大家面前，如一头真正的骆驼，自然缓慢地伸向天空。

塔通道

“骆驼”四条弯曲的腿带来一种奇特的视觉效果——人们可以从各种角度去观赏这座塔，而且，不同的角度与距离会带来完全不同的结构效果。因此塔通道围绕“骆驼”呈螺旋状分布。螺旋通道由主干道开始——因为那里建筑物稀少，而且塔楼远处也能清晰可见——穿越小路与停车场，然后是公园入口、公园小径，最后是一个斜面坡道，通向离地面5米高的入口平台。为让游客在进入塔楼之前可以观赏到塔楼周围与下面的景致，斜面坡道围绕“骆驼腿”弯曲，以入口平台为中心，形成一个八爪鱼的样子。这些人行步道为塔楼提供了一个很私密的通道，并很自然地与公园分离，慢慢将绿地踩在脚下。

结构描述

塔楼由四根立柱（骆驼腿）与顶端网状结构（骆驼头）组合而成。该建筑最引人注目的地方就是其主框架结构灵活的设计在抗震与防风方面是无可比拟的。由于后腿极细，所以该结构的首要的问题是要保持建筑物的稳定性，其次，才可灵活考虑其抗震与防风的功能。

细细的后腿是由不同直径与厚度的钢管沿Z轴制造而成。钢管腿前面30米填充了高阻抗混凝土与1,100绕的水平电磁线圈。较高的部分则是坚硬的内部网状结构。两条前腿由圆柱形钢框架、圈梁与倾斜支架构成。

塔楼所有的建设均采用钢作原材料。由于机械应力的缘故，不同的部位将采用不同种类的钢。前腿支架材料选用235硫化亚铁，楼层内部圈梁采用355硫化亚铁。后腿底部与前腿圈梁则采用一种特殊的高性能可塑性钢460硫化亚铁。这对于主框架结构的灵活性与稳固性非常重要。

依据欧洲建筑规范1、3、4，我们做了一项结构评估。依据UBC1997年第二版关于土壤E，zone2A，对抗震设计作了评估，系数Z=0.15，重要系数 I=1；在更多警告位移测算中Z= 0.2。防风设计依据BS 6399-2，风速45米/秒，与动态压力压强=1,240 N/mp相符。压力分析通过抗震设计动态模型分析，呈线状走向。非线性图形表现为变形分析。

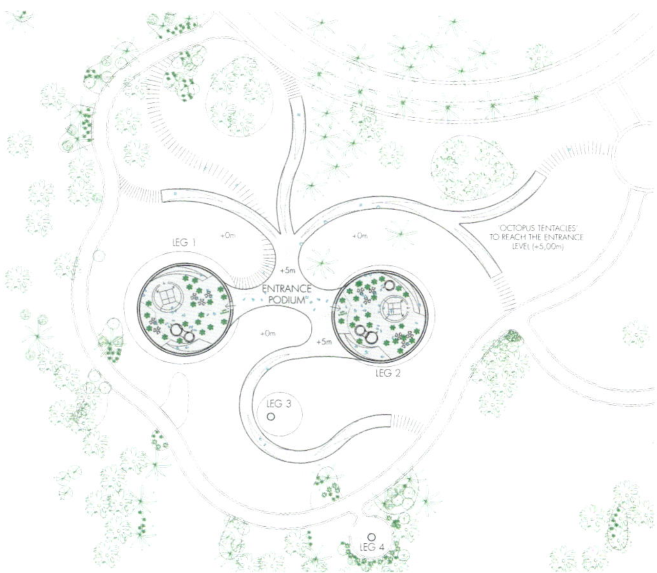
LEG 1
+0m
+0m
'OCTOPUS TENTACLES'
TO REACH THE ENTRANCE
LEVEL (+5,00m)
+5m
ENTRANCE
"PODIUM"
+0m
+5m
LEG 2
LEG 3
LEG 4

BOATING LAKE
RESTAURANT
THE CAMEL
ACCESS TO
UNDERGROUND
PARKING

SITE PLAN

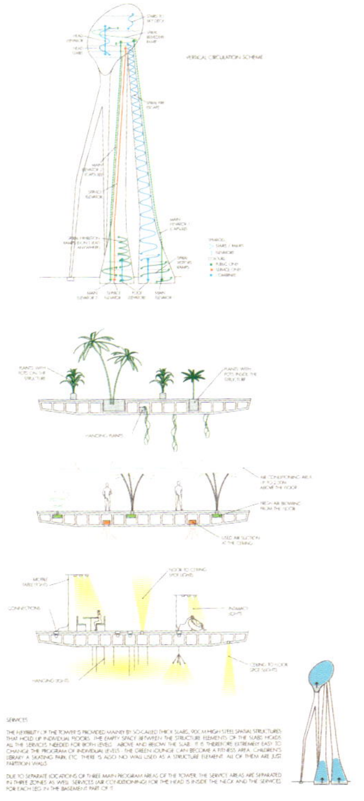

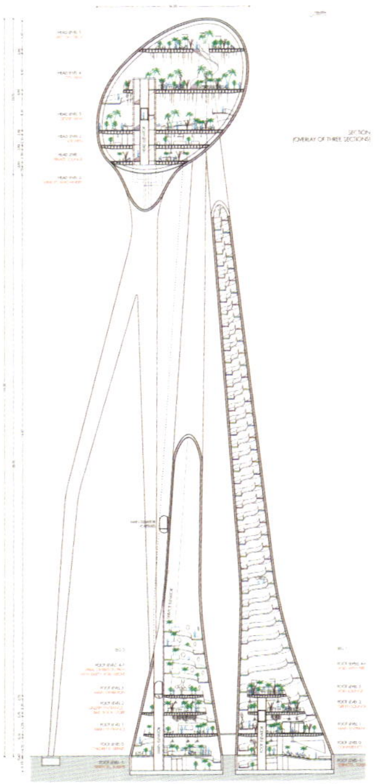

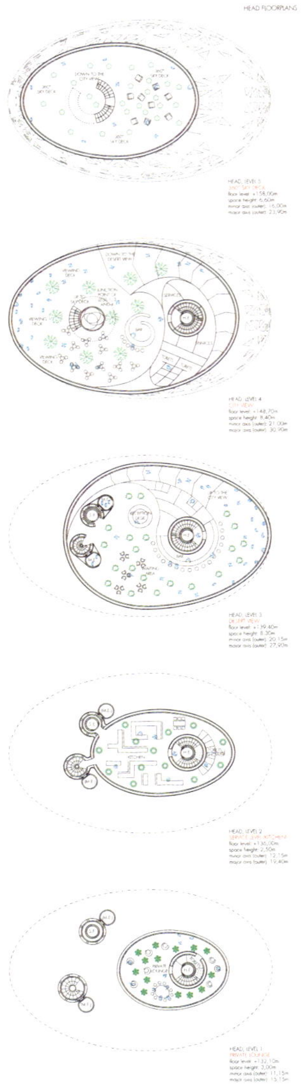

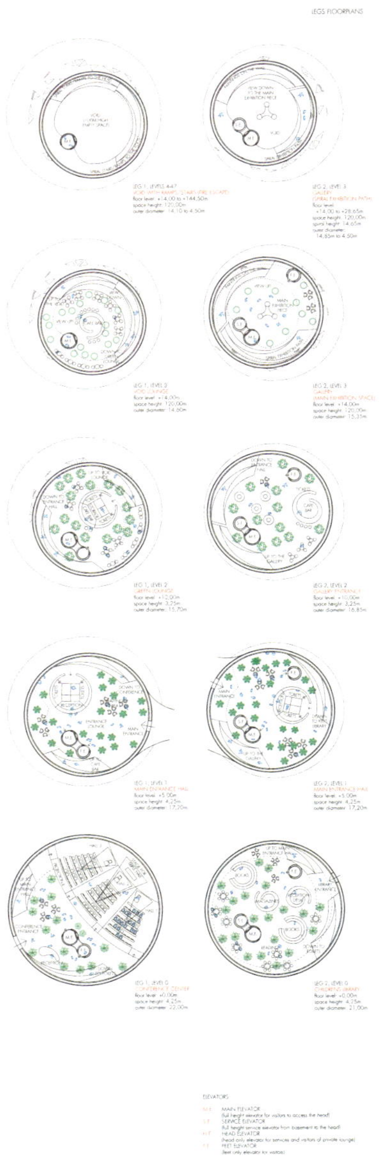

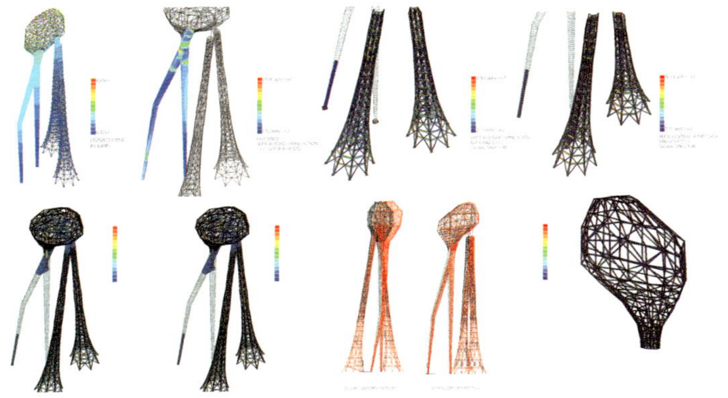

Context and Design Strategy

The audacity of a cosmopolitan city of Dubai is comparable to the courage of caravans crossing the desert. The proposal is homage to this audacity, bringing the image of the caravan to the city, a camel that comes to an oasis to drink water. The tower aims to capture the essence of this journey through the desert, the will of the pioneers to explore the unknown land and the joy in an adventure in the nature. Dubai is an adventure in the desert and an oasis with fresh water at the same time. It is represented in an organic form that elegantly reaches to the sky and at the same time keeps the natural shape of the slow movement of a camel.

Approach

A strange perspective of four distorted legs invites the visitor to explore the tower from all possible angled around it, as the image of the structure is completely different seen from different sides or even distances. Therefore the approach to the tower follows a spiral line with tower in the center. It begins already on the highway - since there are not many buildings around, the tower is well-visible from a distance. The spiral continues through the local road and parking lot, then the entrance to the park, pathways, and finally the ramps that lead to an entrance podium 5m above the ground. To offer a visitor a final tour around (and under) the building before entering it, the ramps are curved around the legs of the tower, creating octopus shape with entrance podium in the center. These walkways provide a more private and intimate approach to the building and establish a spatial separation from the park and the green areas that are left slowly below.

Structure Description

The structure consists of four big piers, the legs, and one reticular superposed structure, the head. Conceiving the structure the most attention was paid to realize the elastic behavior of main structure that must have very little displacements as regard to seismic and wind actions. This condition is very important for buckling stability, the most important problem of this structure, because hind legs are extremely slender. Only secondary elements are allowed to have ductility to adsorb and dump energy of seismic and wind vibration.

The hind slim legs are made with a steel pipe having variable diameters with different thickness along the z axis. For the first 30m of height pipe legs are filled with high-resistance concrete and HEM 1100 column. In the higher part of the pipe there is reticular stiffening structure inside. Two fore big legs are constituted by cylindrical steel brace frame with columns, ring beam and diagonal braces.

All structural elements are steel. Different kind of steel is used in different parts in consequence of mechanical stress. Columns and braces element of big legs are Fe S 235, beams inside the floors are Fe S 355. For base portions of slim legs and ring beams of thick legs a special high yielding steel Fe S 460 is being used. This is necessary to guarantee the elastic behavior of the main structure.

An assessment structure is done in according with Eurocodes 1, 3, 4. Seismic actions are evaluated in according with UBC-1997 volume2, for soil E, zone 2A, coefficient Z= 0.15, importance coefficient I=1; for more cautions displacements are calculated by Z= 0.2 like zone 2b for SLD assessment. Wind action is evaluated in according to BS 6399-2, wind velocity 45 m/s who corresponds to dynamic pressure ps= 1240 N/mp. Stress analysis is developed in linear phase, by dynamic modal analysis for seismic actions. Non-linear geometric behavior is considered for buckling analysis.

University Library Zadar

扎达尔大学图书馆 Studio 3LHD

Credits

Program: public, culture, education, urbanism
Time: 2009
Country: Croatia
Type: public competition
Area: 5,000 m²
Gross Floor Surface: 18,633 m²
Volume: 92,000 m³

Level: 7.00 m
Client: City of Zadar
Team: Sasa Begovic, Marko Dabrovic, Silvije Novak, Tatjana Grozdanic Begovic, Josko Kotula, Vibor Granic, Zeljko Mohorovic, Sanja Jasika, Tin Kavuric
Collaborators: Mateo Bilus (BMP) building physics, Alan Soric (Tehnicke konstrukcije doo) structural engineering, Boris Goreta 3D

扎达尔城市规划的一个突出特征就是规则的城市光栅，这也是该区域规划中的一个元素，同时也是这个城市的典型标志。项目所在区域的电网机关则被委以重任，负责未来整体校园的光栅规划。

拟建于新扎达尔校区的图书馆是第一个意义重大的新建筑，其功能必须要满足当下其所在区域的要求，还要满足未来校园规划的需要。这个现代图书馆的一个主要任务就是其社会责任，即不仅要有藏书，还要有信息媒体，同时还要满足藏书量成倍增加、信息媒体不断升级的空间需要。

建筑体东南边的城市性设置以及其坚定的城市特性完整地定义了建筑的特色。

主要的通道设置成从公园到街区的东西走向，这也明确了公园方向图书馆主入口的朝向。然后作为主要会议及交流的宽阔的大厅完全敞开并垂直上升。

每一层的宽敞的空间和开阔的视野随着我们称之为"左兰尼奇"的象征性的公共楼梯攀爬到具有象征意义的"知识山脉"，如同彼特·左兰尼奇在寓言史诗《山脉》中求索。

鉴于地中海气候及对馆内藏品的保护，内部简洁的空间更需要一个简洁的外部环境和外立面。外立面层层堆叠，每层之间留有空隙，使得图书馆看起来比较内敛。

除了堆叠的书籍，对当地材料的使用也体现了外立面设计的灵感。当地气候下最典型的石头是本科维石，纯天然，非常受欢迎，已经在当地使用了几个世纪。最常见的用法就是自由的堆叠。

facade south

facade north

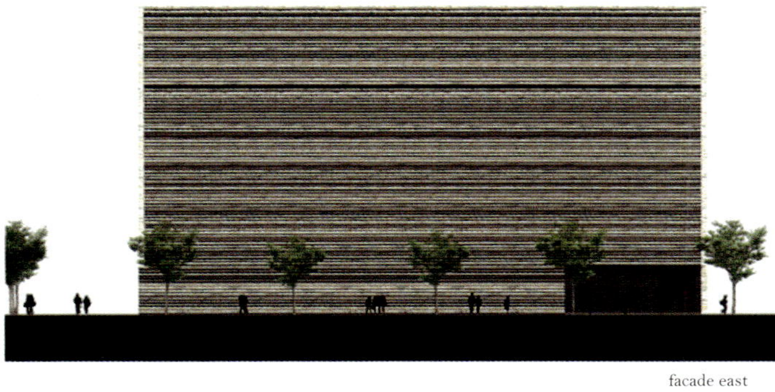

facade east

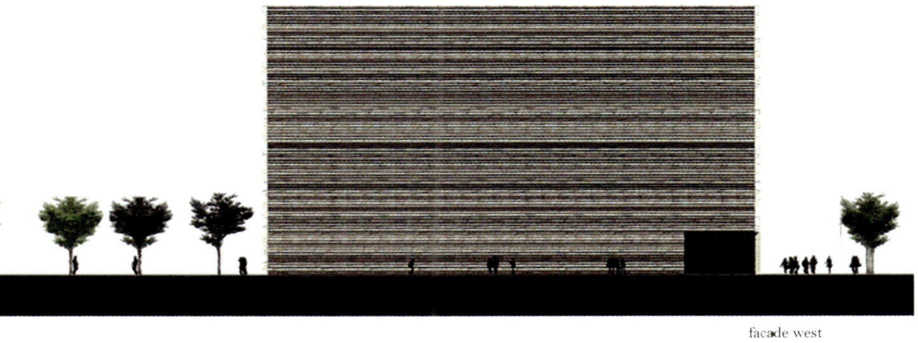

facade west

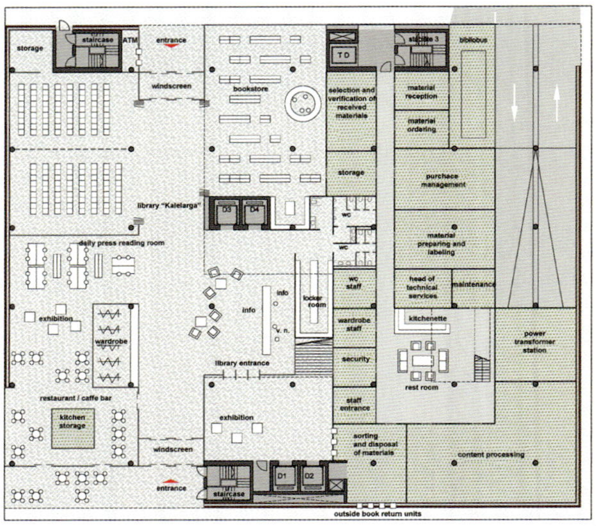

ground floor

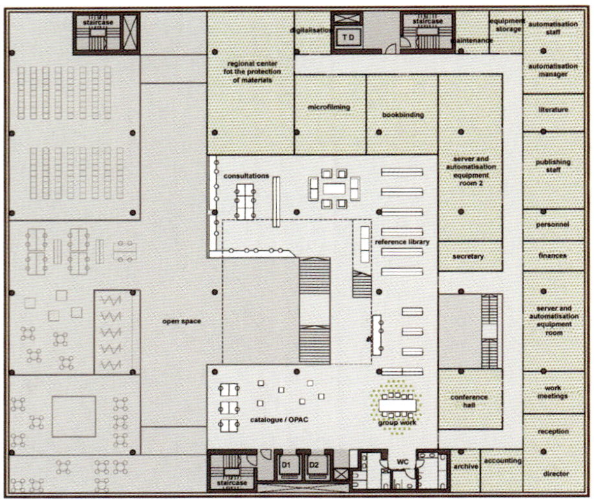

first floor

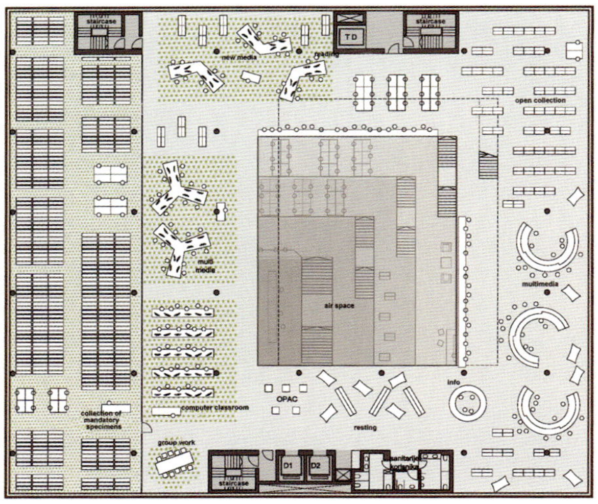

fourth floor

A regular urban raster has always been the dominant motif in the planning of Zadar, an element of regulation for the whole area and the distinctive symbol of the city. Authority of that existing grid in the context of this project is imposed as an essential element in the development of the entire area of the future campus.

The University library is the first significant new building, which plans to build in the area of the new Zadar campus. It must function in the surroundings it will be located today, not only in the ideal scheme conceived in the future. One of the main tasks of a modern library is the social obligation to house not only books but also the information media and allow them space for explosive multiplication.

The urban placement of the building volume in the south-eastern edge of the parcel and the determined urban parameters completely define the buildings character.

The main communication axis is set in an east – west direction, from the park to the street and that clearly defines the main entrance (from the park) and opens the first floor space in that direction. After that large passage which is the main meeting and communication place completely opens and expands vertically upwards.

The openness and visual accessibility of each gallery floor and the representative public staircase, which we called Zoranic stairs, climb to the symbolic "mountain of knowledge", as Petar Zoranic climbs in the allegorical epic Mountains.

Due to the Mediterranean climate, but also because of the physical protection of the library content, the simple inner volume of the introverted house required a simple expression of the external volume and the library facade. Vertical stacking of the facade layers, with a space between them, makes the library more closed from the outside.

Except in a pile of stacked books, the inspiration for the façade was found in local materials. One of the most characteristic stones in the local climate is Benkovac stone. It is very specific, simple, popular and in use for centuries. Its most frequent use is in the form of a panel, usually randomly stacked, and it can always be found stacked on a pile.

Jose Vasconcelos Public Library

巴斯孔塞洛斯公共图书馆 Pascal Arquitectos

Credits
Location: Mexico
Design: Pascal Arquitectos

在当今社会建图书馆？听起来似乎没必要——博物馆藏书不是更合时宜。

所建的这个图书馆看起来像一个大厦，是上一任总裁应一位老顾客的要求而设计的精品。它是从近700个提交项目中选出的前50强。顾客想要在设计方面模仿毕尔巴鄂古根海姆博物馆的建筑结构：通过创造邻近土地的附加值激发房地产的复兴。最终，使该建筑超越物质理念，成为人们记忆的一部分。

直到互联网时代，人们看书的过程依然是认认真真地从故事的开篇开始，到故事的结尾结束。虽然，你可以一下翻到最后一页——没有人在开始之前看故事的结尾。

当今社会一切都是同步的，人们认知的过程已经发生了改变。在互联网的作用下，没有开始也没有终结，只有通过超链接建立的超文本——没有时空序列的限制，反而更像是按照几何分布来排列的。

知识也不再那么具体了，例如，没有基因工程，如今的药品就不存在，空间上也是如此。在这个项目中，由于功能的不断变化，形式已经不再与功能相符。没有了空间的具体化，只剩下一系列发展进化空间上的超链接。

建筑物大部分都建在地下：你看不到它的开始，也看不到它的终结。它和周边城市景观融为一体，那里边界在蔓延，不再有范围的界定，只有冰山一角微微露出，其他一概不知。

摒弃了传统的模型，建筑不再是墙壁、立柱、地板与窗户的组合，它已是一个不能分离的整体。

▲ 建筑摒弃了传统的墙壁、地板与窗户的组合模型，创造出精致完整的空间 Paradigms are broken, architecture is not framed out of walls, floors and windows

▲ 室内效果 interior view

▲ 室内效果 interior view

Why build a library today? Seems absurd, it would be more logical to build a museum for books.
This would be like a mausoleum, a masterpiece of our last president who in a patron's act wanted to be remembered and transcend through its materiality. Our project was amongst the top fifty that were selected from the almost seven hundred that were submitted. His intentions were to mimic what was done with the construction of the Guggenheim Museum in Bilbao; to trigger the real estate recovery of the zone by creating added value to the adjacent land.
Until the age of the internet the process of reading a book would be to start from the beginning and finish at the end, and even though the end was there no one would dare to read it before the beginning.
Nowadays everything is simultaneous, the cognitive process has changed, with the internet there is no beginning and no end, only hypertext and hypervinculation, there is no linear sequence in time and space, it is more in the order of the fractal.
Knowledge is not specific either, as an example, you cannot see today medicine, without relating it to genetics engineering, etc, and same should apply to spaces. In this project as form does not follow function any more because function is continually changing, there would be no specificitv of spaces just hypervinculation of a series of changing and evolving spaces.
As there is no beginning and no end, and most of the building is underground: you cannot see, its beginning or its end. It also melts into the urban landscape, there are no limits but diffused frontiers, you just see the tip of the iceberg everything else is unknown.
Paradigms are broken, architecture is not formed out of walls, columns, floors, windows, and it is a whole that cannot be broken down in parts anymore.

Holocaust Memorial

大屠杀纪念馆 Tomohiro Hata Architect and Associates

Atlantic City Boardwalk Holocaust Memorial Competition: Final 6
大西洋城犹太人大屠杀纪念馆六强项目

这座纪念馆设有真实的窗子，这些窗象征着这个世界不同的时期、地域、文明与发展。窗子只是以简单而抽象的开口方式呈现在建筑的外墙上。这样的设计是为了暗示我们，当代的生活是建立在各民族、地区、文明以及其牺牲者的基础之上的，否则它便不复存在。

一般的窗子是向外开的，而该设计却恰恰相反，内外互置，有种将世界的各个部分都容纳到其内部的意味，纪念馆的内部空间反而成为“世界之外的地方”。

总而言之，透过窗子所看到的，不仅是纪念馆外部的景象，更是围绕纪念馆的整个世界的景象。

我们希望可以通过这座纪念馆看见我们所希望看到的东西，同时也透过表象看到其所象征的时代、地点以及文明。同时，我们也不能忘了馆外的世界也正看着我们。

这个地方不仅为祈祷者提供了为特殊事件做祷告的场所，同时为我们提供了一个思考世界与人类问题的广阔空间。

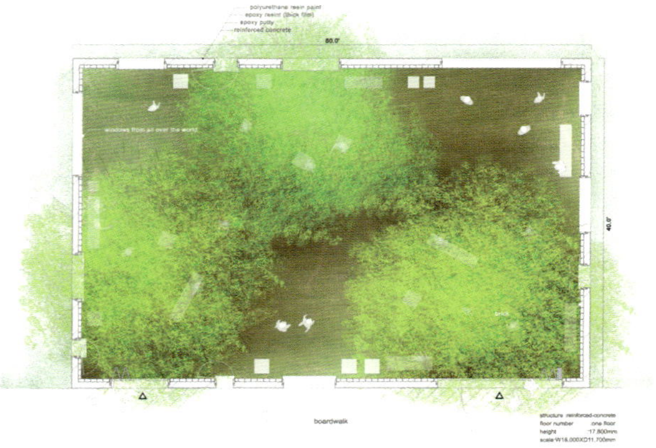

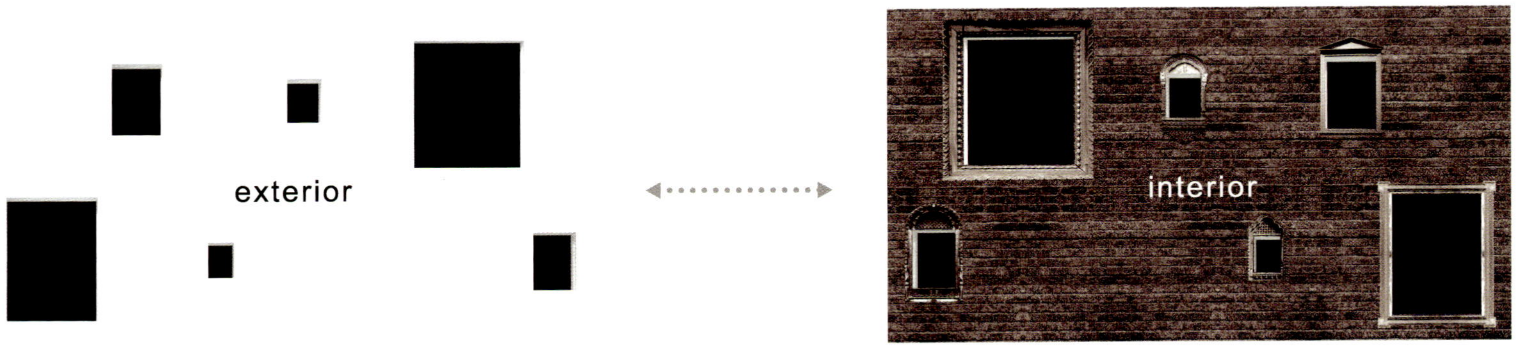

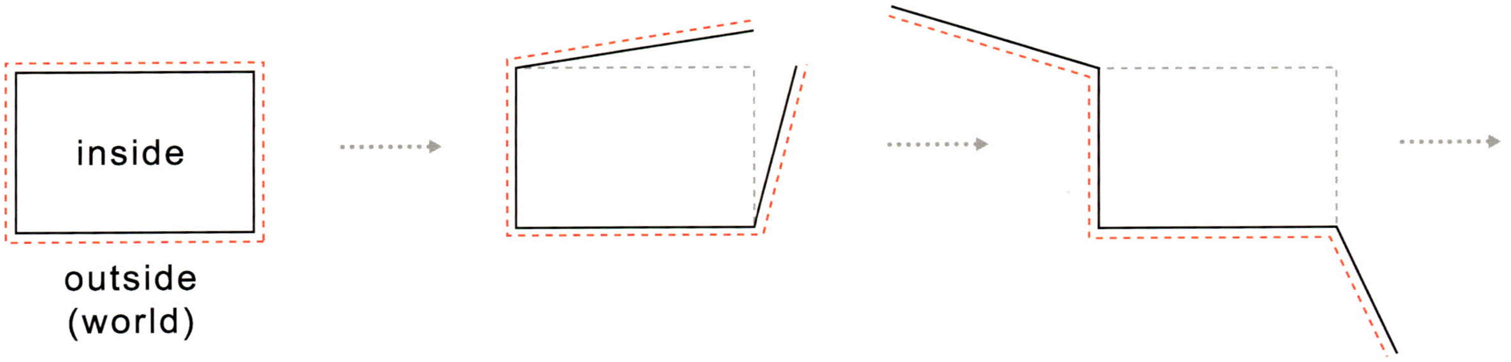

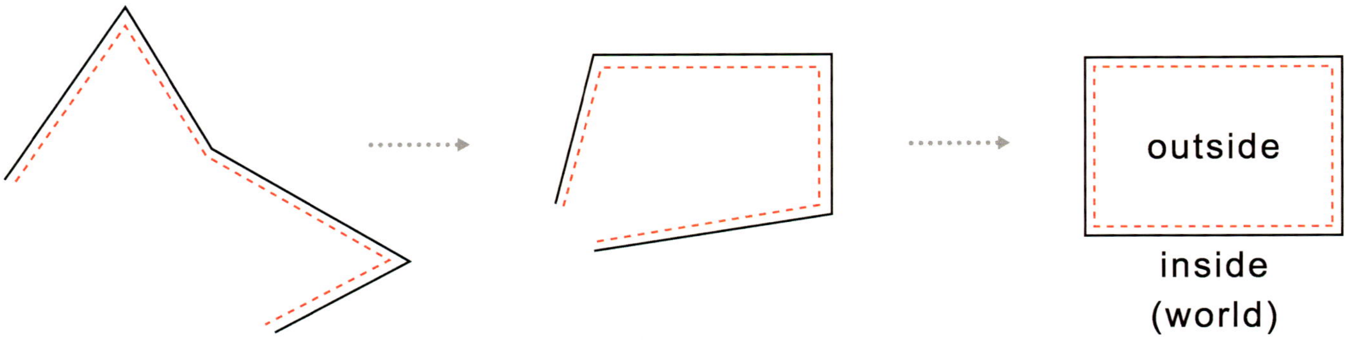

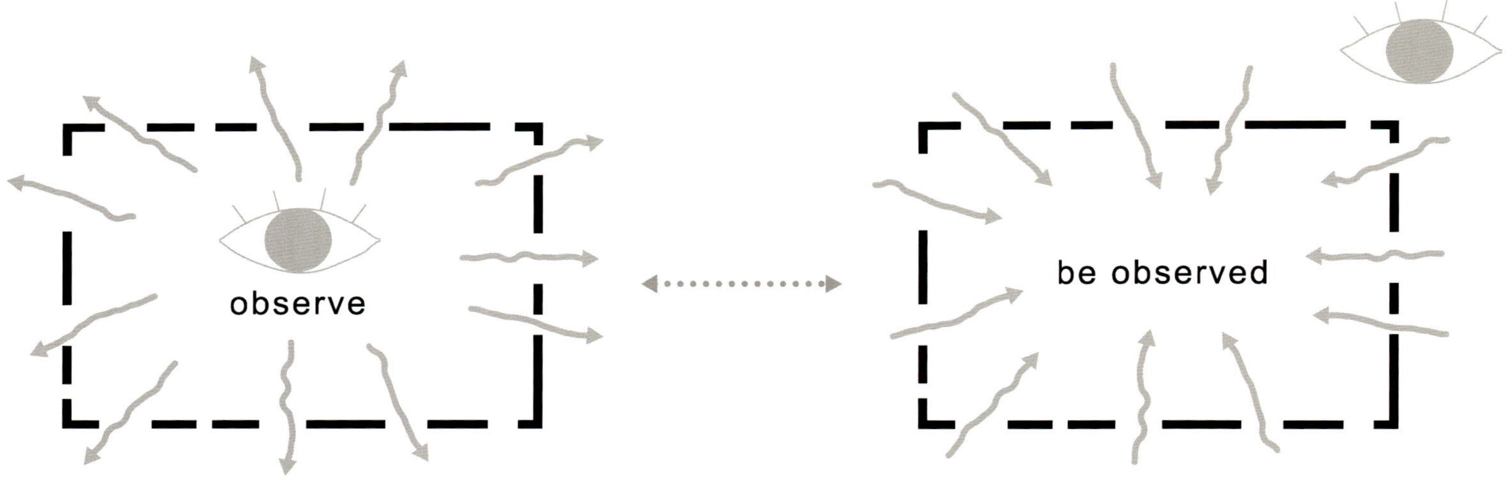

This memorial has real windows, collected for symbolizing various times, places, civilizations and pieces of the world. These are presented as simple and abstract openings on exterior. This process reveals hidden mechanism – our time cannot exist except the fact that it has been standing on various races, regions, civilizations or victims of them.

To arrange the windows opening to the inside instead of originally opening direction to the outside, is folding the pieces of the world up into its inside. That makes turning outside to inside and the place "outside of the world" in a sense.

After all, the scene from the windows is not only the outside of the memorial but also the inside as the world that surrounding the memorial.

We hope this pavilion to be the place, we-ourselves think of what we can see, yet, we must see through the "windows" symbolize various times, places or civilizations. At the same time, we must not forget that we are observed by the pieces of the world through the windows.

This memorial is not only the place of prayer for particular cases but also is the place we take a broad view of think and consider the world, furthermore, consider how we-human beings should be.

ARCHITECTURE COMPETITION WORKS

Office & Commercial
办公与商业

Middle East Technical University MODSIM Modelling & Simulation Research Center

中东技术大学模拟与仿真研究中心

Yazgan Design Architecture

2011 Nominee for European Union Prize for Contemporary Architecture Mies van der Rohe Award

2011密斯·凡·德罗欧洲当代建筑奖提名奖

Credits

Location: Middle East Technical University Technopark, Inonu Blvd., Ankara, Turkey

Design Team: Kerem Yazgan, Burak Turgutoğlu

Employer: Middle East Technical University Presidency

Construction Date: 2008-2009

Total Construction Area: 4,100 m^2

Structural Project: Aydin Pelin - Can Binzet Consultancy Engineering

Mechanical Project: Unlu Engineering Mechanical Industry Trading Co. Ltd.

Electrical Project: Nuve Lighting Industry and Trading Co. Ltd. - Mustafa Karakaş

Interior Design Project: Yazgan Design Architecture

Landscape Project: Yazgan Design Architecture Co. Ltd., Middle East Technical University Construction Services Department

Contractor: Mervecan Construction

Construction Type: Reinforced Concrete, Steel

Project Awards: World Architecture 7th Cycle Chosen Project

中东技术大学的模拟与仿真研究中心大楼坐落在一块5,200平方米的场地上，与周围环境保持连续性的同时，也有自己的特点。从外立面的颜色、楼层的高度、日光的照射方向以及人行道等方面来看，大楼与周围环境保持了协调一致。大楼坐落在一片杨木林之中，外立面的颜色充分体现了其所处的绿色环境。

另外，通过形式、空间、环境创意以及材料的选择，这栋4,200平方米的大楼以特殊的身份向人们展示着一种不同的内涵。大楼严密的矩形结构使其在绿色森林的万千形态之中显而易见。这种矩形几何形态也揭示了其平面布局的独特性。平面图上的条状构成部分都是组成建筑的元素，由外至内排列，将中庭包围起来。这些条状结构主要包括：1.外立面的遮阳钢架 ；2.五种不同绿色的釉彩玻璃制成的外立面；3.工作室；4.储藏室、通风井及烟囱等服务设施；5.走廊；6.中庭；7.中心的镜面泳池。各个元素具有自己的独特性，同时也融为一体。

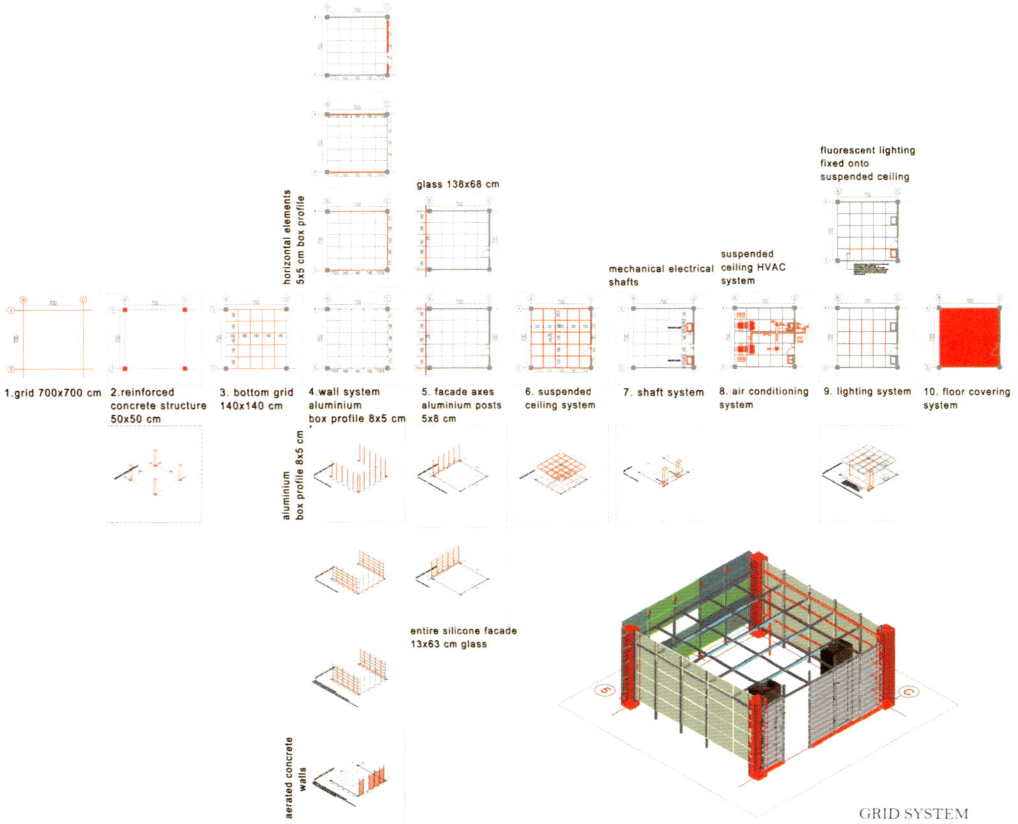

GRID SYSTEM

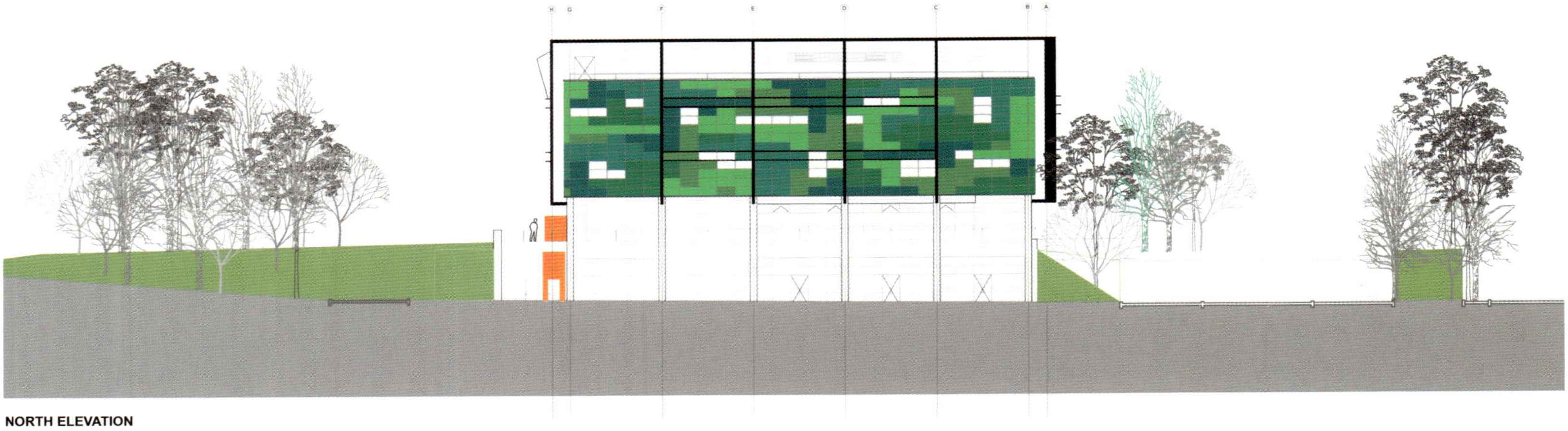

NORTH ELEVATION

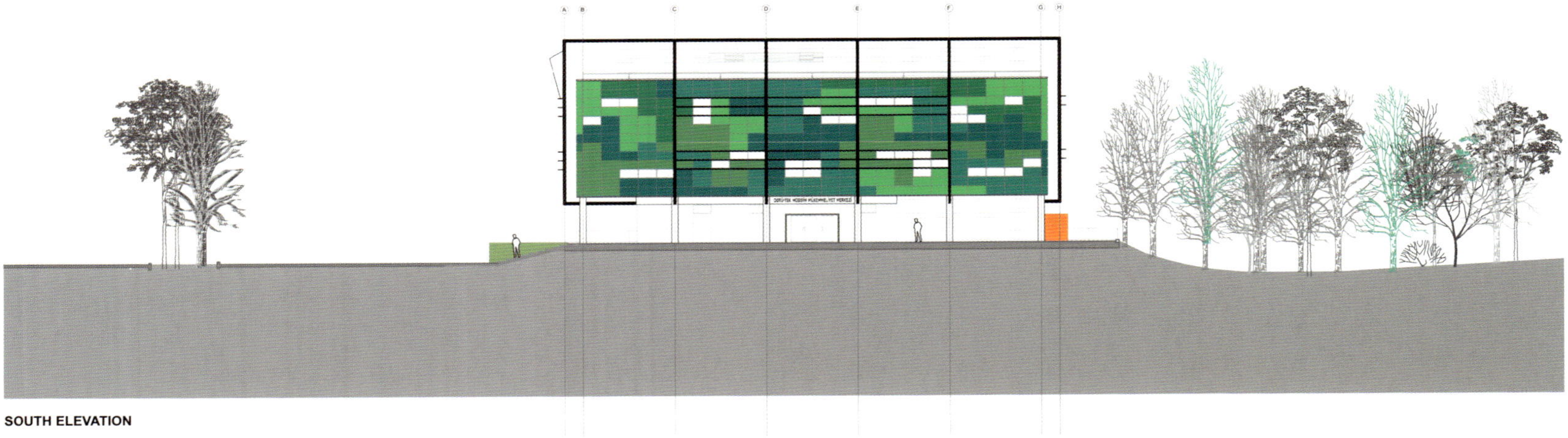

SOUTH ELEVATION

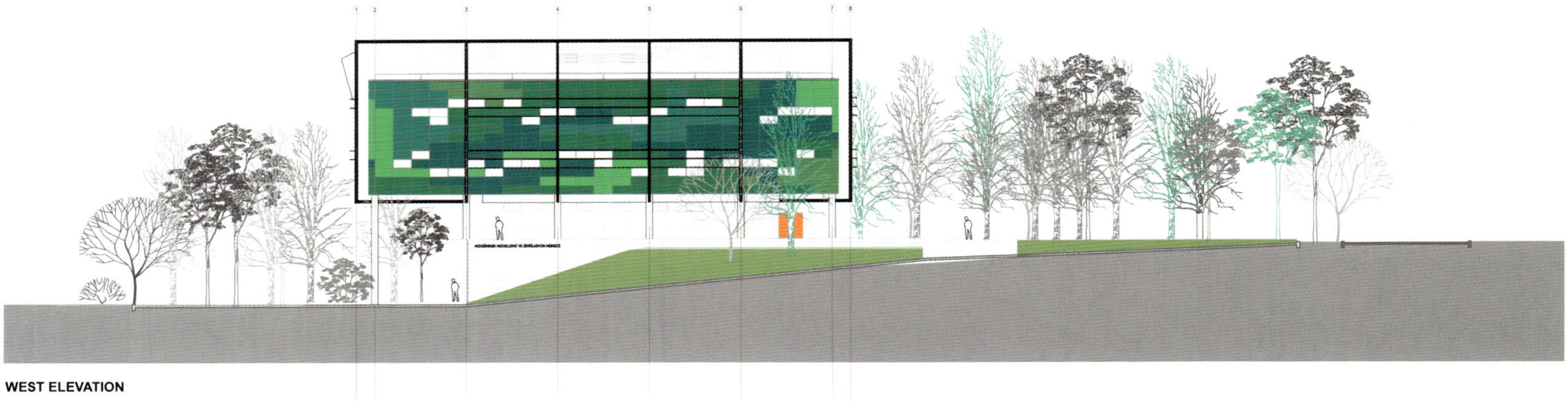

WEST ELEVATION

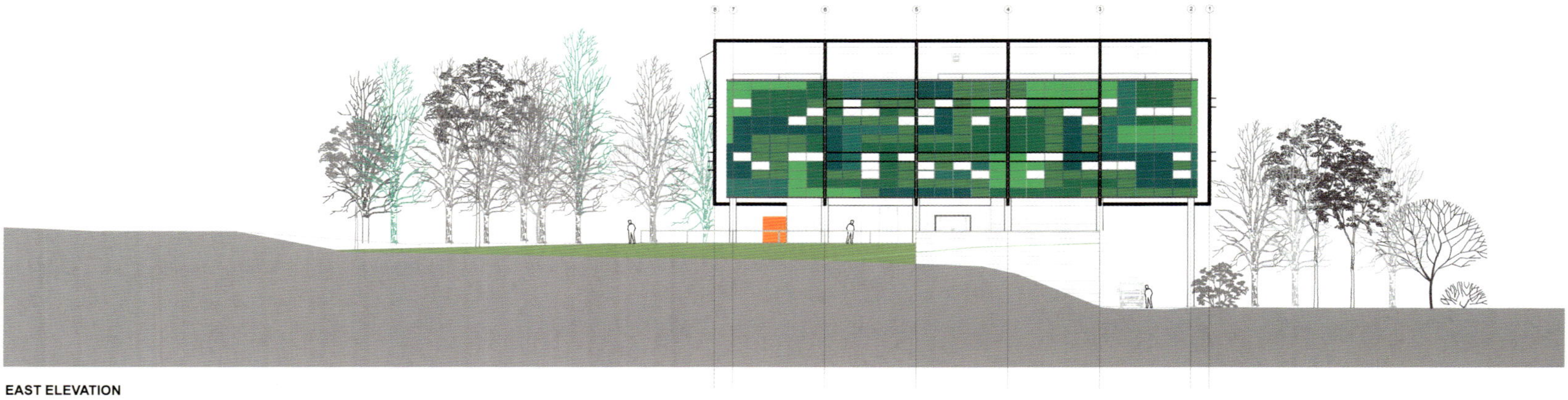

EAST ELEVATION

根据规划，建筑内部将设有模拟与仿真研究中心，虽然两者实际上并没有建设在一起，但是该方案的设计将使之成为可能。为了实现这一灵活性，设计模仿了土耳其地毯的构成方式，将条状的层叠结构与第三维度的网状结构相结合。7米x7米的主网状结构又被进一步分为多个1.4米x1.4米的副网状结构，形成一个矩阵，使得建筑结构的各个组成部分得以定位，包括里面的墙壁、窗子、天花板、柱身、照明设备、管道、栏杆、开关等以及外面的顶棚、广场上楼面形成的覆盖层以及入口大门。建造元素都留有极大的空间供修整、拆卸、重组装以及重建。在今后，网状结构中的任何一部分都有可能经历整修改装。

该方案是由建筑师与客户和工程师共同合作完成的。建造过程中，建筑师、工程师和工艺家在和谐的工作环境中设计出了这样一种矩阵结构，加上各种组成元素及相互之间的联系，成功地建造出了这座大厦。

METU MODSIM is a building situated in a land of 5,200 m^2 that is both bound to and diverts from its context. On one hand, it greets contextual values considering the façade color, the building height, levels of the site, the direction of the sun and the pedestrian axis. Being at the midst of a poplar forest, the building façade is a reflection of all greens of the surrounding.

On the other hand, the 4,200 m^2 building displays a unique identity by exhibiting a different knowledge in terms of both the formal, spatial and contextual means of production and the choice of materials. Its strict, rectangular-shaped geometry renders the building to be seen from every point within the diversity of forms in the green forested context. This rectangular geometry also harbors a distinctive means of plan development. Basically, the plan construct is composed of strips that are made up of architectural elements. They are juxtaposed layer by layer, from the outside to the inside and surround the inner atrium. These strips are: 1. Sun shading steel frame of the outer façade, 2. Facade itself, being composed of enamel painted glass of five different green colors 3. Working rooms 4. Service band that is made up of elements like, storage, shafts and chimneys etc. 5. Circulation corridor 6. Atrium void 7. Mirror coated pool base in the middle. Every architectural element has autonomy while they exist within a totality.

The building has a double program; simulation and research center. Although these two different programs do not mix in terms of circulation, they are designed in such a way that they can be integrated in the future. In order to gain this flexibility in the programmatic development, the layered strip pattern, which resemble Turkish carpet designs' configurations, are integrated with a grid in the third dimension. The whole structural system and the components that make up of the building are located through a 7m x 7m main grid and a 1.4m x 1.4m secondary grid. A matrix is formed through that development. The components that make up of the building inside, which are, walls, windows, suspended ceilings, shafts, lighting elements, ducts, balustrades, switches, etc., and the outside, which are, canopies, floor coverings of the plaza, entrance doors, etc., find a place in the design through this three dimensional matrix. The constructive elements are brutal, easily demountable and remountable and open to re-structuring. Every future changes in the design, may take place within this grid respectively.

Furthermore, the design was not developed through the architect's hand alone; it gives chance to a collaborative work with the client and the engineers. The designed matrix with the strips and the grid gave way to a peaceful working environment between different actors, such as architects, engineers and craftsmen etc., in the process of construction. The building is the result of a three dimensional reference system through matrix and its constitutional elements and relations.

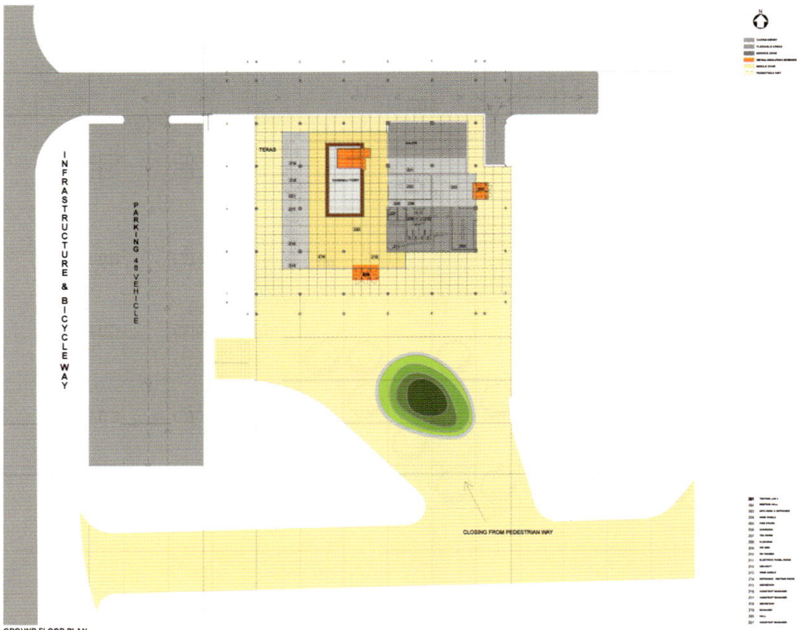

GROUND FLOOR PLAN

Enea Headquarters

景观设计公司Enea总部

Oppenheim Architecture + Design

Competition Entry, First Place
竞赛一等奖

Credits
Scope: 2,787m² Corporate Headquarters
Location: Jona, Switzerland

这个位于瑞士海岸边的建筑是全球著名的景观设计公司Enea的总部。它拔地而起，以木质体量呈现于世，严谨优雅，与场地和景观完美融合，与自然环境相呼应，与边界线和不同的空间条件相匹配。

项目包括温室、仓库、工厂、展示间及行政办公区域。不同区域的排列顺序有利于建筑内部空间的界定，因为所有的区域都由一个服务性的走廊相连接。该走廊沿着建筑将不同的区域连接起来，并将公司的生产线连接起来。

建筑的形式取自于程式化的设计和当地的环境。一方面，程式化以及建筑样式需要一个带有入口点的长而矮的楼房，另一方面，自然环境如光照、风，服务设施及区分前面入口与后面服务区域的必要性使得建筑采用了长而大的合适的形状。建筑所必需的严肃特性以及与周围环境融合的特性增强了简约的特质。

从美学上讲，建筑使用的材料会使其外观融入当地的自然环境；从功能方面考虑，其也会融入建筑内部的空间环境。能量消耗方面，设计果断地采用了节能的措施，整个项目采用如阳光，自然隔音系统的自然资源和诸如地热交换、绿色屋顶等的自然能源保护系统。用料方面还考虑利用诸如再生木材及高效节能的固定装置等，用以扩大建筑的性能空间。

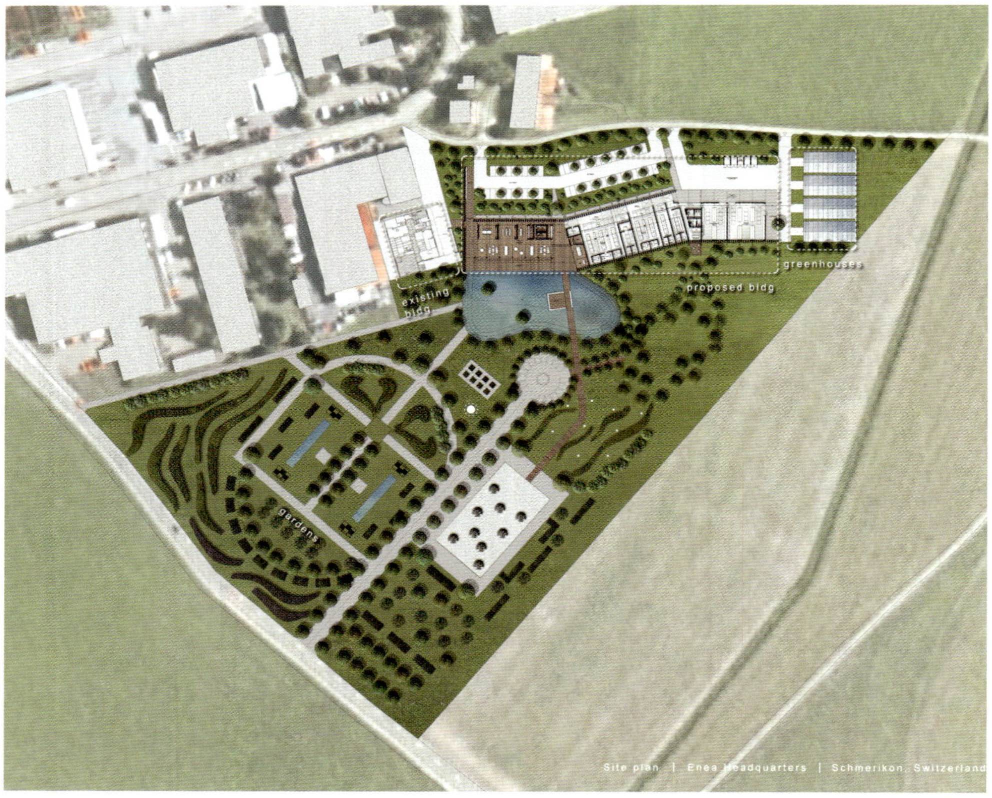

Site plan | Enea Headquarters | Schmerikon, Switzerland

The project, located at the shores of the Obersee in Switzerland, will house the headquarters of the internationally recognized landscape architecture firm Enea Garden Design. It emerges from the site as a wooden volume, discrete and elegant, and blends with the landscape as it responds to the natural environment, boundary lines and zoning conditions.

The program includes greenhouses, warehouses, workshops, showrooms and administrative areas. The sequence of these components contributes to the volumetric arrangement of the building - all areas are connected by a service corridor that runs along the building and organizes not only the circulation between spaces, but the production line of the company itself.

The building derives its form from programmable and site conditions – on one hand, the flow of the processes and program required in the program and their architectural resolution generate a long and low building broken where access points are needed, on the other hand, natural site conditions such as illumination and wind, the service access and the necessity of differentiating front and back - entrance and service areas - make the long mass the appropriate response. Its architectural simplicity is reinforced by the necessity of architectural sobriety and respect towards the elements and nature at the site.

The materials to be used look forward to merge with the natural environment externally - in aesthetical terms, as well as internally - in functional terms. Energy efficiency is decisive in the design process, so the project, consequently, is based entirely on the idea of taking advantage of natural resources such as natural light, natural insulating systems and natural energy conservation systems such as geothermal exchange, green roofs, etc. It is also planned that 'green' materials such as reclaimed wood and energy efficient fixtures will be used for a greater building performance.

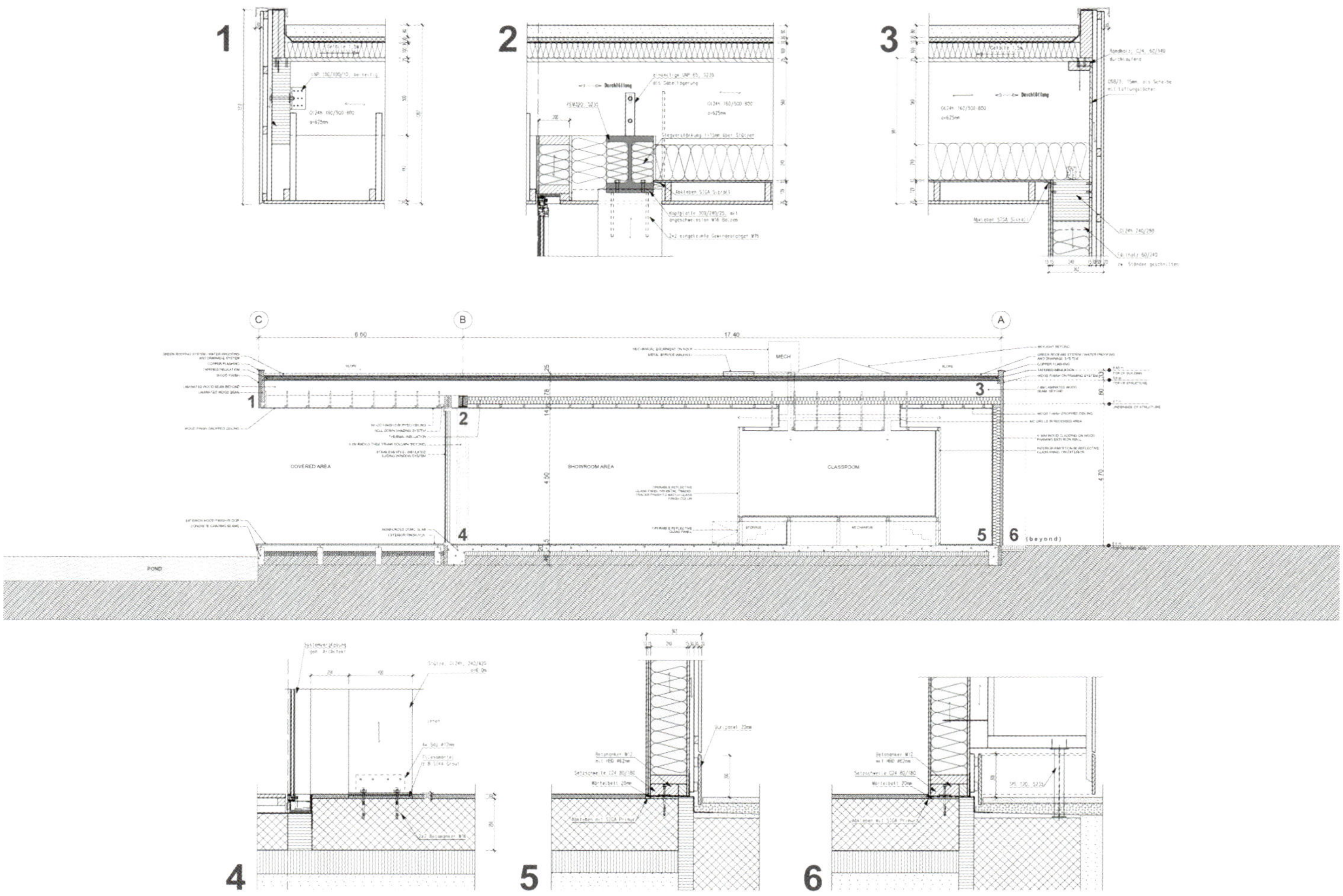

Sectional studies | Enea Headquarters | Schmerikon, Switzerland

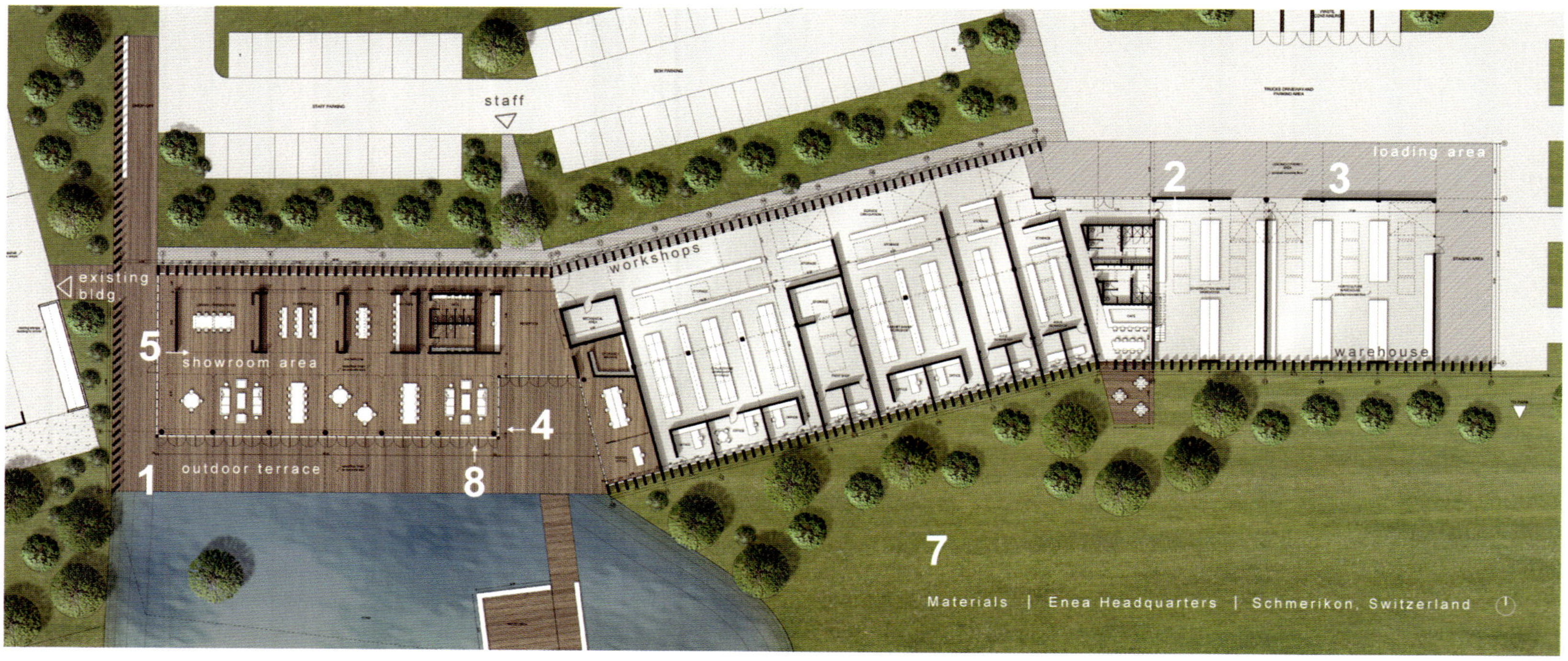

Materials | Enea Headquarters | Schmerikon, Switzerland

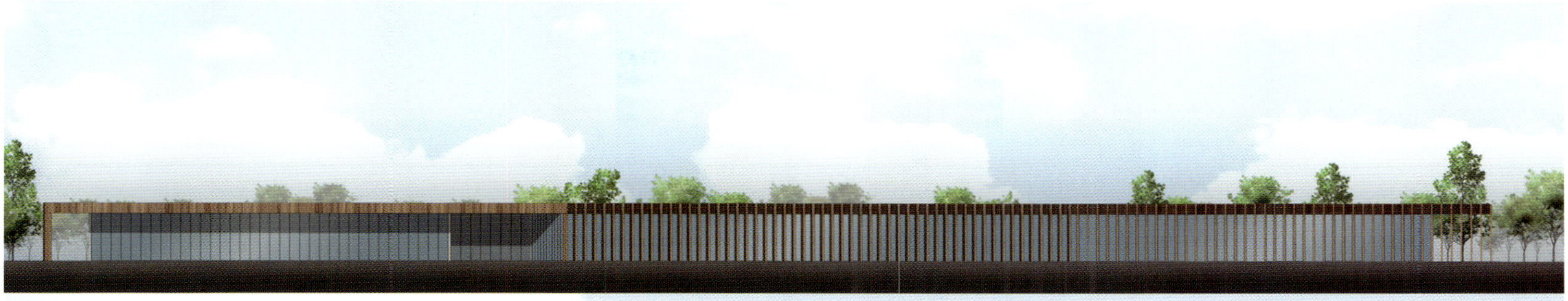
ELEVATION SOUTH

ELEVATION NORTH

ELEVATION EAST

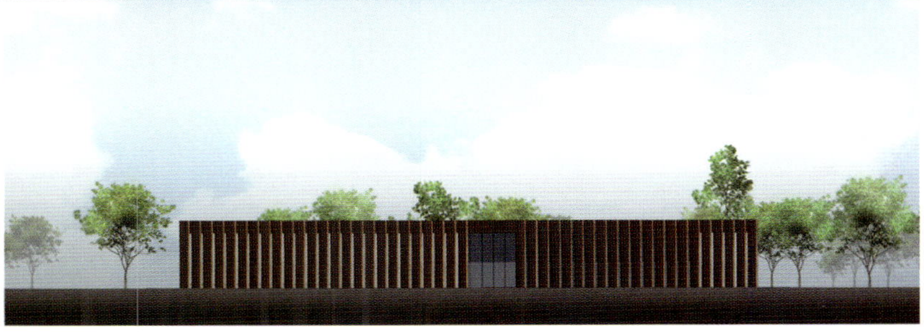
ELEVATION WEST

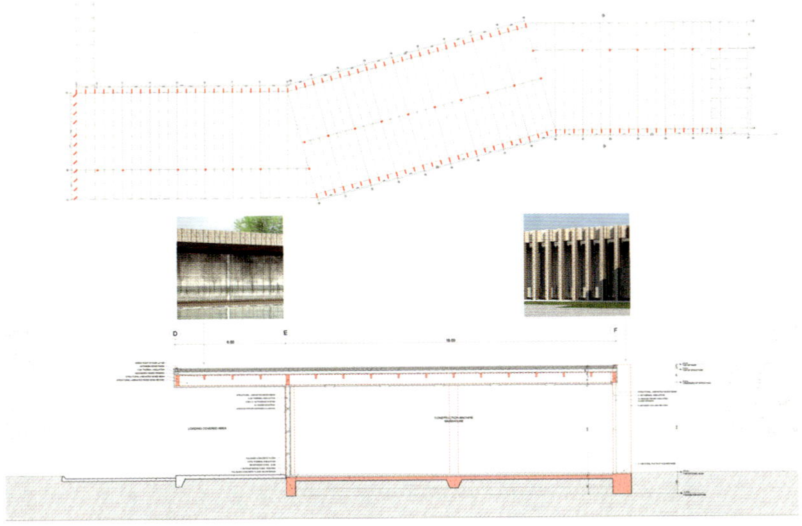

Structural system | Enea Headquarters | Schmerikon, Switzerland

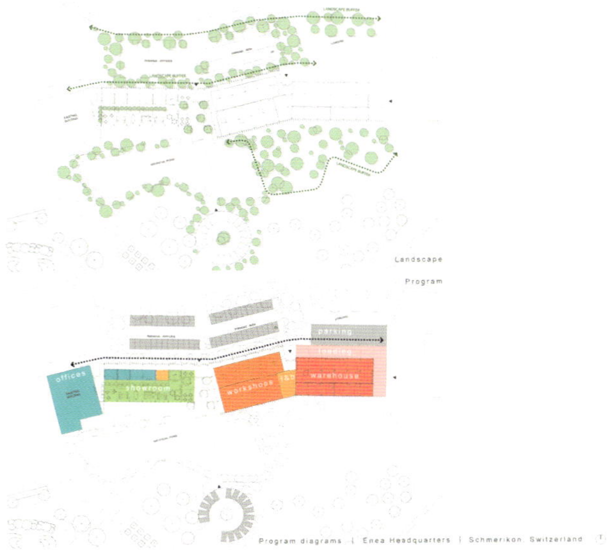

Program diagrams | Enea Headquarters | Schmerikon, Switzerland

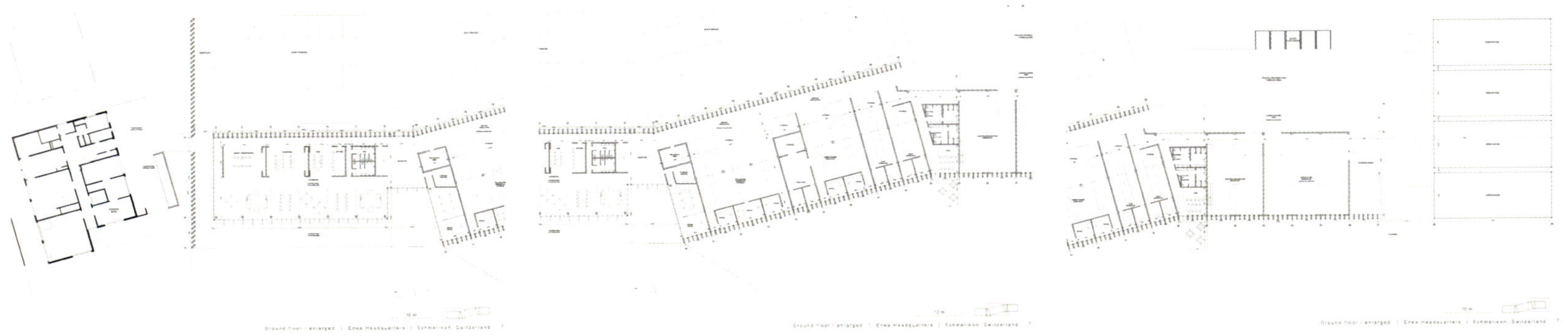

Massing studies | Enea Headquarters | Schmerikon, Switzerland

Interior studies | Enea Headquarters | Schmerikon, Switzerland

Court of Justice Madrid

马德里法院

Ivo Buda Architetto

Competition for the Design of Court of Justice Madrid.
马德里法院竞标方案

Credits

Design Team: Ivo Buda
Status: Competition (Unbuilt)
Client: Court of Justice Madrid
Location: Madrid, Spain
Height: 20 m

Levels: 3 Basements + Ground Floor + 4 Levels
Total Floor Area above Grade: 7,033m^2
Structure: Steel Concrete Structure
Parking Spaces: 70

设计理念

这栋大楼外形如一只海龟的外壳——此设计刚好与该地不规则的地形吻合。它矗立在人行步道中间，"海龟"外壳将进入法院的人群罩在其中，形成公共入口大厅。

"海龟"外壳的移动为大楼创造出更多空间，使光线也能照到入口大厅与地下室。也正由于它的移动，可以在楼顶表面空间建造一个平台。项目设计在不改变圆柱的基本设计理念前提下，尽可能地开拓建筑物空间。"海龟"的外壳作为法院的象征——这个悬空外壳的叠加创造出一个公共的入口大厅。随着人们进入这栋大楼，"海龟"变成了一个气势壮观的透明建筑。圆柱设计理念使"海龟"与周边圆柱形建筑物融为一体，但同时又以其耀眼的外观从中脱颖而出。双层外立面像是在尽其保护建筑物的责任。

整体结构

大楼的整体布局分为公共区域、评审与行政区域、关押与安保人员区域三块，由三条独立显眼的过道划分开。这三块是各自独立的区域，以避免不必要的突发事件。

公共区域

公共区域位于一楼，主要包括导向区域（咨询台、来访登记、咖啡馆等）、通道与等候室。这使通往大楼的公共线路最短，同时也保证了安全。

评审与行政区域

所有的法庭都位于一层与二层。这两层的会议室和法官、评审员的住房与周围走廊相通。法庭为法官、官员、听审人、被保护的目击证人与被拘留者分别设有单独通道。

关押与安保人员区域

该区设在地下室，关押室直接连接通往一、二层法庭的电梯。法庭安保室位于地下一层，拘留处位于地下二层。地下二层与被告席直接相连，与连通该地所有建筑的隧道相通。

这个方案达到了功能明确、安全有保证、环境可持续发展与经济可行的目标，满足了马德里法院的要求。

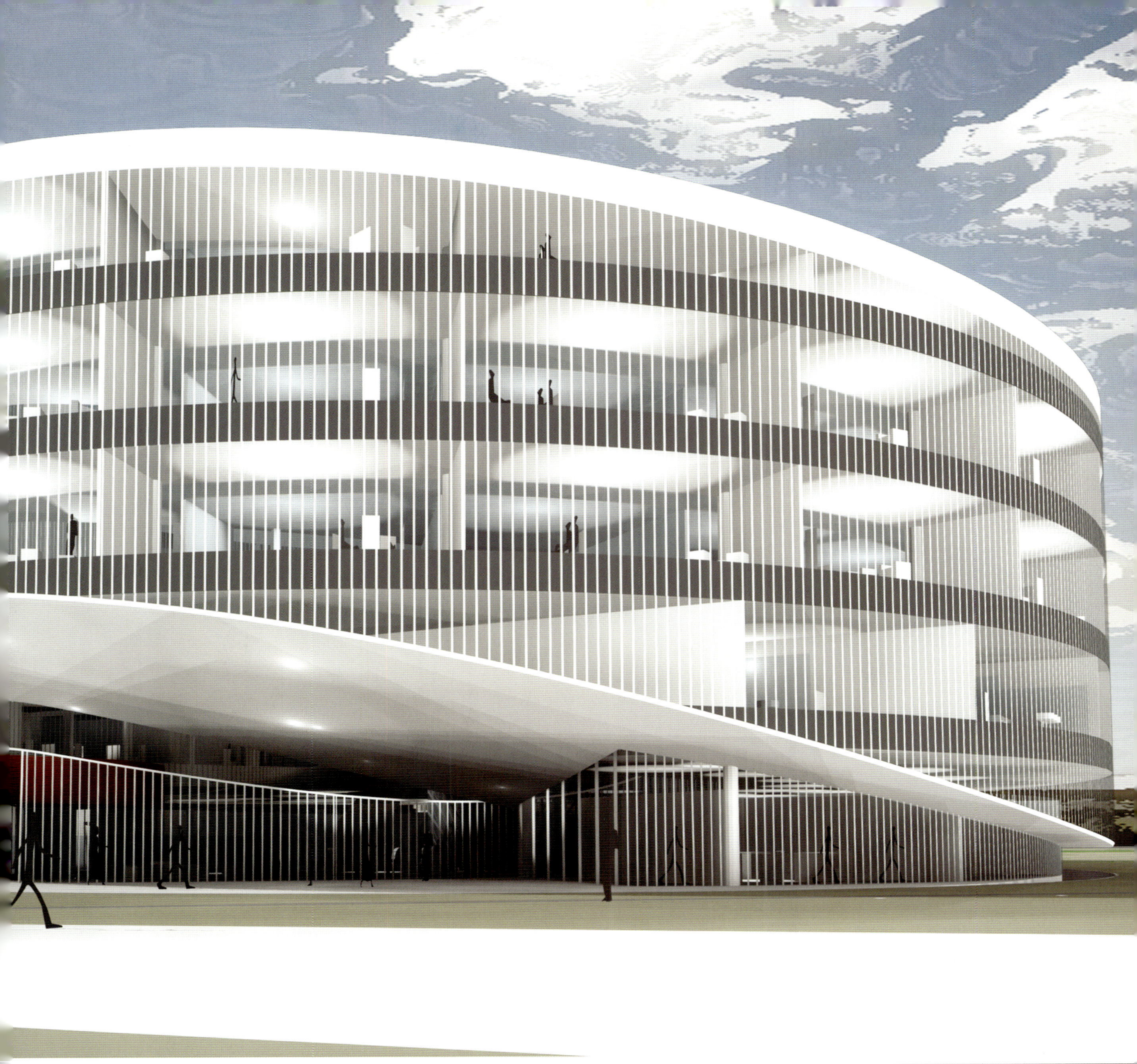

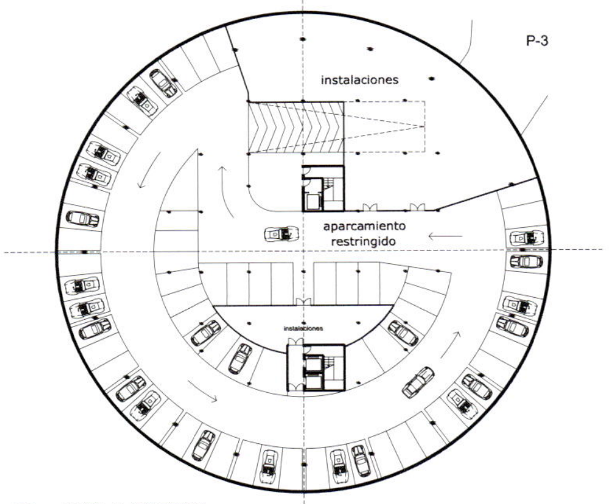

NIVEL -3

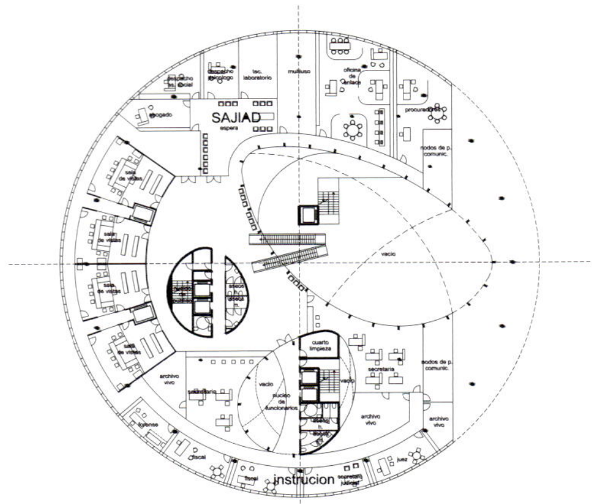

NIVEL 1

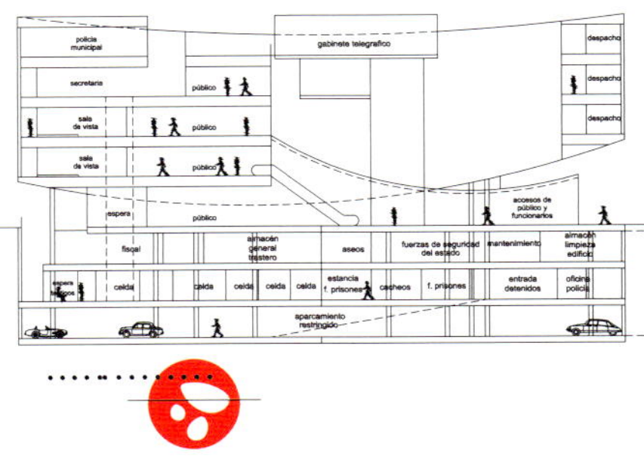

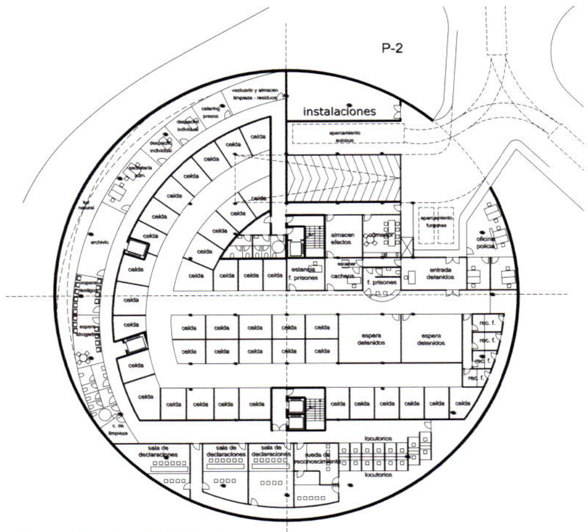

NIVEL -2

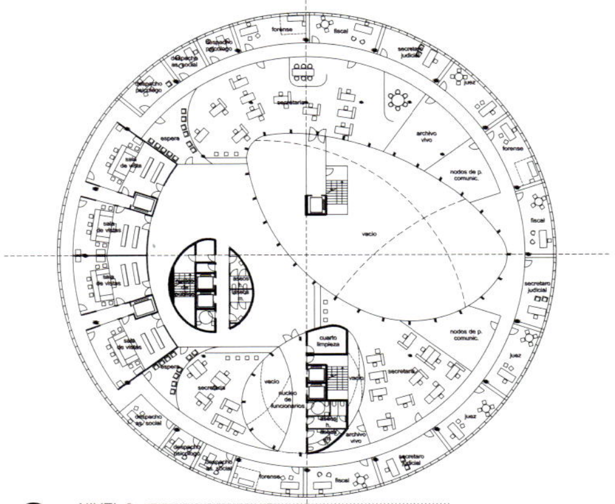

NIVEL 2

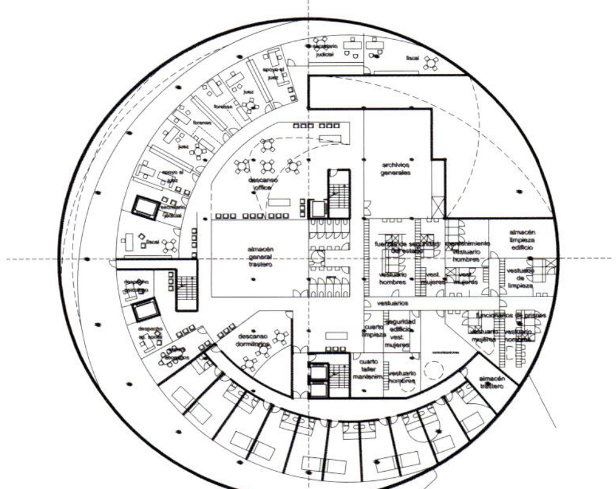

NIVEL -1

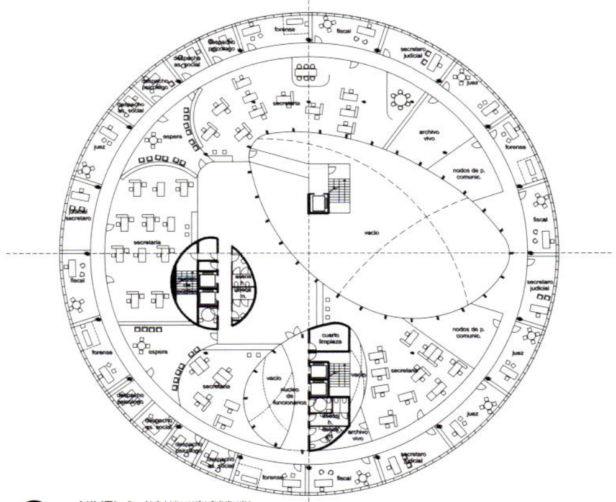

NIVEL 3

Architectural Design

The building is connected to the ground of the campus as a shell of a turtle, with an irregular basis. In this way, fits the plans of the campus, standing on the pedestrian path and stop people under the shell, extending their role to covering element of the public square-entrance of the building. The spaces are created through the movement of the shell, which also allows light into the entrance (hall) and in the basement levels (rooms and bedrooms). Thanks to this movement, it is possible to use the roof surface, creating a terrace. The project is developing the spatial possibilities without altering the basic concept of the cylinder. This research aims to study the shell as symbolic form. The superposition of the overhanging shall create a public square-entrance, resulting in a spectacular crystalline building that is transformed as we move through its surroundings. The basic concept of the cylinder maintains a dialogue with the other circular buildings and at the same time is attractive and recognizable. The facade is composed by a double skin, responsible for protecting the building.

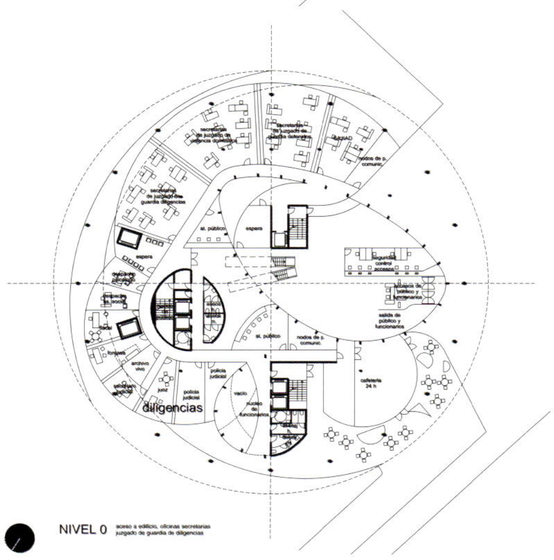

NIVEL 0

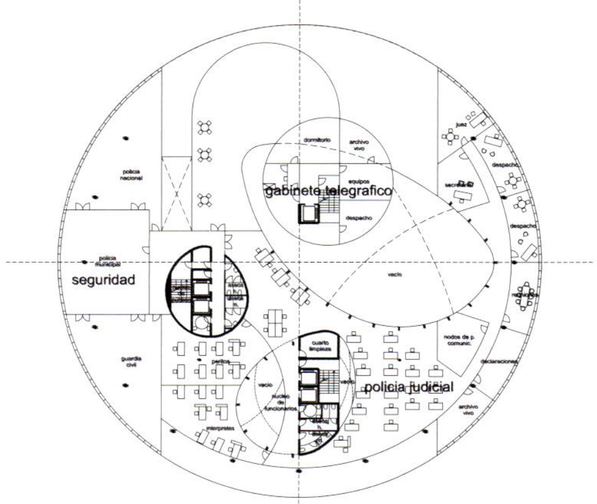

NIVEL 4

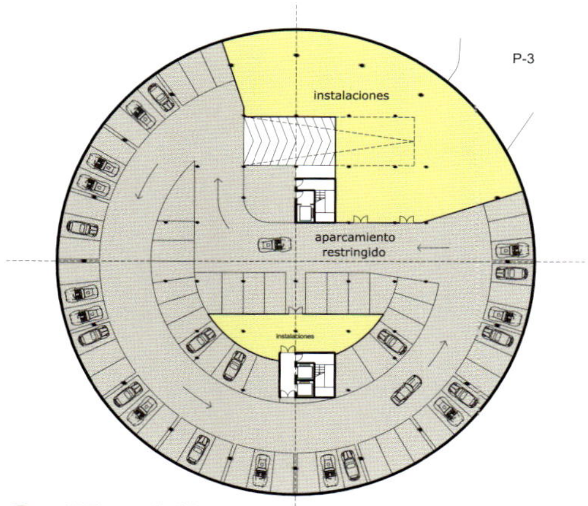

NIVEL -3

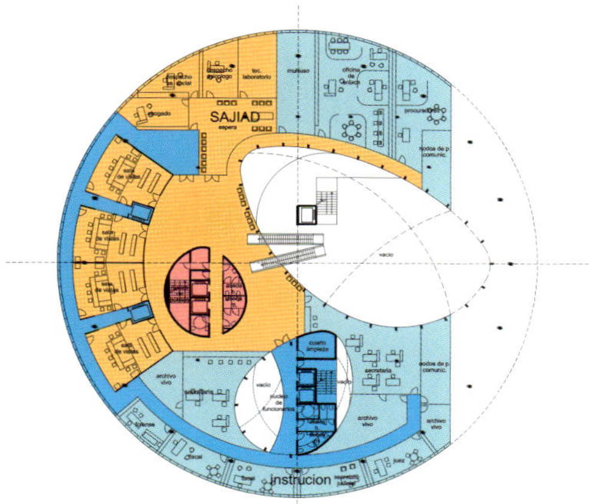

NIVEL 1

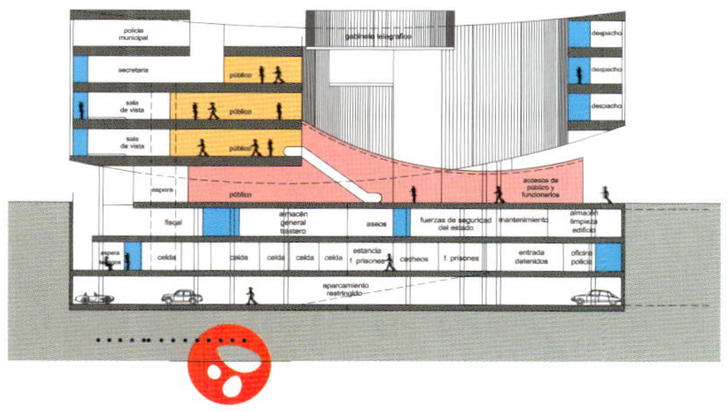

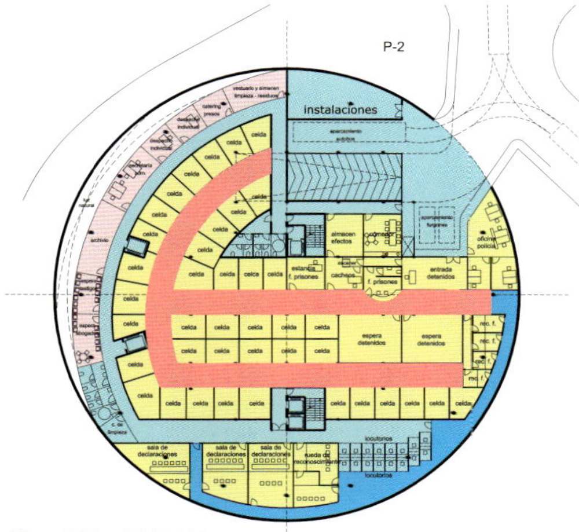

NIVEL -2

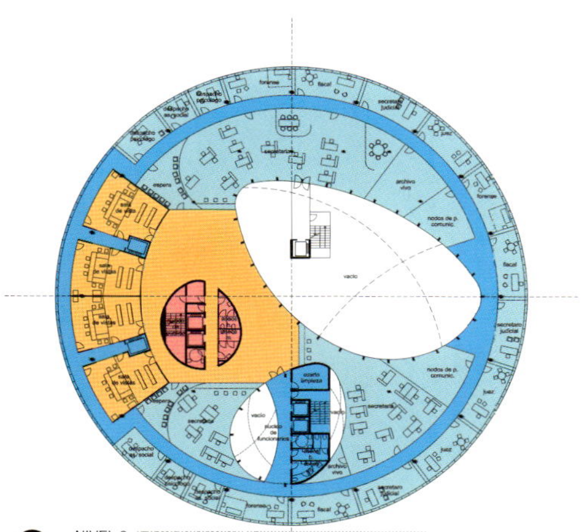

NIVEL 2

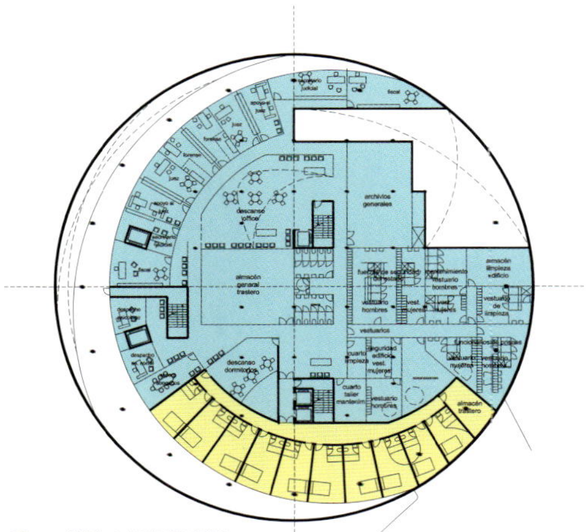

NIVEL -1

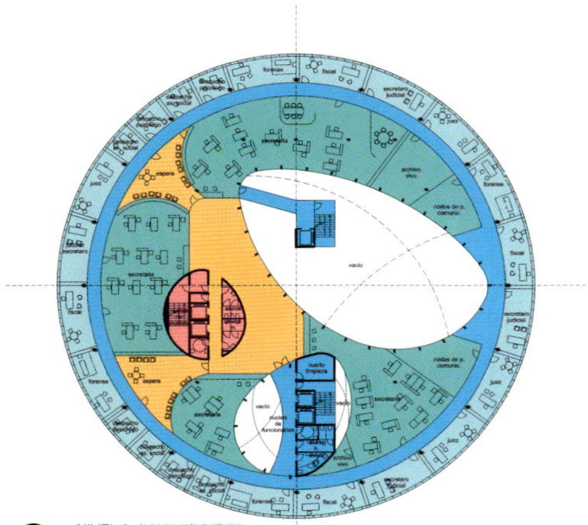

NIVEL 3

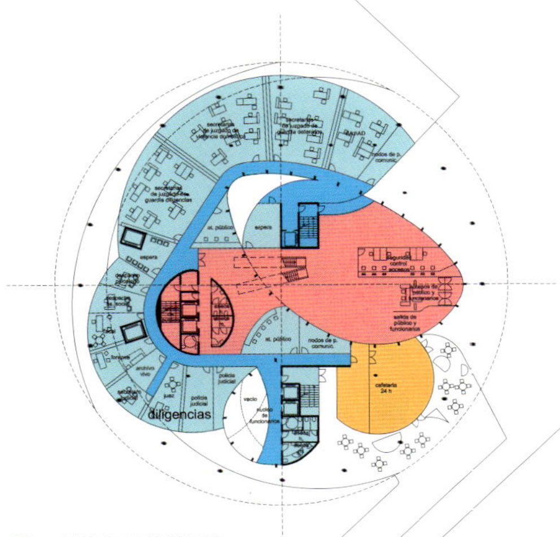

NIVEL 0

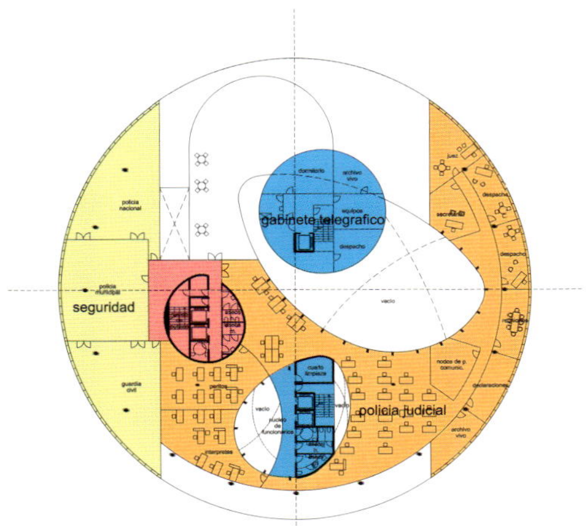

NIVEL 4

General Organization

The overall organization of the building derives from the consideration of three types of spaces linked by three separate and distinct paths: the space for the public, the space for judges and persons belonging to the administration of justice, and the space for the security and arrested. All these spaces are independent paths avoiding unwanted encounters.

Public Space

The public is concentrated in areas of floor care (information, registration, cafeteria, etc.) and access areas, waiting rooms and stay views, all contained on the first floor. In this way the public route through the building is minimized and greatly increasing security.

The Space of Judges and People in the Administration of Justice

All courtrooms are located on the first and second floor. On this floor meeting spaces and accommodation for judges and staff of the administration of justice are connected to the perimeter corridors. The courtrooms have separate entrances for judges, officials, public, protected witnesses and detainees.

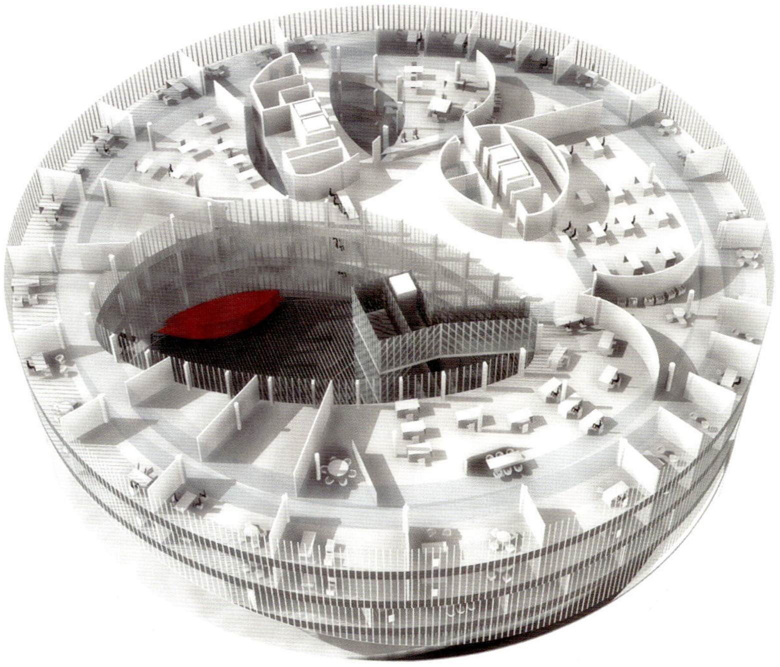

The Space for the Security and Arrested

This space is situated in the basement plants. The space for arrested is directly connected with elevators to the Courts on the first and second floors. Courts guard are located on the first basement and detainees are located on the second basement. The second basement is connected directly to the dock, which connects to the tunnels that connect all campus buildings, which are performed by the work of loading and unloading.

The proposed building meets the goals of functionality, security, environmental sustainability and economic viability required by the Justice Campus Madrid.

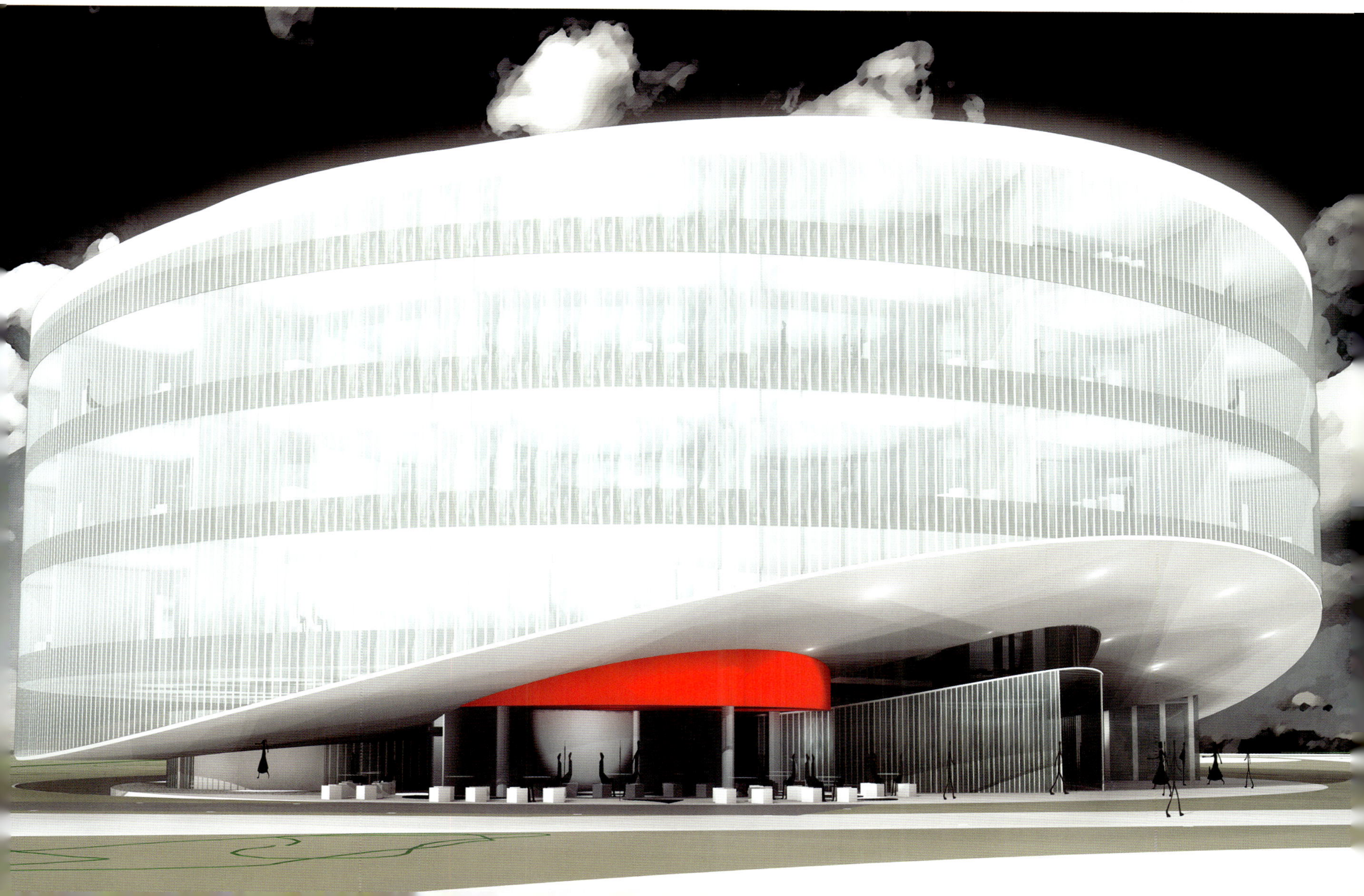

Hasselt Court of Justice, Hasselt, Belgium

哈瑟尔特法院

J. Mayer H. Architects　a2o-architecten　Lens°ass architecten

International Competition, 2005, First Prize
2005年国际竞赛一等奖

Credits

Client: Stationsomgeving NV Hasselt, Hasselt, Belgium
Completion: February 2011
Location: Monseigneur Broekxplein, 3500 Hasselt
Function: Courthouse including Offices
Site Area: 3,640 m^2

Building Area: 3,200 m^2
Total Floor Area: 21,600 m^2
Stories: 13
Structure: Concrete and Steel
Announcement Date: February 2007

一期：根据West 8事务所的规划，对哈瑟尔特火车站区域重新开发，包括两幢大楼以及由酒店、办公楼和住宅组成的低层街区。

二期：哈瑟尔特法院。

该法院是围绕火车站的新城区开发计划的两个标志性项目之一。一座法庭的后勤保障、地理位置，以及它的安检设施，三者共同造就了高度集中的内在联系。参考依据包括之前占据并界定该地点范围的老旧工业钢制结构，它们有机的新比利时艺术风格形成了哈瑟尔特文化遗产的组成部分。除了以实体存在，还有一棵树与之呼应，成为哈瑟尔特小镇的一个象征，唤醒了中世纪欧洲之前的历史传统——在一处居所的中央，一棵大树下，设立一处特别的“表明公平正义的场所”。

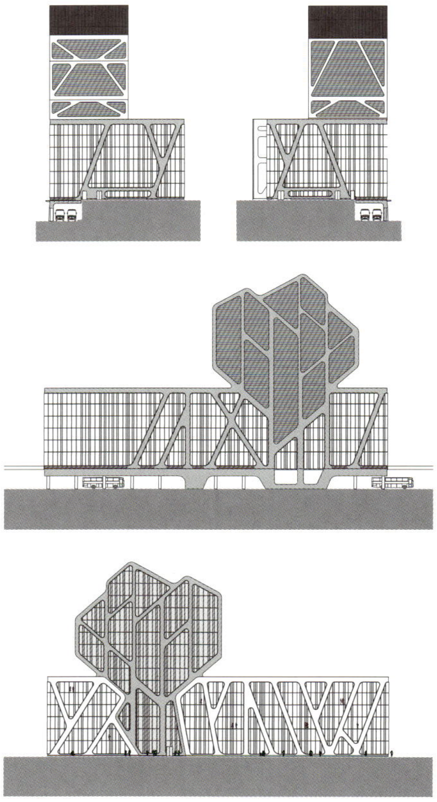

Phase 1: Redevelopment of the railway-station area in Hasselt based on the West 8 masterplan, including two iconic buildings and low-rise blocks for hotels, offices, and housing.
Phase 2: Court of Justice, Hasselt.
The Court of Justice is one of two iconic projects within the new urban development around the main rail station. The logistics and siting of a courthouse with multiple security barriers of security results in a massing composed of three interconnected volumes. References include the old industrial steel structures that formerly occupied and defined the site, their organic Belgian Art Nouveau forms constituting part of the cultural heritage of Hasselt. There are also echoes of a tree, which, in addition to being is the Hasselt town emblem, as well as also harks back to the historic pre-medieval European tradition of holding a special “place of speaking justice” underneath a large tree in the center of a dwelling.

CSAI Spa Office Building

CSAI Spa办公大楼 Modostudio

Restricted Architectural Competition, Second Prize

建筑设计方案邀请赛二等奖

Credits

Location: Terranuova Bracciolini, Italy

Client: Centro Servizi Ambiente Impianti Spa

Area: 1,366 m^2

Year: 2010

Design Team: Modostudio (Fabio Cibinel, Roberto Laurenti, Giorgio Martocchia, Javier Esquembre, Julie Carlier) with CCDP

Consultants: Sustainable Systems; Ing. Michele De Beni

这是一幢位于美丽的托斯卡纳的生态服务办公楼。该地地处山谷，景色宜人，高低起伏的山脉在每一个山谷与山脊处都呈现出异样但却相似的美丽。

新的办公大楼鹤立于群峰，但又没有打破山谷原本和谐的美丽。周边山脉别样的空间格局与颜色衬托着大楼，既与之相融，又形成极致的对比。

借助于地理位置的优势，办公楼的形象如中世纪的观景塔——它曾经是周边景观的标志性建筑。

该办公楼密实的结构与高技术含量的设计，与自然的景色完美地结合在一起——不同的材料，融为一致的美。

另外，其外表皮采用可持续材料设计，尽可能地利用该地天然资源，非常环保。

自然光线

窗户玻璃的面积不超过建筑物外立面的40%，太阳光线的照射率因窗户的差异、墙壁的厚度与窗户外层的遮帘而不断变化。

健康的工作环境

设计去除了天花板与浮动地板的装饰，这样使工作人员最小限度地接触恶劣的环境。

室内灵活设计

室内设计了可移动隔板，可以根据实际需要灵活改变办公室的大小。

可持续设计

设计全部采用木料建造，最大化地实现环保，这使该项目获得Klimahaus认证。

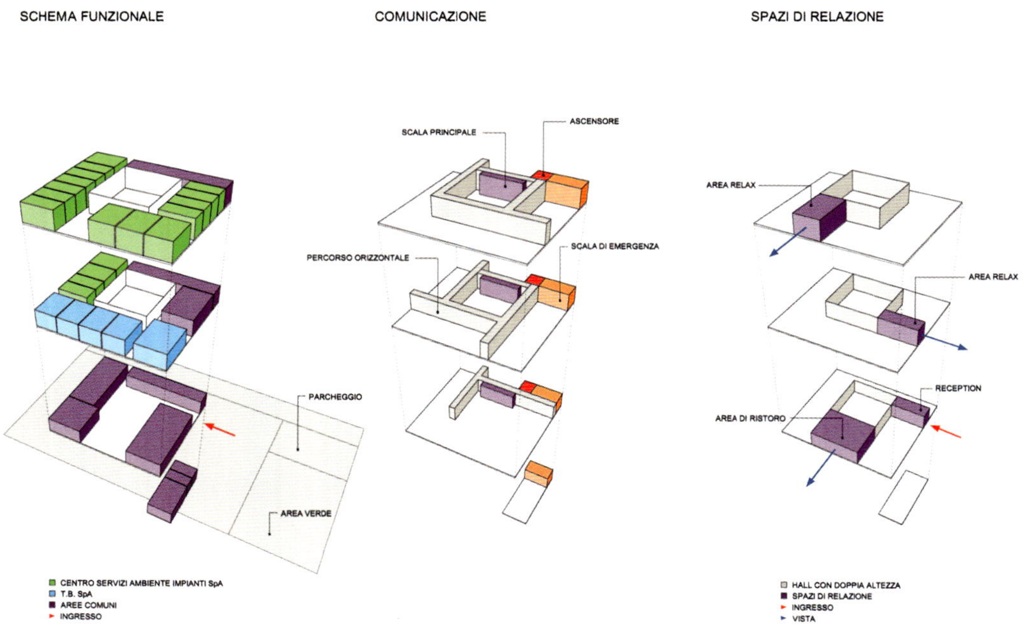

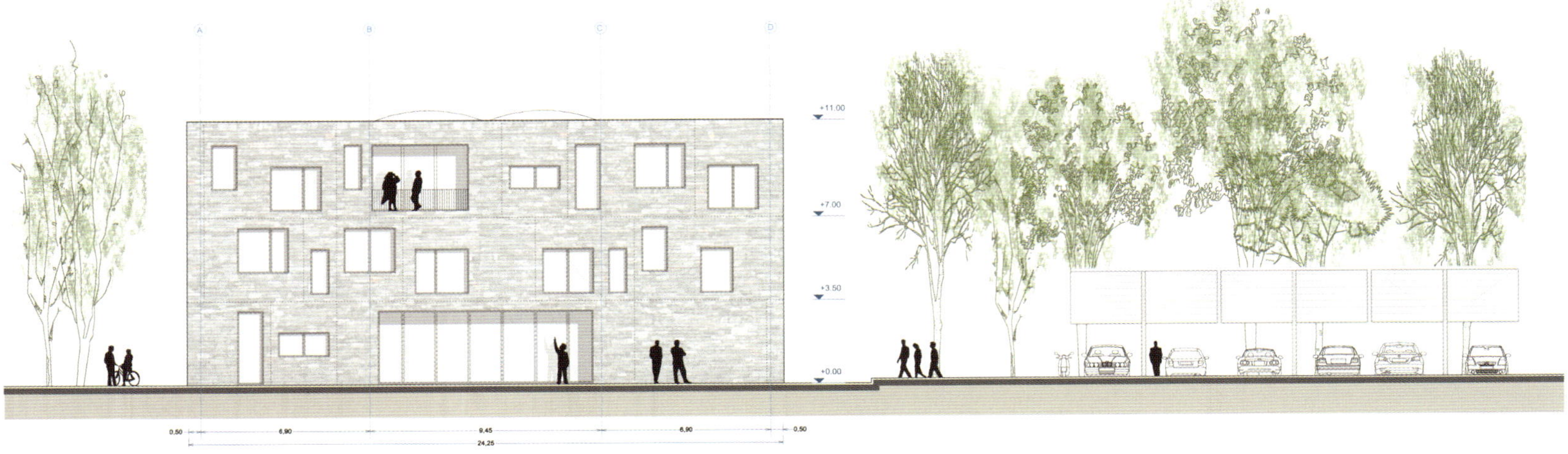

PROSPETTO SUD B-B

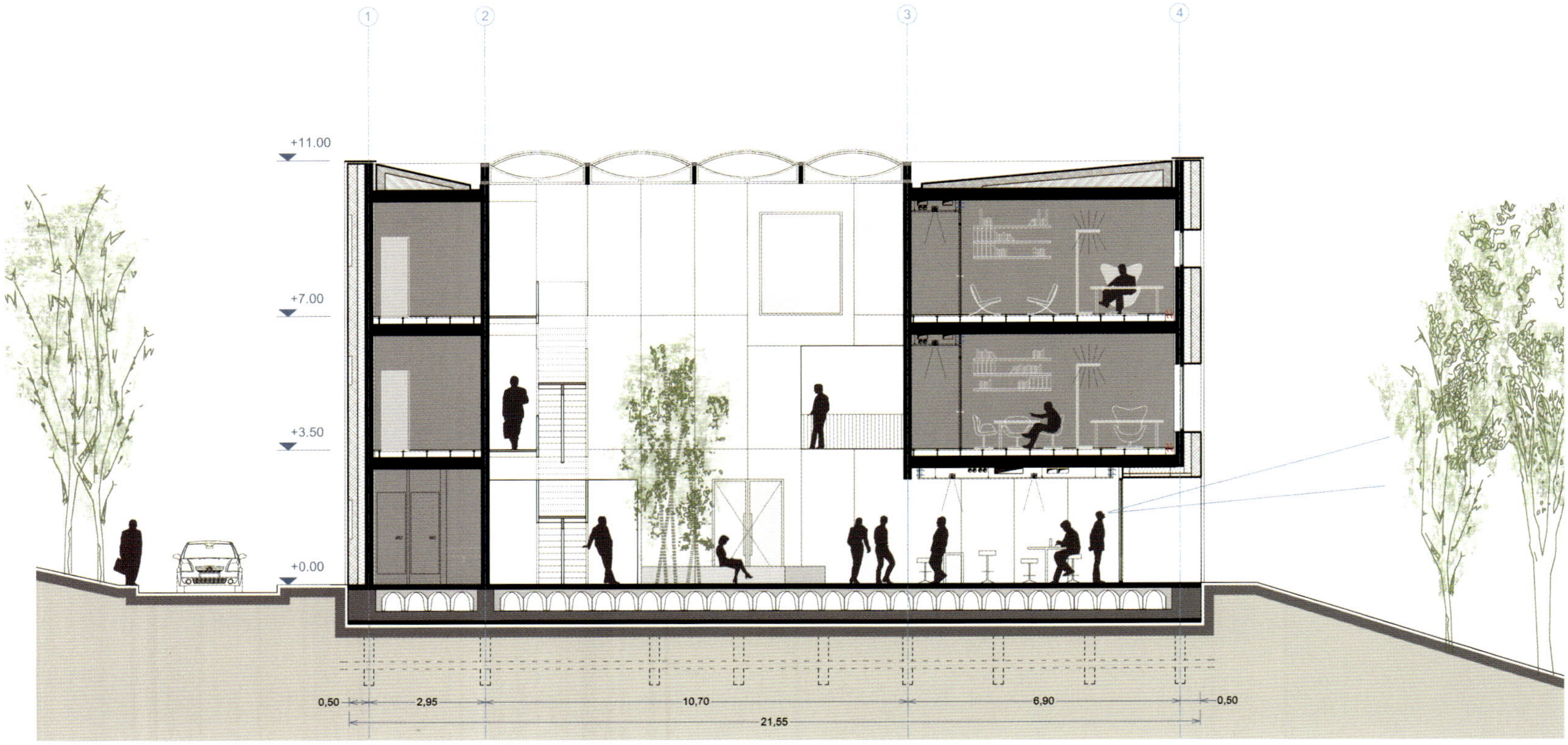

SEZIONE A-A

SEZIONE C-C

SCHEMA STRUTTURALE

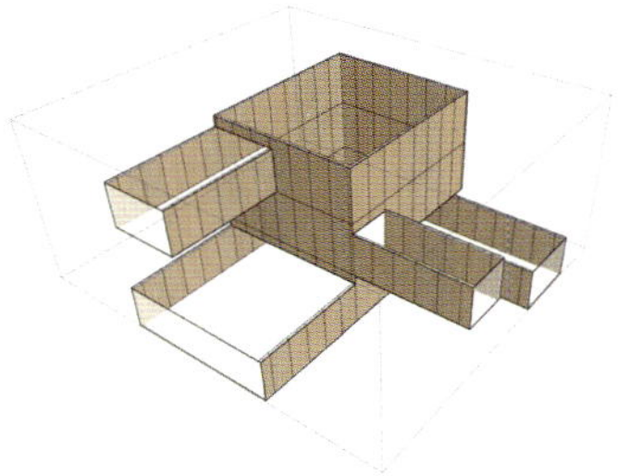

STRUTTURA INTERNA PORTANTE IN PANNELLI DI LEGNO TIPOLOGIA XLAM

+

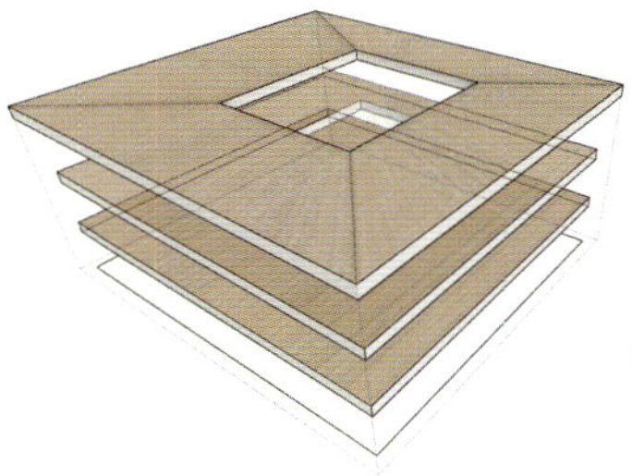

SOLAI AUTOPORTANTI IN LEGNO TIPOLOGIA XLAM

+

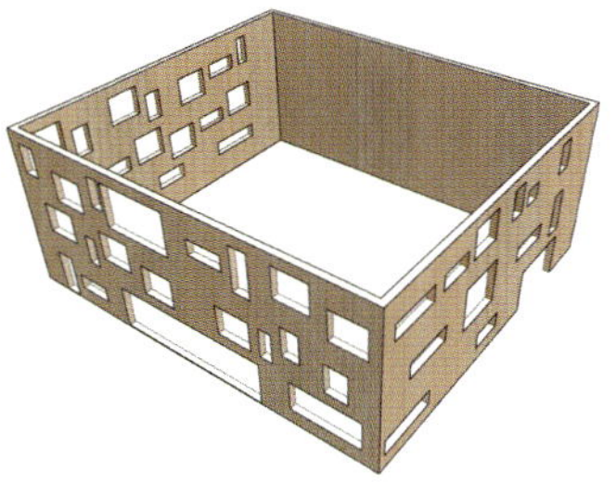

INVOLUCRO ESTERNO PORTANTE IN PANNELLI DI LEGNO TIPOLOGIA XLAM

+

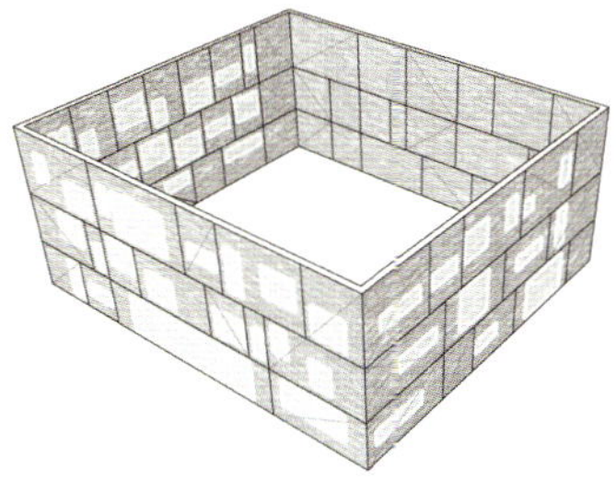

RIVESTIMENTO IN PIETRAME DI RECUPERO CON FISSAGGIO A STRUTTURA IN ACCIAIO CONTROVENTATA

ILLUMINAZIONE NATURALE

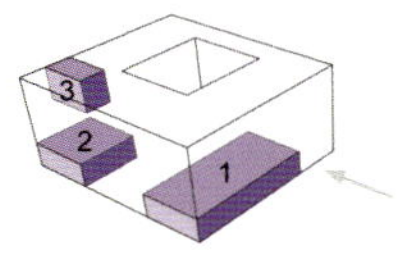

1

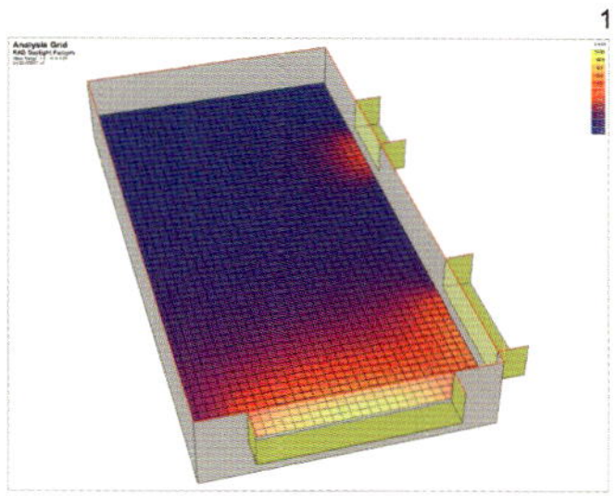

FATTORE MEDIO DI LUCE DIURNA ALL'INTERNO DELLA SALA FORMAZIONE: 3,22% (valutato con il metodo BRE - building research establishment "split flux")

2

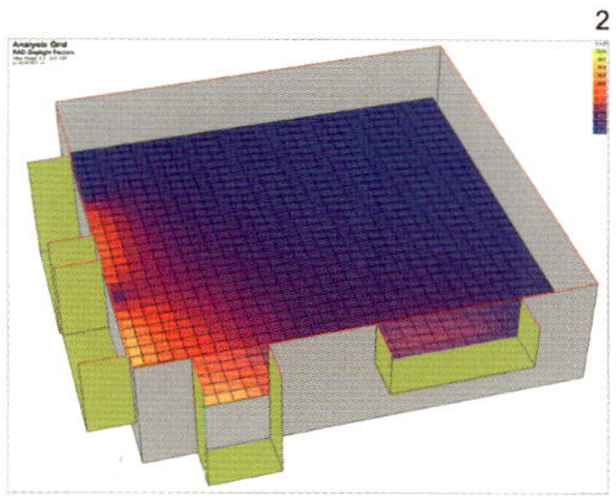

FATTORE MEDIO DI LUCE DIURNA ALL'INTERNO DELLA SALA RIUNIONI: 5,67% (valutato con il metodo BRE - building research establishment "split flux")

3

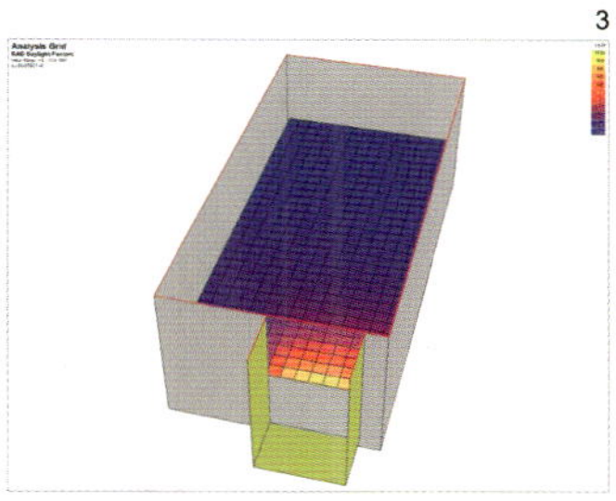

FATTORE MEDIO DI LUCE DIURNA ALL'INTERNO DELL'UFFICIO PIU' SFAVORITO : 2,04% (valutato con il metodo BRE - building research establishment "split flux")

$$\eta_m = 2,04\%$$

FATTORE MEDIO DI LUCE DIURNA DEL FABBRICATO UNI EN ISO 10840:

ORIENTAMENTO VISIVO AREE UFFICI

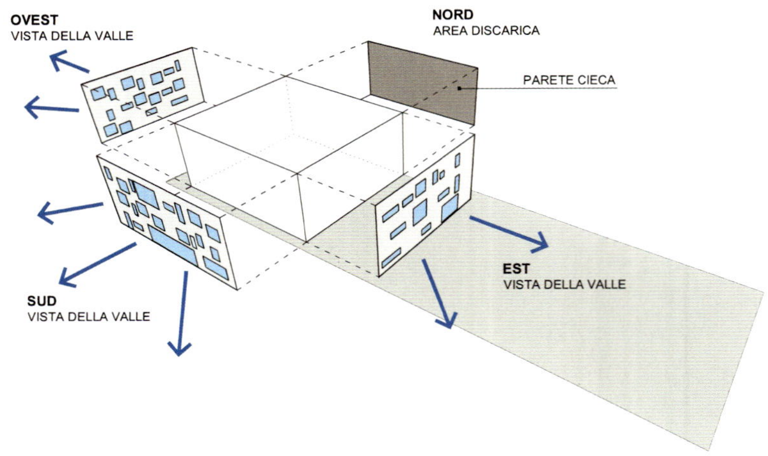

PERMEABILITA' DELLE AREE ESTERNE

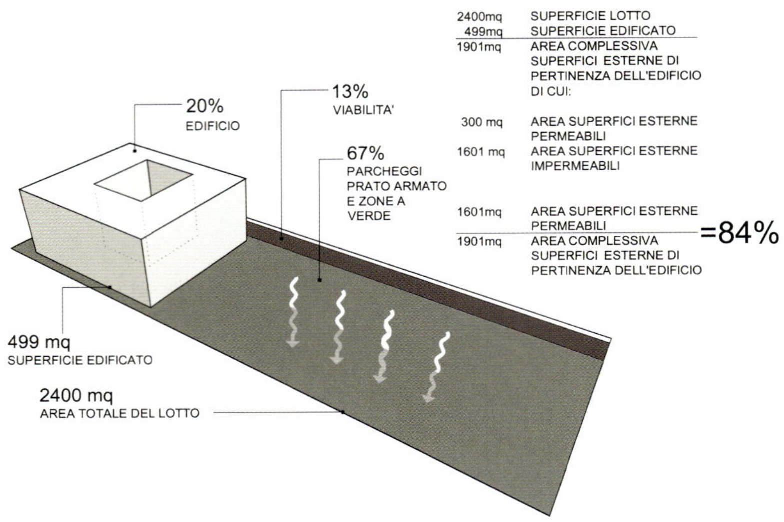

COEFFICIENTE DI FORMA OTTIMIZZATO
AI FINI DELLE PRESTAZIONI ENERGETICHE

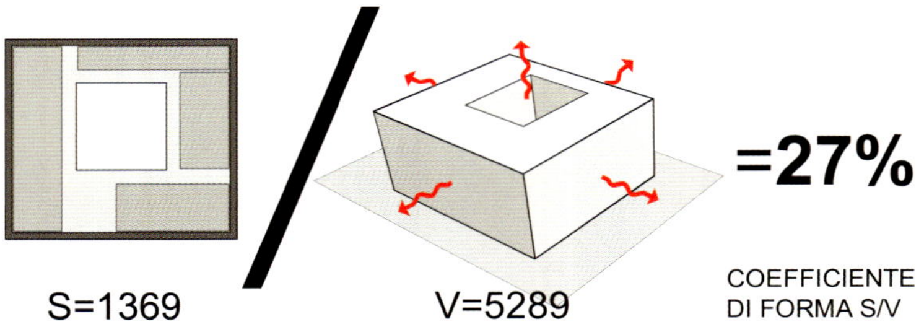

RECUPERO ACQUE PIOVANE

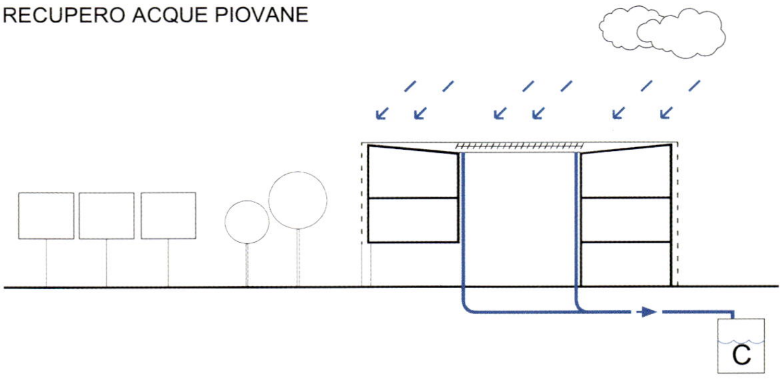

INVERNO → RISCALDAMENTO

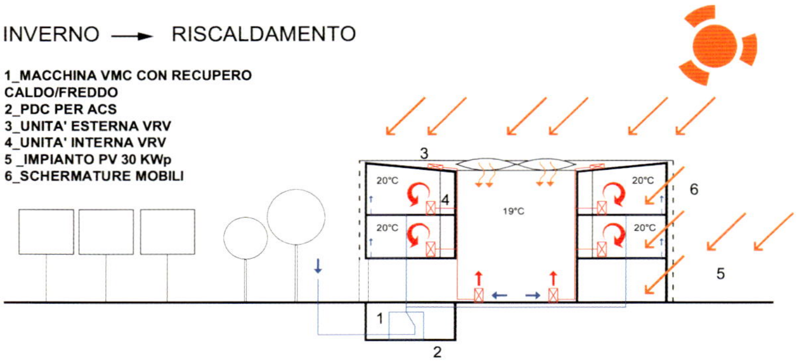

ESTATE → RAFFRESCAMNETO

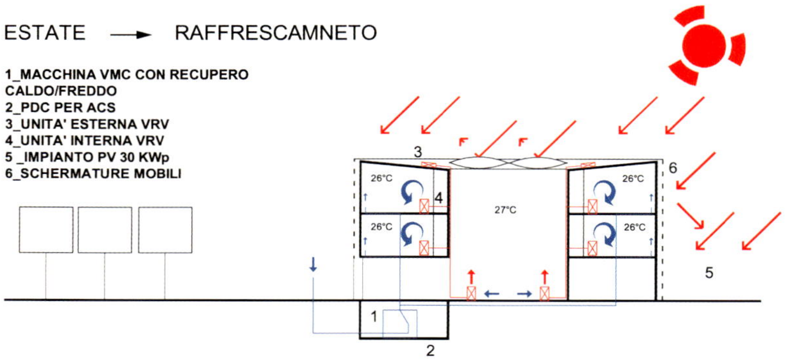

BILANCIO ENERGETICO INVERNALE

The building will host offices for a company which deals with ecological services in the beautiful Tuscany landscape, a place which is a valley from connotations typically Tuscan. The colors of the rolling hills created at every turn, every small ridge, a view which is always the same and always different. The design of the new building stands just as single element, although a highly characterized, not altering the balance of colors and materials of the valley in which it resides. The impact of the environment is of utmost respect and stands as counterpoint to a place characterized by space and colors of the hills.

Its image is connected to the old middle age sighting towers which is a reference to the landscape of these places. The new buildings identify the site where it is built as a privileged place for its position.

Its compact size and high technological content, blends with the natural look of his prospects that result assembly of different materials by type and consistency.

The building also features in its outer skin as message of thought where the need for sustainable build the new is resolved with the use of materials recovery and also exploiting the natural resources of the place.

Natural Light

The amount of sun radiation is constantly adjusted through the choice of windows, whose glass surface does not exceed 40% of the entire building, and by shading systems from both the thickness of the walls and external blinds.

Healthy Work Environment

The possibility of eliminating the use of ceiling and the provision of floating floor reduces to workers the risk of exposure to bad environments.

Flexibility of Interior Spaces

The plans show different ways of partitioning spaces that demonstrate the

PIANO TERRA

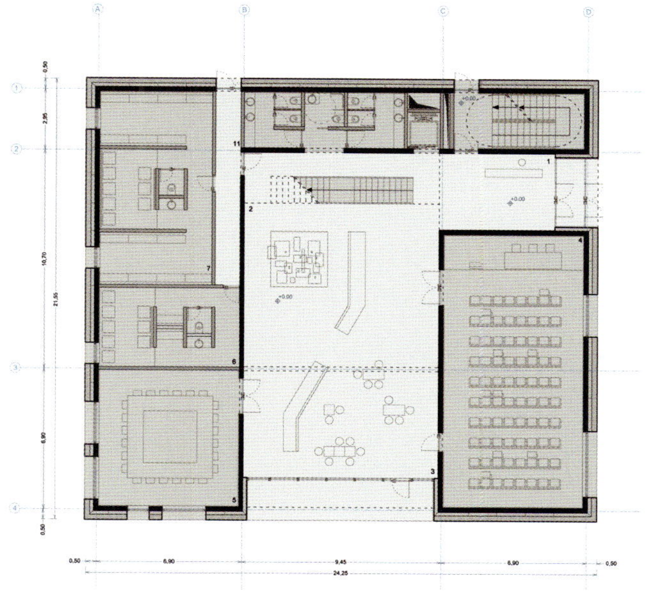

SECONDO PIANO

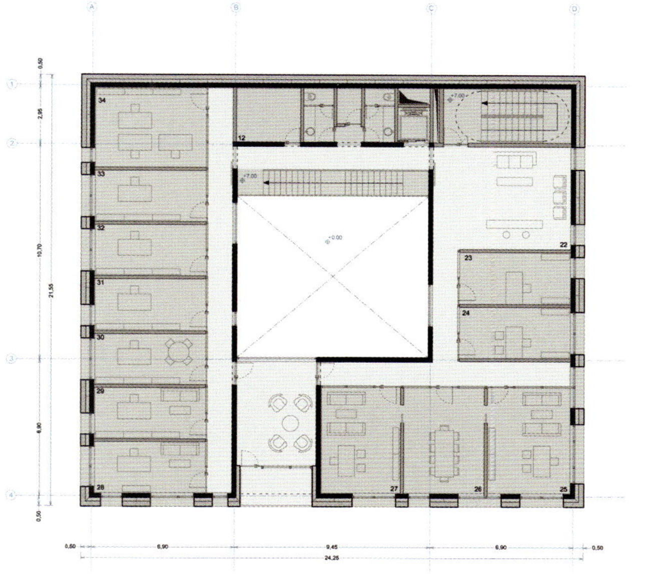

flexibility of the building at different deployment configurations of offices, which can expand or shrink by moving the partitions.

Sustainable Design

Thanks to the entire wooden structure, which give great environment benefit, the building will be totally Klimahaus certified. The cladding will use recycled stone in order to have a minimum ecological impact.

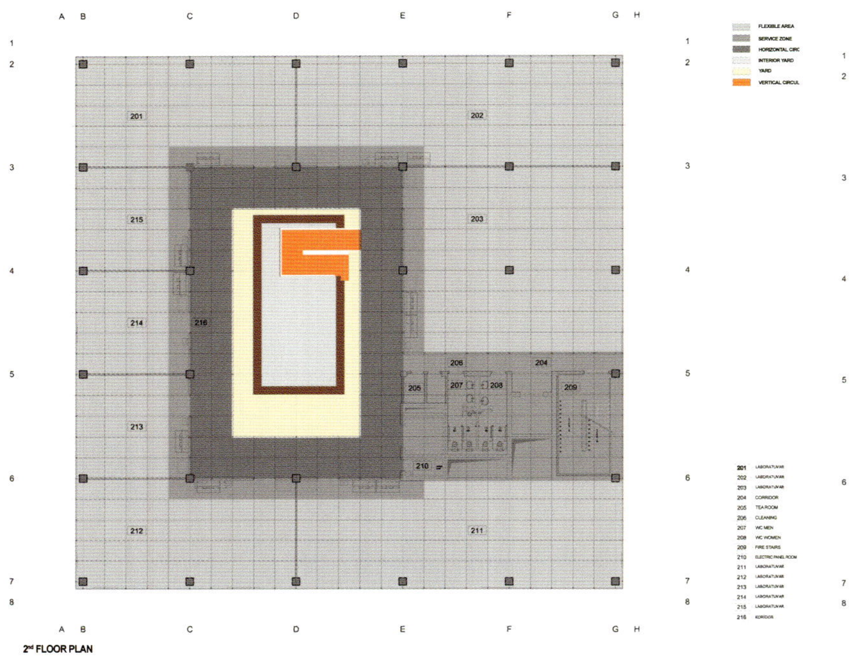

2nd FLOOR PLAN

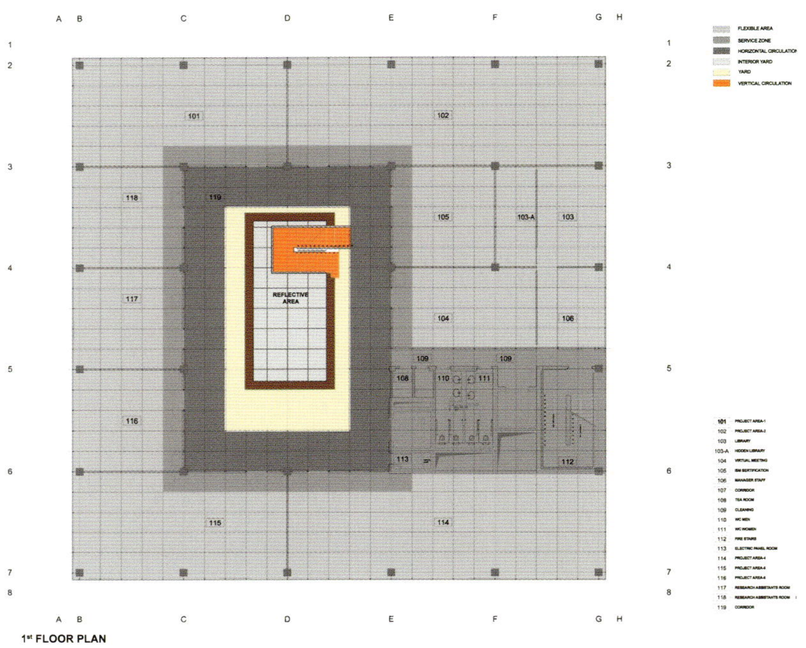

1st FLOOR PLAN

Science Tower for The Panum Complex, Part of the University of Copenhagen

哥本哈根大学帕纳姆中心科学大厦

C. F. Møller Architects

First 1st Prize in International Architectural Competition, 2010
2010年国际建筑竞赛一等奖

Credits
Client: The Danish University and Property Agency
Area: 35,000 m²
Address : Blegdamsvej, Copenhagen, Denmark
Expected Completion: 2014
Architect: C. F. Møller Architects
Landscape: SLA
Engineer: Rambøll
Collaborators : aggebo&henriksen, Cenergia, Gordon Farquharson, and Innovation Lab

帕纳姆中心扩建工程以创建适合现代教研的最佳环境为目标而规划设计。另一目标是，建造一座帕纳姆中心和大学北校区一体化的、同质化的、关键性的、雕刻般的大楼。同时，新中心还将扮演一个与毗邻街区和整座城市有关的开发规划的动力源泉。

16层高的科学大厦将为中心以一种清晰可辨识的形式提供一个统一的、动态的教研焦点。正如树木有它的树根网，大厦也是建立在一群包含日常功能的小型建筑之上，包括大礼堂、教室、餐厅、实验室、会议室以及阅览咖啡馆。最激动人心的部分莫过于，扩展性的科学广场将会形成中心的新社交中枢。广场作为整个中心的主要入口，还将服务定位于主要社交见面场所，把新旧功能区连接起来。

新帕纳姆中心看起来将会是开放的和外向的，透明的底层楼面将帮助淡化大楼与城市之间的边界。公众也会被邀请参观大厦的顶层，那里有咖啡馆、休息室和观光点。

在楼宇之间，新广场将会与配套有凉亭和座椅的内部花园空间一起浮现出来。它们除了将成为教室和办公室的延伸拓展外，还将为城市增添新的绿洲。一条校园中央大道贯穿整个区域，旁边有人行道和自行车道，将会与周边城市区亲密接触，创造出一个生意盎然的城市公园。

建筑的有机形式表达了象征能力和创造力，同时它又通过色彩搭配、节奏变化和沉重感来迎合已有的帕纳姆中心。

建筑正立面以分解大楼外观的层高落地窗网格状式结构建造起来。

使用丹麦能源最大程度有效利用的实验室，将使项目在能源利用方面成为开拓先锋。从通风系统中排出的废弃能源，将以迄今为止前所未有的程度被循环利用，用来平衡整个建筑的所有能源消耗。

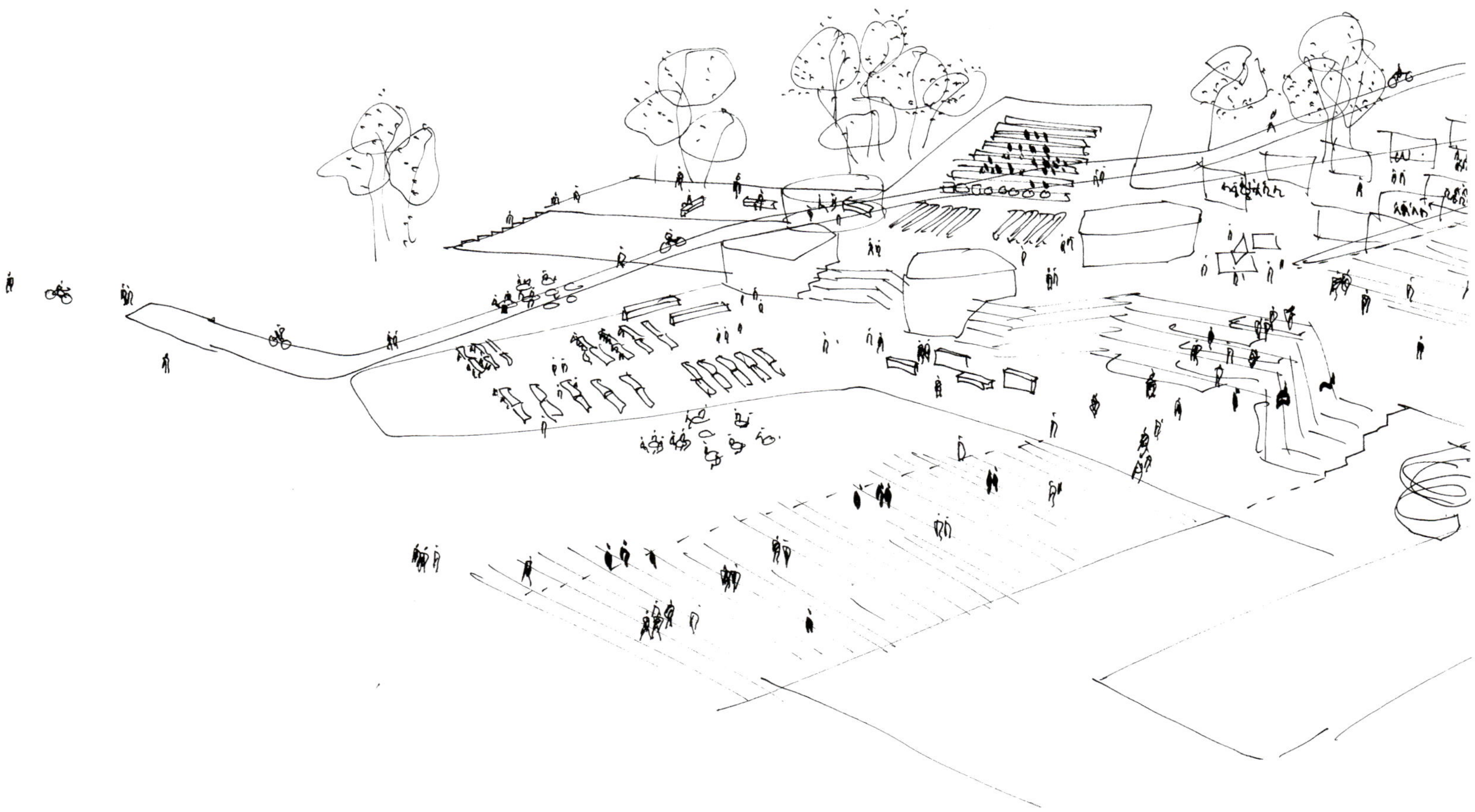

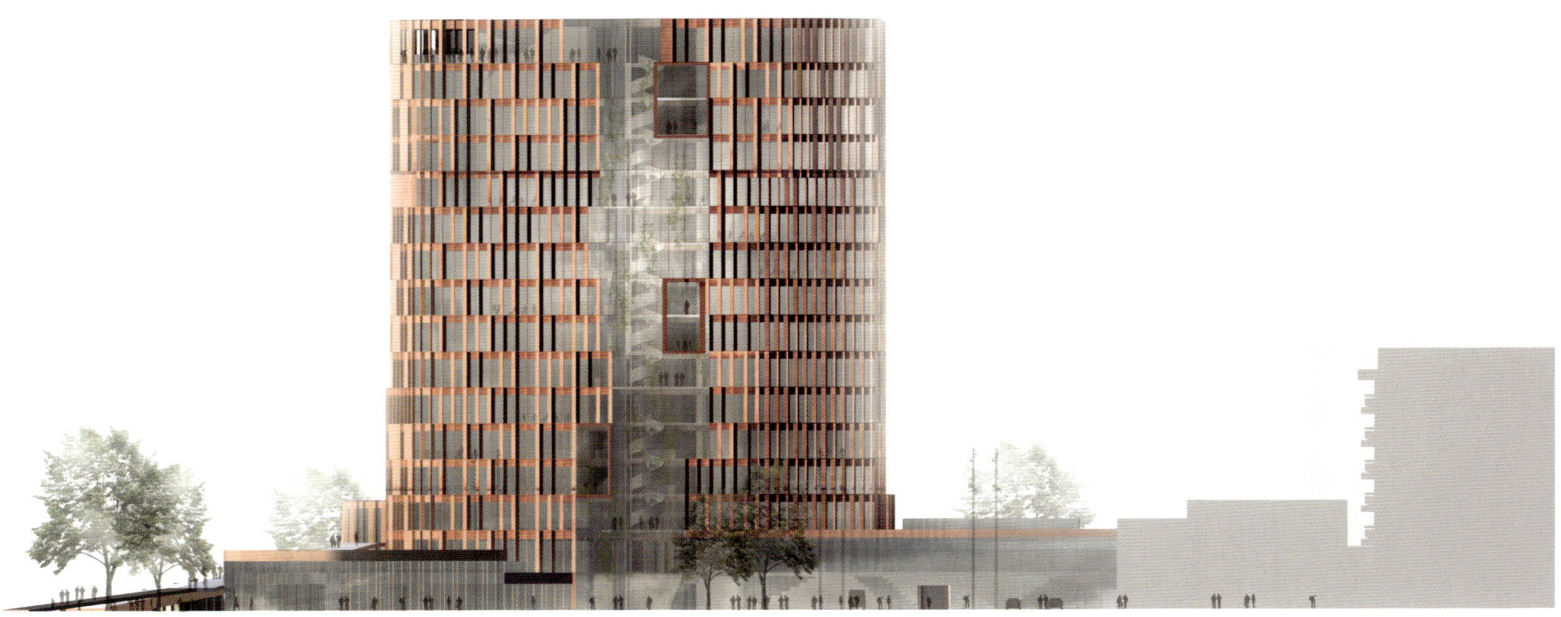

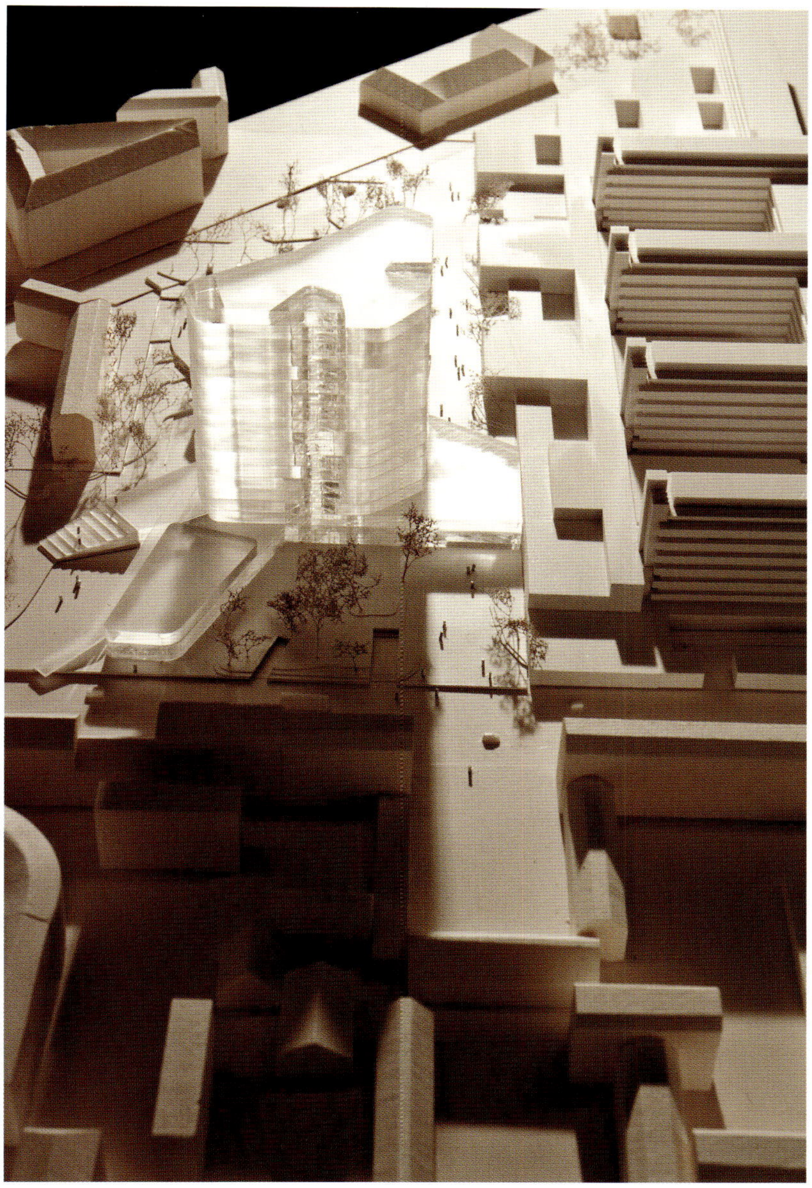

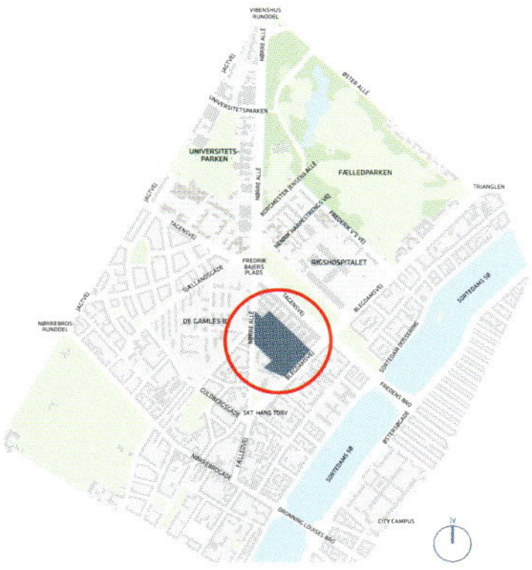

URBAN CONTEXT ANALYSIS
The site of the new Panum extension is located in central inner Nørrebro - one of the countries most dense urban areas. The surroundings are characterised by diverse ethnicity, some large institutions and few large/scale green spaces.

PATCHWORK
Spatially, programmatically and typologically the area is characterised by diversity and juxtaposition. The new landscape aims to enhance this by introducing a new urban nature, combining urban space with the natural.

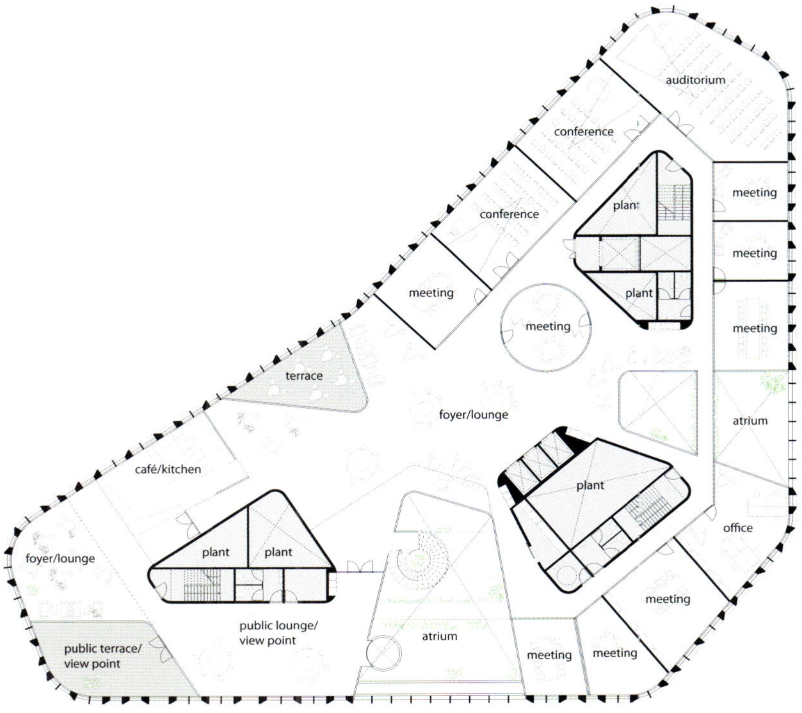

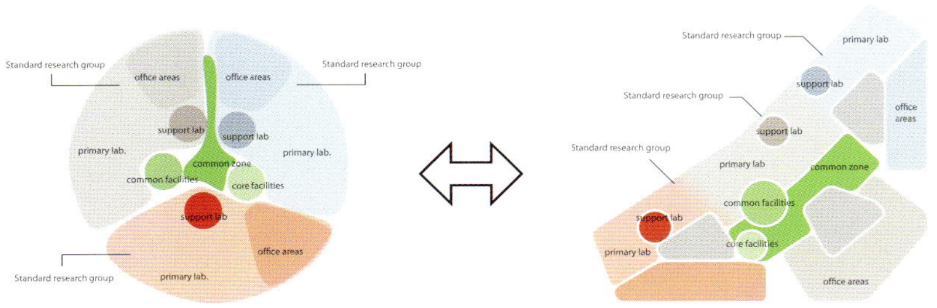

FUNCTIONAL DIAGRAM OF RESEARCH UNITS & FLOOR PLAN LAYOUT

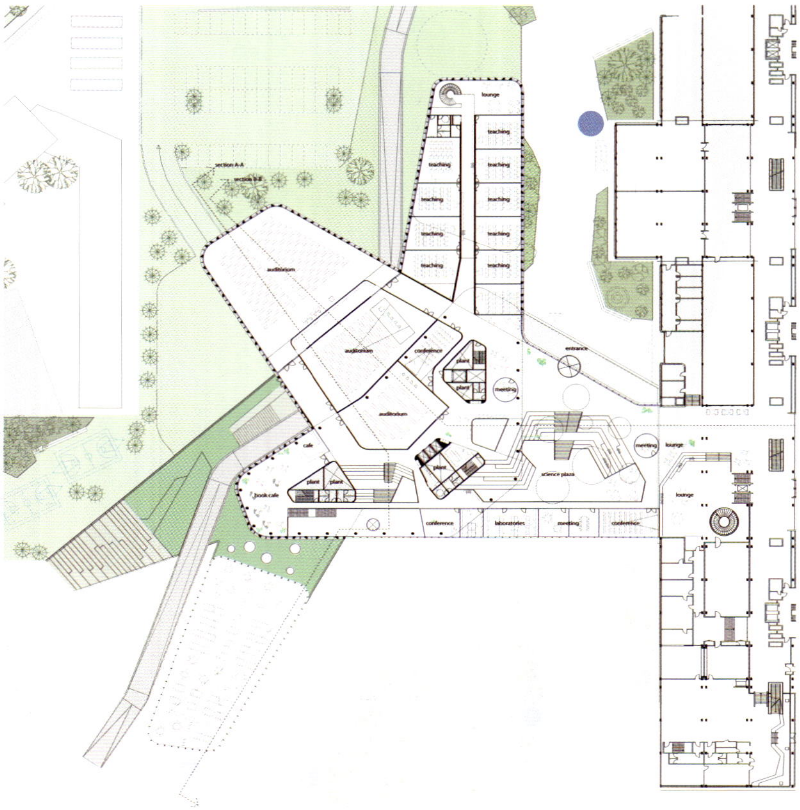

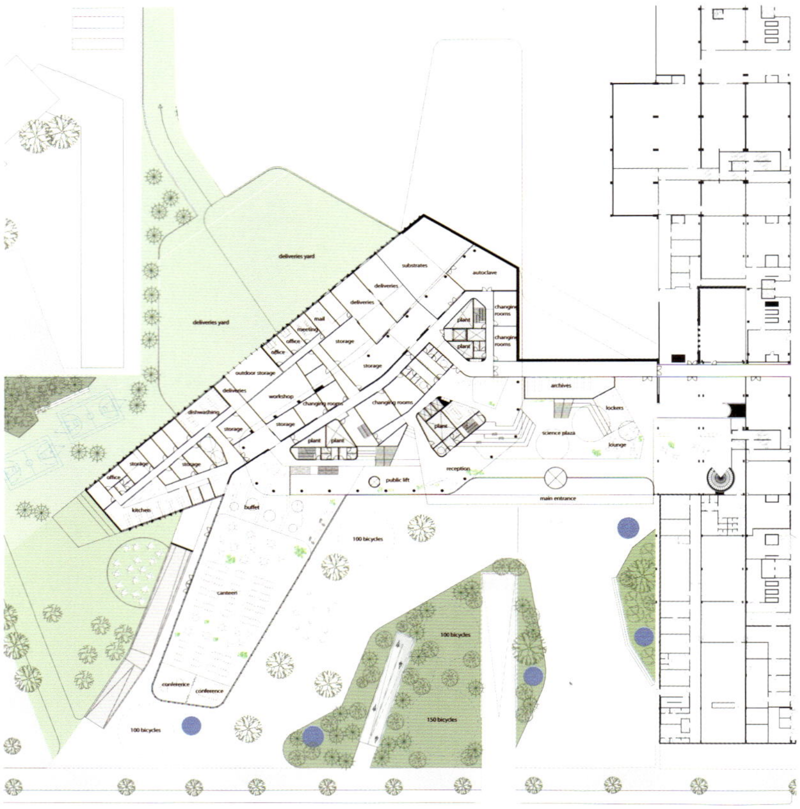

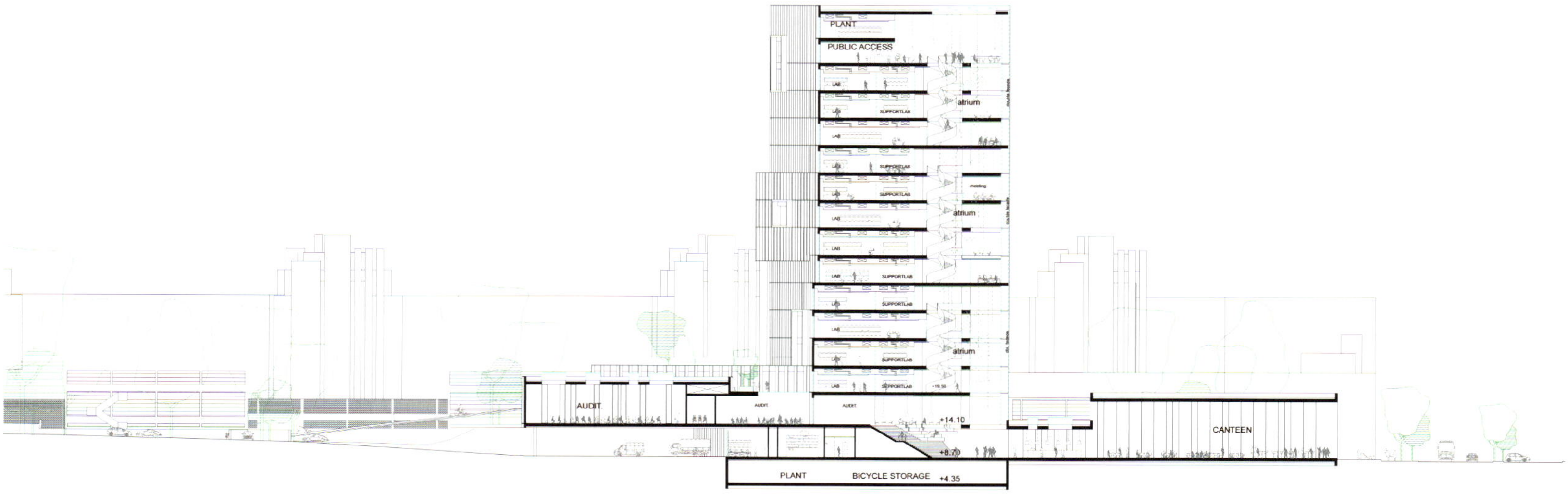
PLANT
PUBLIC ACCESS
atrium
atrium
atrium
AUDIT
+14.10
+8.70
CANTEEN
PLANT
BICYCLE STORAGE
+4.35

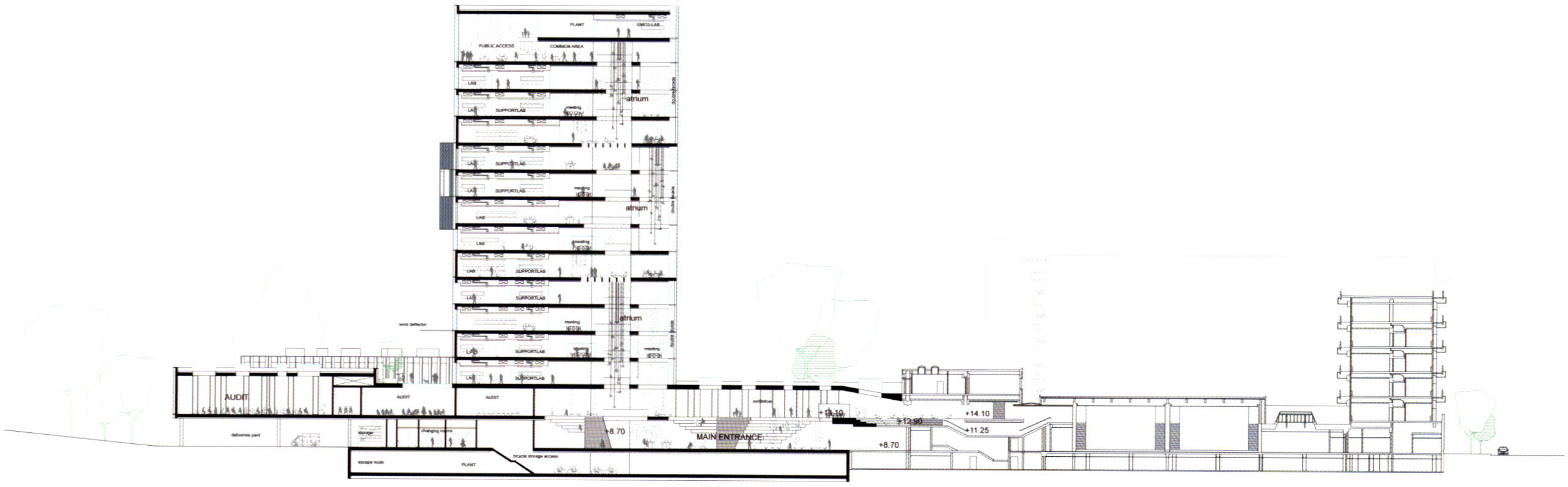
atrium
atrium
atrium
AUDIT
+8.70
MAIN ENTRANCE
+12.90
+14.10
+11.25
+8.70
PLANT

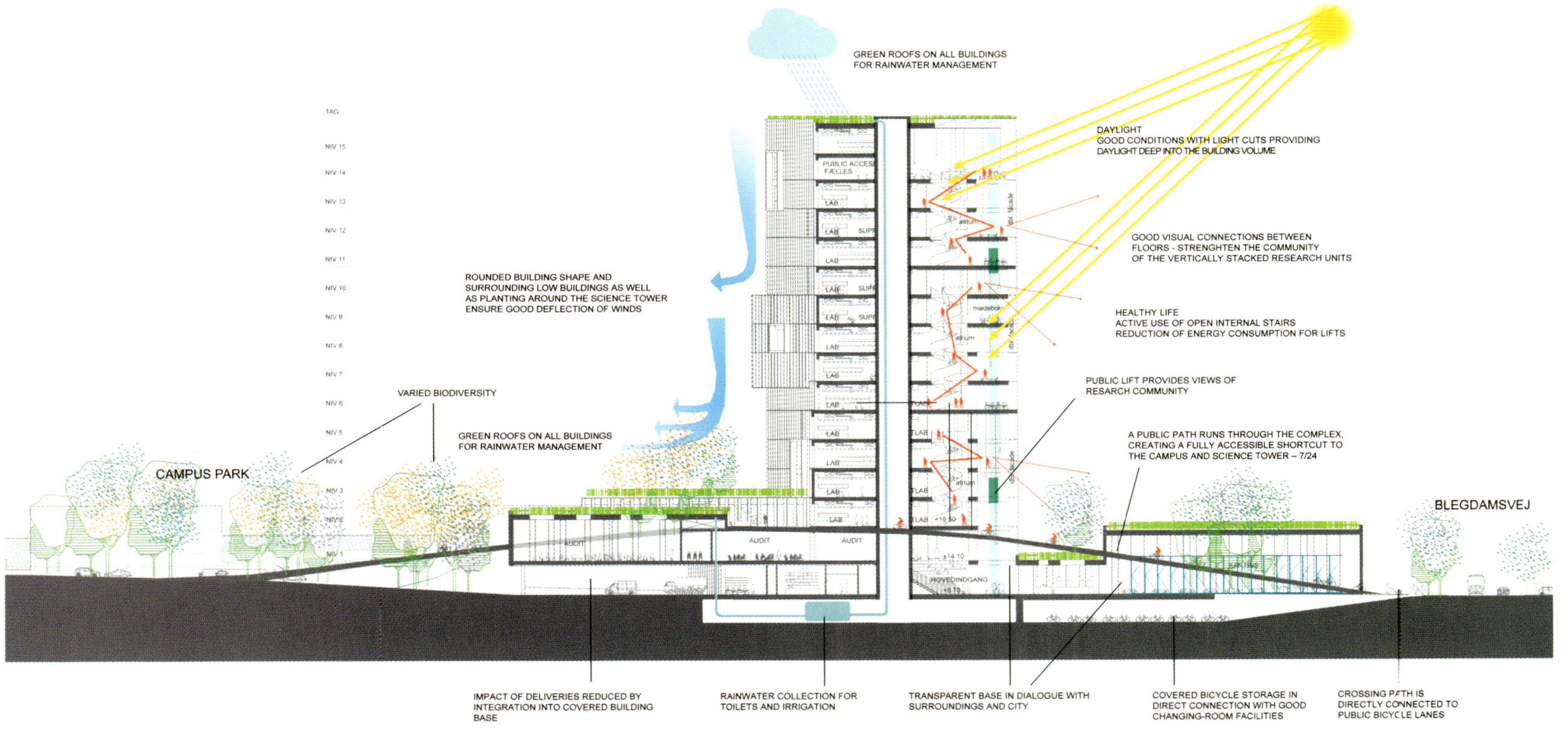

SUSTAINABILITY DIAGRAM

ENERGY EFFICIENCY DIAGRAM

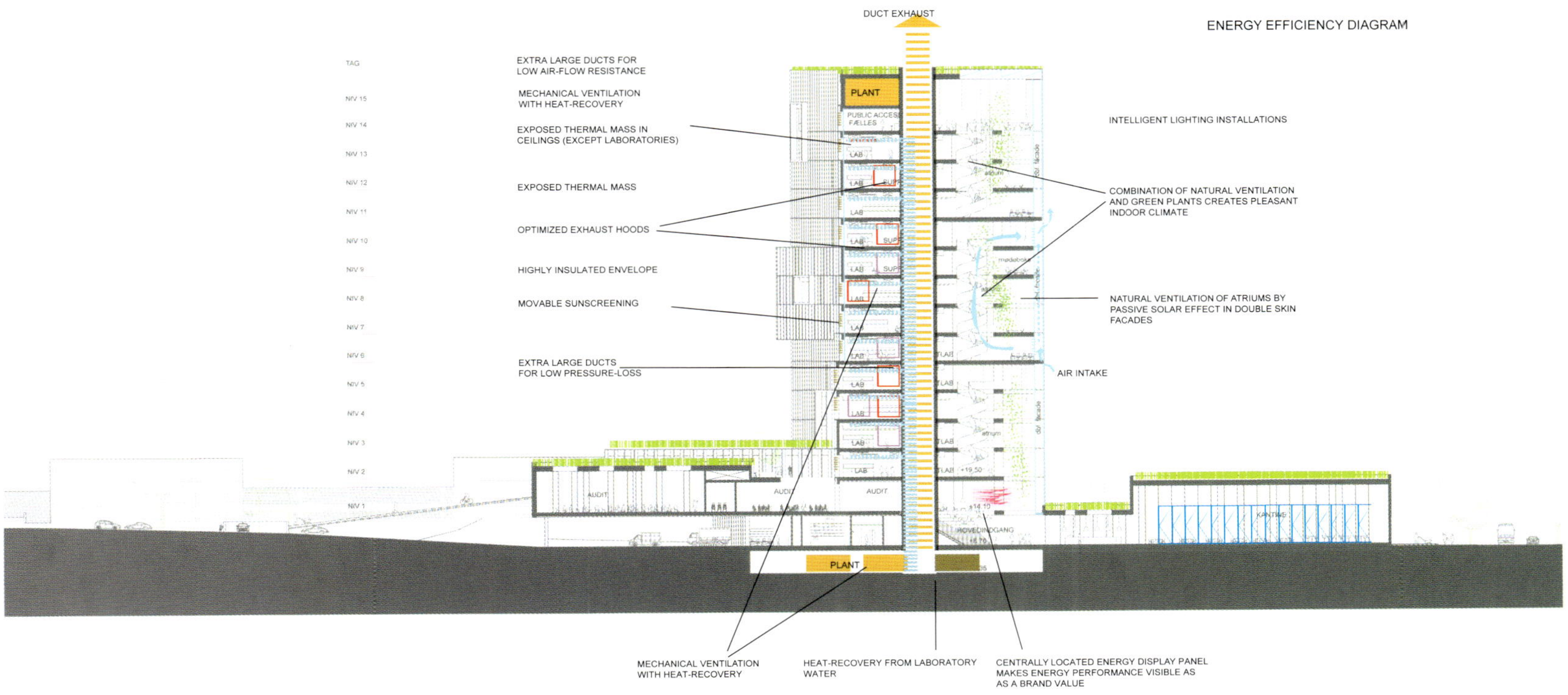

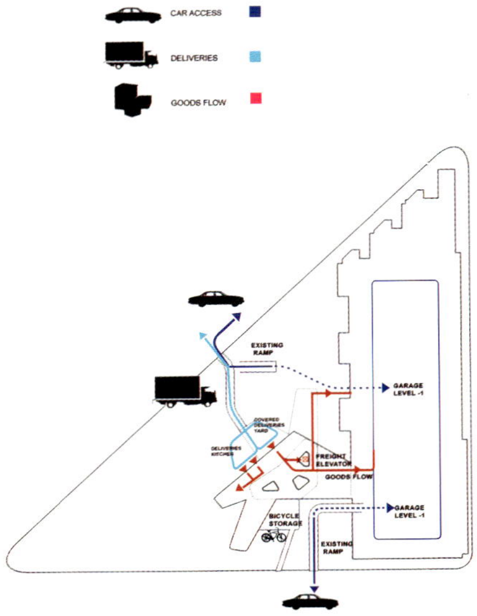

ACCESS DIAGRAM: DELIVERIES & PARKING GARAGE

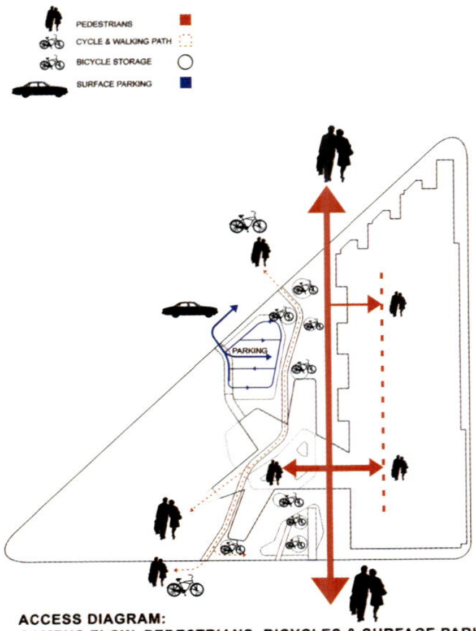

ACCESS DIAGRAM: CAMPUS FLOW, PEDESTRIANS, BICYCLES & SURFACE PARKING

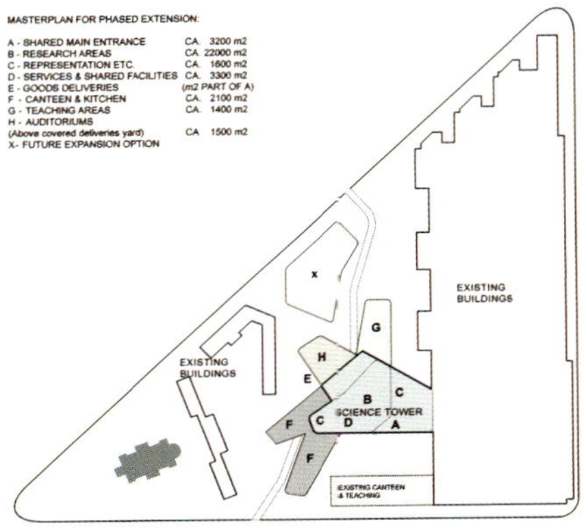

PHASES

The extension of the Panum complex has been designed with the aim of creating the best possible environment for modern research and teaching. A parallel objective has been to create a building which will stand out as an identity-creating, sculptural linchpin for the entire Panum complex and the university's Nørre Campus (i.e. the North Campus) as a whole. The new complex is also intended to act as the generator of a positive urban development in its immediate neighbourhood and in relation to the entire city.

At sixteen storeys tall, the Science Tower will provide the complex with a unifying and dynamic focal point in a clear and readable form. But just as a tree has its root network, the tower rests upon on a series of smaller buildings which contain the common functions: the auditoriums, classrooms, canteen, show lab, conference rooms and book café. The most striking part of the root network is the extensive science plaza, which will form the new social hub of the complex. The plaza accommodates the main entrance and will serve as the main social meeting-place, linking all functions between the new and the existing Panum complex.

The new Panum complex will have an open and outward-looking appearance, with a transparent ground floor that will help to blur the boundaries between the building and the city. The public will also be invited to visit the top of the tower, where there will be a café, lounge and viewing points.

Between the buildings, new plazas will arise, together with internal garden spaces equipped with alcoves and seating. These will function as an extension of the study rooms and offices, but will also add new green oases to the city. A campus thoroughfare passing through the area, together with pedestrian and cycle paths, will create a vibrant urban park with intimate links to the surrounding city.

With its organic forms, the building expresses signal power and innovation, but is also adapted to the existing Panum complex through its colour scheme, rhythm and gravity.

The facade is built up in the form of a grid structure of storey-high window fields that break up the building's large scale.

The project will be pioneering in energy usage, with Denmark's most energy-efficient laboratories, in which waste energy from the ventilation system will be recycled in the overall energy balance of the building to a hitherto unprecedented degree.

Elisabeth and Helmut Uhl Foundation

Elisabeth and Helmut Uhl基金会

Modostudio

Architectural Restricted Competition, Winning Project

建筑设计方案邀请赛获奖项目

Credits

Location: Laives, Italy

Client: Elisabeth and Helmut Uhl Foundation

Area: 1, 150 m²

Year: 2009

Design Team: Modostudio (Fabio Cibinel, Roberto Laurenti, Giorgio Martocchia)

Structural Engineer: KHing Knippers Helbig Beratende Ingenieure Consulting Engineers

这个项目是由来自于意大利、瑞士、奥地利与德国的五个设计团队参加的竞赛获奖项目。

竞赛的目的是要设计基金会大楼。大楼将建在山顶，面向博尔扎诺地区山谷，周围环境优美，景色宜人。

项目将按照原有建筑的地基建造，以最大程度地保护环境。这里将会建造一个玻璃房，里面设有实验工厂培养基地、研究人员办公室、一个餐厅以及其他的设施，以服务基金会活动。

设计展示了建筑功能性与地理环境之间的紧张关系：室内空间与室外环境的关系、传统地理位置与创新基金研究之间的关系、工作空间与生活空间之间的关系。项目想要充分利用周边的环境来保护人类免受环境的侵害。

在环境可持续发展方面，新的建设材料大部分来自原有建筑拆毁后的重复利用，以及最新的光伏太阳能系统（将用于首次实验）。这样的设计使项目获得Klimahouse的认证。

“Elisabeth and Helmut Uhl 基金会”

2008年成立于慕尼黑附近的格勒福尔福，“Elisabeth and Helmut Uhl Foundation”是一个以创始人Rainer Uhl父母的名字命名的非营利性机构。

基金会成立的目的是通过以下几方面促进科学与艺术的发展：

1. 各个学科与文化之间先进的网络交叉，以达到为社会创新，培养个别有天赋的前沿工作者与思想者；
2. 举办学科间的文化交流会议；
3. 推广结合了时尚美学的可持续生态建筑。

The project is the winning design of a competition by invitation which has seen the participation of five teams from Italy, Switzerland, Austria and Germany.

The goal of the competition was to design a building able to host the activities of the foundation. It will be located on the top of a mountain facing the valley towards Bozen. The site has a spectacular panoramic view and an amazing surrounding.

The project wants to preserve the environment using the same footprint of the existing buildings which will be demolished once the construction will start. It will host a glasshouse for experimental plants colture, researcher's room, a refectory and other spaces to support the foundation activities.

The design wants to express the tension generated from the particular site and from the specific functions of the foundation building. A tension generated from the relation between the indoor spaces and the outdoor environment, between the traditions of the site and the innovation of the foundation research, between the space for working and for living. The project wants to protect its guests from the particular climatic conditions but at the same time to take advantages from the spectacular surroundings and environment.

The design will take care about sustainability. Most of the used materials will be recovered from the demolishing buildings, an innovative photovoltaic+solar system will be experimented for the first time. The building will be Klimahouse certified.

"Elisabeth and Helmut Uhl Foundation"

Founded in 2008 and named after the parents of the founder, Rainer Uhl, "Elisabeth and Helmut Uhl Foundation"is a non-profit organization (öffentliche Stiftung des bürgerlichen Rechts) based in Gräfelfing near Munich.

Aim of the foundation is to advance sciences and arts by:

1. Advancing networking between disciplines and cultures in order to create innovations for the society, fostering individual (highly gifted) frontier runners and lateral thinkers, who exemplify such networking.
2. Organizing interdisciplinary and intercultural meetings.
3. Promoting an architecture, which combines sustained ecological aims with a novel, gentle esthetics.

IDEA!
WINE
VIEW

SKY
EARTH

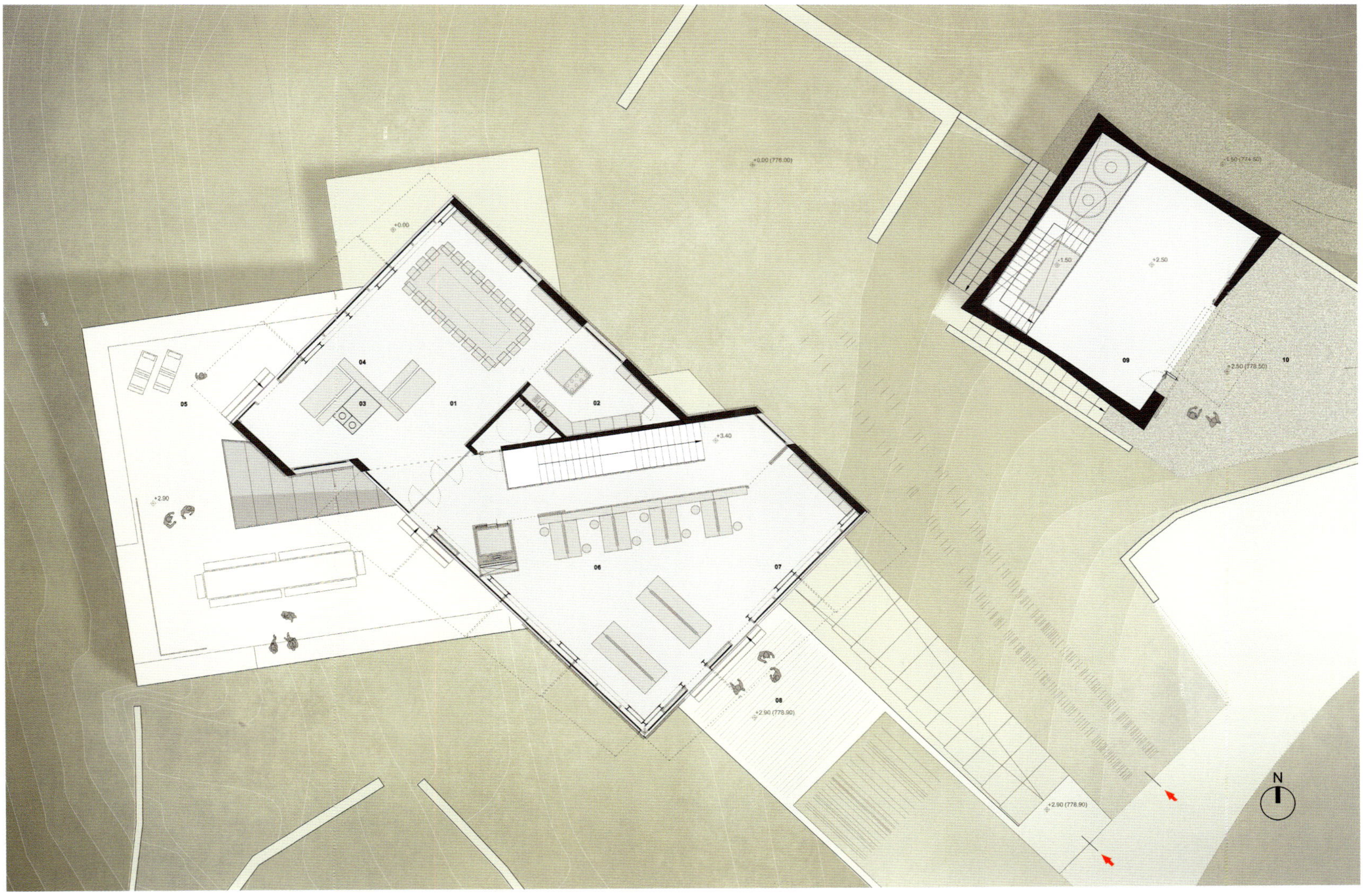
N

Energie Steiermark Administration Headquarters

能源公司办公总部

Ernst Giselbrecht + Partner ZT GmbH

Competition, September 2006, First Prize

2006年9月竞赛第一名

Credits

Location: Graz, Austria
Client: Energie Steiermark AG
Architect: Ernst Giselbrecht + Partner ZT GmbH
Project Manager: DI Peter Fürnschuss
Collaborators: Renate Mussbacher, DI Mirela Seva, DI Christina Giselbrecht, DI Christian Liegl, Christoph Wagner, Hannes Walzl, Philipp Erkinger (renderings)
Structural Consultant: Freller ZT GmbH, 9020 Klagenfurt
Electrical Consultant: Feistritzwerke GmbH
Construction Physics Consultant: Rosenfelder & Höfler Consulting
Landscaping Consultant: Büro ko a la
Light Planning Consultant: Vedder Lichtmanagement
Main Contractor: Strabag AG
Roof Contractor: Buchacher GmbH
Facade / Metalwork Contractor: SFL GmbH
Carpenter Contractor: Kompacher GmbH
Design & Documentation: 2006-2009
Construction Year: 2008- 2010
Built-up Area: 1,200 m²
Net Area: 14,000 m²
Total Floor Area: 16,900 m²
Cubage: 58,500 m³

项目描述

这栋20世纪五六十年代的建筑显然已经无法满足当代节能结构的最基本要求，如何处理该建筑，修复还是拆除，这已经成为迫切需要解决的问题。尤其是，该建筑是能源公司Energie Steiermark的办公楼，而作为能源公司，无论如何也应该在其所处领域起到模范作用。研究表明，相对于建造一些新的建筑，在这栋9层的长形建筑上做一些延伸反而更为有利。

考虑到安全法律的要求，建筑上的附加部分最多只能有5个楼层，而若在建筑两侧进行延伸的话，延伸部分则可以达到现有结构的高度。另外，延伸的部分可以作为一个新的整体结构，并且避免造成新建筑与旧建筑内部的差异。短而清晰的路线也属于该设计理念的一项优势，同时，设计不仅利用了安全技术环境中的协同效应，也极好地改良了旧楼层中的疏散路线。最后，但也很重要的是，楼上的绿色空间会被保留，而现有附加建筑则被移除，从而留出更多的空间，使得绿色区域融合为一个连贯的整体。

这样一个经过扩展的密实空间拥有优越的先决条件来满足被动房的能量标准。所有立面都会安装附加的隔热层、特殊的玻璃板、光伏发电设施以及太阳能热水装置。巨大的立面即使在远处也清晰可见，它不仅是这个著名的能源供应商的旗舰标志，同时也作为其传播媒介和身份象征，通过建立在前沿技术之上的设计展现公司的经济与生态实力。

设计理念

对项目场地的合理利用及现有建筑的巧妙扩展；

在格拉茨市建立起一个自信而清晰的建筑形象；

立面配以折叠式百叶窗，每天都呈现出不同的样式；

LED立面在夜晚展现给城市的是一系列与能量有关的情节景象（延续了白天的日出、降雨、瀑布和森林，所有这些无不体现出生命的活力）；

每一层采用不同的色彩，使得雇员能够清楚地辨认出自己的办公楼层所在；

室内设计涵盖了各个独立办公室的设计以及整体解决方案。

Northeast elevation

Southwest elevation

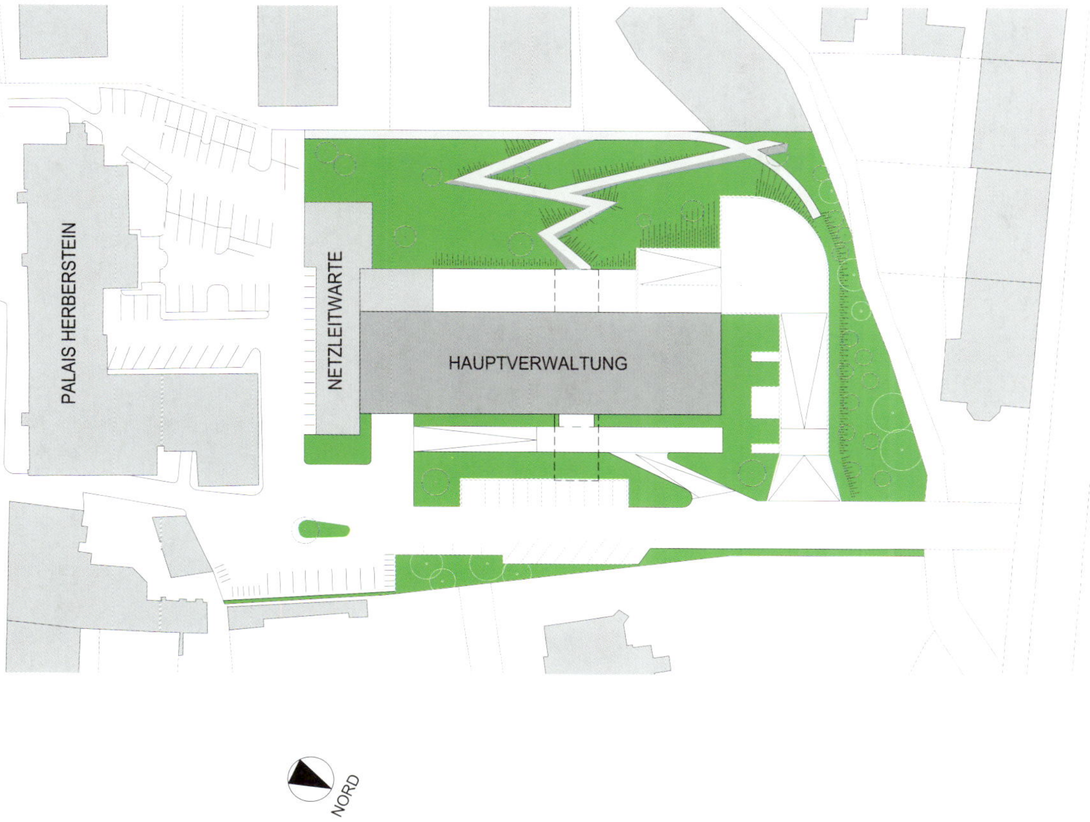

Site plan

Competition plans

Project Description

To renovate or to demolish? The question as to how to handle buildings from the 1950s and 1960s that no longer fulfill even the most basic requirements of modern energy-economical structures is becoming increasingly pressing - particularly for an energy company that should, after all, also function as a role model in this field. The project "Energie Steiermark - Neu" is based on the results of a study that showed that it would be more advantageous to extend the existing administrative building - an elongated plate construction with nine storeys - than to construct additional buildings.

Due to legal safety requirements, an annex could only have a maximum of five storeys, while a continuation of the building on both sides could keep up the level of the existing structure. Another advantage is that a new whole will be created and internal differentiations between the old and the new building substance be avoided. Short, clear structured routes are another benefit of such a construction concept. By extending the building on its two narrow sides, synergies in the context of safety technology can also be utilized and the escape routes of the old building be significantly improved. Last but not least, the green spaces on the facility will not be truncated, as existing annexes will be removed, thus freeing up more space and allowing the green areas to be integrated into one coherent park.

The compact enlarged volume has excellent prerequisites for fulfilling the Passivhaus energy standards in the new future. All façades will be equipped with an additional insulation layer and special glass panels as well as a photovoltaic power generation facility and a solar water heating installation. The enormous façade will be visible for miles, not only as the flagship of this particular energy provider, but also as its media carrier and corporate identity reflection, underlining its economic and ecological competence with a design that relies on state-of-the-art technology.

Architectural Concept

-Good usage of construction site and extension of the existing building
-A confident and unmistakable profile within the image of the city of Graz
-Façade concept with folding shutters, thus the façade will look different every day
-LED-façade - presenting a profile to the city at night-time too with a scenario that relates to energy (extending the day with a rising sun, rain, waterfall and forest – those very energies from which all life springs)
-Interior color concept - unmistakable colors for each storey, helping employees to identify with 'their' floor
-Interior design ranging from single offices to group office solutions

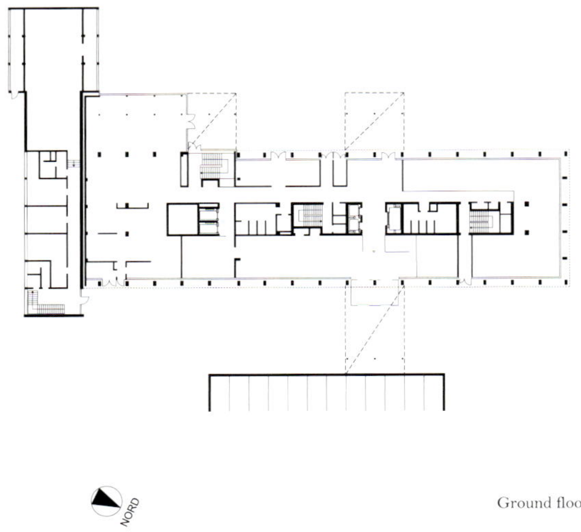

Ground floor plan

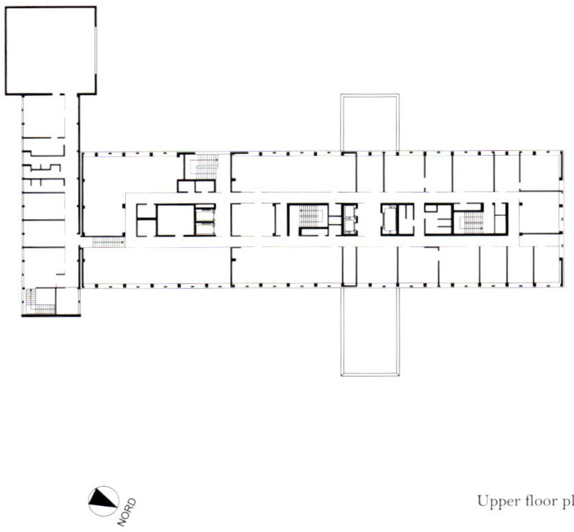

Upper floor plan

Lujiazui-Itochu Development Building
陆家嘴伊藤忠商事株式会社发展大楼

Sydness Architects

Credits
Location: Shanghai, China
Type: Corporate Headquarters

该项目是由分别来自于日本，德国，英国的建筑事务所和美国Sydness 建筑事务所参与国际竞赛而共同设计的作品，获得了设计大奖，同时也完成了设计施工的相关文件。这座34层高的建筑位于陆家嘴金融贸易区中心一隅，不仅与其所处地形相呼应，也有效地呼应了其地理位置。这座建筑从平面图来看是两个沿直径稍有偏移的半圆的奇妙组合，然后转化成标准的办公室模型，边缘呈锯齿状，前后边长有1.2米的递增或递减。建筑的垂直花岗岩肋部更突出了表面起伏的闪亮幕墙和花岗岩，节节攀高，与天空交相辉映。低一些的楼层覆盖了水平石材窗框及灰色透视玻璃和拱肩玻璃。西南面的大型广场通向圆形转门入口。三层的地下室包含了一个停车场和机械设备。该楼的业主是上海市陆家嘴发展公司和日本伊藤忠商事株式会社跨国公司。

This commission is the result of an international design competition which included four architects from Japan, Germany, Great Britain and the U.S. Sydness Architects won the design competition and completed the construction documents for the tower. The thirty-four story building is a response to the configuration of the site and to its location on a prominent corner in the center of the Lujiazui Finance and Trade Zone. The tower's plan is two semi-circles which are shifted along the diameter to fit snugly within the property lines. The building is then modified in deference to standard office modules and building products and has a serrated edge on a shifting 1.2-meter module. The undulating perimeter curtain wall of flamed and polished granite is accented by granite vertical ribs which rise up the tower to a stepped top that echoes the plan against the sky. The lower floors are clad with horizontal stone mullions and gray vision glass and spandrel glass. A large plaza facing southwest leads to a recessed circular glazed entry. Three levels below grade contain a parking garage and mechanical equipment. The owners are the Lujiazui Development Company of Shanghai, and a Japanese multi-national company, Itochu.

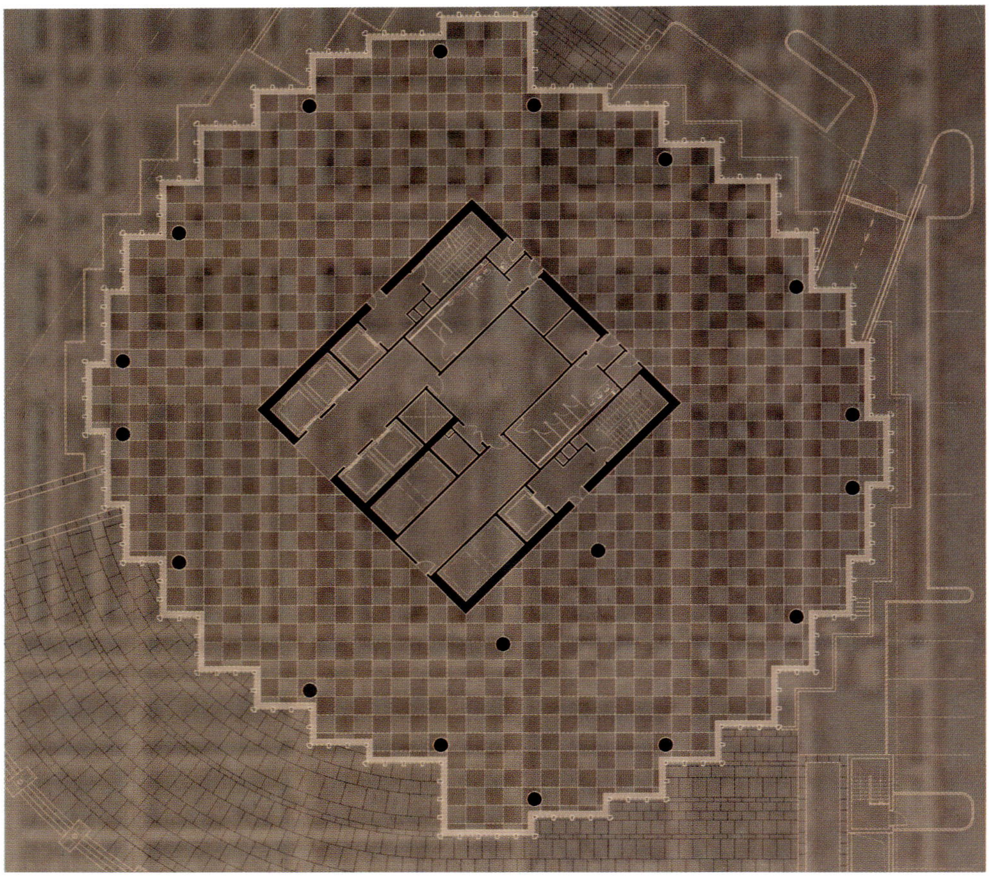

▼陆家嘴伊藤忠发展大楼平面图
Photo of the plan of the Lujiazui-Itochu Development Building

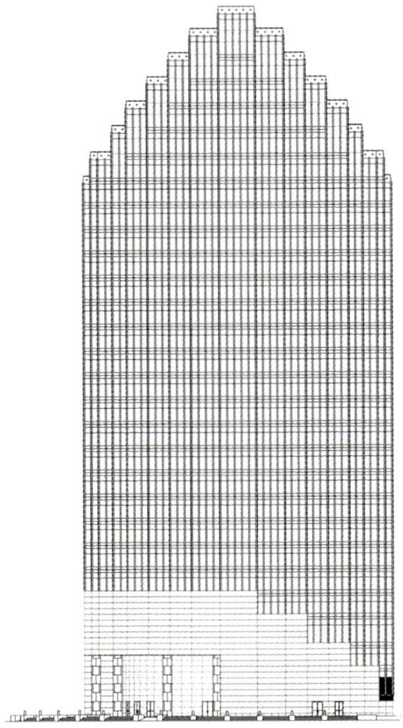

▲陆家嘴伊藤忠发展大楼南面和西面的立面图
Plans of the south and west views of Lujiazui-Itochu Development Building

▲陆家嘴伊藤忠发展大楼内部大堂表现图
Professional rendering of the interior lobby of the Lujiazui-Itochu Development Building

▶陆家嘴伊藤忠发展大楼表现图 1
Professional rendering of the Lujiazui-Itochu Development Building

▼陆家嘴伊藤忠发展大楼外部广场表现图
Professional rendering of the exterior lobby of the Lujiazui-Itochu Development Building

▼陆家嘴伊藤忠发展大楼表现图 2
Professional rendering of the Lujiazui-Itochu Development Building

Bading Development Tower

八顶发展大厦

Sydness Architects

Invited Competition
邀请赛参赛方案

Credits
Location: Shanghai, China
Type: 41 Story Office Tower

设计旨在将这座41层高的八顶发展大厦建设成一个世界级的金融中心。位于黄浦江畔浦东新区滨江地带的中心，这座大楼面朝浦西的外滩，将成为陆家嘴金融中心展现实力的新焦点。

大楼被设计成八角形，使其从城市的任何方位看上去都优雅地矗立着。每一面都有玻璃幕墙配合垂直花岗岩使建筑给人一种强烈的通往天空的立体感。每个立面都有轻微凹陷和旋转，像一个盘旋而上的玻璃王冠，使站在最上面的人可以从独特的视角观看浦西外滩。

The forty-one-story Bading Development Finance Mansion was designed as a world-class financial center. Positioned on the Huang Pu River at the heart of Shanghai's Pudong New Area riverfront Bund, the building will face the Puxi Bund and stand as a symbol for the strength and security of the new focal point that is the Lujiazui Financial Center.
The plan of the tower is an octagon, creating a tower that stands elegantly when viewed from any position around the city. The facade features vertical granite ribs extending up the central portion of each face giving the building a strong vertical expression as it rises skyward. Each face recesses slightly and rotates, spiraling toward the glass crown, which provides the upper floors with unequaled views of the Puxi Bund from this unique vantage point.

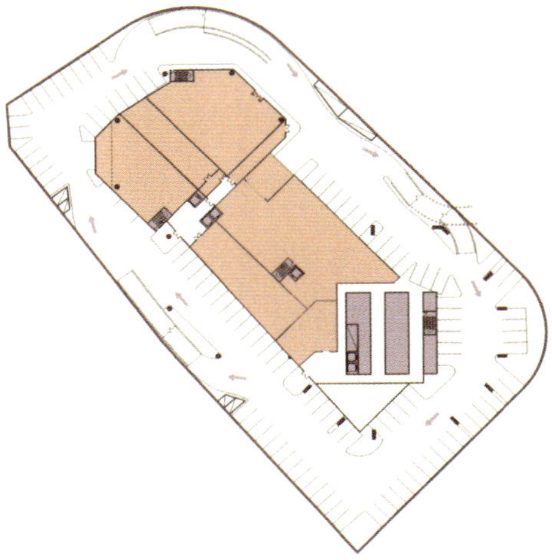

▲ 八顶发展大厦地下室平面图
Professional drawings of floor plan of the Bading Development Tower

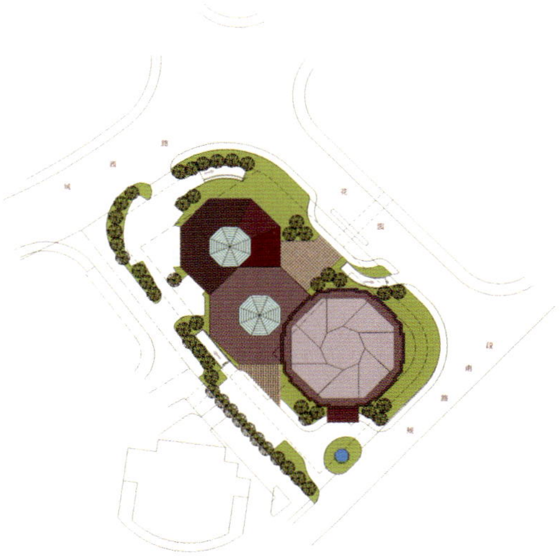

▼八顶发展大厦截面图
Professional drawings of section 1 plan of the Bading Development Tower

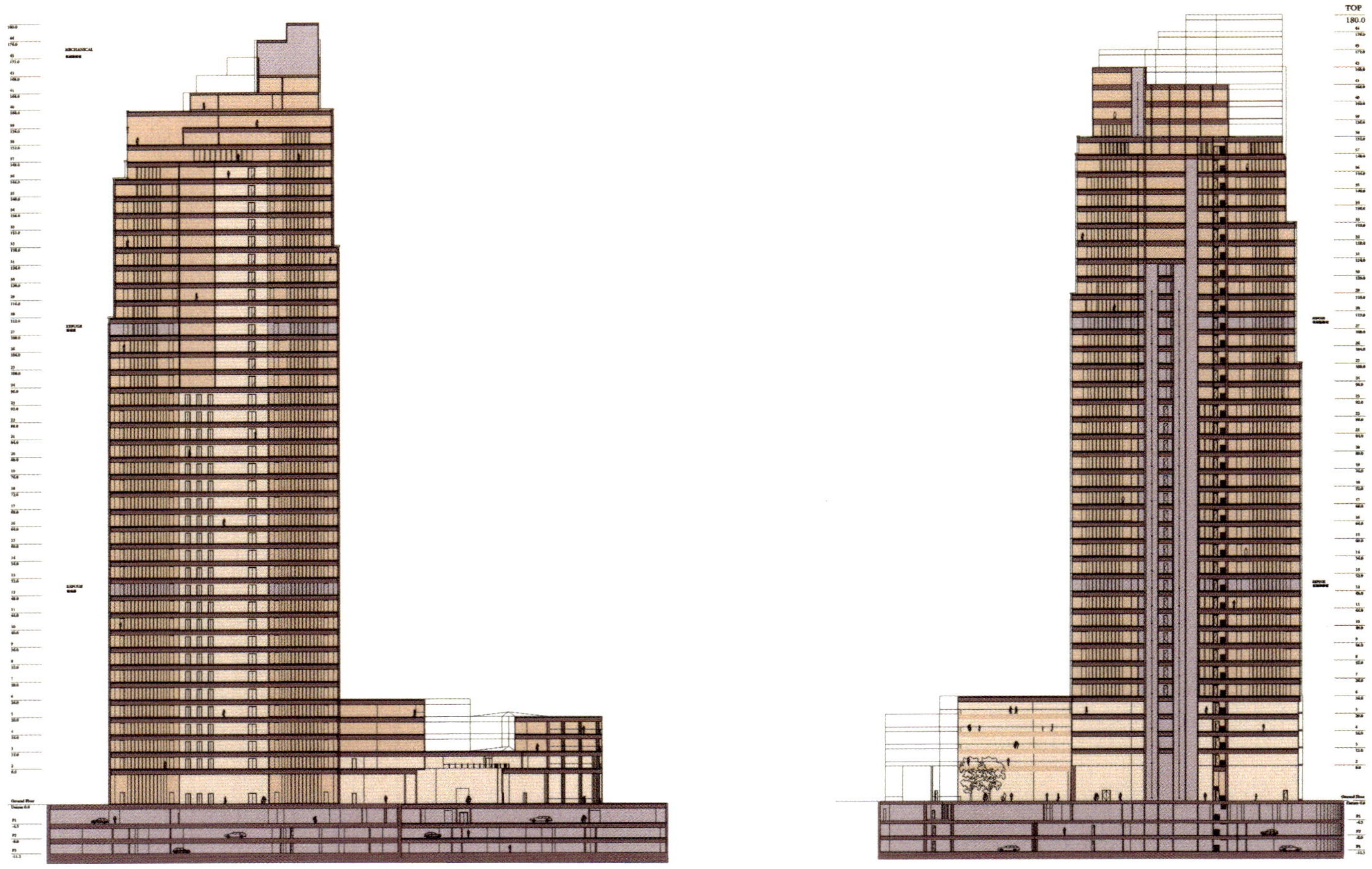

▲ 八顶发展大厦表现图 1
Professional rendering of the Bading Development Tower

▲ 八顶发展大厦表现图 2
Professional rendering of the Bading Development Tower

▲ 八顶发展大厦表现图 3
Professional rendering of the Bading Development Tower

▼ 八顶发展大厦表现图 4
Professional rendering of the Bading Development Tower

▼ 八顶发展大厦表现图 5
Professional rendering of the Bading Development Tower

Turkey's New Embassy in Berlin

土耳其驻柏林大使馆

J. Mayer H. Architects

International Competition 2007
2007年国际竞赛参赛方案

Credits
Team – J Mayer H, Andre Santer, Alex Arnold, Steve Molloy

土耳其驻柏林新使馆建筑向世界展示了土耳其的过去、现在以及未来。

设计重新诠释了土耳其的传统建筑形式和盖瓦贴砖的图案，以创造一个全新的、开放的、孔状的建筑立面，一种被置于建筑内部的中庭和庭院的结构。

大使馆的功能区是根据水平方向划分的，最敏感区域在最上层，公共空间则在入口周围。

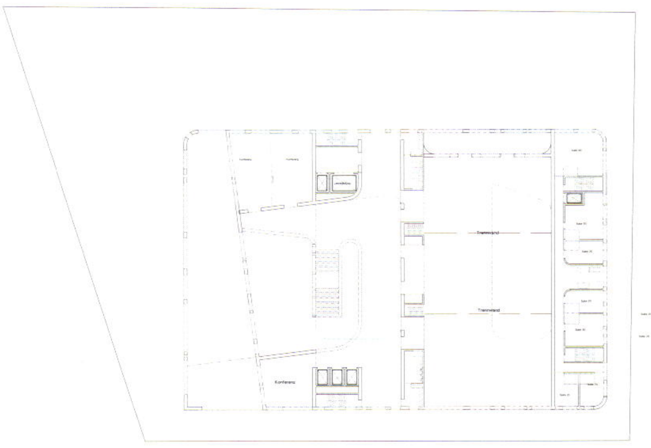

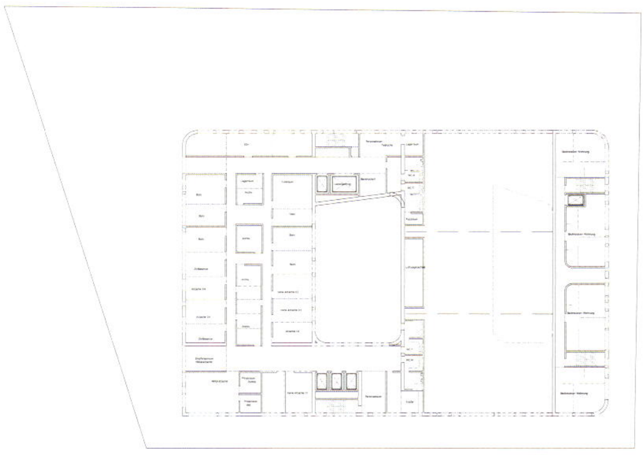

Turkey´s new embassy building in Berlin is to represent Turkey´s past present and future faces to the world.

The design reinterprets traditional Turkish building forms and tiling patterns to create a new, open, perforated facade, a structure carried through into the atria and courtyards of the interior.

The functions of the embassy are layered horizontally, with the most sensitive areas to the top and the public spaces grouped around the entrance.

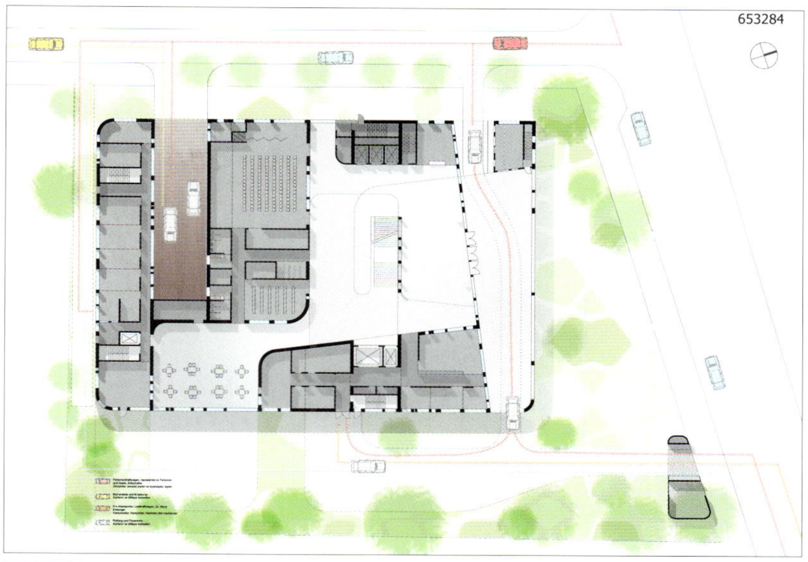

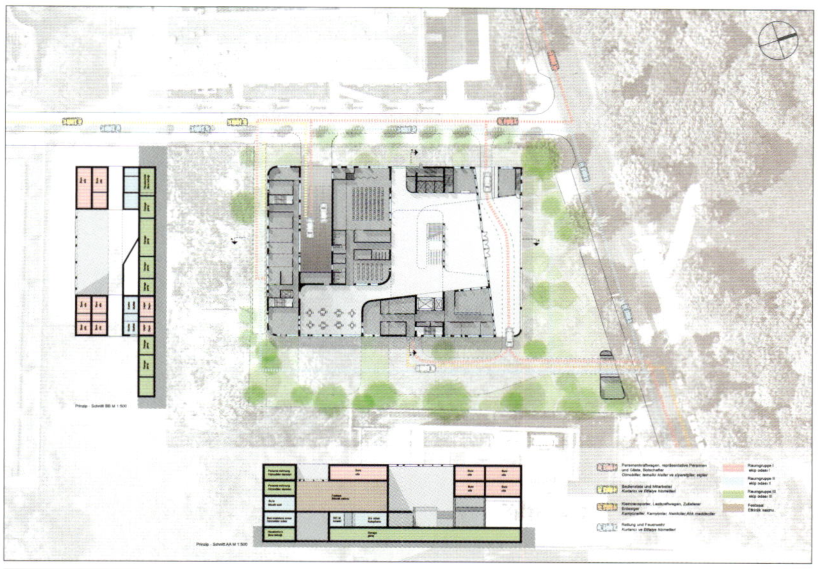

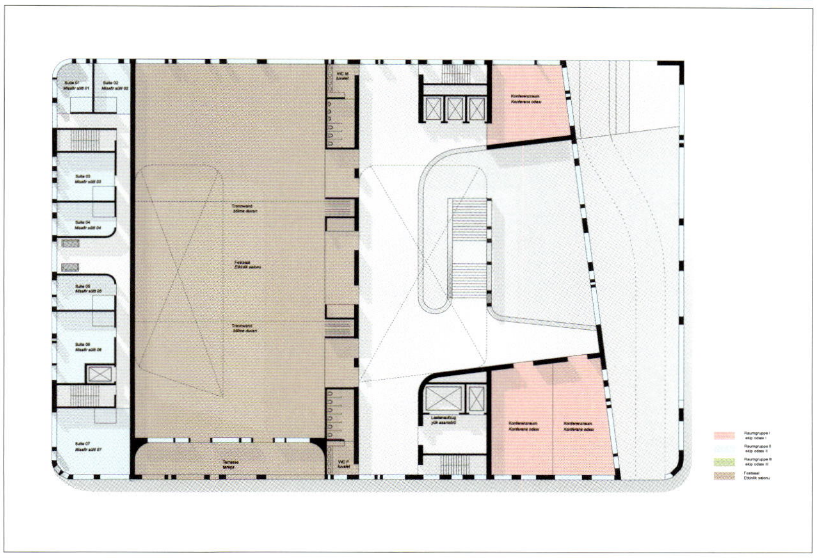

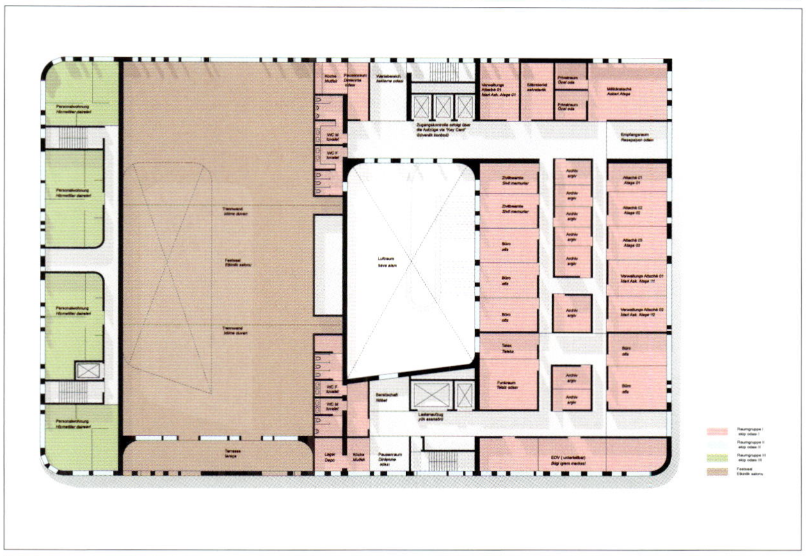

The White Peacock
"白孔雀"多功能建筑体

Ivo Buda Architetto

Designing in Teheran
德黑兰设计竞赛参赛方案

Credits
Project Design: Ivo Buda
Project Team: Ivo Buda , Manfredo Bianchi, Tomaz Kristof
Client: Benetton
Location: Teheran, Iran
Year: 2009
Program: Commercial, Office and Residential

Height: 31m
Levels: Basement (4 levels) + Ground l.+7 levels
Total Floor Area above Grade: 7,125 m^2
Structure: Steel-concrete Structure
Parking Spaces: 114

白孔雀

这个项目的设计基于人类多样性的理念，这不由让人想起孔雀开屏时美丽的羽毛。

孔雀代表着荣耀、不朽与清廉。她是人类最值得崇敬的性格的化身，是正直与美丽的象征。在巴比伦王国与波斯，孔雀被当做王族的守卫者。

建筑设计

孔雀尾巴运用了钢斜撑技术，特殊的透视带领人们领略建筑物各个角度的景致。尽管外形简洁，这个建筑看上去有很多神秘之处——它没有棱角，原本应有的静止被动态栅格代替。这些栅格给人视觉上不断生长的感觉，并包围着一系列透明与半透明的空间。

这个建筑功能多样化且灵活，包括商业、办公与住宅。不同的楼层由一个电梯与楼道系统连接。商业、办公与住宅楼的入口设计采用热情的颜色，在视觉上，这些入口与建筑物楼层垂直"切口"相连。这些"切口"使商业区呈现多层，从办公楼看去，甚为壮观。

结构描述

建筑结构由钢斜撑 "幕帘" 圆周、混凝土台阶、核心区与6个圆柱组成。幕帘是一个三维的框架，在水平与垂直方向都具有强大的支撑力。易受荷载影响的硬质材料可以分散地震能量。这些材料组合在对角线之间交叉点与楼层钢板贴近位置，这使其具有异乎寻常的强大水平力，在抗震测试时具有较高数值的结构系数。混凝土核心区域保证了建筑防风抗震的高强度稳固性。

楼层底板由高黏性钢板与混凝土厚板组合而成，黏土减轻了承重压力，为Rck=300千克／平方厘米，最大厚度下承重1600千克／立方米。横梁为HEB型，斜撑与立柱为盒状。钢斜撑幕帘由S355钢热轧断面制成，接口处是焊接并用螺栓加固的。尽管构造相当复杂，钢结构依然按标准框架设计，接口在车间焊接而成，横梁接合处在现场完成。玻璃铝制不透风外墙构成外立面。

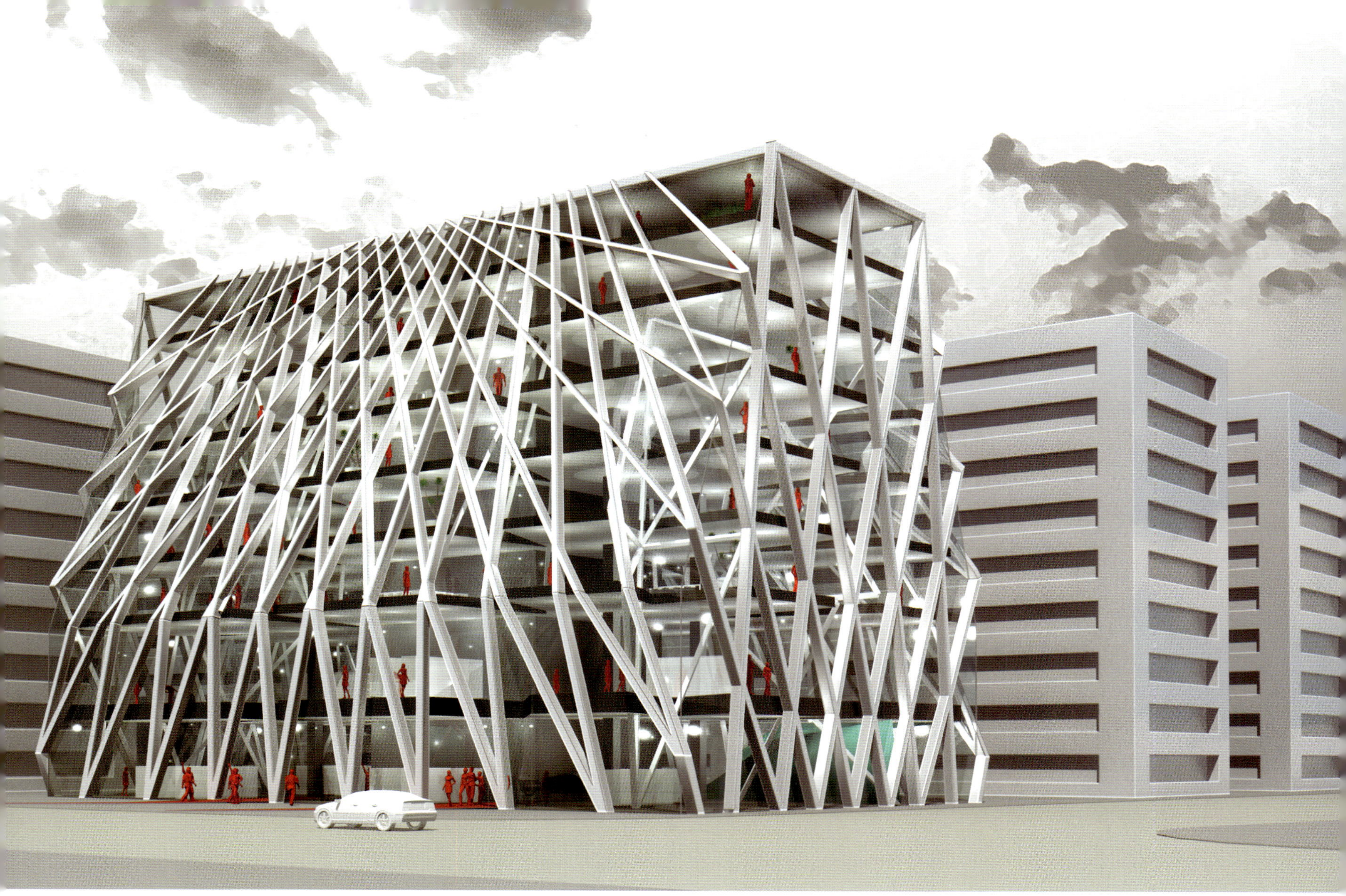

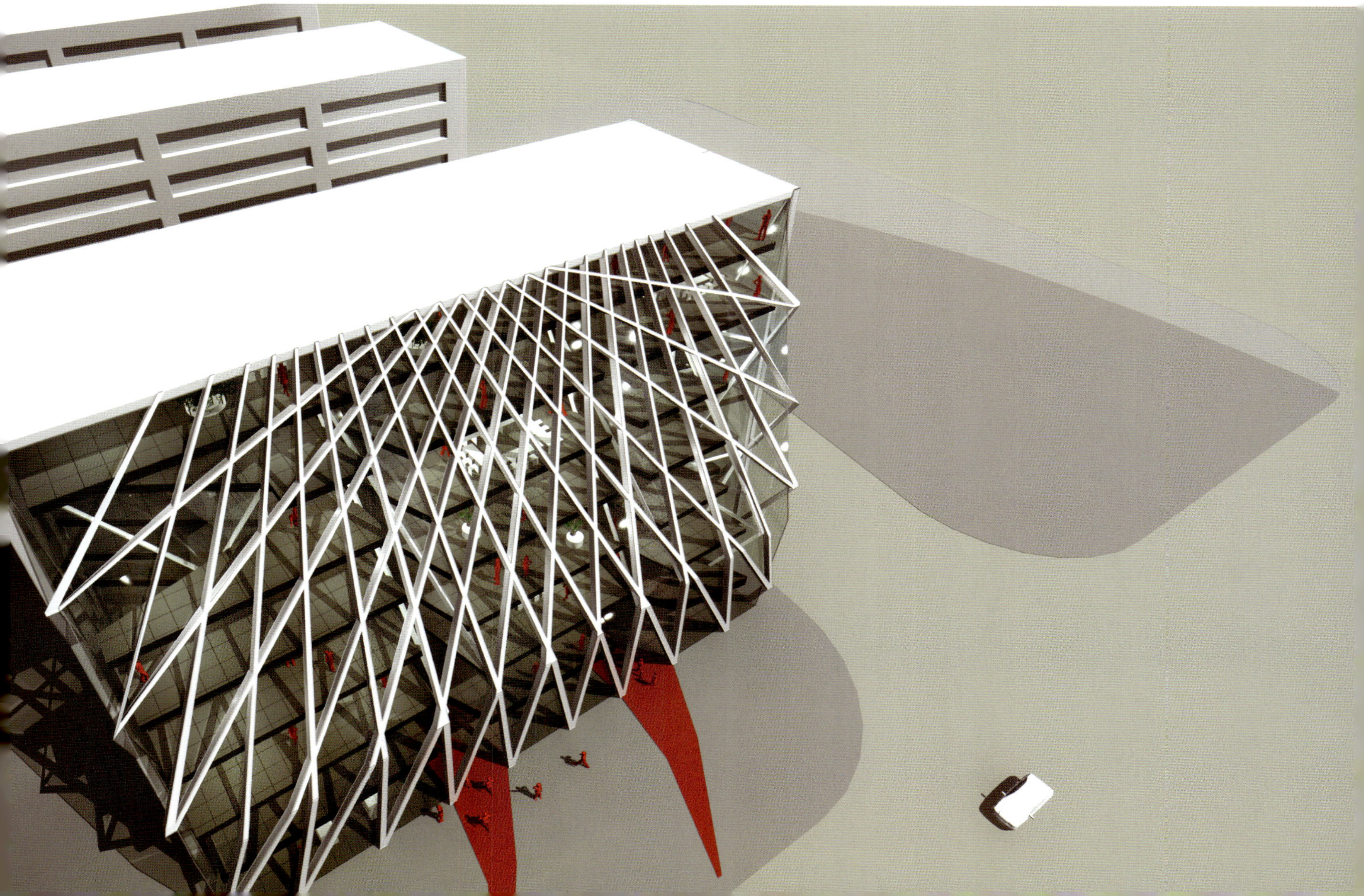

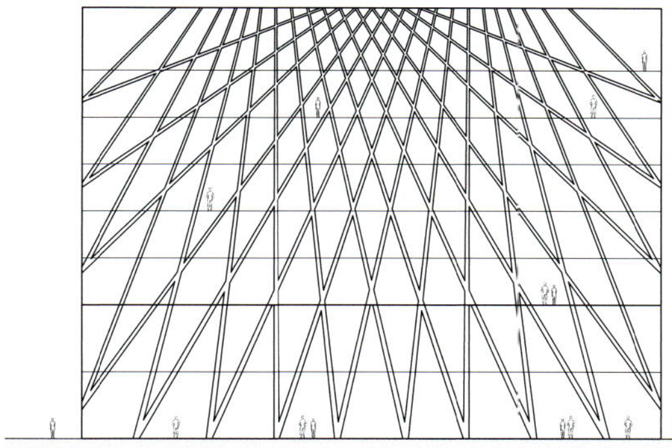

ENTRANCE FACADE

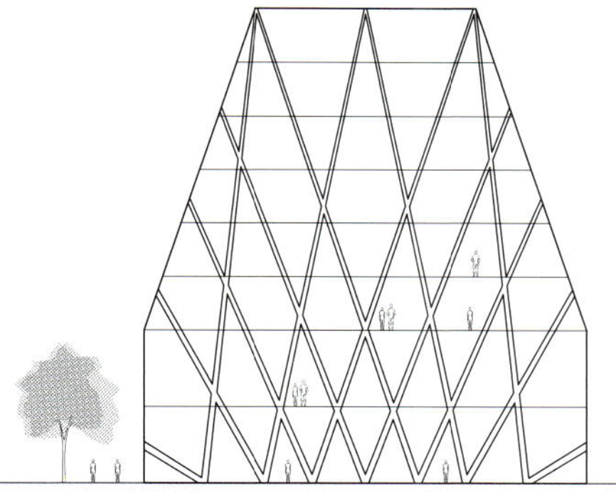

SIDE FACADE

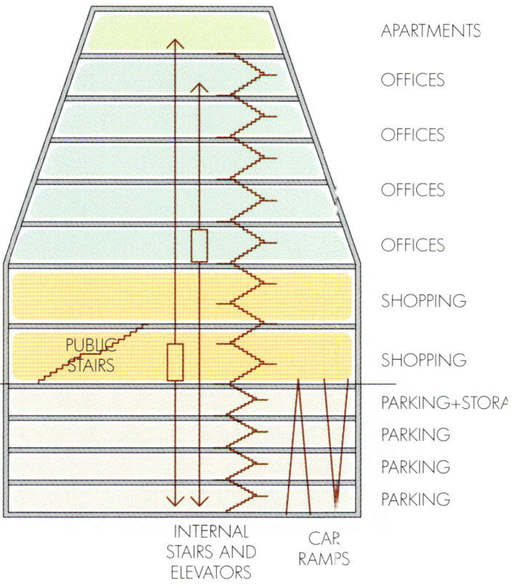

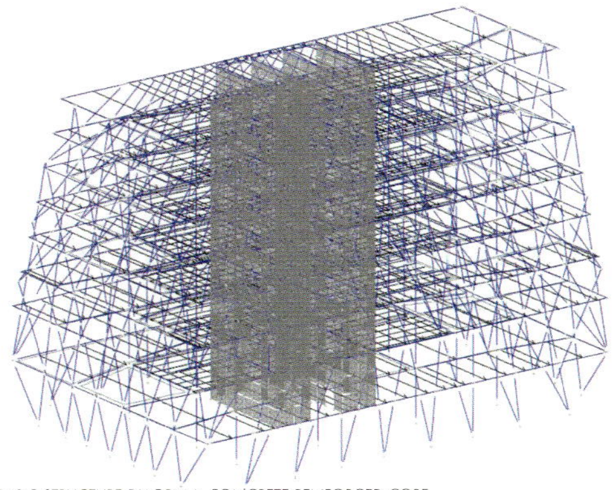

BASIC STRUCTURE DIAGRAM: CONCRETE RENFORCED CORE, STEEL EXOSKELETON, STEEL SLABS

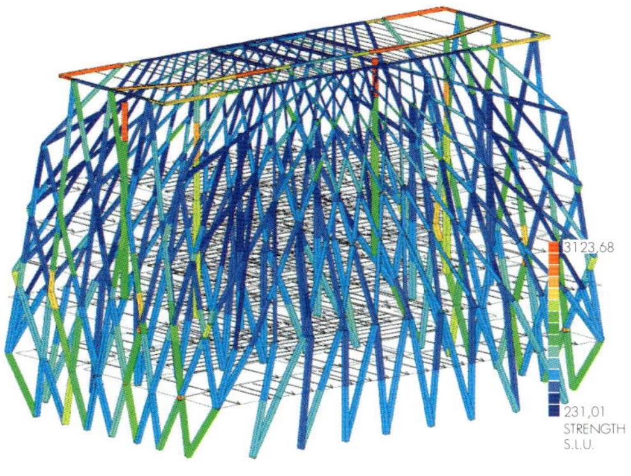

STEEL STRUCTURE, VERTICAL AND SEISMIC LOADS

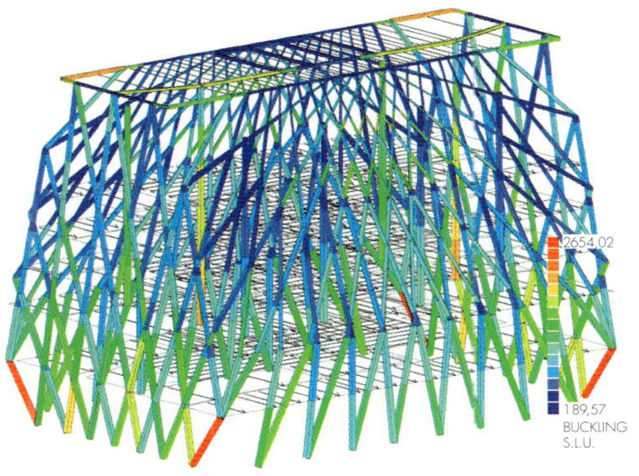

STEEL STRUCTURE, VERTICAL AND SEISMIC LOADS

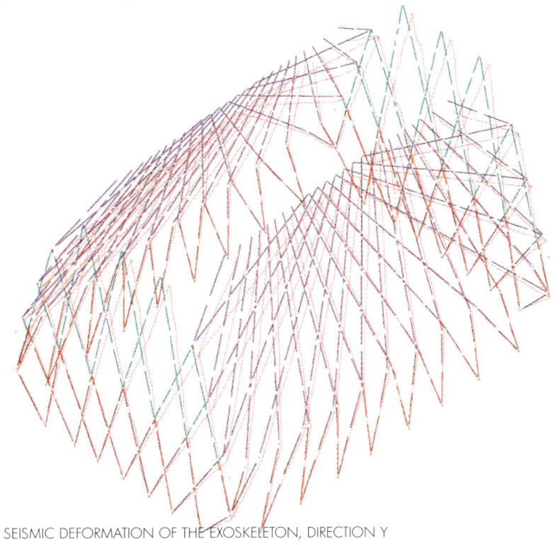

SEISMIC DEFORMATION OF THE EXOSKELETON, DIRECTION Y

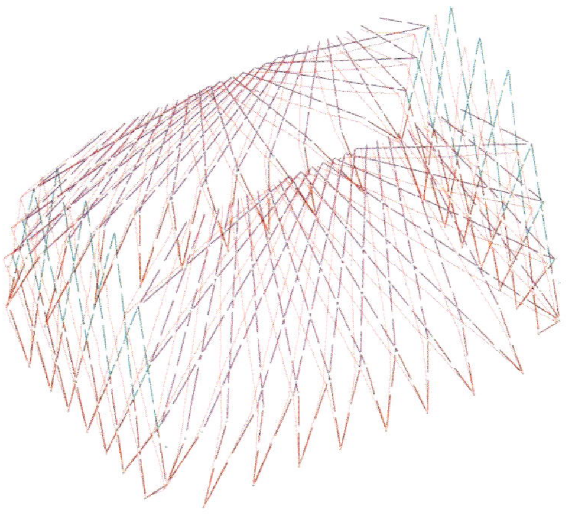

SEISMIC DEFORMATION OF THE EXOSKELETON, DIRECTION X

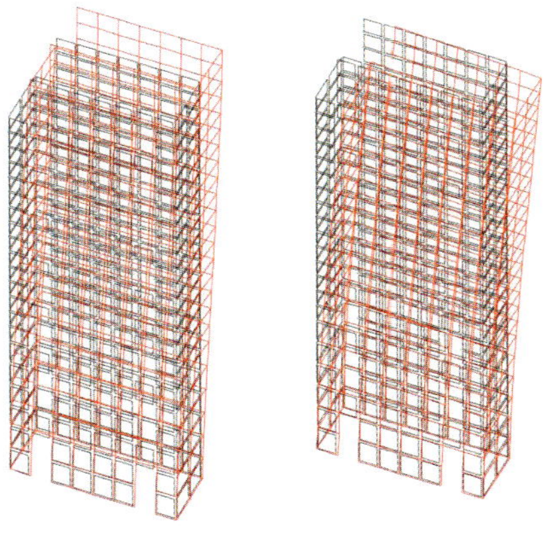

SEISMIC DEFORMATION OF THE CORE, DIRECTIONS Y AND X

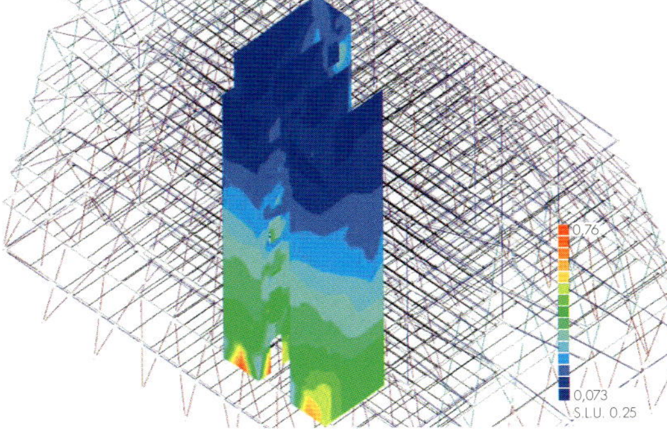

CONCRETE RENFORCED CORE, VERTICAL AND SEISMIC LOADS

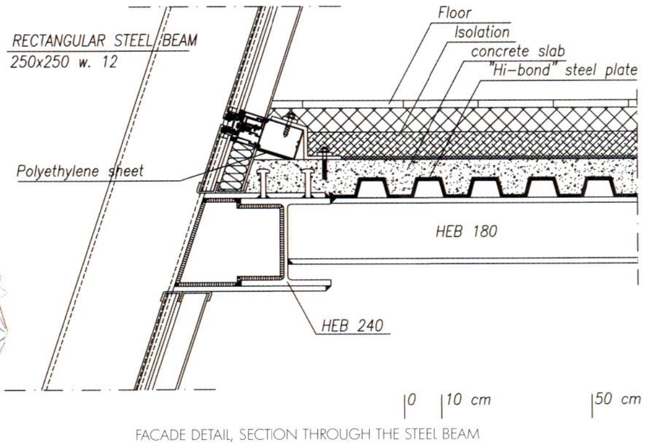

FACADE DETAIL, SECTION THROUGH THE STEEL BEAM

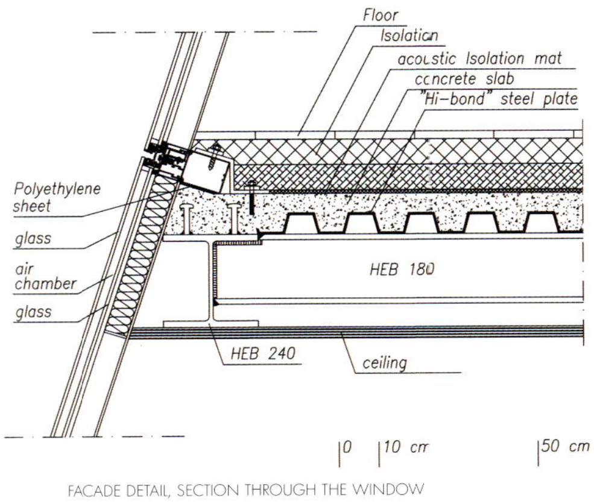

FACADE DETAIL, SECTION THROUGH THE WINDOW

The White Peacock

Today we are called upon to dedicate our resources and capacities towards a society built upon the richness of human diversity that echoes the beauty of the unfolded plumage of the peacock.

Peacock is representative of glory, immortality and incorruptibility. It is a possessor of some of the most admired human characteristics, and is a symbol of integrity and beauty.

In Babylonia and Persia the peacock is seen as a guardian of royalty.

Architecture Description

A strange perspective of the "peacock tale" made of steel diagonal elements invites the visitor to explore the building from all possible angles around it. Although its shape is very simple the building appears to have many mysteries, the corners disappears and the perception of a static volume is substituted by a dynamic grid that seems to grow and to wrap a series of transparent and translucent spaces.

The building is a flexible container for commercial, office and residential activities. Different levels are strongly connected by a system of elevators and stairs. The visitors of commercial, office and residential spaces have dedicated entrances, which are visually connected through vertical "cuts" in the building slabs. This "cuts" provides fascinating multistory commercial spaces and gives attractive views from the offices.

Structure Description

The structure is consisting of the "curtain" of steel diagonal elements on the perimeter, concrete stairs and lift cores, and 6 internal columns. The curtain is a three-dimensional frame with a strong horizontal strength and high capacity for vertical support loads. The presence of many stiff elements subject to cutting loads provides the dissipation of seismic energy. These elements are formed when the intersection between the diagonals are close to the slabs, which makes them eccentric with a strong horizontal strength which for seismic testing allows a high value of the coefficient of structure q. The concrete core guarantees a high stiffness of the building against wind and earthquake. The slabs are of steel plate type hi-bond, cooperating with the concrete slab, lightened with clay expanded Rck = 300Kg/cm^2 and max density 1600 kg / m^3. The floor beams are of the type HEB. Diagonal elements and the columns are of box type. The steel diagonal curtains are made with hot-rolled sections of steel S355, the links are partly welded partly bolted. Although a relatively complex geometry, steel structure is achieved through standard profiles, with the nodes fully welded in the workshop, and the junction of the beams to be executed on site.

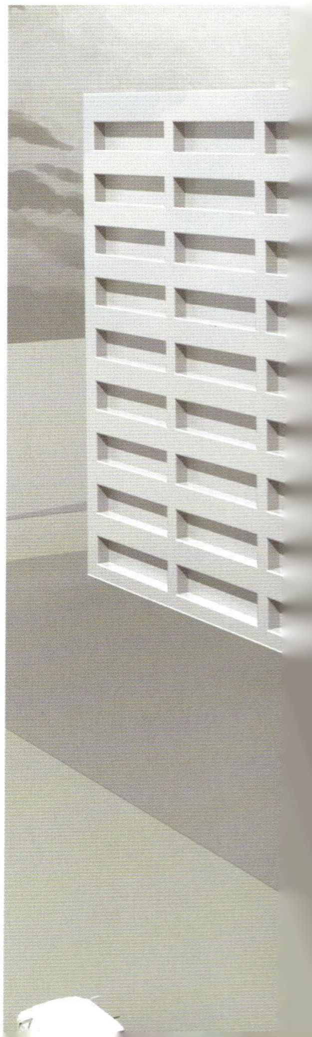

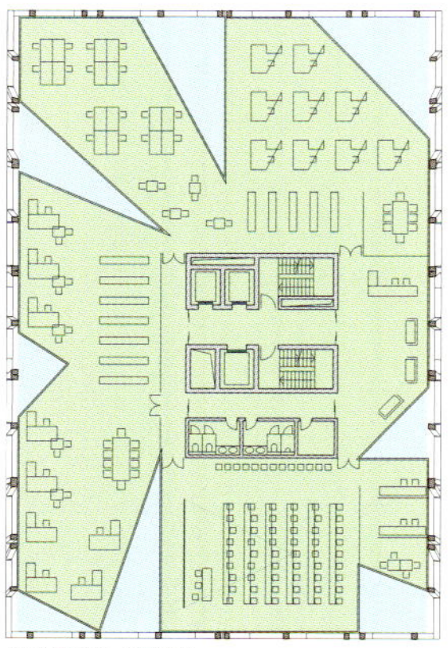
2ND FLOOR - OFFICES

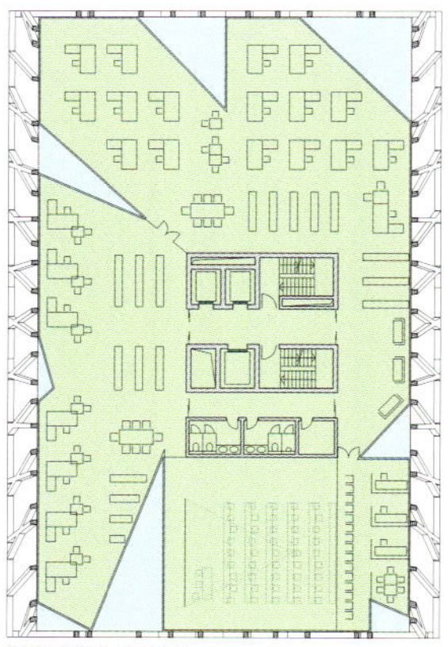
3RD FLOOR - OFFICES

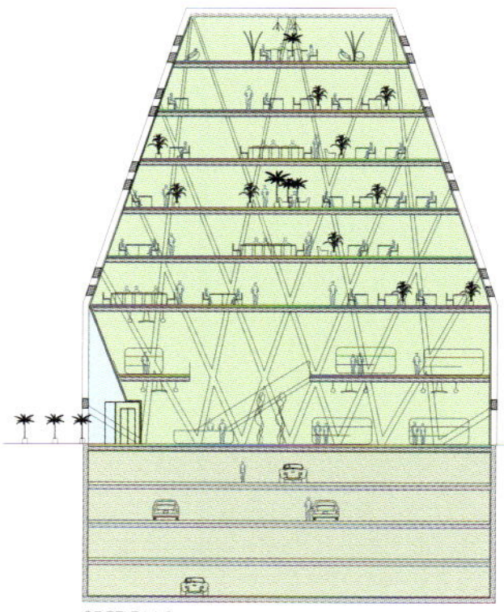
SECTION 1

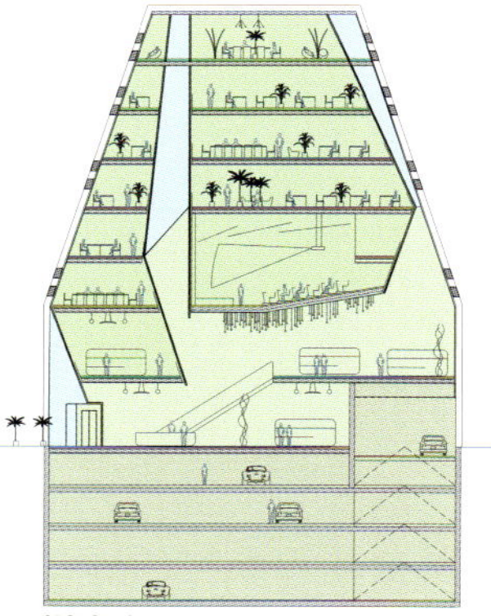
SECTION 2

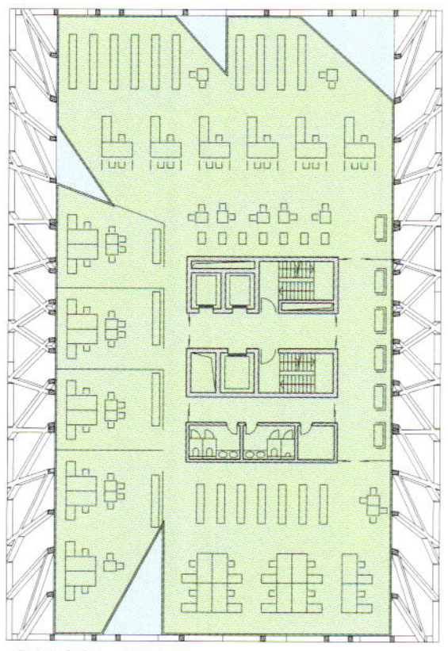
4TH FLOOR - OFFICES

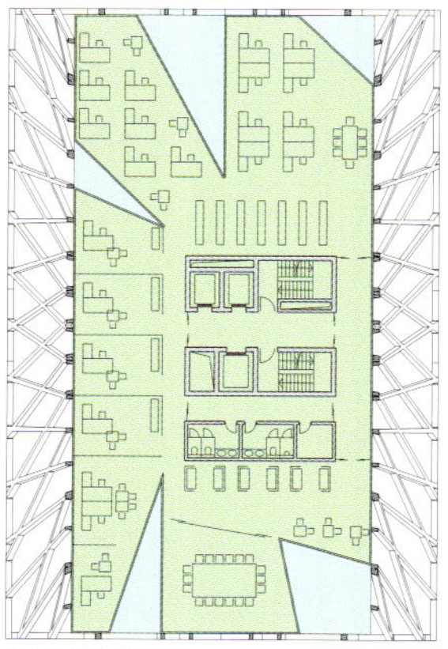
5TH FLOOR - OFFICES

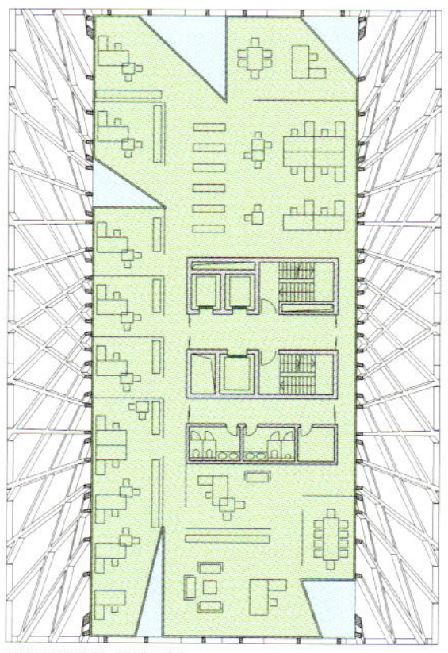
6TH FLOOR - OFFICES

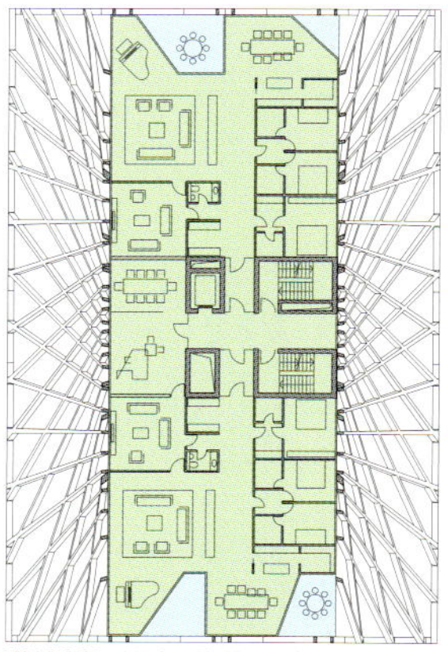
7TH FLOOR - LUXURY APARTMENTS

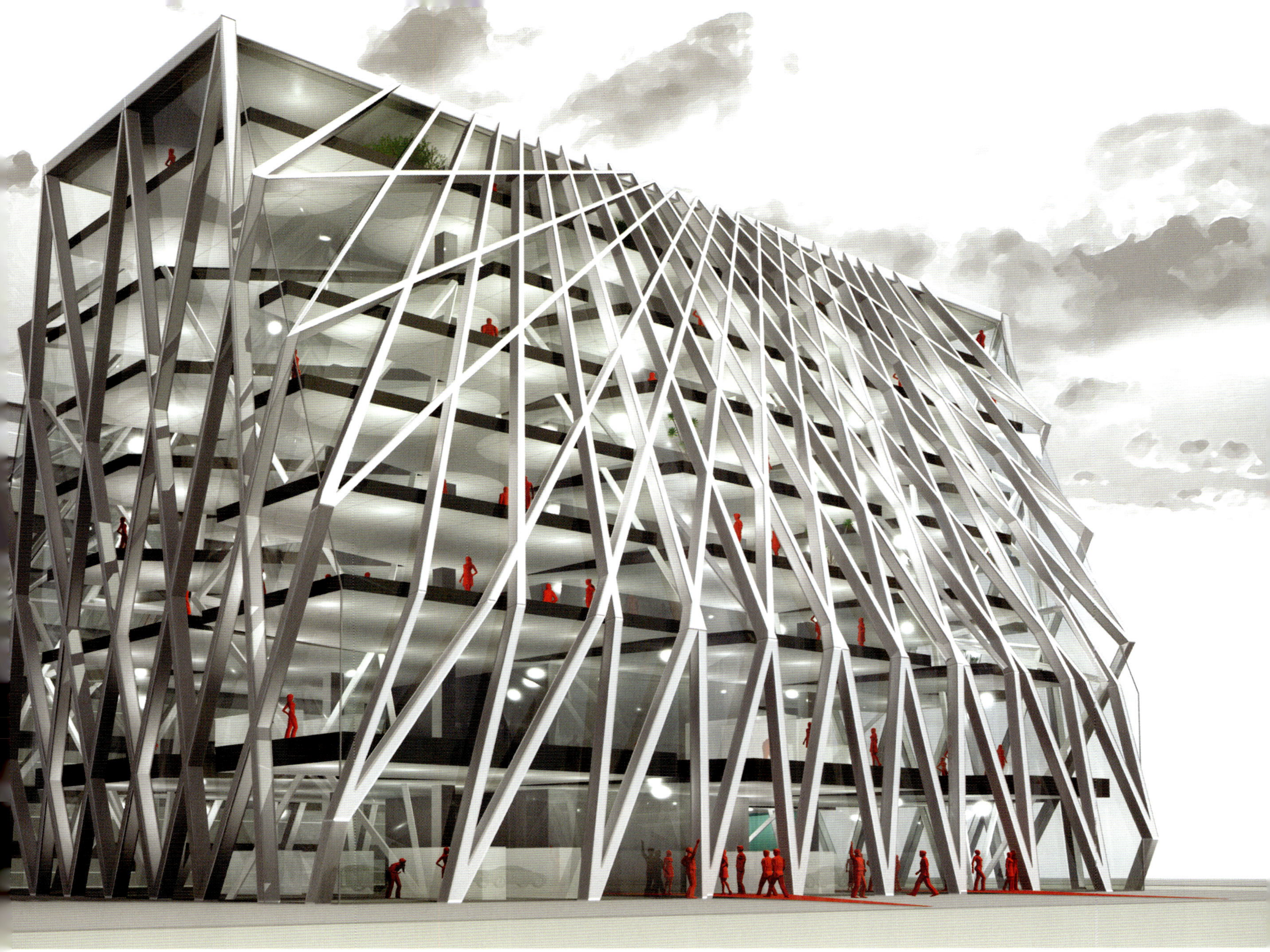

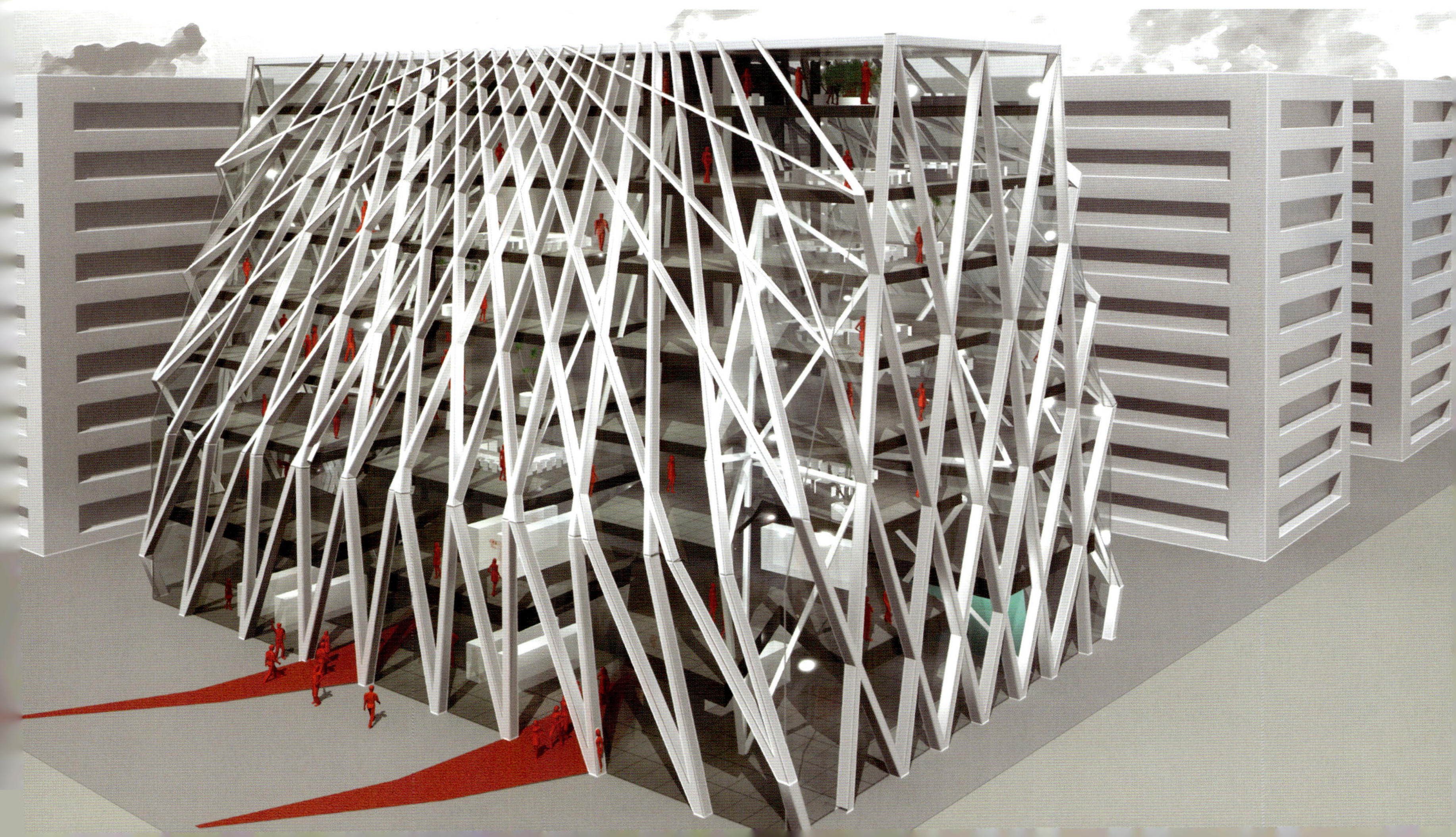

KRYSTALKLAR ("Crystal Clear")

"透明的水晶"雕塑建筑综合体

C. F. Møller Architects

First Prize in Architectural Competition, 2009
2009年建筑设计竞赛一等奖

Credits
Location: Oslo, Norway
Architect: C. F. Møller Architects in collaboration with Kristin Jarmund Arkitekter
Landscape: C. F. Møller Architects in collaboration with Kristin Jarmund Arkitekter
Competition Collaborators: ATKINS, Erichsen & Horgen AS, MIR (illustrations), Oslo Modellverksted (model)
Area: 92,000 m² (75,000 m² above ground)

挪威最重要的交通枢纽位于奥斯陆市中心，而这座高层建筑就坐落于此，可以欣赏到海滨和峡湾的美丽景色。设计是要建立一个地标性的雕塑建筑综合体，与首府低层建筑和谐共存。三座建筑的高度分别约为110米、65米和55米，沿着枢纽区域的边缘依次排列，最高的一栋与现存的奥斯陆广场和波斯格瑞白格特塔一排，而另两座较低的与城市连接。

三座建筑外部都具有清晰垂直的轮廓，大范围的开放区域以及凹进的区域形成窗户，可以观赏到不同的景色。相反地，朝里的立面由堆叠的釉面体组成，自由排列，形成一个棱镜一般的水晶外观。这种布局不仅保证了从这三座建筑可以欣赏到水边的景色，还保证了从远处也可以欣赏到水边的景色。

在建筑之间，一个包含商店和餐厅的两层的底座通过斜路、平路以及楼梯与街区连接，形成跌宕起伏的景观。这个底座被高层建筑围起，创造了一个平静的城市花园，为城市及居住者提供休闲空间和咖啡厅。这三栋建筑经过高度灵活的设计，可以作为办公、酒店甚至住宅之用。

A high-rise development, located at Norway's most important traffic hub in central Oslo, and with fantastic views of the waterfront and fjord-landscape beyond. The idea is to create a landmark sculptural ensemble of towers, yet observe the harmony with the surrounding, low-rise urban fabric of the capital. The three towers of approx. 110m 65m and 55m height, are arranged along the edges of the site, and the tallest tower is aligned with the existing nearby Oslo Plaza and Postgirobygget towers, while the lower buildings form the link to the city.

The three towers have clear-cut and vertical elevations to the exterior of the site, with large openings and setbacks forming windows to selected viewpoints. In contrast, the elevations towards the interior of the site are composed of stacked, glazed volumes, freely arranged to form a prismatic and crystalline appearance. The layout secures the views over the water, not only for the three new buildings but also the city beyond.

In between the towers, a two-storey base containing shops and restaurants forms an undulating landscape that connects to street level via ramps, plateaus and stairs. This base creates a calm urban garden, framed by the tall buildings, with recreational space and cafes for the city and the buildings occupants. The towers are designed with a high degree of flexibility to house offices, hotels and possibly housing.

entra
posten
steria
oslo
Radisson SAS

+ 100 m
+ 70 m
+ 50 m
+ 15 m

entra
EIENDOM
posten
steria
CopyCat
Radisson SAS

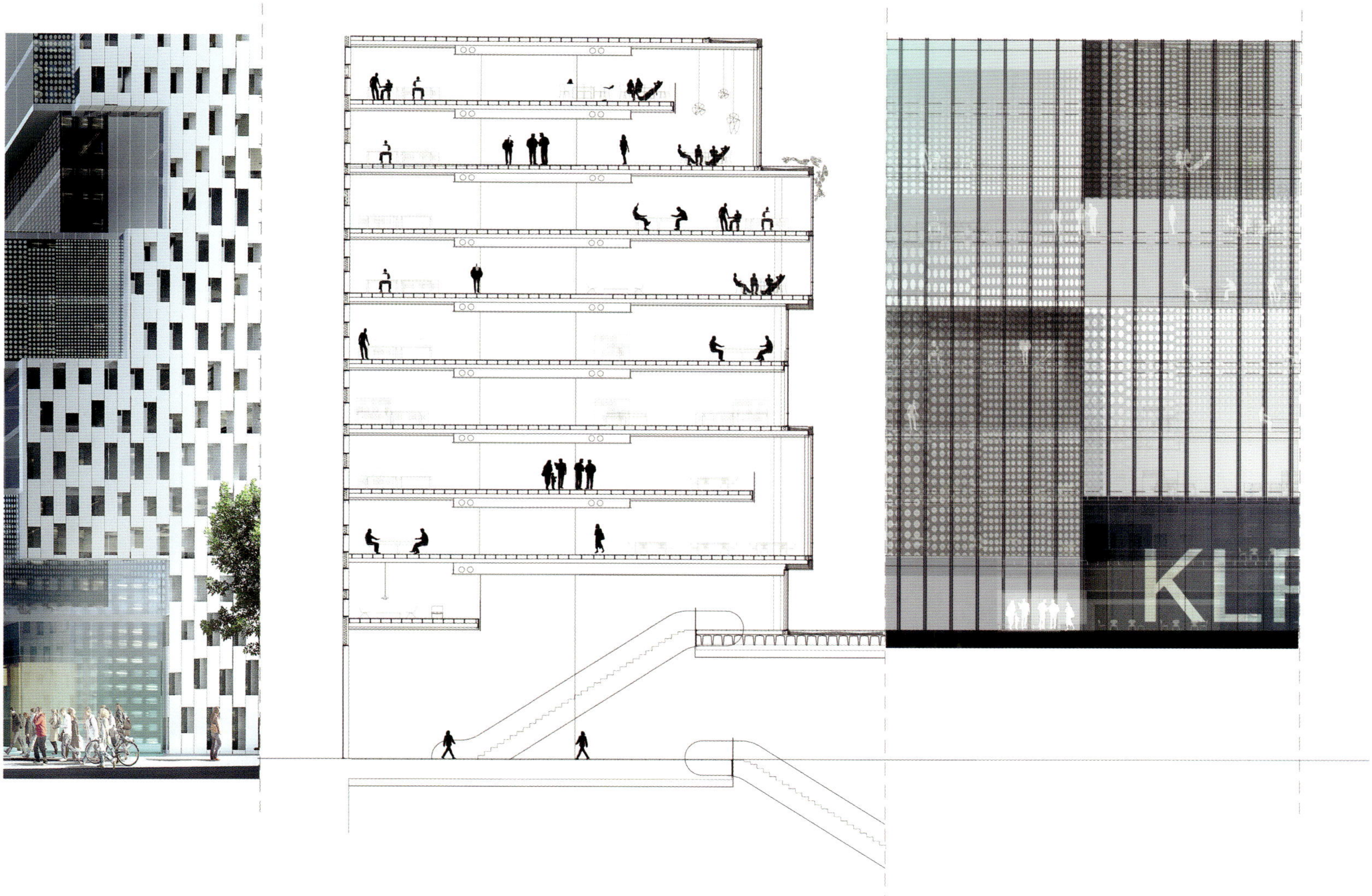
KLP

CAFE BABYLON

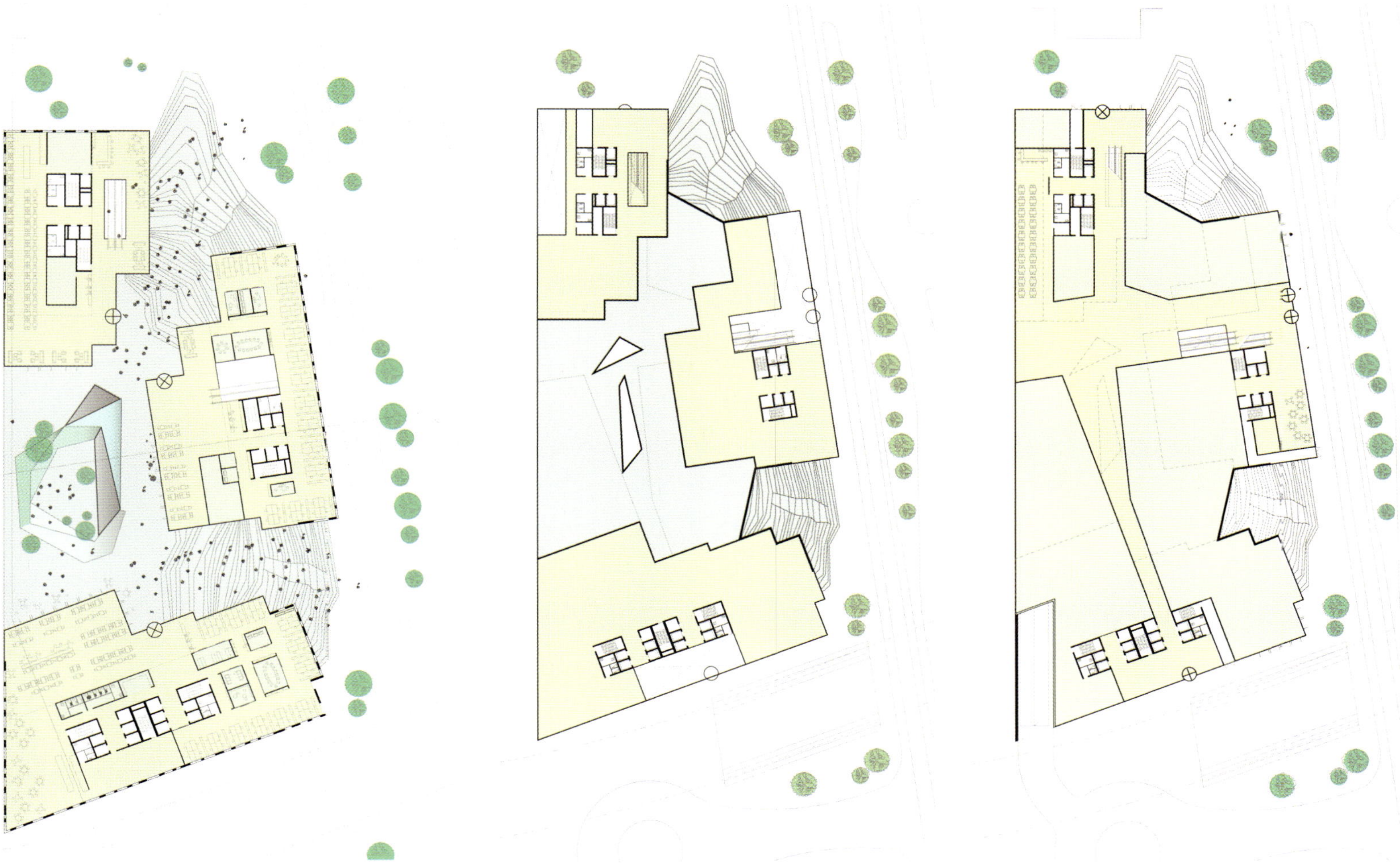

Continental Tower, Stockholm

斯德哥尔摩大陆塔

C. F. Møller Architects

Credits

Location: Centralstationen, Stockholm, Sweden

Landscape: C. F. Møller Architects

Area: 30,000 m² (including the existing hotel)

Year: 2009-Now

Images: C. F. Møller Architects

斯德哥尔摩建筑区建设是与其城市轨道交通的扩充建设相联系的。

这个100米高，约30层的建筑综合体将包括一个酒店，若干办公区及附带的小部分住宅区。在该城市主要的旧地铁站森特林的对面，建有一个新地铁站，而该建筑便设有直接到达新地铁站的入口，由此这个建筑也将成为斯德哥尔摩新地铁的一座里程碑。

C. F. Møller 的设计基于一个多功能的理念，意在建立一座同时具有多项用途的建筑。建筑从斯德哥尔摩这座城市中获取灵感，如坐落在该建筑右侧的斯德哥尔摩最大的教堂Klara教堂，就是该城市中建筑的典型代表。

该项目重新诠释了城市美丽的市政大厅和经典的教堂尖塔，它们一同构成了该城市的历史性地标，也创造了斯德哥尔摩的历史天际线，同时使森特林周围城区的独特的城市感集中体现。

The tower block construction in Stockholm has been planned in connection with the extension of the City Line rail link.

The complex of 100 meters, the equivalent of approximately 30 stories, will include a hotel, offices and possibly housing. The building will be equipped with an entrance to a new station at street level directly opposite the city's old main metro station, Centralen, and will therefore be a landmark for Stockholm's new metro.

C. F. Møller's design is based on a concept of multi-function, a building where many varying uses can take place simultaneously. The tower draws inspiration from the city of Stockholm, an example of which is Stockholm's largest church Klara kyrka situated right next to the site.

The project reinterprets the city's beautiful town hall and classic church steeples, which combined make up the city's historic landmarks and create Stockholm's historic skyline, but at the same time integrates the unique urban feeling defining the area around Centralstationen.

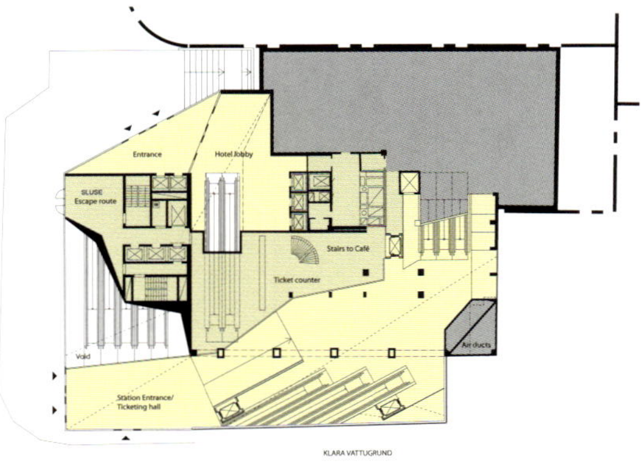

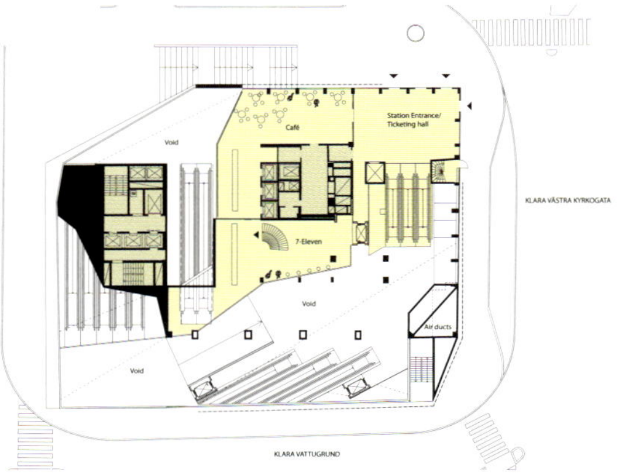

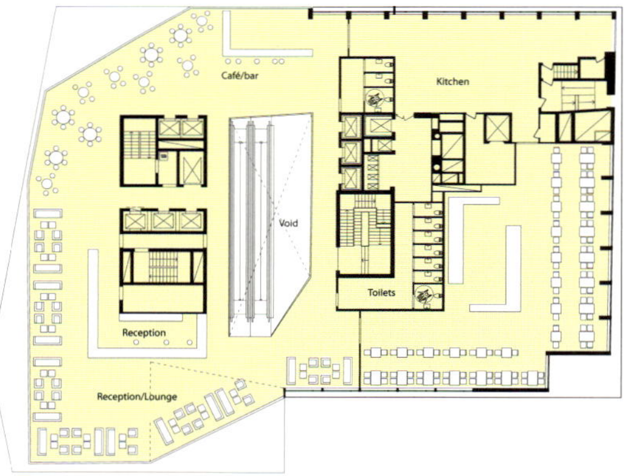

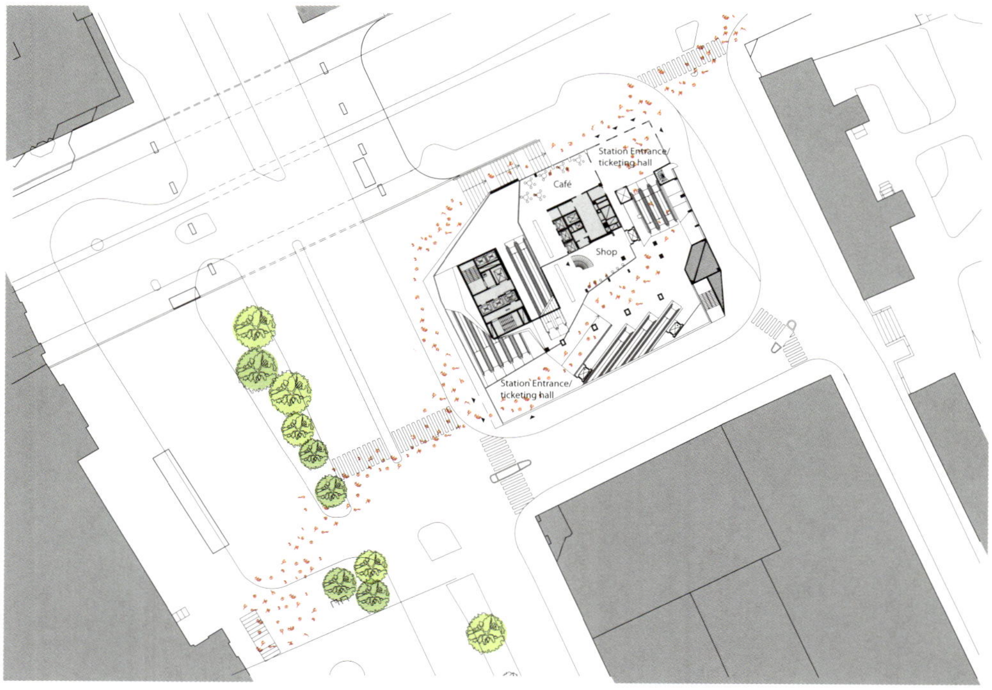

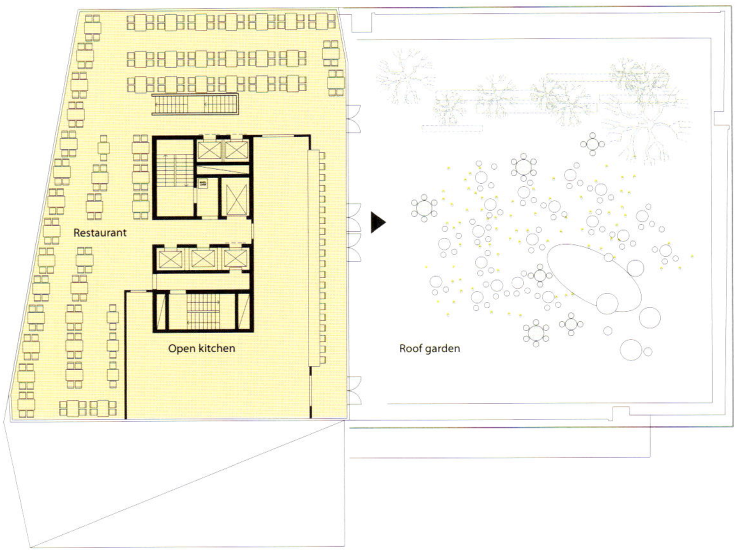
Restaurant
Open kitchen
Roof garden

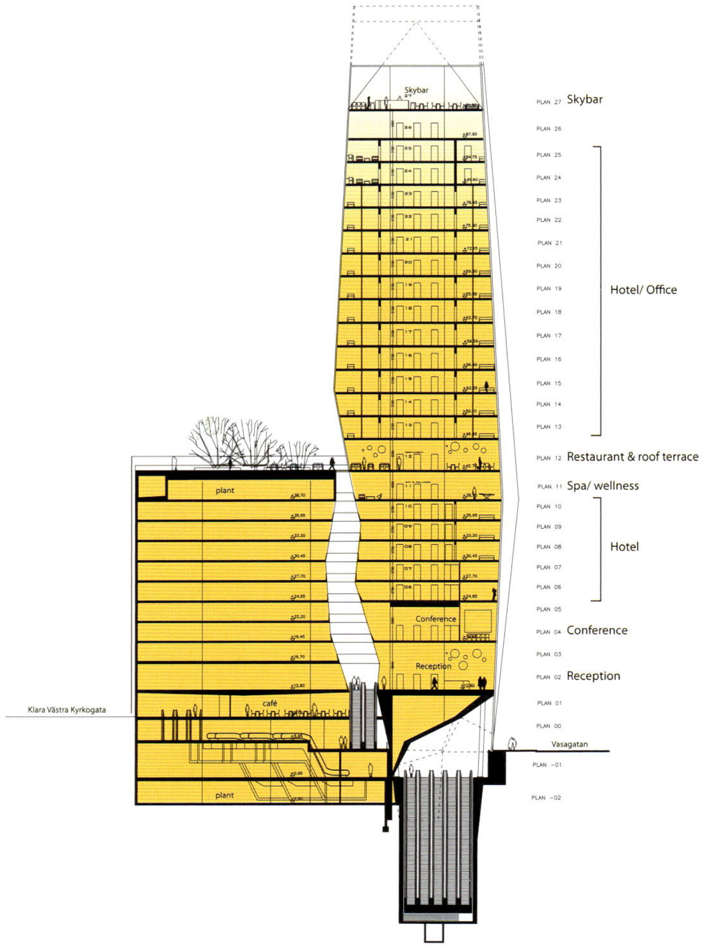
Skybar
Hotel/ Office
Restaurant & roof terrace
Spa/ wellness
Hotel
Conference
Reception
Klara Västra Kyrkogata
Vasagatan
plant
café
plant

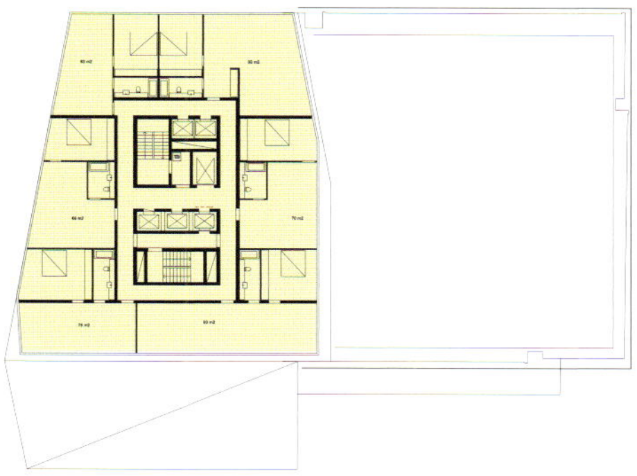

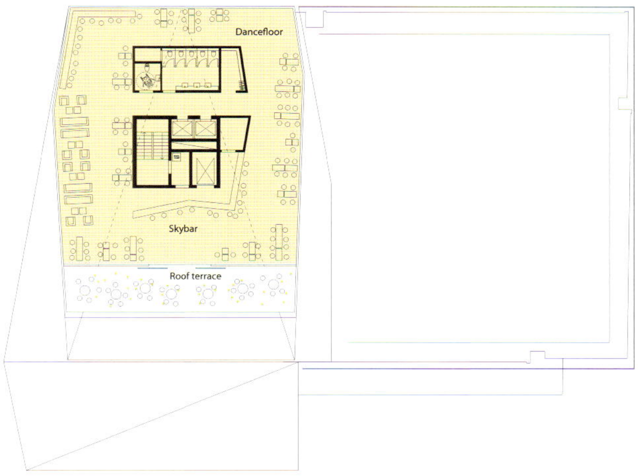
Dancefloor
Skybar
Roof terrace

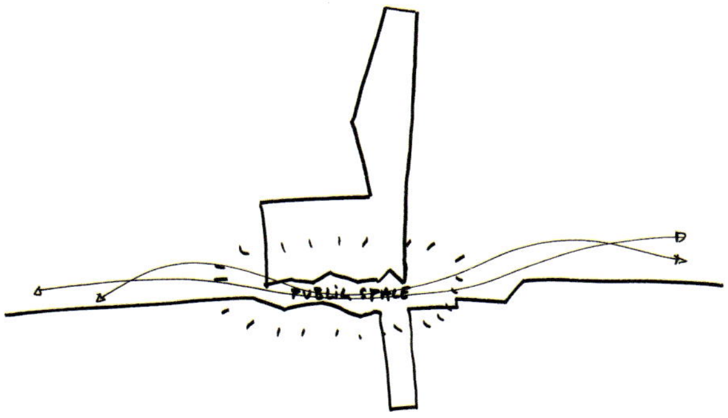
PUBLIC SPACE

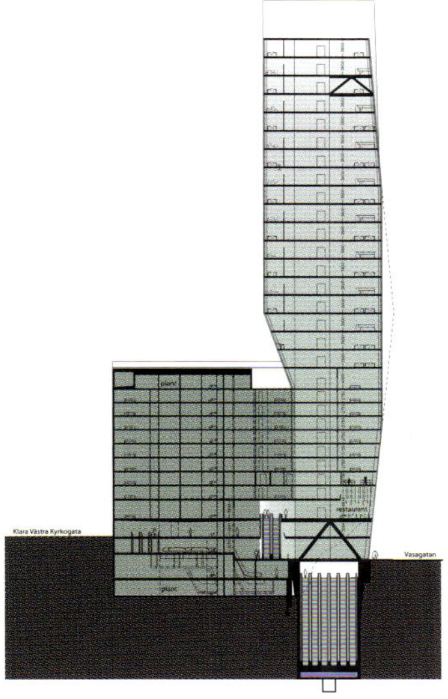

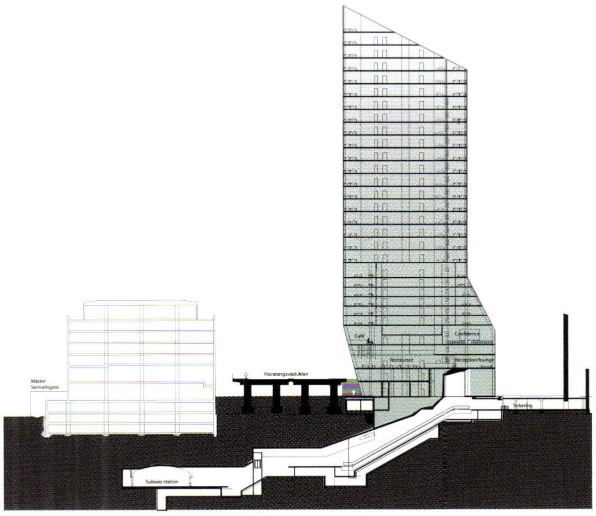

Manifold Dubai

多样的迪拜——蒂森克虏伯集团大楼

International Competition, 2009
2009年国际设计方案赛参赛方案

Credits

Project Type: Observation Tower
Location: Dubai, United Arab Emirates
Area: 1,000 m^2
Client: Thyssen Krupp Company

Principals in Charge: Marcelo Spina and Georgina Huljich
Project Designers: Hironori Kamizono and Geoffrey Elander
Project Assistants: Jeeyea Kim and Nathaniel Moore

迪拜的新标志？

设计师对于该项目的设计基于一种区别性与独立性的角度，表达了一种与当前设计完全不同的理念。

对于蒂森克虏伯集团大楼的设计，设计师提出了开放性策略：将主功能区域与其他相关区域隔开，使其更加灵活。不再将又长又厚重的核心部分与观景平台牵强地结合在一起，从而突出大楼的不同——风格上，全球的大同与当地的特殊恰当融合；颜色上，淡紫与深紫、青灰与蒙胧自动感结合；材料上，普通与复杂适宜对比；技术上，低端与高端相互补充，从而消除对立。

项目主要由三个独特的部分组成：立柱、五个核心区与桥梁骨架。

立柱位于角落里，从扎比尔公园两侧都能到达，这样可以缓解附近公园的人流拥挤的状况。立柱里面前厅位置设有一个小小的礼堂、礼品店，与一个大型的等候区融合在一起。

五个核心区包含大楼主要设施，构成大楼脊柱与运输系统。核心区以占用平台为基准等距离分布，三个出口楼梯井道占领了整个核心区域的延伸部分，供失火或者机械故障时使用。另外，空气调节管道与电力系统也设在核心区域。大型的制冷设施设在桥梁骨架服务区域，可直接通往核心区。

核心区采用可重复利用底层模块化金属框架而制成的预应力现浇混凝土建造。大楼内墙为30～50厘米厚。所有用到混凝土的地方都将用飞灰代替水泥，以减弱对环境的污染。桥梁骨架稳定着外形窈窕的大楼，呈对角斜向移动，与核心区域拓扑连接。

桥梁骨架主要包含一些间接连接的方案性设施。这些设施均向中间聚拢，同时，休息室与憩坐处自然地分布在岔道上。桥梁骨架的外缘面积较小，只容小数目人群，可欣赏到美景，并为建筑物提供自然通风。内部设有自动扶梯，作为最重要的交通与流通工具；它连通所有楼层，并且可以欣赏到迪拜那美妙绝伦的景致，从沙漠到小溪，从海岸线与岛屿到高楼大厦。每一层都设有一个阶梯式的观景平台，由于是斜坡式的，平台呈现出完美私密的口袋状空间，可以体验欣赏到全景的愉悦。

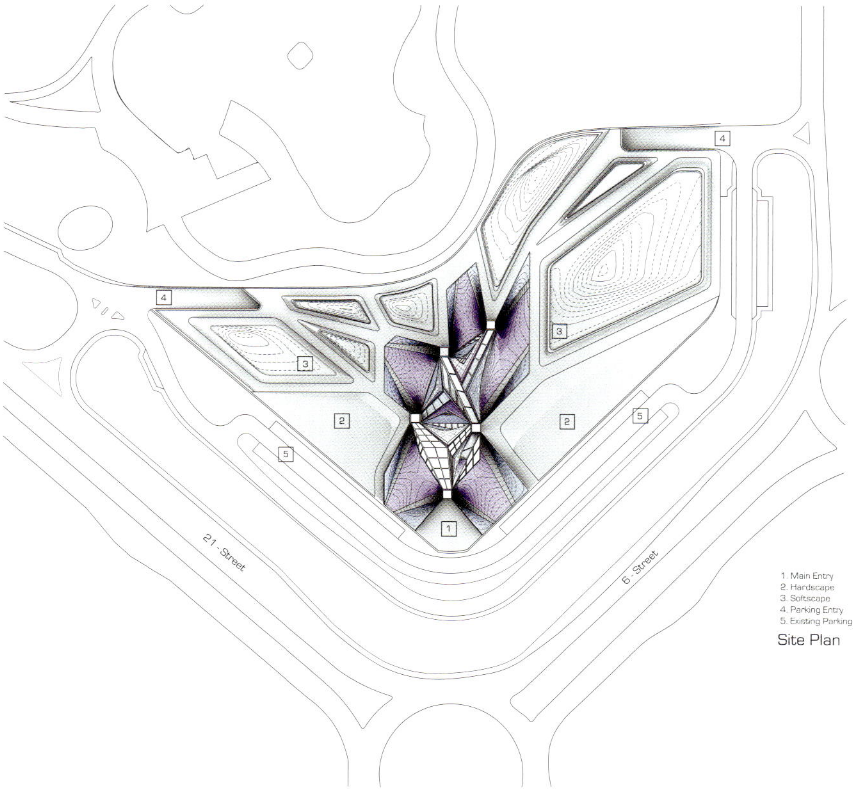

1. Main Entry
2. Hardscape
3. Softscape
4. Parking Entry
5. Existing Parking

Site Plan

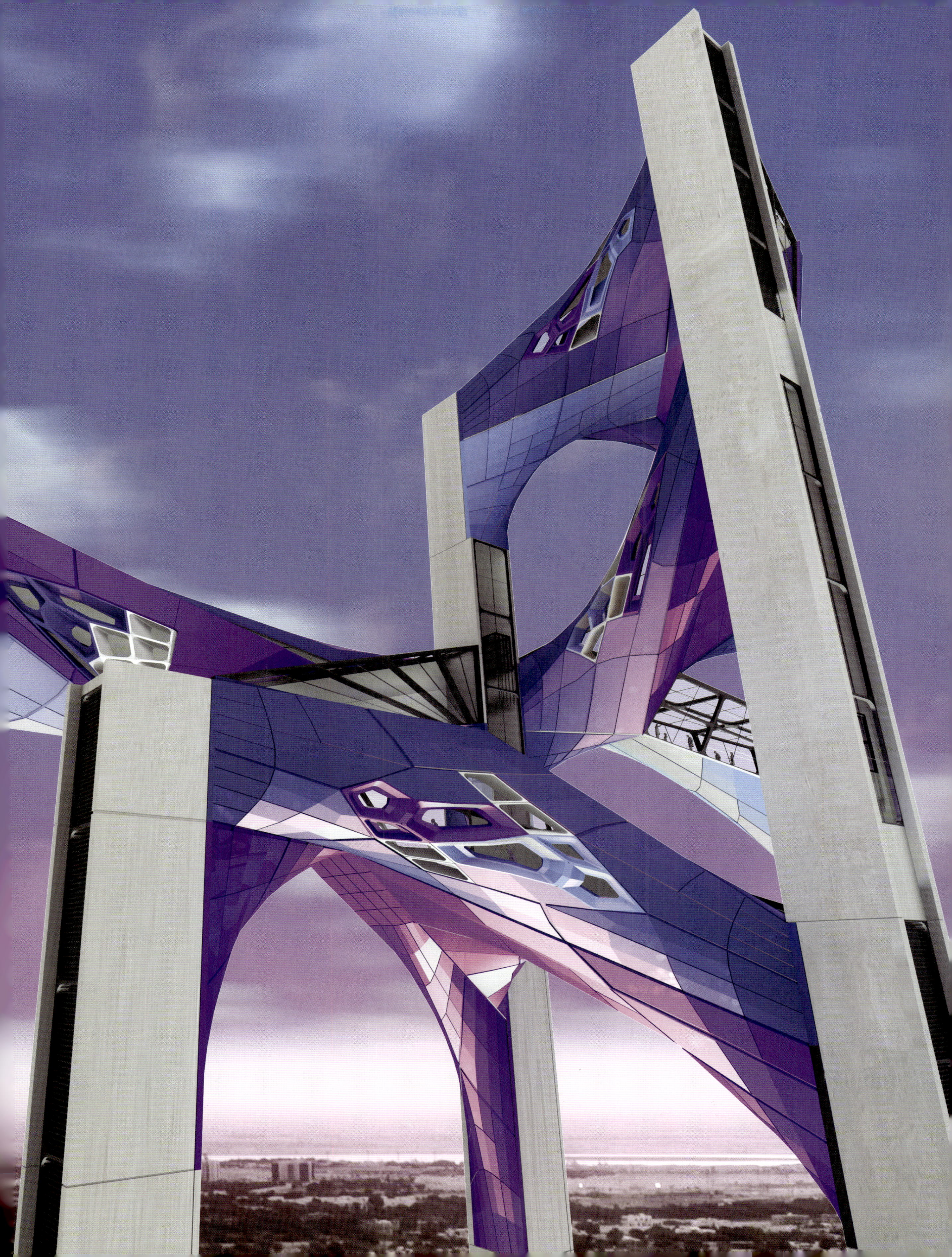

Sequence Diagram

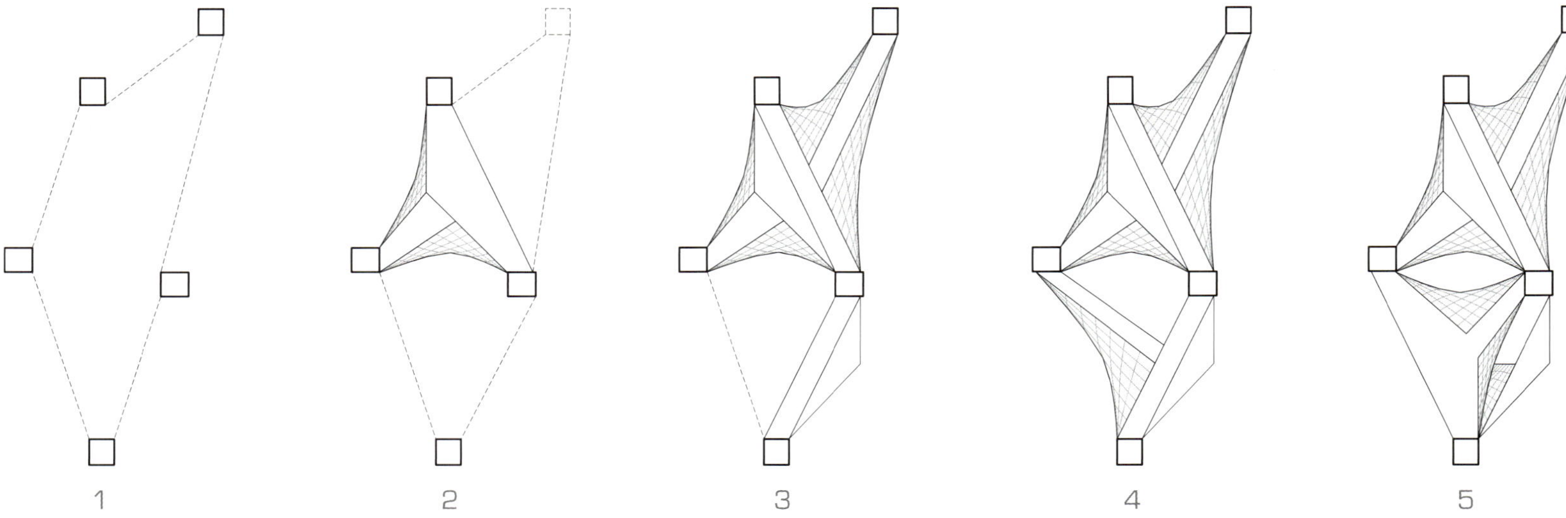

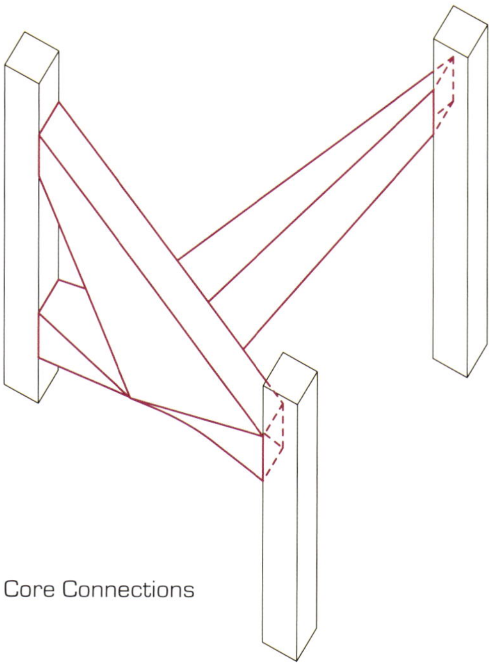

Core Connections

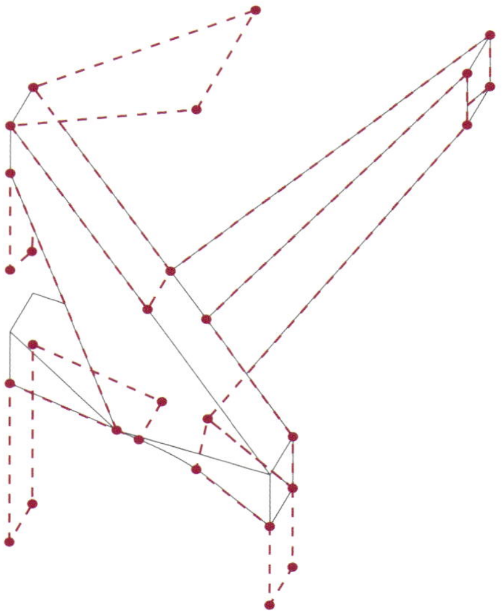

Shell Extrapolation Network

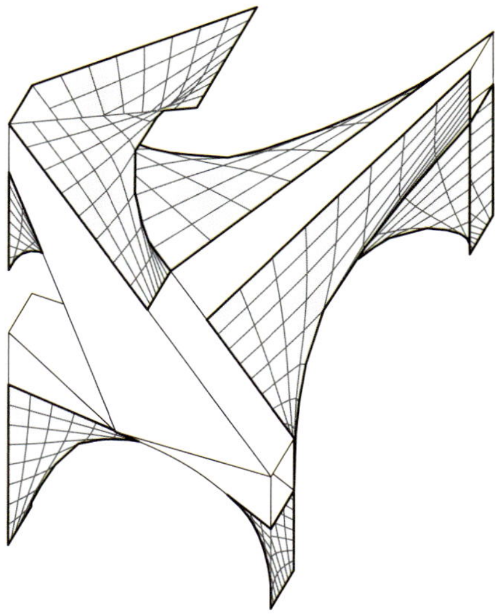

Minimal Surface Shell Geometry

PALAZZOLO

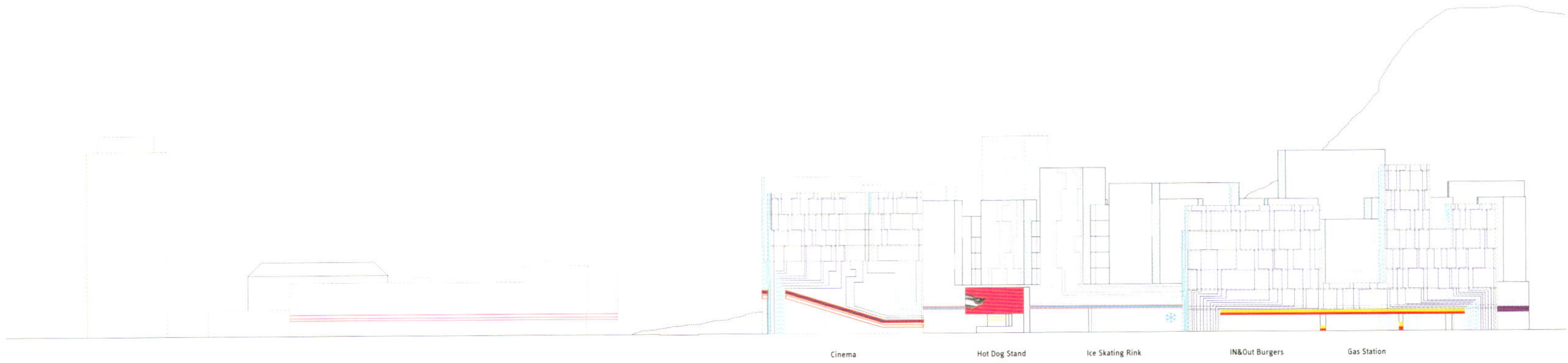

Cinema
Hot Dog Stand
Ice Skating Rink
IN&Out Burgers
Gas Station

Rain Water Collection Surfaces
Rain Water Collection Surfaces
Rain Water Collection Surfaces
Communal Spaces
Rain Water Collection Surfaces
Rain Water Collection Surfaces
Rain Water Collection Surfaces
Communal Spaces
Bus Terminal
Travel Agency
Super Market
Ice Skating Rink
Cinema

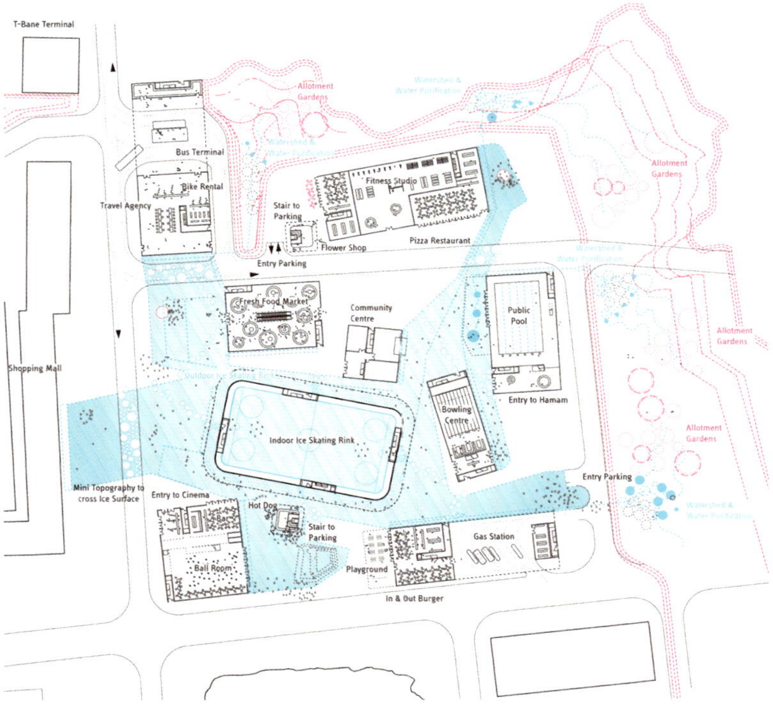

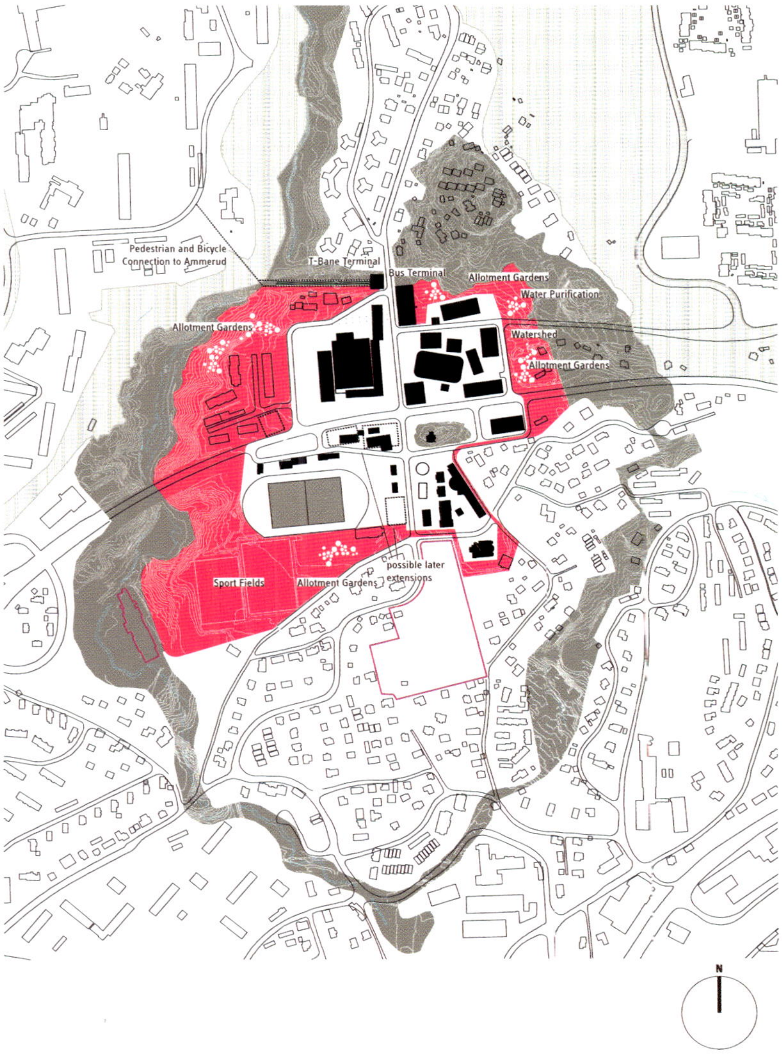

MOSAIC OF ENCLAVES
Grorud Senter potentially linking the surrounding neigborhoods

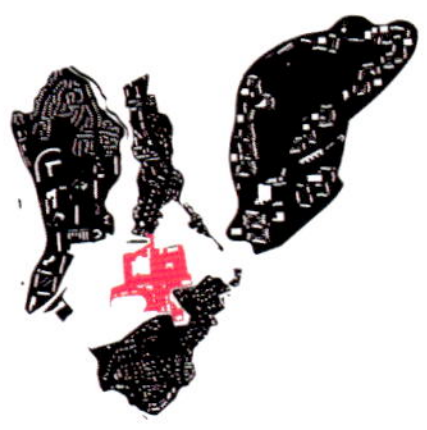

EXISTING ENCLAVES
Grorud, Ammerud, Romsas and Grorud Senter

PROMOTED ECOLOGIES
defining and connecting the enclave trough adjacencies

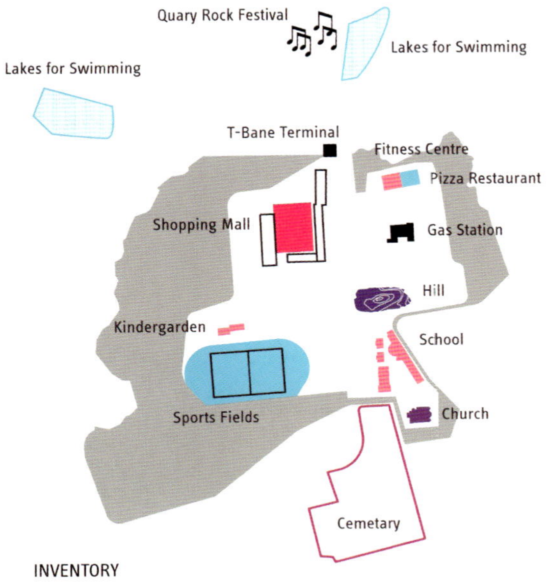

INVENTORY
it's all there

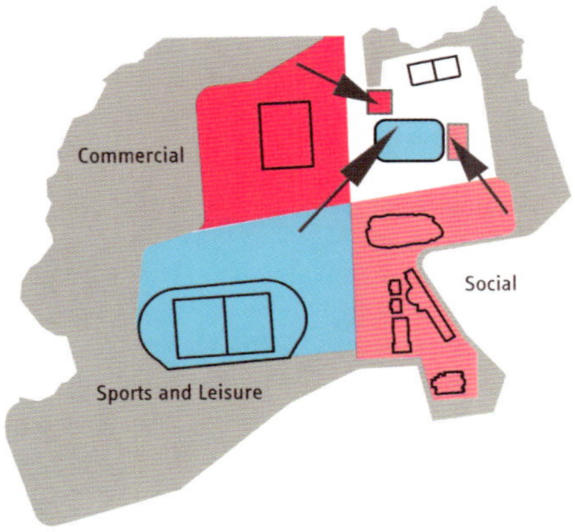

4 PROGRAMMATIC FIELDS AND CROSSOVER AS INITIAL

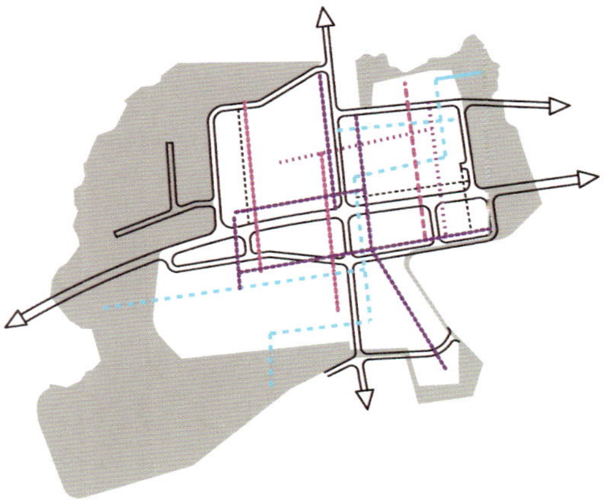

WEB OF URBAN ROADS AND PEDESTRIAN PATH
supports horizontallity of movements

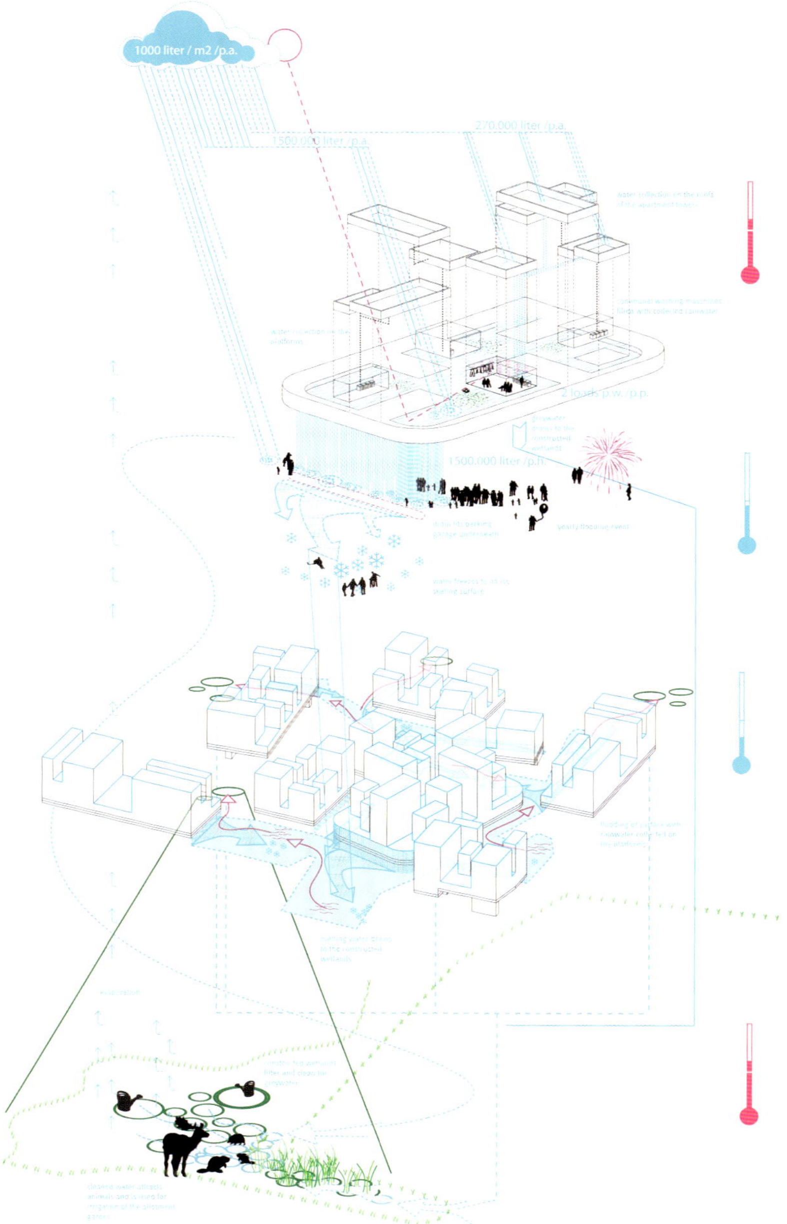

Oslo's population is expected to double within the next ten years. The outer city center Grorud is part of Oslo's densification strategy and therefore aims at bringing in new inhabitants while consolidating the existing. Diversifying social structure comes along with the sustainability imperative – what is the public space to project from these demands?

Cumulus is a proposal for Grorud that believes in the latent potentials of the site: its latent ecologies and orders, the environmental presence in outer city low density urbanism, the mix and publicness of the blunt peripheral programmes and the underlying ideas of modern light and lit dwelling manifest in the existing housing slabs of Grorud center.

Cumulus thus is a strategy to re-cycle, de-sectorise and confront these latencies and interweave them through a concept of (water) flow: The project conceives social spaces as related to environmental dynamics and develops a close - but not too close - public sphere inherent to the loose aggregate at hand.

Cloud! The white and heavy cumulus congestus cloud serves as performative model to develop an architectural typology for this synthesising approach: undetermined and open on the ground, heavy and stable on first floor, a spongy heap in the sky. It floats above and fixates rudimentary every day programmes, it enables an intimate and lit inhabitation of the skies, it collects and stores water while the release of this water becomes a social event – the day in nearing winter when it is cold enough to freeze the flooded surfaces to an ice skating plane.

Morphologically the typology enables desired shifts in scale: On ground the crowding of programmes assembles towards a patchy canopy rather than a homogenising cover. Above ground miniature skyscrapers at the scale of single rooms produce a multiplicity of spatial situations and light conditions. The aim is to fabricate a sense of urban "inside" at a small scale while keeping it open to the environment. Social densities and encounters merge with openness and porosity.

Cumulus builds on an accumulation of public programmes within a defined area framed by an ecology of gardening and sports. It strives for a packed mix and the adjacencies of different users: Close, but not to close. Above this social field hover private dwellings, which are developed as a means to live in the sky: You can wander from one part of your L – shaped apartment to the other by crossing a glass bridge: Air and the outdoors become part of the house.

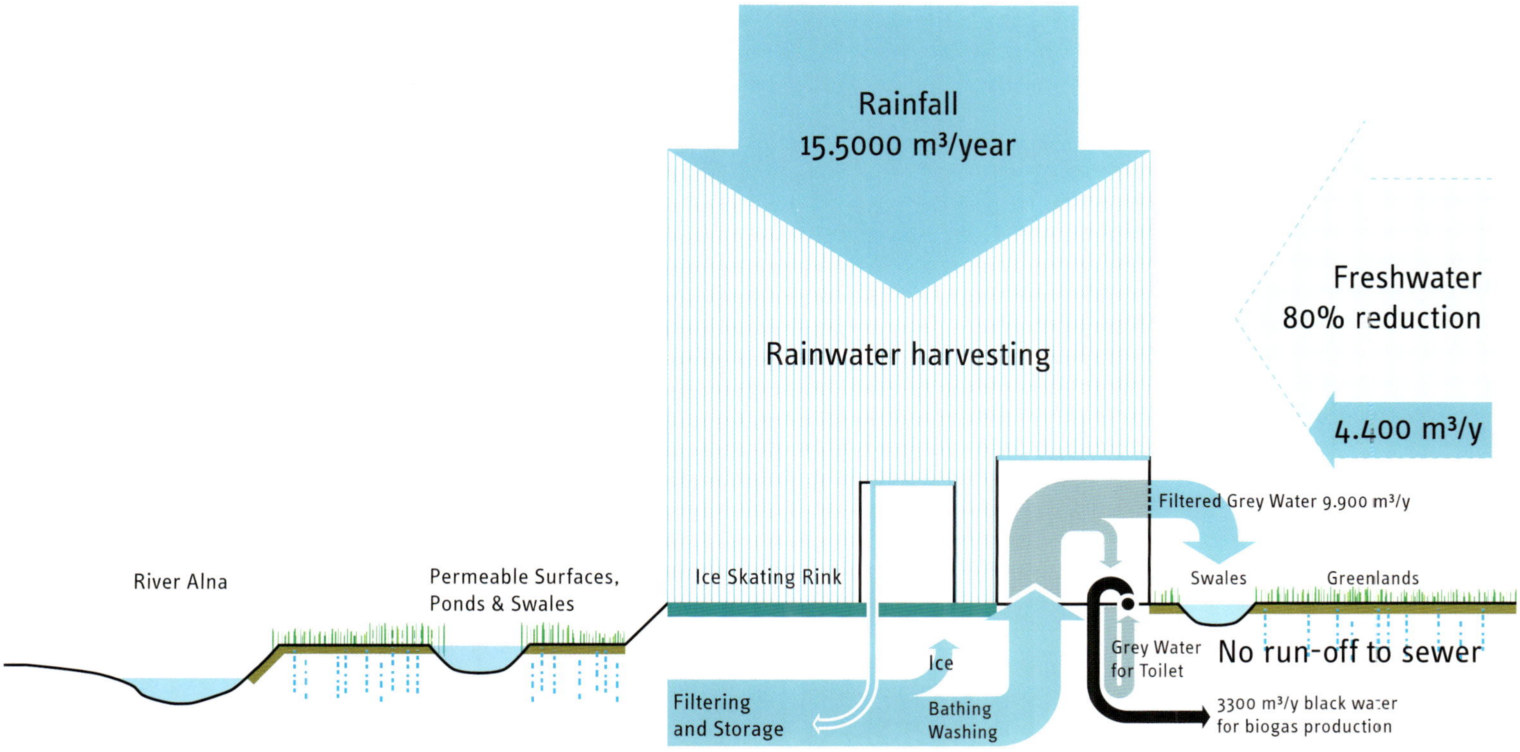

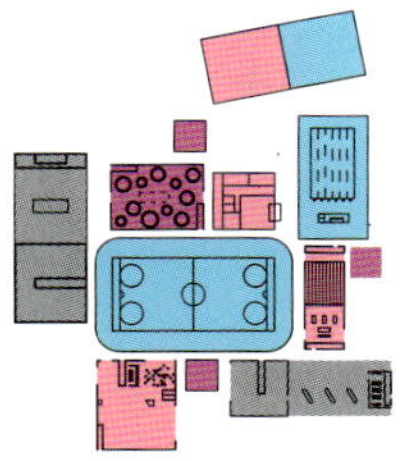

DENSE PACKING OF PROGRAMMES

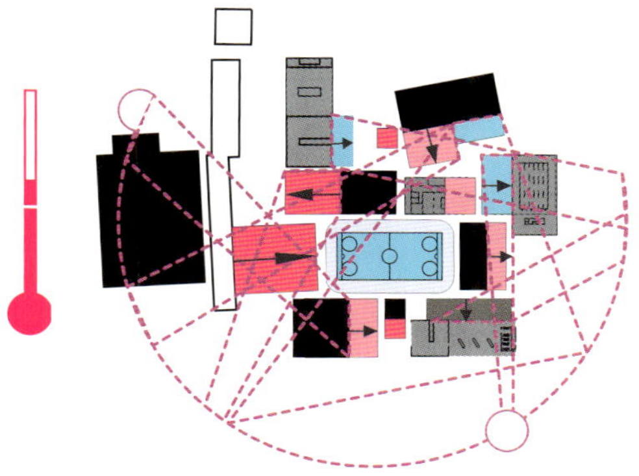

SUMMER OUTDOOR PROGRAMS
rearranging the packing according to sun exposure:
close but not too close relationships

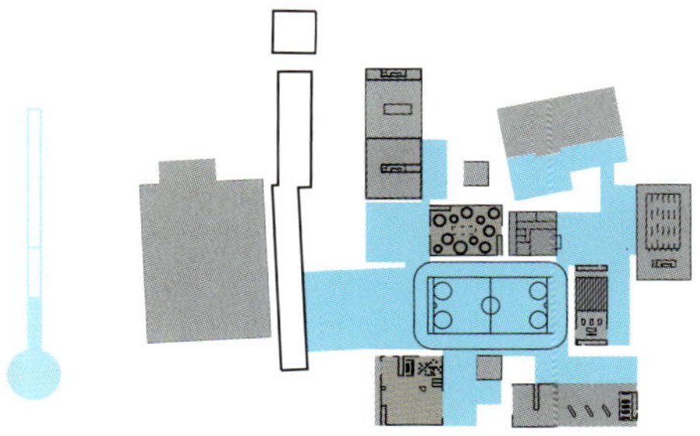

WINTER OUTDOOR PROGRAMS
ice skating surface connects all public programmes

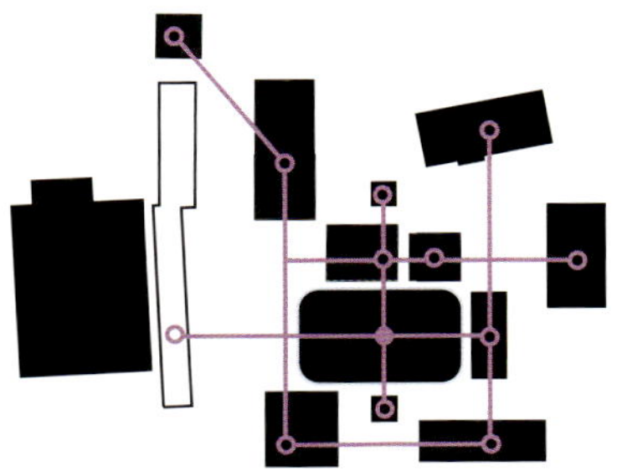

CONNECTIONS AND RELATIONS OF PROGRAMMES

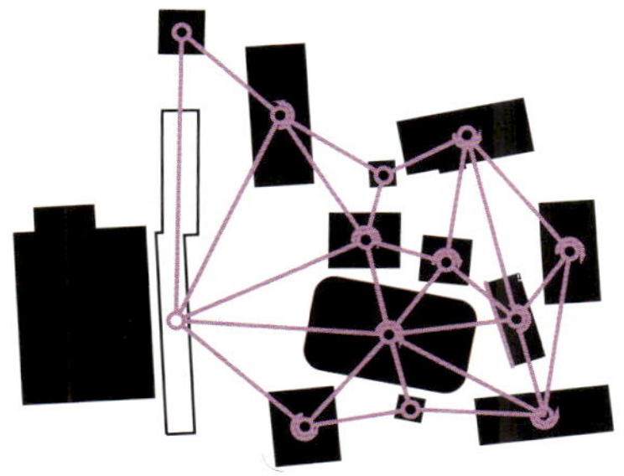

ADJUSTMENTS OF PROGRAMMES
to site boundaries and through rotation of the central ice skating rink for better connectivity

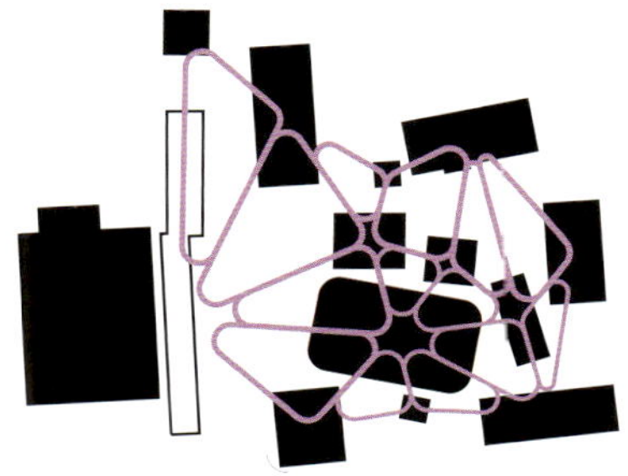

RESULTING SPACES OF INTERACTION

CROWNING ORDINARY PROGRAMS OF THE PERIFERY WITH HOUSING
a cloud inhabited by air dwellers is defining a continuous horizontal public field on ground level

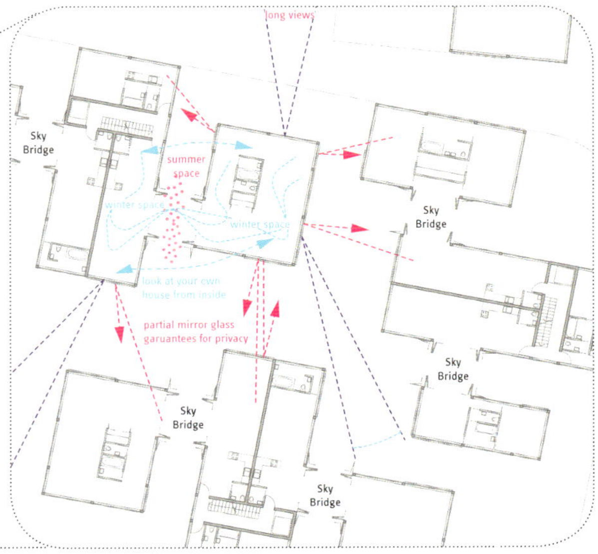

APARTMENTS INHABITING THE SKY

HABITATION LEVEL

Wadi Resort

旱谷度假区

Oppenheim Architecture + Design (OAD)

国际知名 “绿色” 建筑师乍得·奥本海姆的大作旱谷度假区位于约旦的月亮谷，预计于2014年完工，为设计和生态理念树立了新的基准。奥本海姆建筑设计事务所在一次全球的竞争中脱颖而出，执行设计这个史无前例的项目，包含47个沙漠小旅馆，在古城佩特拉外围为狂热的世界旅行观光者们提供一次未来的原始体验。

奥本海姆的获奖方案企图通过充分利用沙漠中沙石会合的神秘谷重新诠释社会与自然的关系。该项目与周边完美的景致默默地融合，开发并增强沙漠的自然之美，从而建立独一无二、简洁而奢华的住宿设施。项目所处位置不仅引人注目，沙漠小旅馆和别墅的外形也展现出不同的形态，与文化和地域紧密相连。由此游客将产生一次丰富而奢华的体验。

沙漠小旅馆的结构将雕刻成砂岩悬崖，利用现有的岩石几何结构设计其外形。其他结构包含夯土和与当地红沙混合的水泥。无论是建设的还是雕刻的，这种极小但又很强大的建筑可以创造和谐与平衡，同时健全环境。 内部与外部都有意地设计成点状，从而通过最小的努力建立最大的影响。从原始环境中得到灵感，奥本海姆运用其可持续发展设计的专业技能创造出被动的通风方式，充分利用岩石的自然冷却效果而适当地定位，使该项目减少能源消耗，最大限度地提供舒适健康的环境。

“经过训练，我们已经提高了在视觉、嗅觉、味觉、听觉及触觉方面欣赏月亮谷神秘之美的能力。基于对大自然过去、现在和将来的力量、韵律和模式的理解，我们通过原始本能的设计触碰了沙漠固有的动力。” 奥本海姆如此解释他这个项目的创造过程。

项目采用的策略是经过千百年来证实的。我们已经从那些数千年美丽而神奇的沙漠文明中学到了许多。设计十分注意利用当地材料和节水措施，建立一个相对封闭的集雨水窖，通过自然中植物和生物的作用再收集灰色和黑色的水。所有的系统和服务都将是不可或缺的设计。

International "green" architect Chad Oppenheim sets a new benchmark for design and ecological sensitivity with the Wadi Resort-located in Wadi Rum, Jordan, set for completion in 2014. Oppenheim Architecture + Design (OAD) beat out a global competition and will execute an unprecedented project comprised of 47 desert lodges, setting forth a future primitive experience for the avid globetrotter, an hour and a half outside of Petra, the ancient city of the Nebataeans carved into the desert rock.

Oppenheim's winning proposal set out to reinterpret the way society deals with surrounding nature by taking full advantage of the mystical valley where desert sand meets desert stone. The project merges silently with its wondrous setting, exploiting and enhancing the natural beauty of the desert to establish accommodations that are uniquely elemental and luxurious. Dramatically situated the lodges and villas in their various incarnations; are all about a visceral connection to culture and place. The resulting experience is a revolutionary notion of opulence that is intentionally reduced to what is essential.

The structure of the lodges will be carved into the sandstone cliffs, utilizing the existing geological geometries of the rock to devise the form. Other structures are comprised of rammed earth and cement mixed with the local red sand. The minimal yet powerful gestures of the architecture both built and carved serves to create harmony and balance while framing and amplifying the surroundings. The interior and exterior are deliberately blurred

establishing maximum impact with minimum effort. Inspired by the primordial, Oppenheim used his expertise in sustainable design to create passive means of cross ventilation, taking full advantage of the natural cooling effect of the rocks, and proper positioning allowing the project to minimize energy consumption and maximize comfortable healthy living.

"We have trained and heightened our senses to see, smell, taste, hear, and touch the mystical beauty of Wadi Rum. We tapped the inherent power of the desert through primal and instinctual design moves, informed by the forces, rhythms and patterns of nature- past, present, and future," says Oppenheim about his creative process for the project.

The strategies employed are those that have been proven over the last thousands of years. We have learned a great deal from the civilizations that have lived in the beautiful and magical desert for millennia. Great care has been given to utilizing local materials as well as various water conservation measures for both human and site irrigation to establish a relatively closed system of harvesting rain water in subterranean cisterns and re-harvesting grey/black water through a living machine of botanical and biological nature. All systems and services will be completely integral to the design.

The 80,000-square-foot architectonic form responds directly to the rich regional cues: an evolutionary process that has established, over millennia, a clear and appropriate identity found in the Middle East.

TERZ

三度空间 SMAQ

Credits

Project Team: Sabine Müller, Andreas Quednau, Matthias Titze, Robert Gorny

Client: Leipziger Wohnungs- und Baugesellschaft mbH (LWB)

Location: Leipzig, Germany

Program: Hotel with spa, pool, conference, bar and fitness; loft building with office space, ateliers, gallery and shops

Site Area: 7,247 m²

Gross Floor Area: 17,500 m²

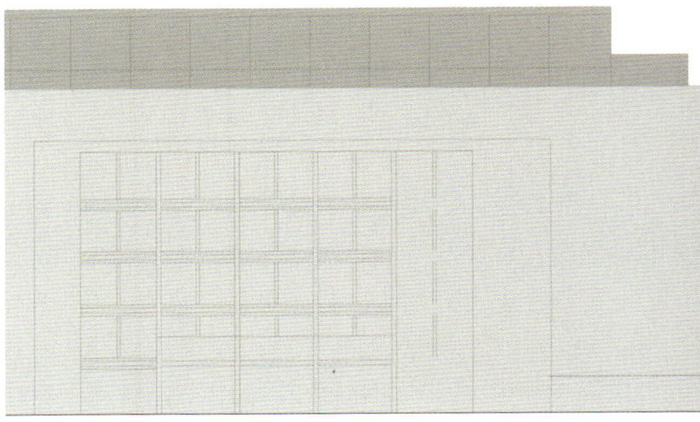

这一被称为“三度空间”的城市新建筑，主要是由两栋新建筑环绕现有的冬季花园大厦而形成。两栋新大楼作为一个整体屹立在城市广场上，处在一个具有高参考点的环境之中。三度空间发掘出城市的三维潜能，并在莱比锡的中央位置建立起一个特别的场所。

建筑师在进行三度空间的设计时充分考虑到该场地的特殊性，它坐落在市中心与“Grafischen Viertel”住宅区之间，一边是有大型项目和部分公共项目的市中心，另一边是有综合性项目的住宅区。

莱比锡的这一区域对该场地的未来发展有着各种不同的甚至有些相互冲突的要求，于是三度空间项目面临了各种挑战，包括最大化楼面面积和渗透性，同时与场地的环境和象征作用相呼应。

为了实现这些目标，两栋L形建筑的设计采用了二重策略。

1.垂直：得益于高参照点，住宅区的建筑密集度得以增加，但是不能过于紧密。这些建筑与高耸的“冬季花园”相呼应，并形成了一个整体。这样该场地便形成了市中心一个显著的场所。建筑保持了较低高度，这样一来就可以欣赏到远处火车站和维克多Residenz酒店的景象。

2.水平：建筑下面的3～4层楼作为基础部分与街道相邻，两栋楼均与社区楼层的轮廓和高度保持一致，形成城市临街建筑群和Wintergartenstrasse的边缘，Querstrasse也因此变得格外醒目。总而言之，两栋楼成了住宅区与市中心之间的一个界面。

这些水平与垂直的策略创造了一个公共的空间，并在住宅区中设有人行道，这样一来，这个空间便形成进入市中心的入口。

两栋楼的建筑面积共17,500平方米，其中一栋有3层和13层楼两部分，而另一栋有4层和8层楼两部分。

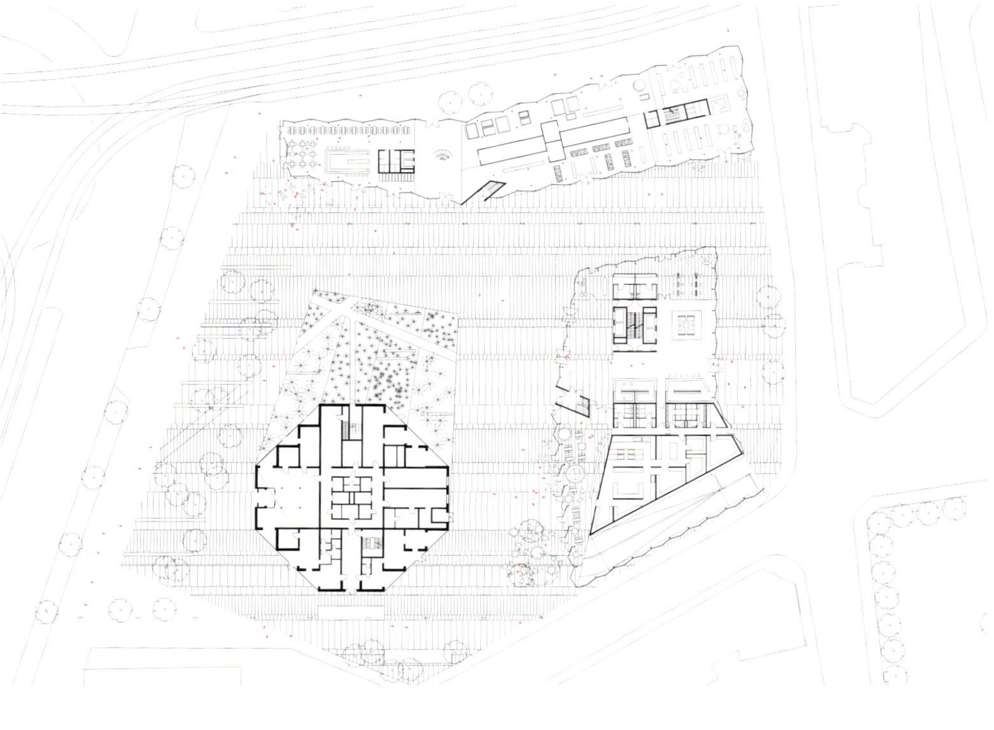

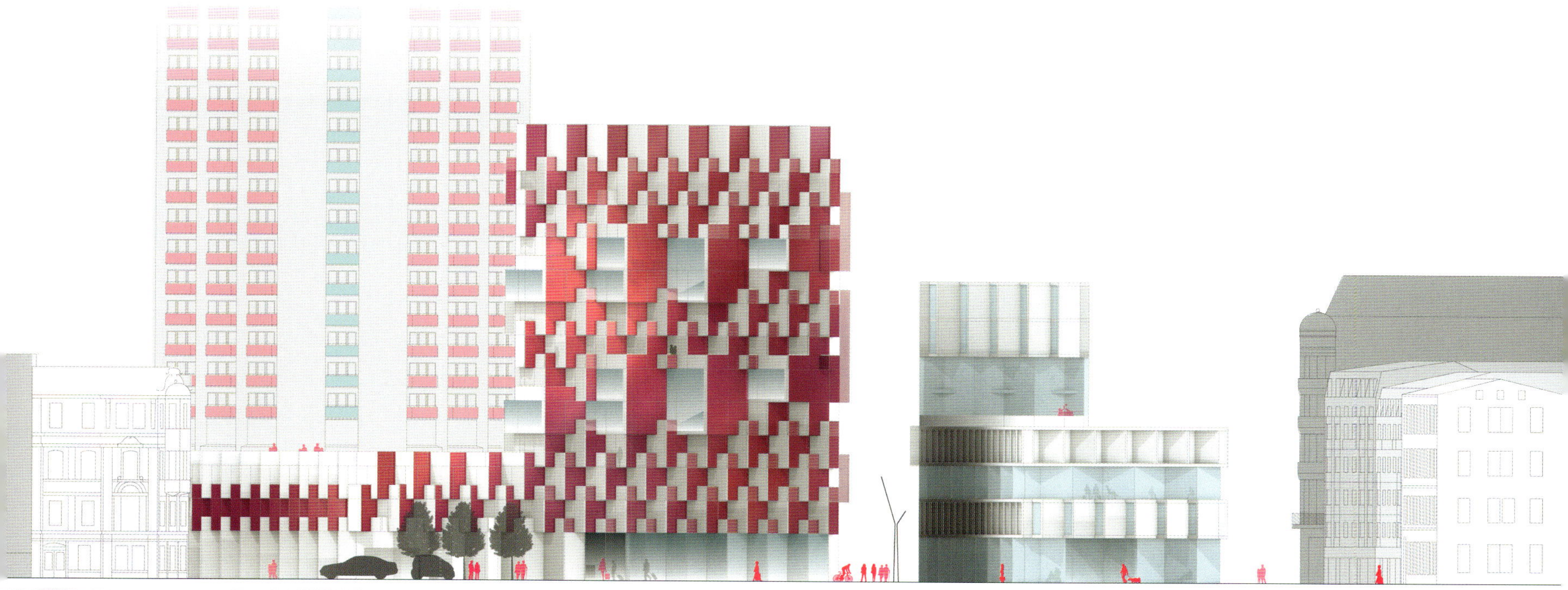

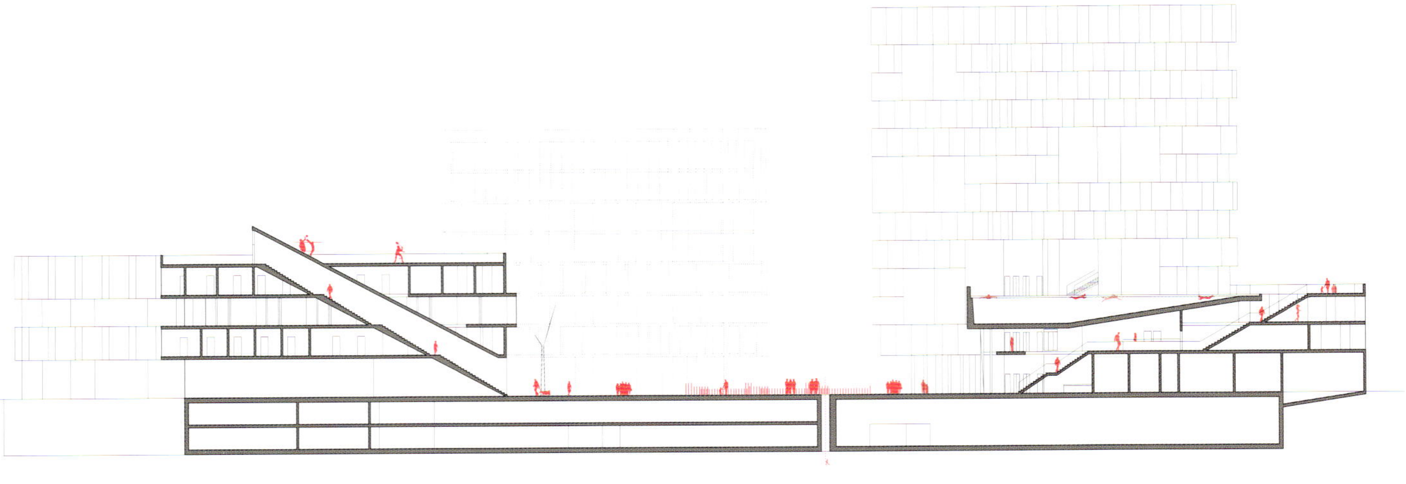

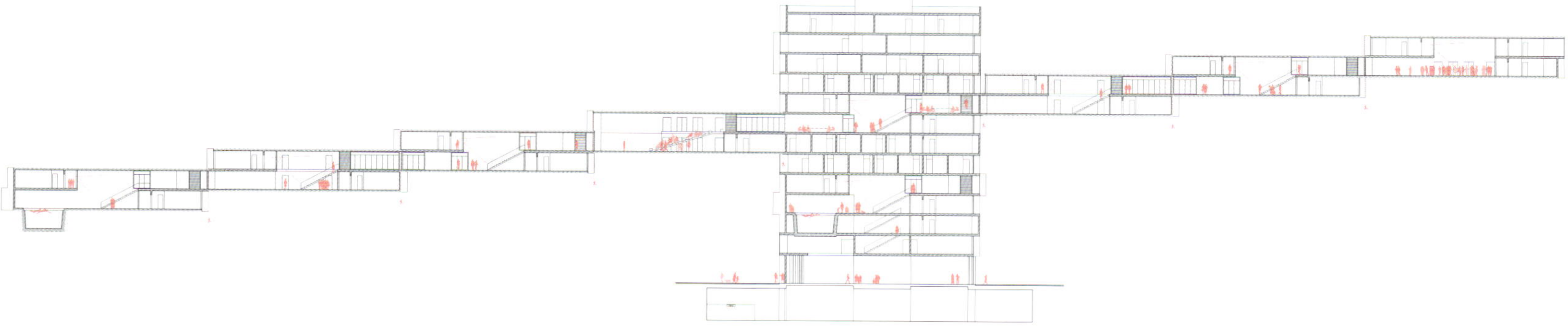

In the urban figure TERZ ("third") two new buildings swirl around the existing "Wintergarden Tower" and meld themselves together as a grouping on an urban square; the ensemble of stacked buildings is sited within a context of high reference points. TERZ draws out the three-dimensional potential of the city - volumetrically and programmatically - and establishes a special place in a central location of Leipzig.

TERZ negotiates the particularity of the site between the inner city, which has large and partially public programs, on the one hand, and the "Grafischen Viertel" neighbourhood, which has mixed-use programs, on the other.

This section of the city has diverse and, in part, conflicting requirements for the future development of this site. The challenges include maximizing floor space and permeability, as well as responding to the context and symbolism of the site.

In response to these goals, the proposed two L-shaped buildings following a two-fold strategy:

1. Verticality: high points allow the project to increase density in the neighbourhood accordingly, but without over-building. These buildings correspond to the verticality of the Wintergarden Tower and create an ensemble. As such, the site is established as a distinct place in the inner city. The building heights are also stepped down in order to frame a perspective view towards the train station and Viktor's Residenz-hotel.

2. Horizontality: The bottom 3 to 4 storeys of the buildings form bases which organize the street space. They are aligned with the contours and heights of the neighbouring, and thus form urban block edges. The street edge of Wintergartenstrasse is recreated and Querstrasse becomes a distinct space of its own. The two proposed buildings create an interface between the "Grafischen Viertel" and the inner city.

These Horizontal and Vertical strategies create public spaces with pedestrian pathways in the "Grafischen Viertel." It is thus simultaneously a gateway into the city and a destination.

A total area of 17,500 m^2 of the gross floor is generated on the site in the two buildings: one with staggered heights of 3 and 13 storeys, the other with staggered heights of 4 and 8 storeys.

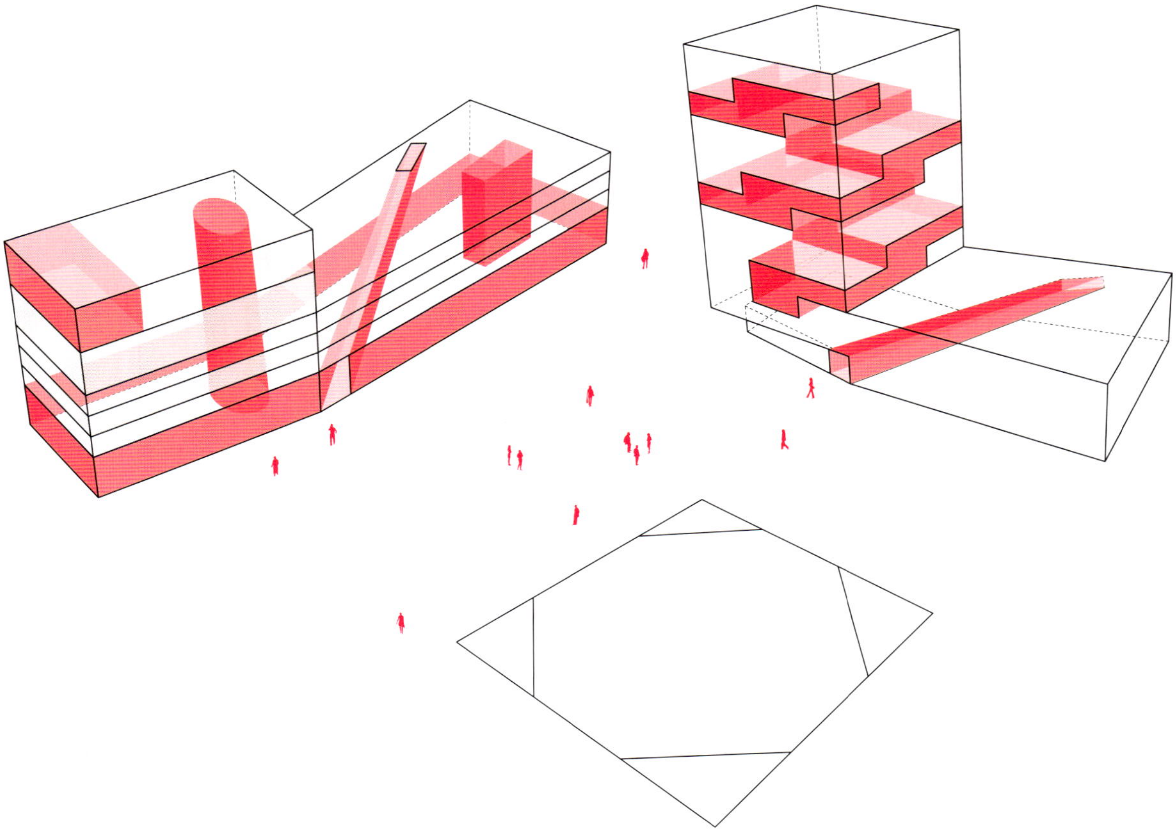

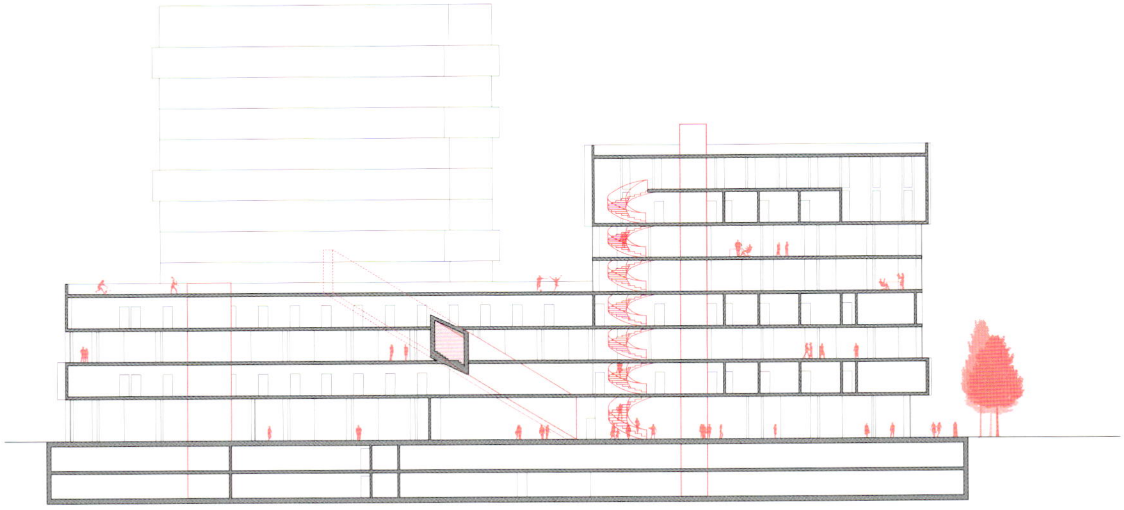

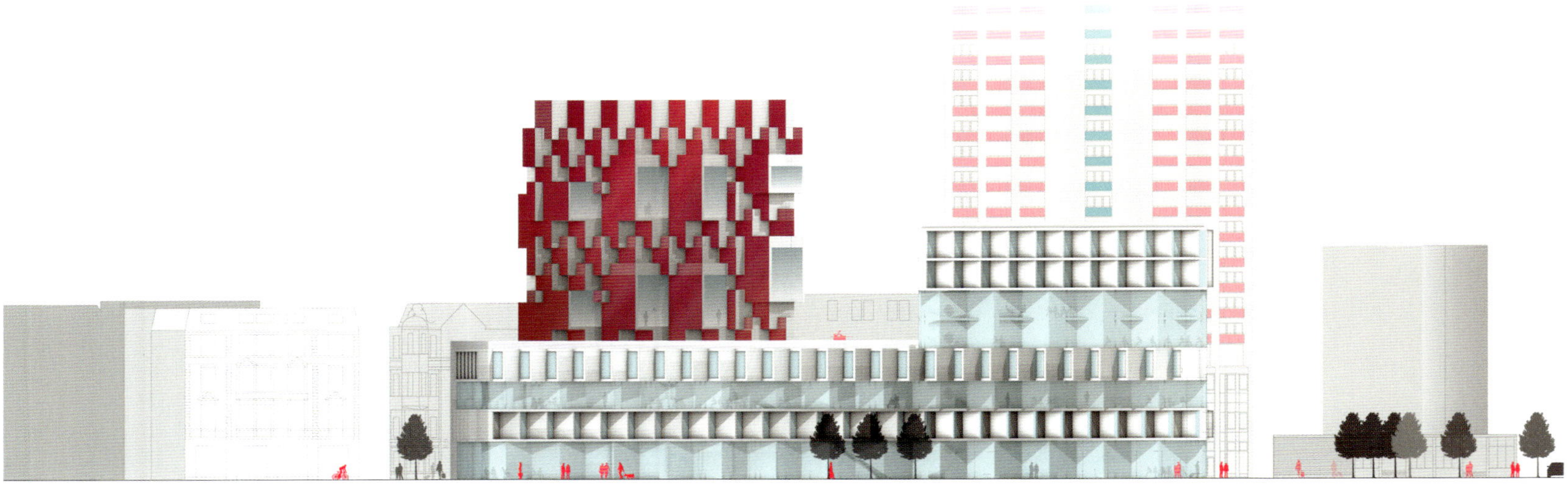

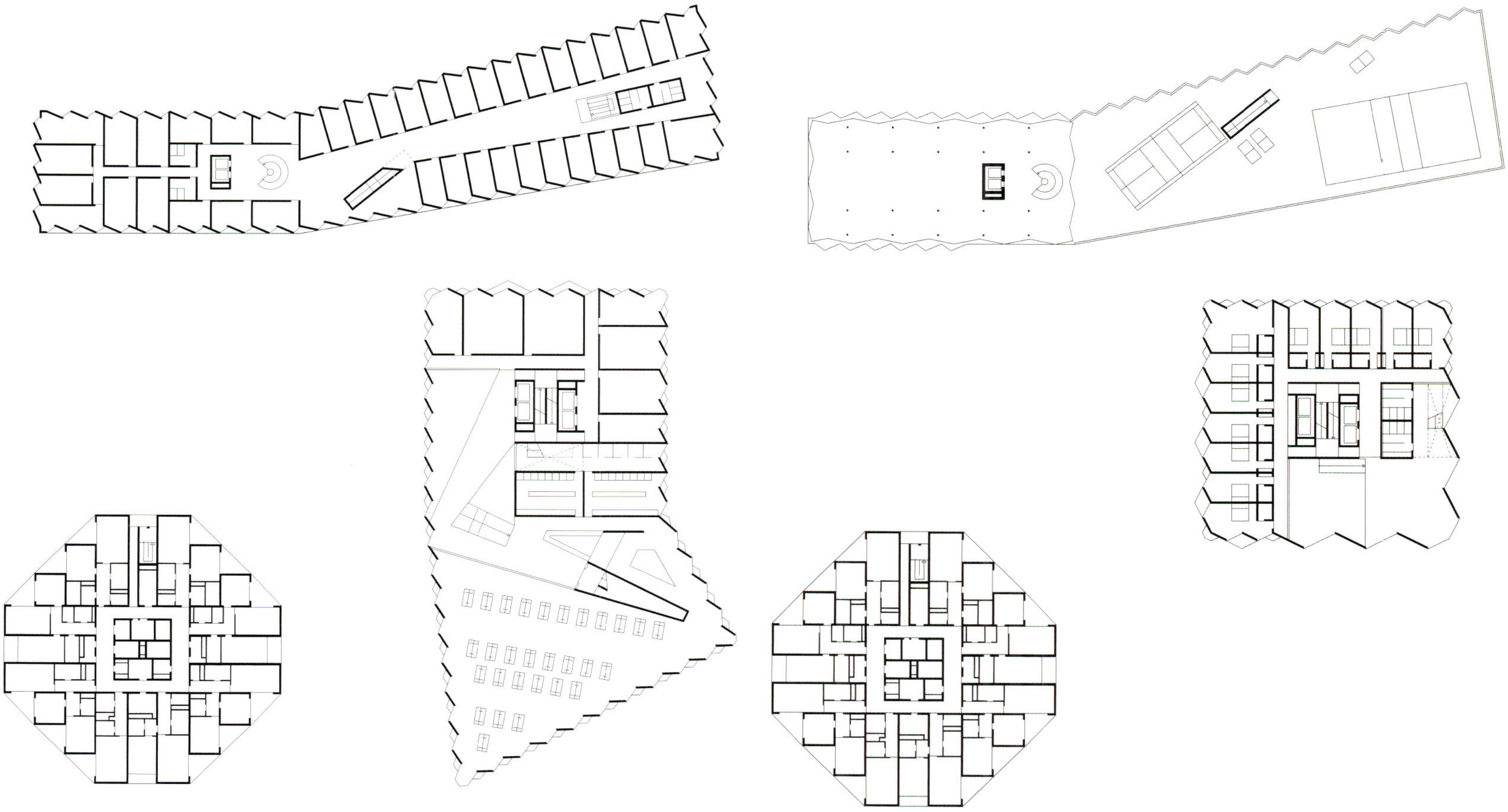

Green Pavilion Restaurant

绿馆餐厅 Studio 3LHD

Competition Prize, 2010
2010 年竞赛获奖项目

Credits

Title: Canteen of Agriculture and Forestry University in Maksimir
Program: public, culture, education, leisure, housing
Year: 2010
Location: Tvrtko Milos Street, Faculty of Agriculture and Forestry campus
City: Zagreb
Country: Croatia
Type: invited competition
Site Area: 4,536 m^2
Size: 1,600 m^2
Footprint: 1,702 m^2
Elevation: 130.20 m
Budget: 1,700,000.00 Euro

Client: University of Zagreb, Faculty of Agriculture and Forestry
Project Team:Sasa Begovic, Marko Dabrovic, Tatjana Grozdanic Begovic, Silvije Novak, Paula Kukuljica, Nevena Kuzmanic
Project Team Collaborators:Leon Lazaneo, Eugen Popovic, Anna Kocsis
Landscape Design: Boris Goreta
3D: Nikola Miletic (Kap4 (Miletić)
Structural Engineering: Zoran Divjak (Dekode D.O.O.)
Specialist Design: Mario Lukenda (Ured 3T)
Hvac: Slavko Simunovic (Hit Projekt)
Plumbing and Drainage: Zlatko Galic (Nova-Lux D.O.O.)
Electrical Engineering: Josip Karaman
Cost Consultant: Mediha Ahmetovic

“绿馆餐厅”位于浓密的树林之中，餐厅最重要的元素和灵感便来自于此，餐厅四周树林密布，只有餐厅东侧树丛相对稀疏。新餐厅意在开辟绿色领地，并由此定义余下的建筑样式。餐厅顶层绿意盎然的“空中花园”也孕育了下层的设计。覆盖于屋顶上的植被既是餐馆最朴实实用的元素，也是天然的光照调节器。

因为餐馆中庭开阔，大楼内光照充足，无论从视觉还是建筑实体角度都与自然融为一体，并使位于树木周围区域的透明度大大增加。偌大的中庭是主要的入口通道，设置了会客区域、开放的露台和一个通向大楼各处的通道。大楼中庭的两旁坐落着两家主要的餐馆(食堂和自助餐厅)。食堂的风格颇具森林公园的味道，打开滑门便可让您置身于大自然的怀抱。学生会办公室和多功能大厅位于东侧的大楼内。两幢大楼与餐厅的朝向相同。

这里还建有临时的小楼和暖房，这样的建筑作为学术文化的象征可以在任何大学内找到，而该项目中10套宿舍公寓也汲取了这种精髓。宿舍楼是独立式的，设计时充分考虑了个人隐私。每幢公寓楼房顶楼的屋檐之上布满了不同的植被，富有启示作用。

餐厅意在秉承可持续性建筑的理念，建筑材料、建筑风格和建筑结构等各个方面都讲求环保。在设计过程中，自然光照和空气流通极为重要，它可以进一步减少人工灯照和空调设备带来的能耗。餐馆的屋檐和特大的植被灯具是天然的温度调节器，植被灯具犹如厚厚的保护层，为餐厅驱寒避暑，使其冬暖夏凉，适宜居住。

园林设计是基础建筑理念和细节中必不可少的一部分。该餐厅屋顶大都布满了茎叶肥壮的植被，形成了别有情趣的屋顶花园。而餐厅其他普通区域拥有特色的植物种植区。餐厅的普通区域间专门为顾客和就餐的人准备了通信线路和小型的休息区。

每天餐厅的厨房需要准备和提供1,500道菜肴，所有的工作区和服务区都在一楼——也就是最下面的楼层，功能齐全且四通八达。厨房最特别之处在于与外界连通。

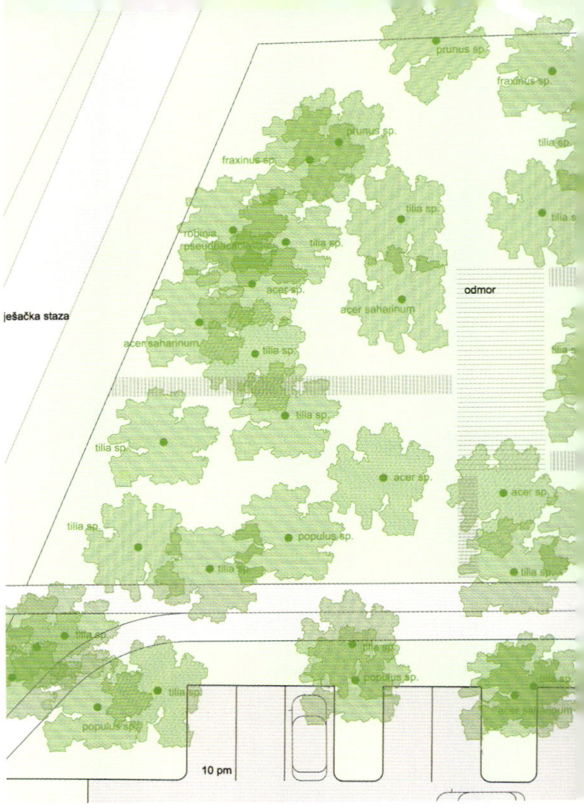

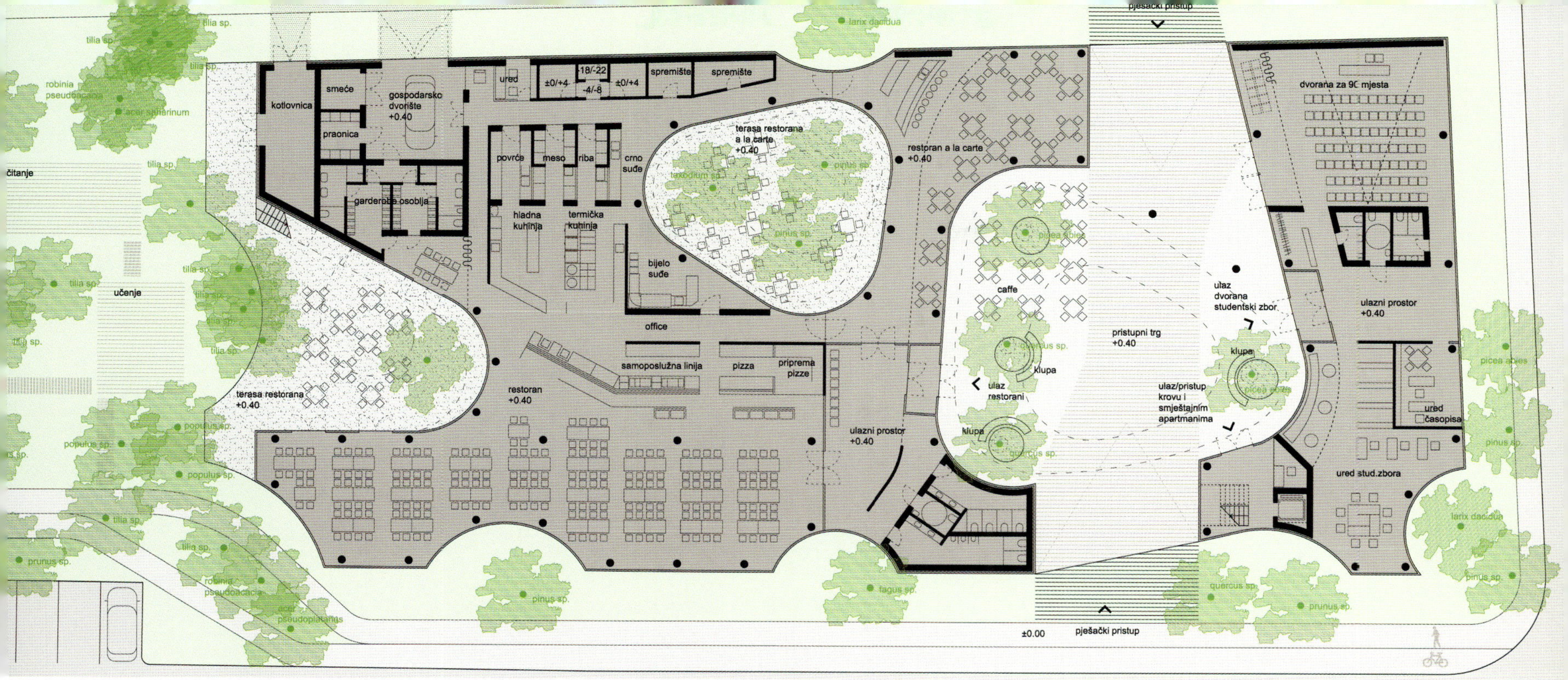
kotlovnica
smeće
gospodarsko dvorište +0.40
praonica
garderoba osoblja
ured
spremište
povrće
meso
riba
crno suđe
hladna kuhinja
termička kuhinja
bijelo suđe
office
samoposlužna linija
pizza
priprema pizze
terasa restorana a la carte +0.40
restoran a la carte +0.40
caffe
pristupni trg +0.40
ulaz dvorana studentski zbor
dvorana za 90 mjesta
ulazni prostor +0.40
ured časopisa
ured stud.zbora
klupa
ulaz restorani
ulaz/pristup krovu i smještajnim apartmanima
terasa restorana +0.40
restoran +0.40
ulazni prostor +0.40
čitanje
učenje
pješački pristup
±0.00

kolni pristup
pješački pristup
pergola
+5.00
2.5%
+4.00
5%
+6.00
tehnički prostor
gredice
P+1
+0.40
±0.00
staza
zeleni krov
pergola
+6.00
5%
+4.00
2.5%
+5.00
pješački pristup ±0.00

+8.20
+8.80
+9.50
+5.00<+7.00
+5.00<+6.00
+0.40
±0.00
1

+4.00<+5.00
+5.00<+6.00
+0.40
±0.00
2

+8.90
+8.30
+4.00<+5.00
+5.00<+7.00
+0.40
±0.00
3

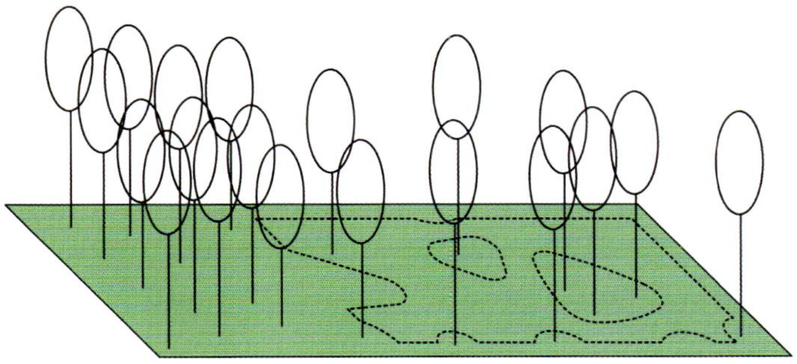

- preserving the trees
- facilities placed between the trees

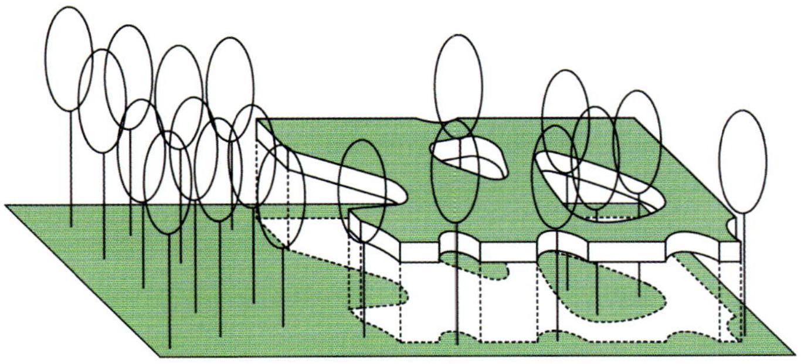

- cutting the territory around the trees
- formation of new surfaces in the air

The most important element and the inspiration for this project is the location itself, with extremely high quality existing vegetation, which is densely distributed throughout the plot and gradually diluted out toward the eastern edge. The idea of the new Pavilion / Restaurant is to construct a space by mapping the area of high quality trees and to define the remaining territory for construction. By raising this "cut" green territory in the air the space "below" is created. The basic formal and functional element and a seasonal regulator of light is so-called "vegetative" canopy that has been formed along the entire roof edge.

Open atrium spaces which let the light enter all parts of the building and provide visual and physical connections with the natural environment are created to enhance transparency in the area around the trees. The large entrance atrium is the main entrance Plaza, a meeting place, an open terrace and a main entrance to all building spaces. Two main restaurants (canteen and a'la carte) are located next to two main building atriums. The canteen space is oriented to the Forest Park and the sliding doors enable its expansion to the outside environment. The Students association offices and the multifunctional Hall are part of the eastern volume, thus allowing the pavilion to be built in phases.

The small temporary buildings and greenhouses that have always been presented throughout the campus as part of the faculty culture have served as inspiration for the accommodation units with ten apartments which have been incorporated into the project concept. The apartments are designed as freestanding volumes raised on the roof for privacy. Zones with different plants which may also serve educational purposes were formed around the apartments, on the roof surface.

The idea is to create a sustainable house, which will contribute to the environment in every respect - with building materials, construction and structure. In designing, a special significance is given to natural lighting and natural space ventilation, which further reduces the energy requirements for artificial lighting, heating and cooling. The roof and its oversized vegetative canopy are excellent regulators of heat, a thick protective cover that keeps the house from cold in winter and heat in summer.

The landscape design is an integral part of the basic architecture concept and the whole plot. The largest part of the roof surface is covered with succulents typical for roof gardens; other areas are organized into regular shapes inspired by plant growing plots. Communication paths and small leisure zones for guests and canteen users are created between the regular shapes.

In the restaurant kitchens approximately 1,500 meals will be prepared and served daily. All working and service spaces are organized on one level - the ground floor, to enable the ideal functionality and access to all units. One of the specifics of the kitchen is in full contact with the outer space.

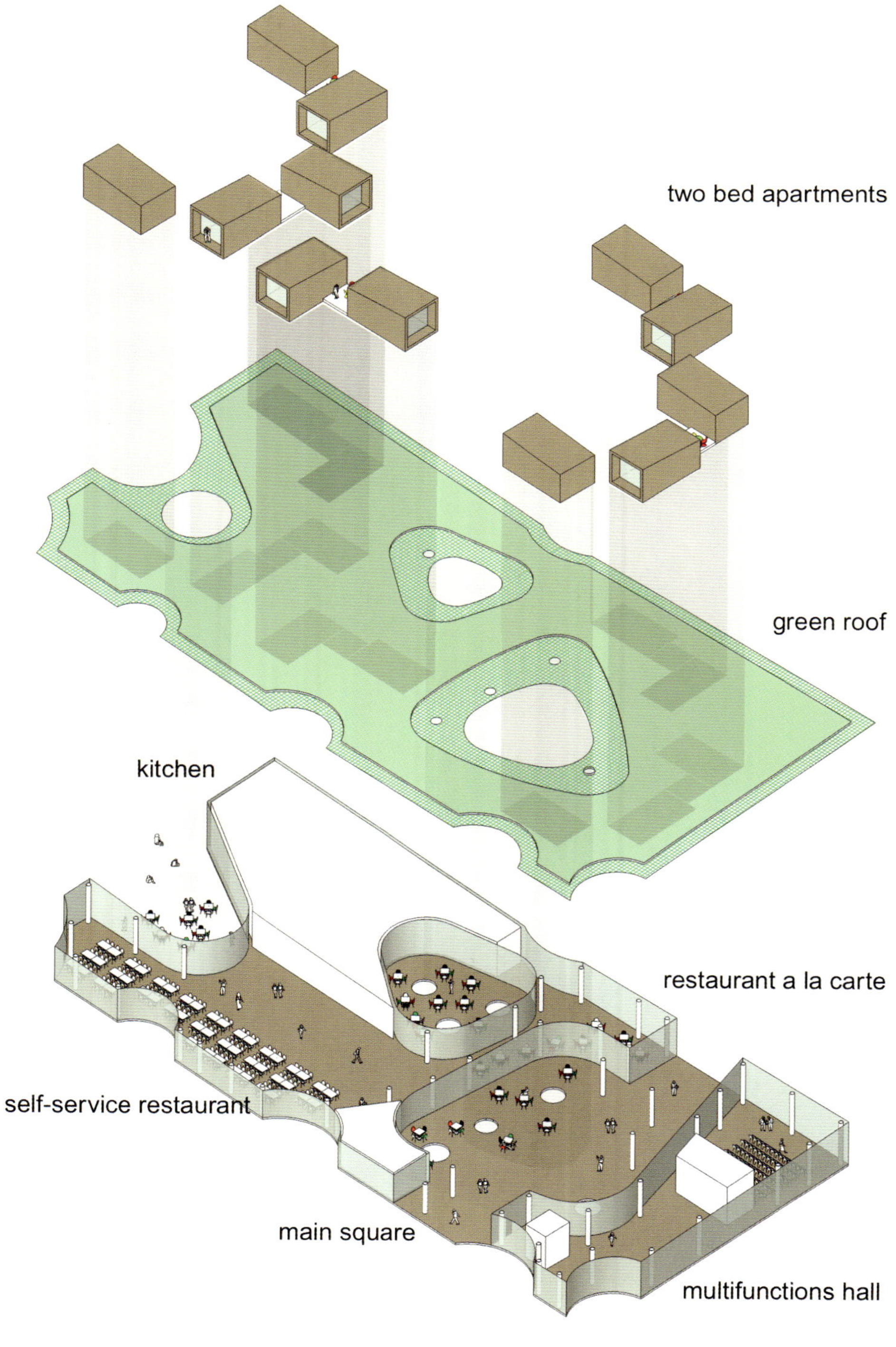

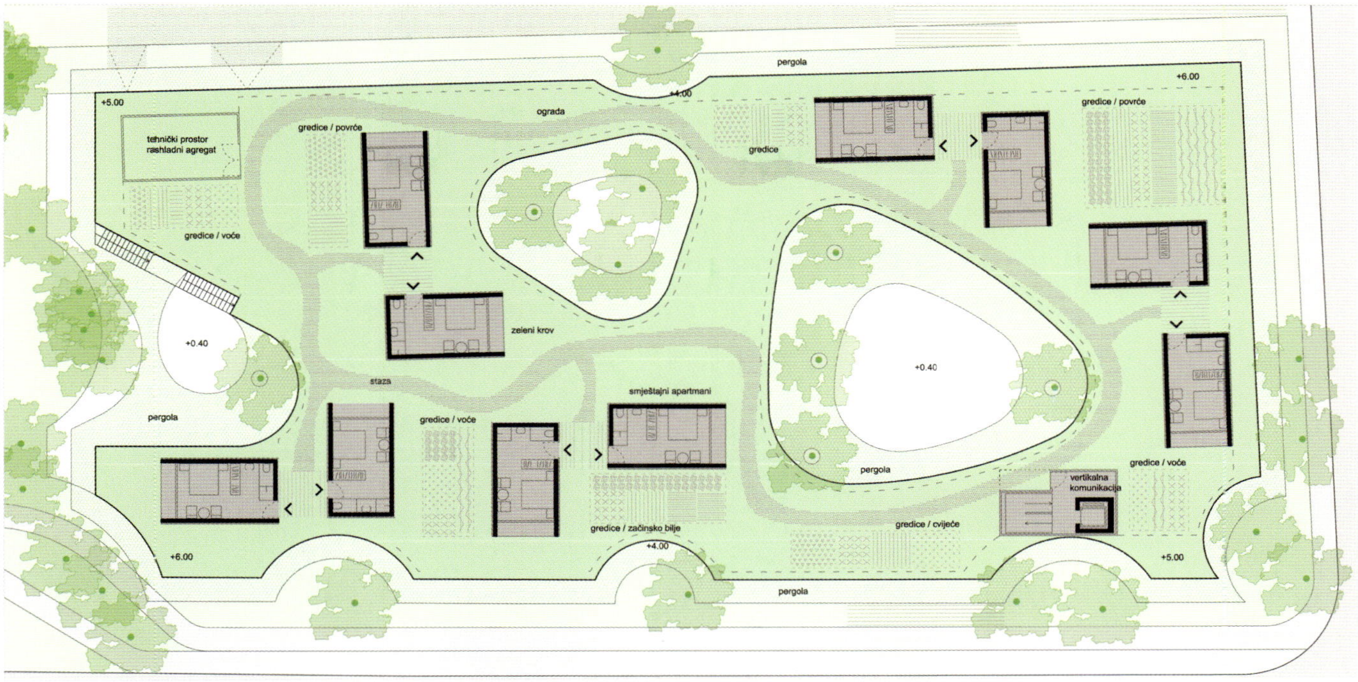
pergola
+5.00
+4.00
ograda
+6.00
gredice / povrće
tehnički prostor
rashladni agregat
gredice
gredice / povrće
gredice / voće
zeleni krov
+0.40
staza
smještajni apartmani
+0.40
pergola
gredice / voće
pergola
vertikalna komunikacija
gredice / voće
+6.00
gredice / začinsko bilje
+4.00
gredice / cvijeće
+5.00
pergola

Edil Tomarchio Commercial Park

Edil Tomarchio 商业区

Modostudio

Architectural Competition, Second Prize
建筑设计方案赛二等奖

Credits
Location: Aci Sant'Antonio - Italy
Client: Edil Tomarchio
Area: 14,950 m²
Year: 2010

该地位于埃特纳火山与海之间，历史悠久，传统文化底蕴丰富，但是最近几年发展混乱——这也是设计的难点。

设计将复兴周边农业，并修筑通往村落中心的步道与自行车道，以及一个大型的地下停车场。这个停车场除了能服务顾客，还可以供公共活动使用，以活跃村落气氛。

外观圆形设计轮廓清晰而兼备现代风格，使其极具代表性——而旁边则是设计简单的建筑，一个外立面与地板是石灰熔岩的仓库，还有两栋用砂岩建造的楼房。

设计功能区域划分明确，其中位于北侧的“大圆”将建造大型停车场，底层为主要商场；商业楼设有管理部门与销售区域。

南面的“小圆”，毗邻环道，和商业楼顶棚相连。这里主要设有咖啡馆、会议室、图书馆、健身房与顾客接待处。这样明确的功能划分可使三个“圆”独立运行，节约能源消耗，避免了公共区域与私人活动区域重叠。

设计将自然材料、低能耗与最小能源的提案放在至关重要的位置。当地的石头与其利用穿孔板的独特运用，为销售与服务楼提供了自然的庇荫设施。另外，建筑物的玻璃都将采用低排放型的。

楼顶将安装一个太阳能热敏光伏系统，便于利用雨水清洁与灌溉，达到能源有效再利用的目的。

It's difficult to deal with an area so rich in history, traditions, an area extremely fascinating, between Etna and the sea. An area in which recent years has had a chaotic development.
The intervention will propose a recovery of the surrounding agricultural tissue and will create a pedestrian and cycling connection the center of the village with the new intervention which will host a big underground parking. The parking, in addition to serving customers company, will be used as parking public by improving the livability of the village.
The circular volumes want to propose an architecture very recognizable, but linked to the contemporary context. If in fact deliberately circular forms identify a representative image of the intervention, detached building low-quality surroundings, the coating materials used are stone local lava for the floor and the facade building a warehouse and for the sandstone two other buildings, material rooted in tradition local building.
The intervention is based on a clear division of functions with the proposed volumes. The bigger and North located circular building in his basement hosts a large parking lot, while the ground floor hosts the main store; the commercial building, hosts the administration and sale area.

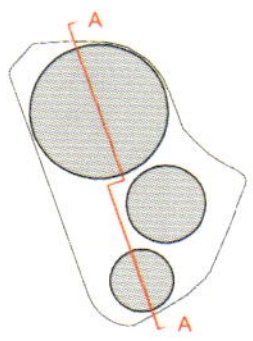

SEZIONE LONGITUDINALE A-A

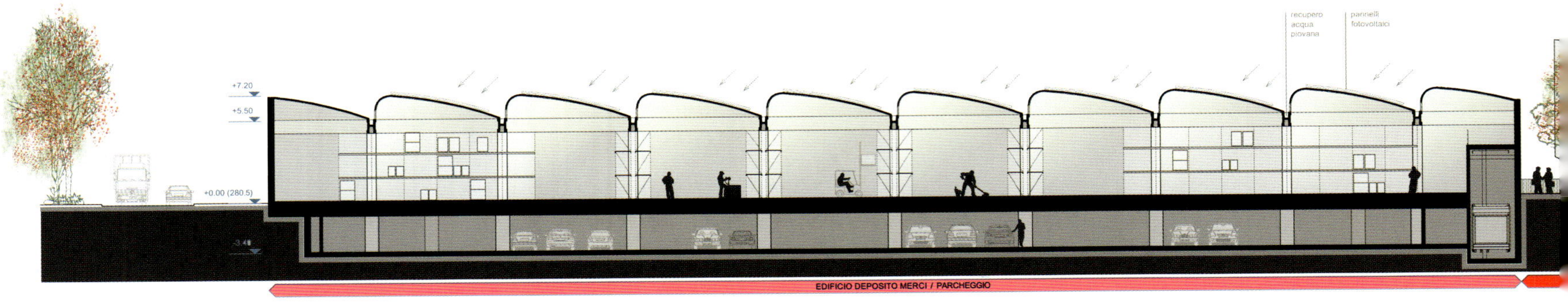

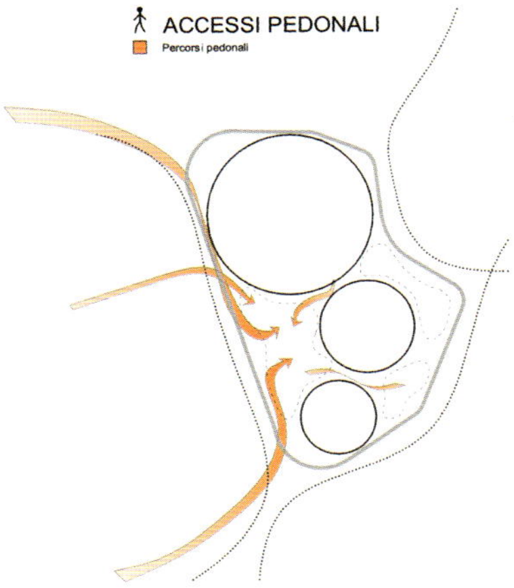

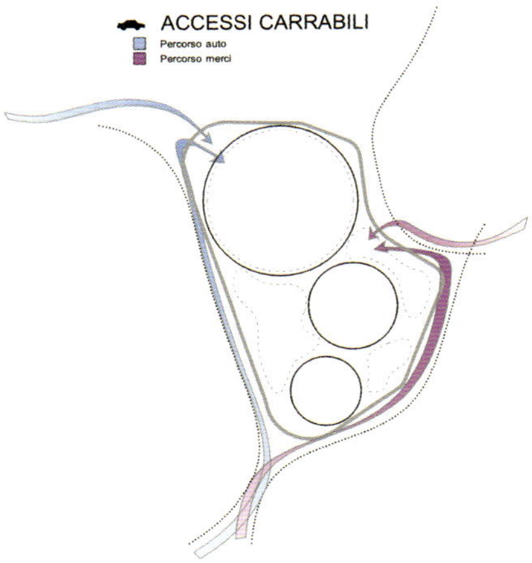

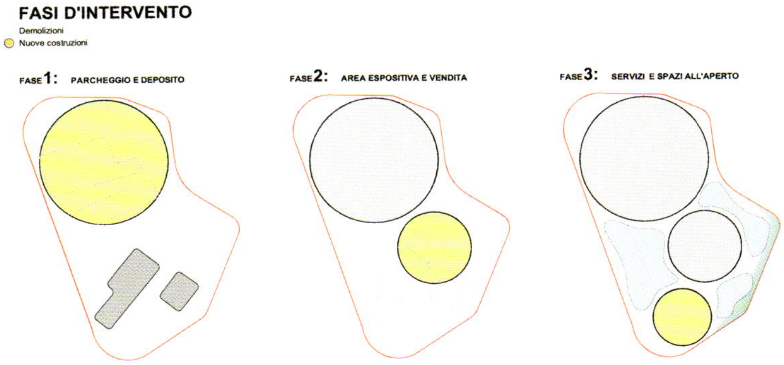

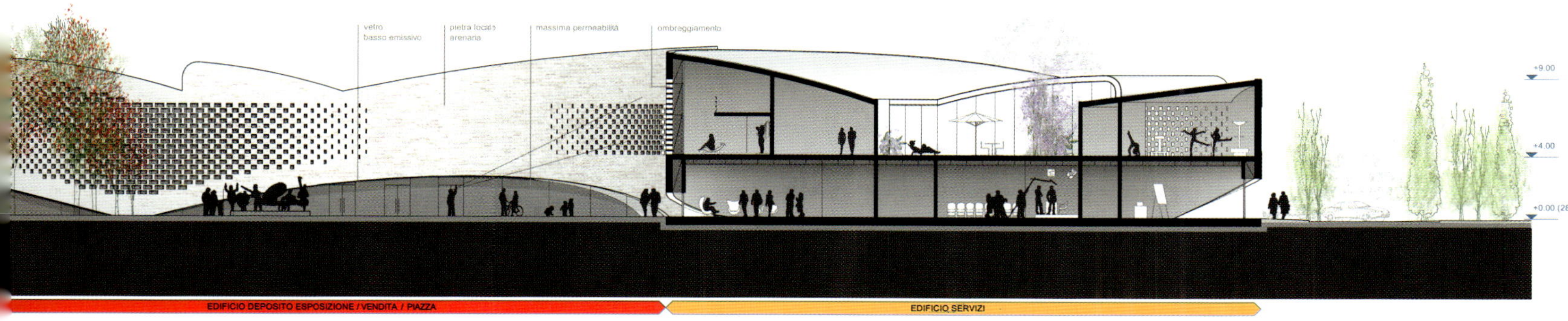

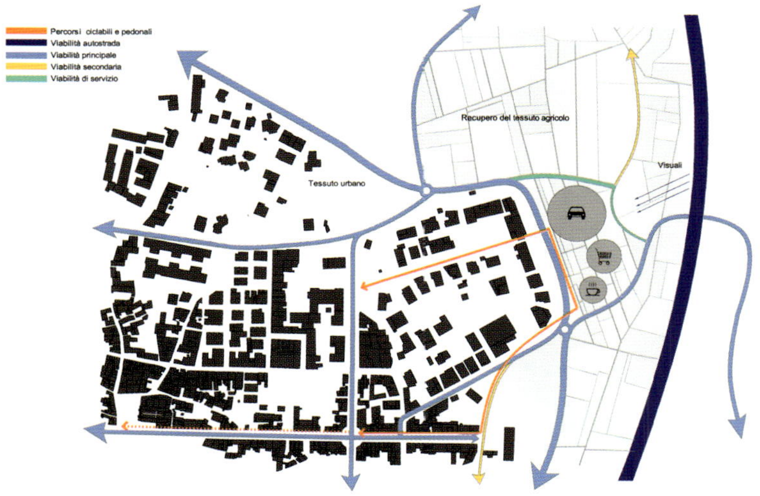

DETTAGLIO SEZIONE VERTICALE
EDIFICIO MOSTRA / VENDITE 1:50

1_Carter in lamiera metallica.
2_Pannello prefabbricato in cls con isolamento termico.
3_Pannello composto da blocchi in pietra arenaria locale.
4_Controsoffitto in cartongesso.
6_Pavimento galleggiante con finitura in gomma naturale.
7_Vano per posizionamento impianti e luci.
8_Pilastro con finitura naturale.
9_Serramento a taglio termico con doppio vetro.
10_Facciata con profilo a taglio termico e vetro basso emissivo.
11_Pavimento in cls con finitura in resina.
12_Pannello di copertura sandwich con doppia coibenza acustica e termica

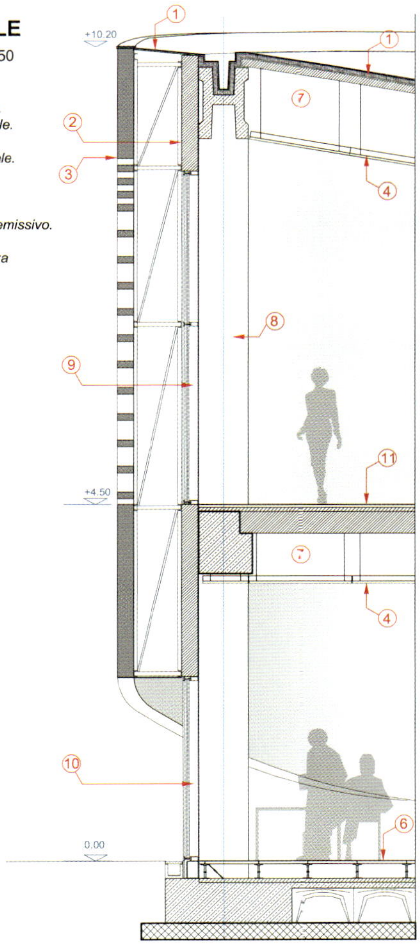

The southern building, close to the roundabout and connected with a canopy to the commercial building, hosts the cafeteria, a conference hall, the library, the gym and the guest rooms. Such clear division of functions allows the possibility of using the buildings individually, saving energy consumption and avoiding overlapping functions without public and private interference.

A great importance is given to the proposition of solutions which ensure natural construction materials, low power consumption and minimum energy supplies. The local stone and its particular implementation through perforated panels will provide a shading natural casting for the for sale and services buildings, which will use low emission glass.

The covers will host an installation of solar thermal and photovoltaic system. The roof will facilitate the recovery of rainwater that will be used for sanitary purposes and irrigation.

Intercontinental Hotel

洲际酒店 Ivo Buda Architetto

Open International Competition on the Development of an Architectural Concept for an International Business Centre with an Intercontinental Hotel in the City of Yerevan

埃里温国际商务中心与洲际酒店设计理念发展方案公开赛参赛方案

Credits

Project Design: Ivo Buda
Project Team: Ivo Buda, Paulo Barbaresi, Manfredo Bianchi
Client: Intercontinental Hotels
Location: Yerevan, Armenia
Year: 2010
Program: Hotel with congress centre, offices and apartments with panoramic view on the city and Mt. Ararat
Height: 92m
Levels: Basement (4 lev.) +Ground l. +23 levels
Total Floor Area: 77 000m^2 + parking area
Structure: Steel-concrete structure
Parking Spaces: 1,394

方案概述

这个方案是关于埃里温国际商务中心（IBC）与洲际酒店的设计方案。

这项可行性研究展示了IBC与洲际酒店的设计细节，以及三栋综合建筑——集住宅、办公及购物为一体。每一部分的具体设计都结合客户要求与实际情况，具体对待。基地延伸区域建造了与地形城市相匹配的楼房，以丰富整个地块的建设。

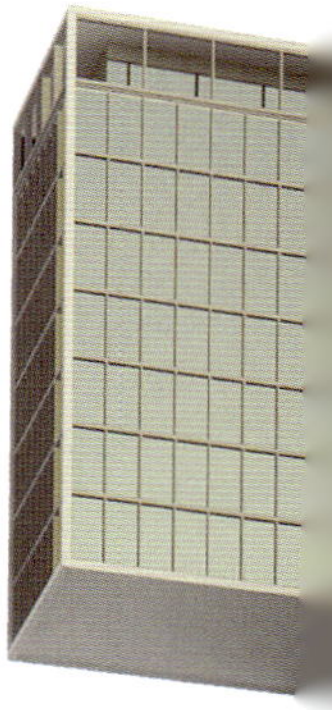

另外，设计也将环保作为重要组成部分。

设计理念

位置

设计充分地利用了该地的地形优势，这里是埃里温城市的一块肥地，绝对有机会开发世界一级的设计方案。这个位置可以俯瞰整个城市，同时在任何一个角落都可以看到这里。因此可以说，这里是“顶端城市”。

堆叠——设计语言

大楼设计成方盒叠方盒的样子，表达堆叠的意思。通过这些堆叠的方形盒子，设计师阐述了建筑空间组合的完美理念。建筑整体高大，结构紧凑，在表达设计理念的同时也为环保作出贡献。

绿色生态

绿色生态也是我们设计理念中的一个很重要的原则。所有的设计部分均向绿色靠拢，设计留出大块绿化区域与公共空间，为埃里温城市新增了绿色的元素；以“城市森林”的概念，为人们提供了散步、骑车的休闲空间，以尽享公共场所的舒适与让人赞不绝口的美景。

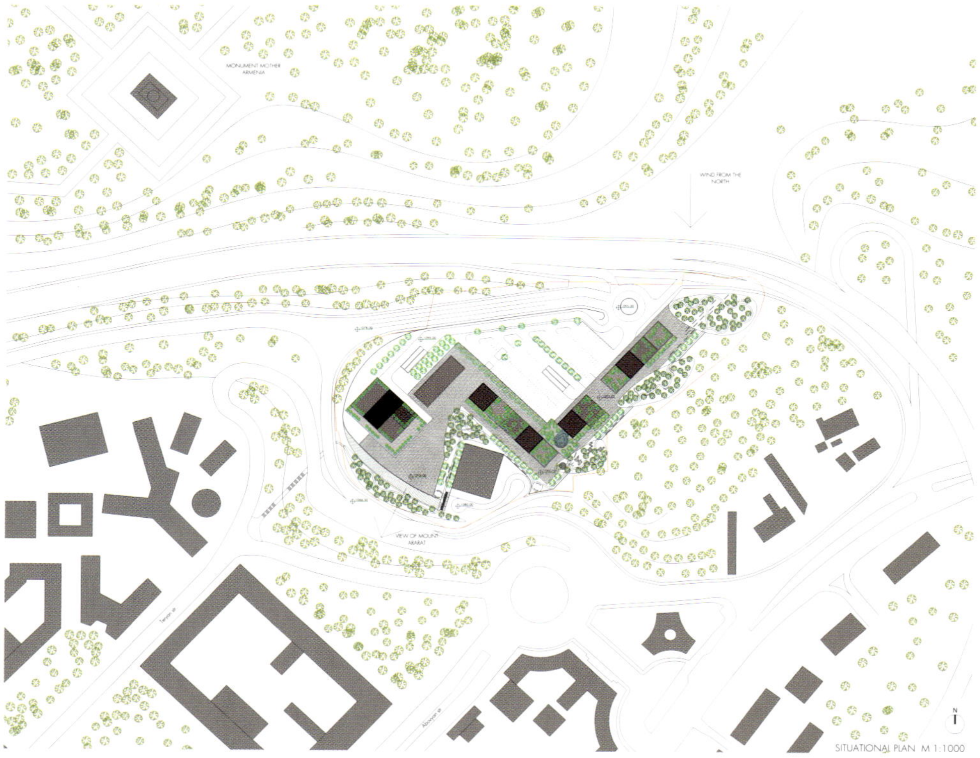

INTEGRATION WITH THE CITY

THE SCHEME OUTLINE FOLLOWS THE URBAN GRID

Hotel, offices and residential buildings are aligned to the main urban grid and the monuments of Mather Armenia

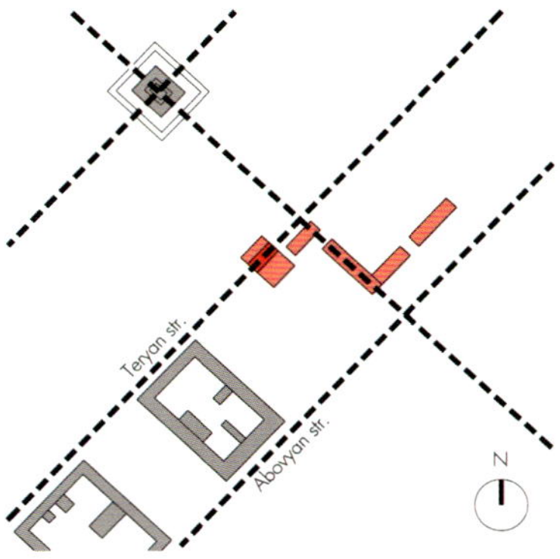

VIEW OF THE CITY AND THE MOUNT ARARAT

The setting out of the hotel's rooms, the apartments and offices were studied in order to face the city and Mount Ararat in the background.

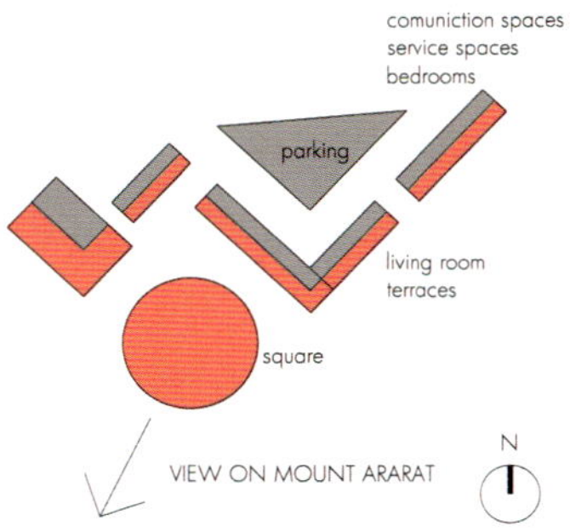

VEHICULAR AND PEDESTRIAN ACCESS

Veichular acces is organized from the northern highway and from the west. Pedestrians can also access directly from Teryan street.

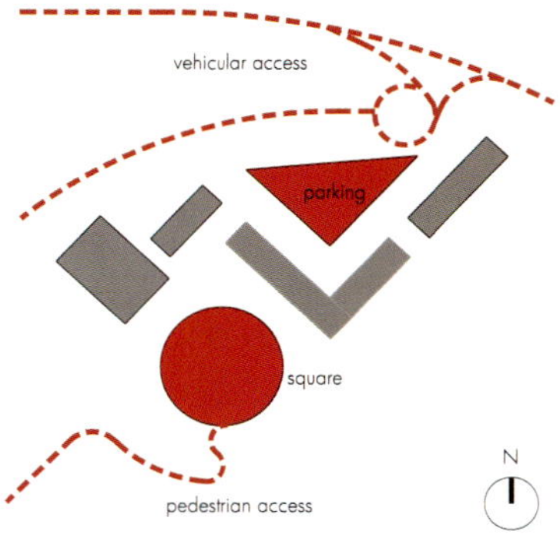

Phase **1a**: Hotel & IBC + Parking 1
Phase **1b**: Serviced Apartments D
Phase **1c**: Serviced Apartments E

Phase **2**: Building A (residences and shops) + Parking 2
Phase **3**: Building B (office-studios and shops) + Parking 3
Phase **4**: Building C (residences and shops)

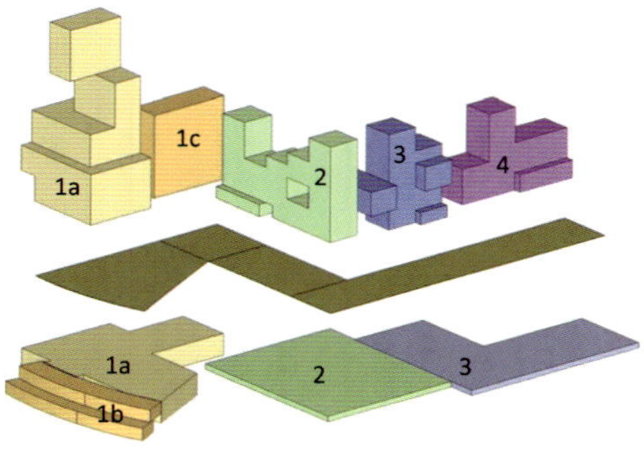

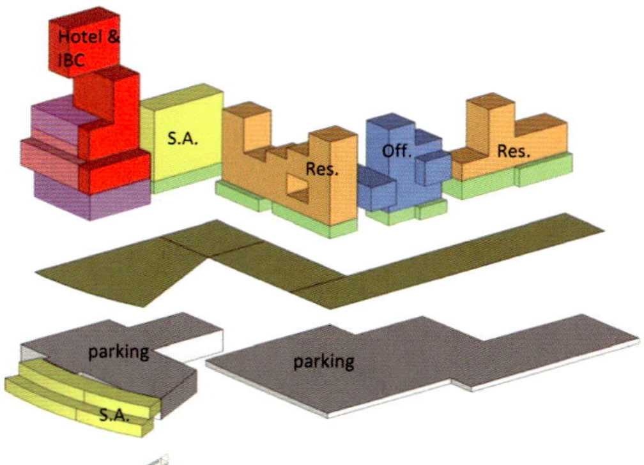

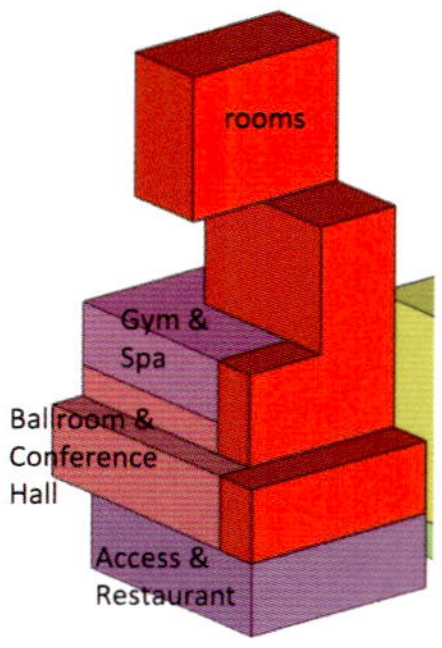

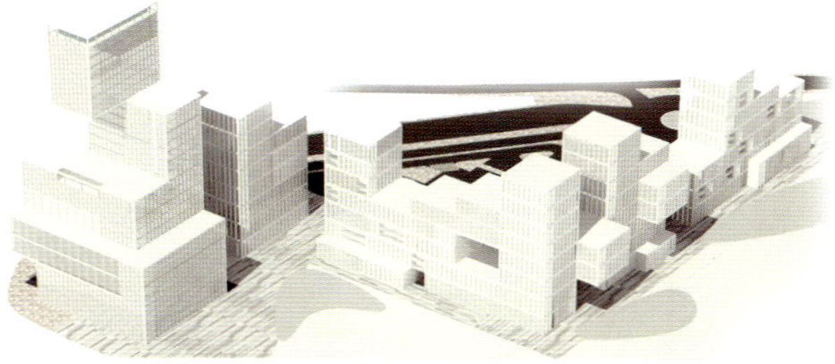

WIND

Buildings face south. North façades are closed and accommodate service spaces. South façades have big glass openings.
The setting out of the buildings acts as a barrier to the northern wind, acclimatizing and offering wind-free open spaces to the public realm and terraces.

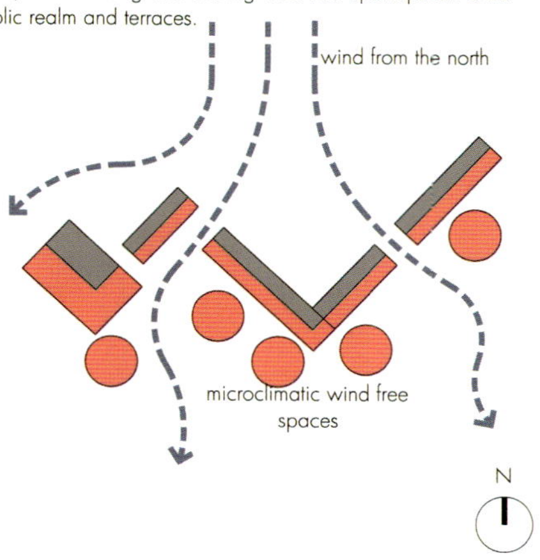

PUBLIC SPACES

The location of the open public areas were arranged to face the city and the Mount Ararat. Private open spaces were incorporated in terraces.

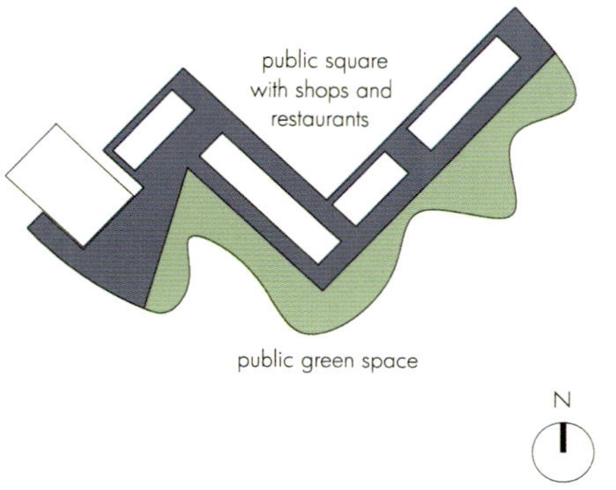

BUILDING PHASES

Phase 1a: Hotel & IBC + Parking 1
Phase 1b: Serviced Apartments D
Phase 1c: Serviced Apartments E
Phase 2: Building A (residences and shops) + Parking 2
Phase 3: Building B (office-studios and shops) + Parking 3
Phase 4: Building C (residences and shops)

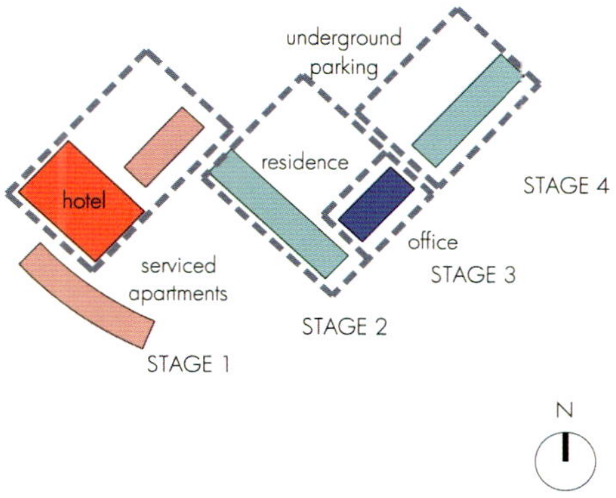

Introduction

The following text briefs and provides useful information about the main items regarding the proposal for the International Business Centre, Intercontinental Hotel in the City of Yerevan.

The following feasibility study shows details of the Intercontinental Hotel & IBC as well as three mixed-use building developments with residential areas, work spaces, and shops. The proposed scheme shows a series of spaces according to the requirements of the brief, but also provides phases of it. This extension is composed by a series of buildings, which fulfill the plot site, according to what we consider proper for the site and the city.

The proposed development takes into account key environmental considerations, explained beneath.

Architectural Concept

Location

The concept takes advantage of the huge possibilities that the site location offers. It is one of the best places in town and gives an extraordinary possibility to develop a first-class world architectural scheme. Because of the location, the proposal has the advantage to be the one which overlooks the whole city. At the same time it would be seen from everywhere. It would somehow be like the "acropolis" of the city, the idea of the city that observes and is seen from everyone.

Massing-Architectural Language

Buildings convey the meaning of mass, boxes located one on top of the other. Going through "white" piled-up boxes, we obtain the right volume composition to get a great architectural scheme. A compact and rising building makes part of the meaning for the concept as well as a contribution to some environmental key points.

The Green

A strong point in the concept attempts to resolve the lack of green areas in Yerevan. It drove us to group and to join all the uses in one strong linear intervention, leaving a big amount of free space for green areas and public spaces. The scheme gives the idea of "forest" to the city, allowing people to walk, cycle, and enjoy the public realm and the stunning views.

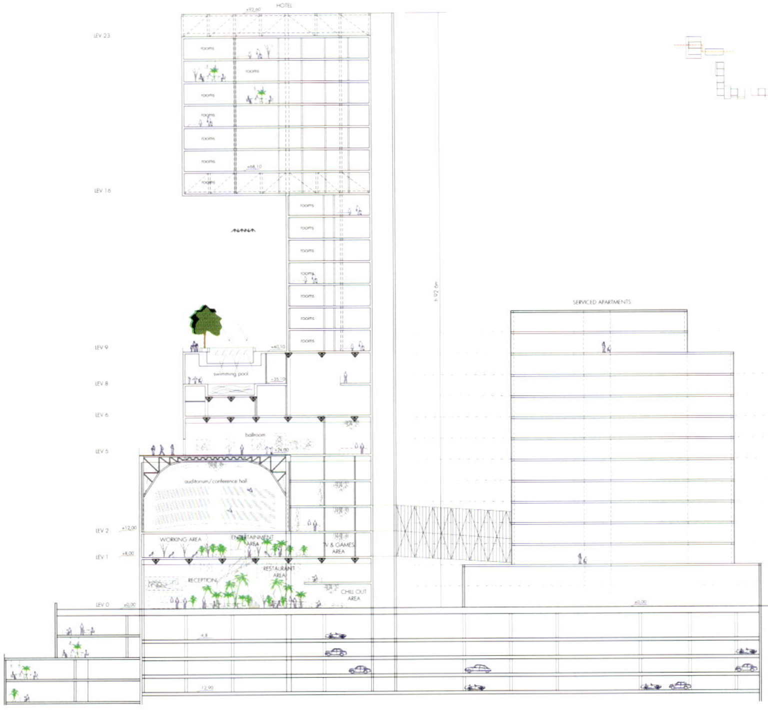

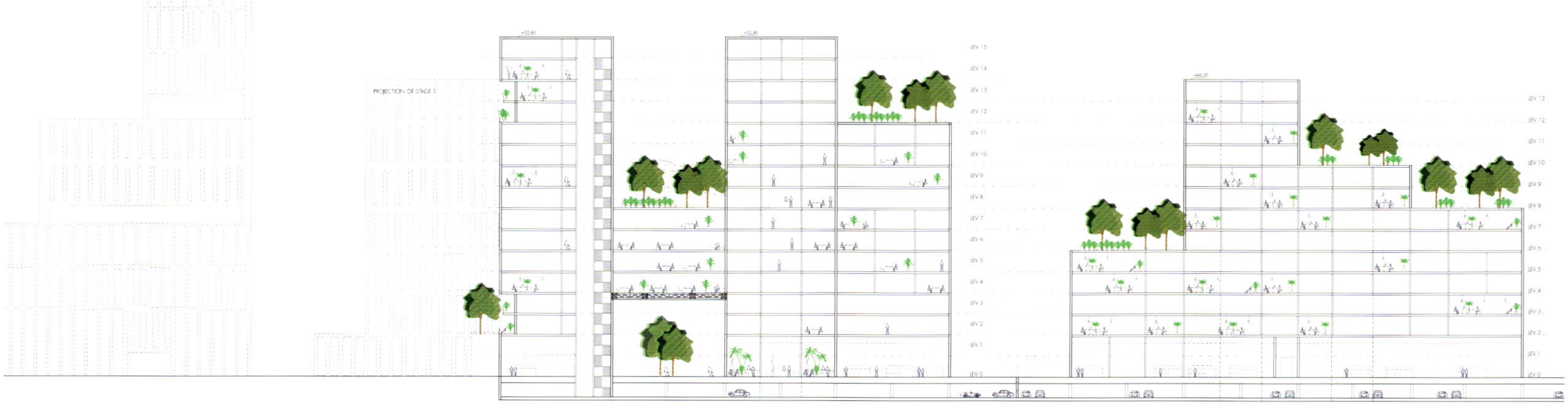

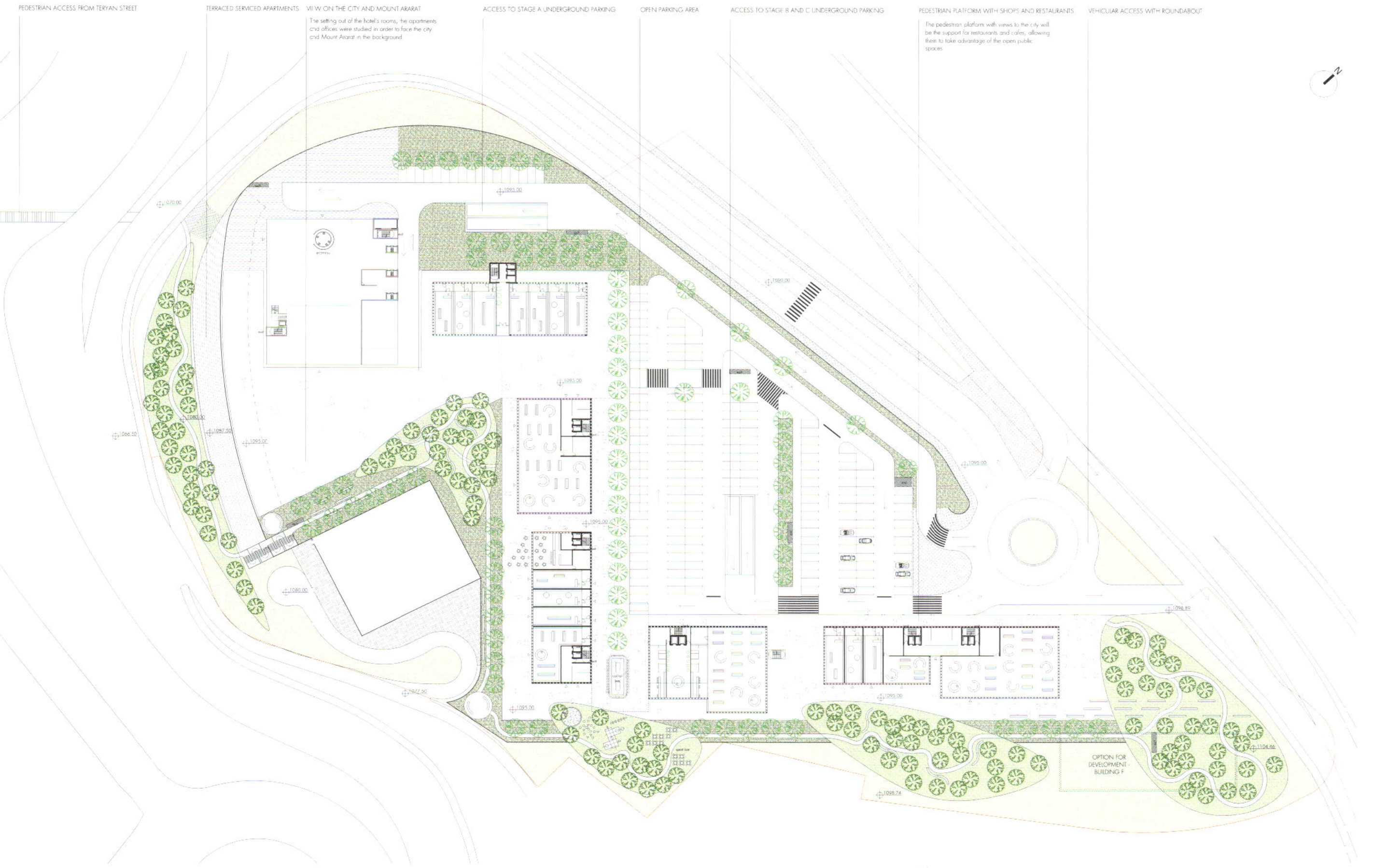
PEDESTRIAN ACCESS FROM TERYAN STREET
TERRACED SERVICED APARTMENTS
VIEW ON THE CITY AND MOUNT ARARAT
ACCESS TO STAGE A UNDERGROUND PARKING
OPEN PARKING AREA
ACCESS TO STAGE B AND C UNDERGROUND PARKING
PEDESTRIAN PLATFORM WITH SHOPS AND RESTAURANTS
VEHICULAR ACCESS WITH ROUNDABOUT

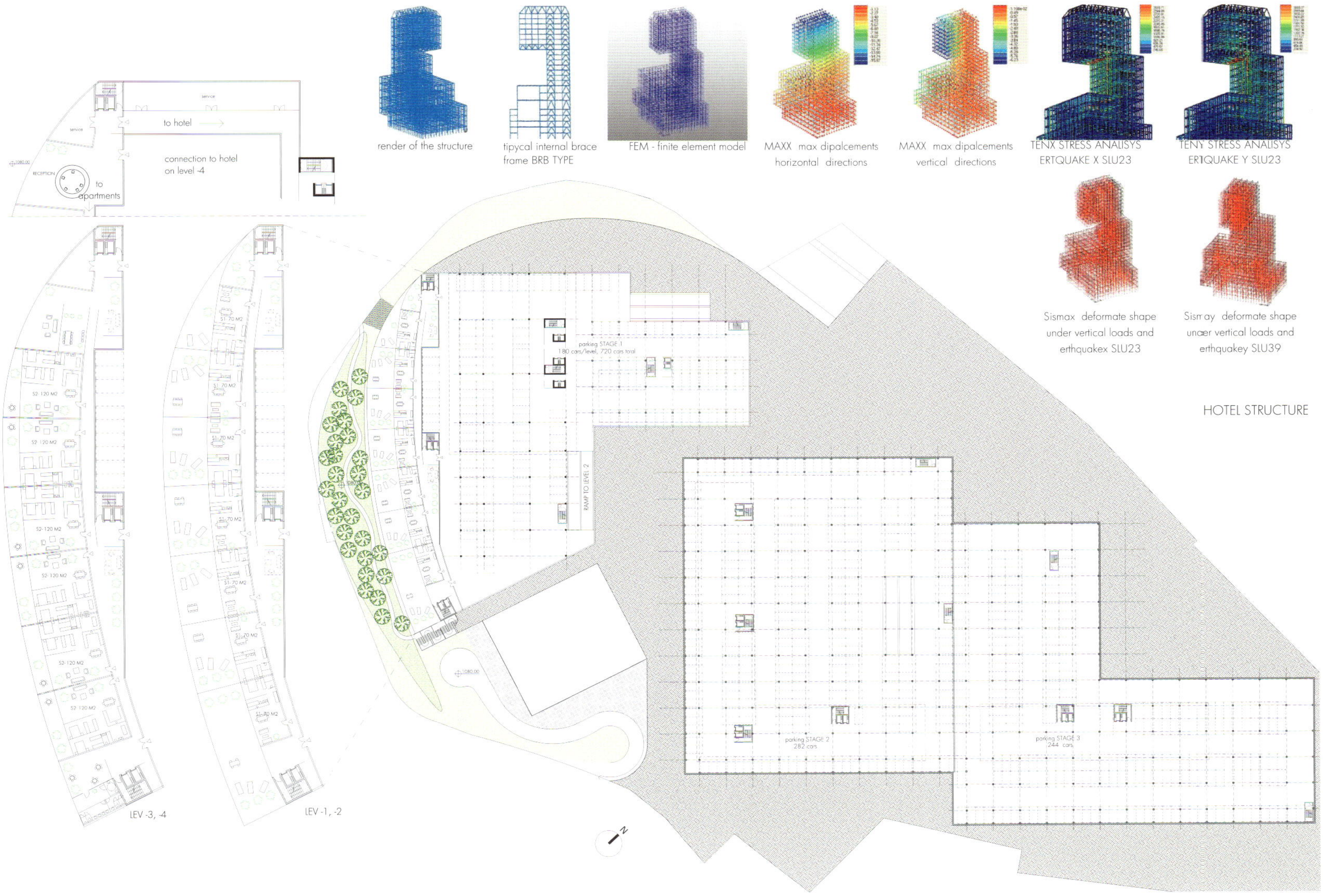
render of the structure
tipycal internal brace frame BRB TYPE
FEM - finite element model
MAXX max dipalcements horizontal directions
MAXX max dipalcements vertical directions
TENX STRESS ANALISYS ERTQUAKE X SLU23
TENY STRESS ANALISYS ERTQUAKE Y SLU23
Sismax deformate shape under vertical loads and erthquakex SLU23
Sismay deformate shape under vertical loads and erthquakey SLU39
HOTEL STRUCTURE
to hotel
connection to hotel on level -4
to apartments
parking STAGE 1
180 cars/level, 720 cars total
RAMP TO LEVEL -2
parking STAGE 2
282 cars
parking STAGE 3
244 cars
LEV -3, -4
LEV -1, -2

Mandarina Duck
鸳鸯品牌店

Arch. Pierluigi Piu

Credits
Year: 2008
Project Description: Concept for Renewal of the Brand's Shops Image
Images: Arch.Antonio Cuccu

意大利鸳鸯——国际知名品牌，1977年成立，主要做箱包与服装配件——产品定位于都市年轻人，设计热情有活力，关注不断变化的新产品，创造旅行的梦想。

“旅行梦想”这个观念，引导着该品牌在世界各地店面的翻新设计理念——与品牌层层涌现的休闲设计相符。

在设计主题众多的旅行选项中（如冒险、航海、工作等），我们选择了朴实的方式——火车或者高速公路——带领旅行者游遍全世界各大城市。最终，我们设计了一个高速公路网；在网络交叉的结点处是世界各地理想的大都市。该设计创造出一种室内低浮雕的效果，又像是放置了无数宽大的格子。而地面上则用透明的树脂，宣扬“意大利鸳鸯”在世界范围内无休止的旅行。

室内设计了许多竖直的展示架，这些展示架沿“旅行路线”布置，由三个简单模型按不同组合设计而成。展示架短短的，但因沿旅行路线摆放，反而显得绵延不断。每个架子都有两块对立平行的玻璃滑板支撑，线条流畅，外形稳固，体态轻盈，视觉上透明化。另外，墙壁上也设计了悬挂式的展示架；墙壁呈白色，背光铝制的方格，交叉点处刻着全世界各大城市的名字。造型像三明治一样，两面白色，中间夹杂着黑色——很容易让人想起地下交通图的样子，这也是该项目的特点，设计如绘图一样的感觉。

除了上面的那些立架，我们在墙壁上清晰地雕刻了一些壁龛，上面置入上釉的无框匣子，可以展示鸳鸯新品——眼镜、钱包、香水等小物件及服饰配件。

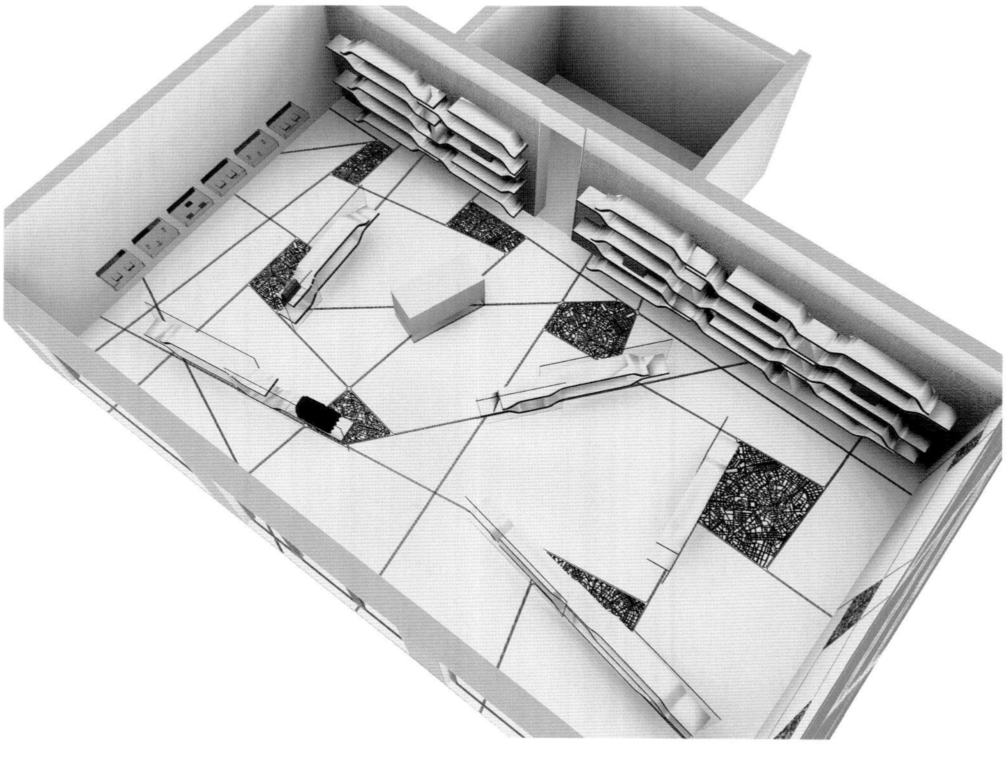

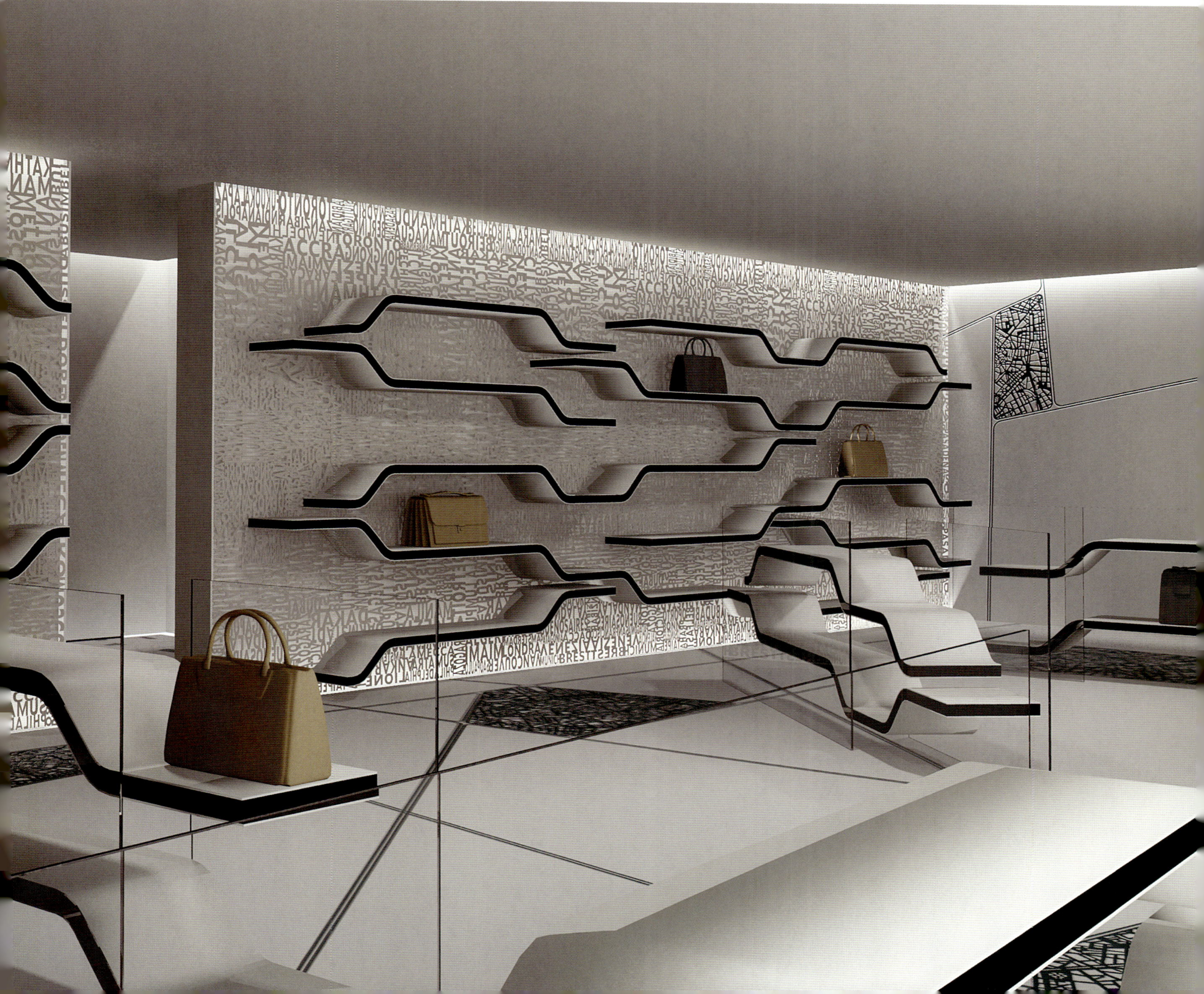

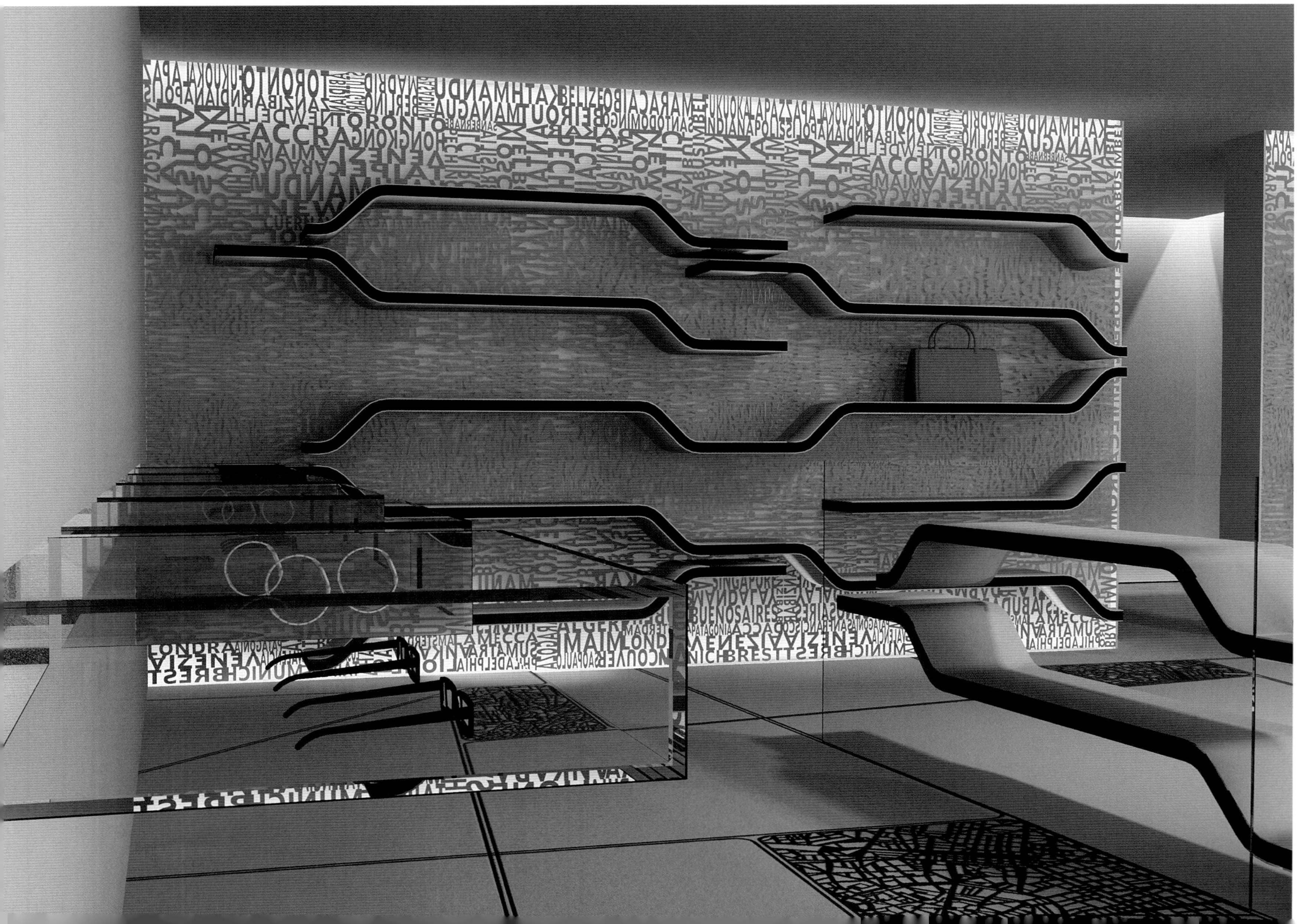

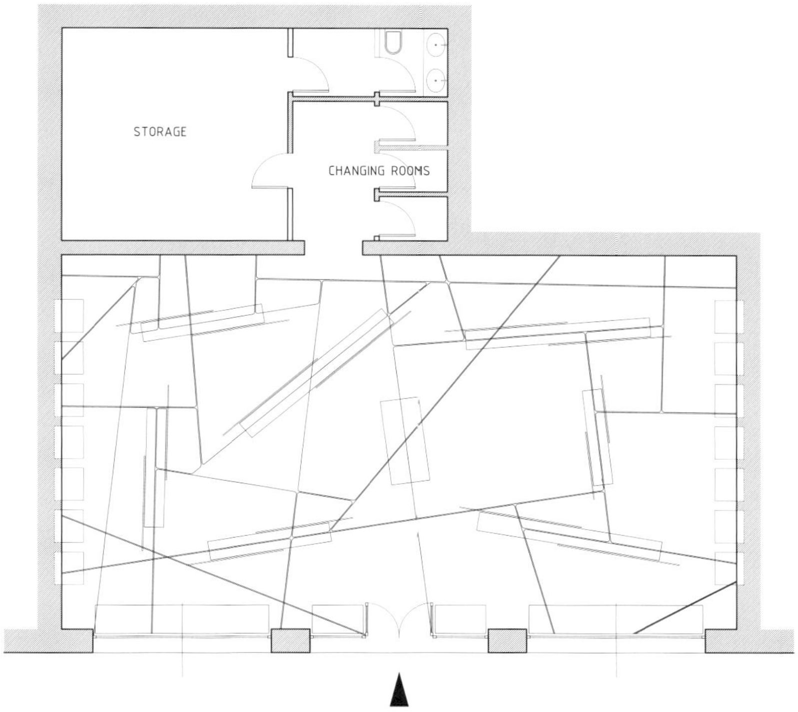
STORAGE
CHANGING ROOMS

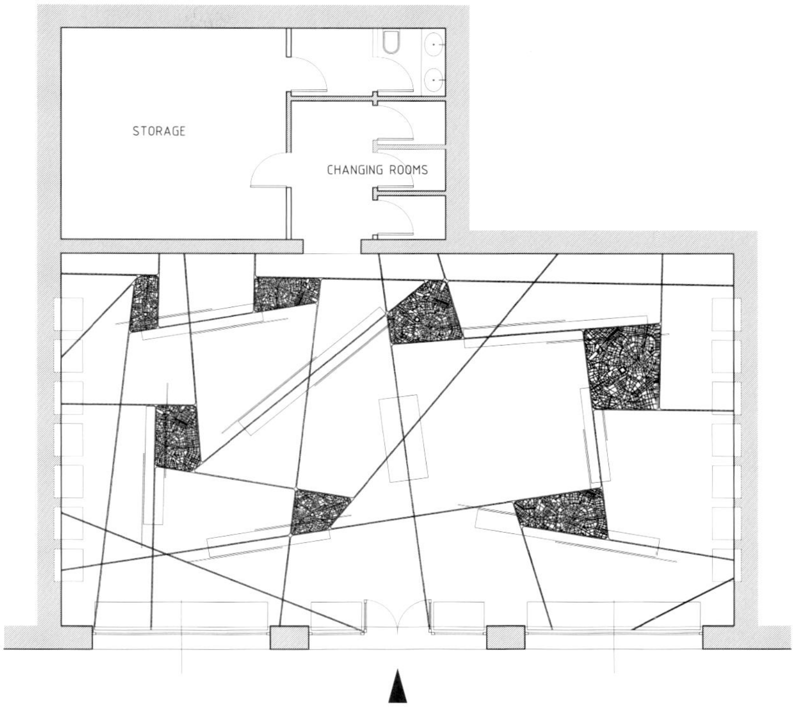
STORAGE
CHANGING ROOMS

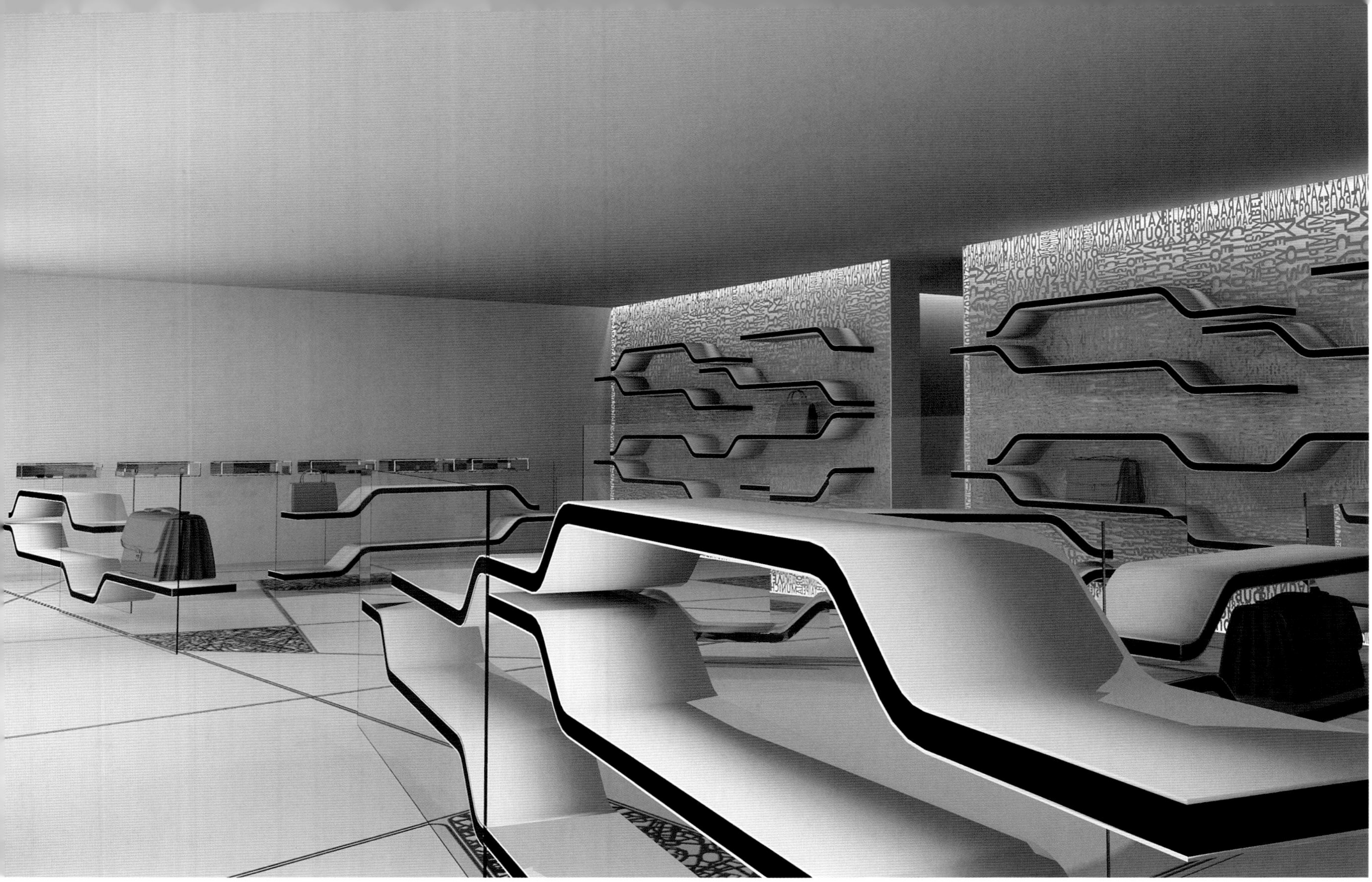

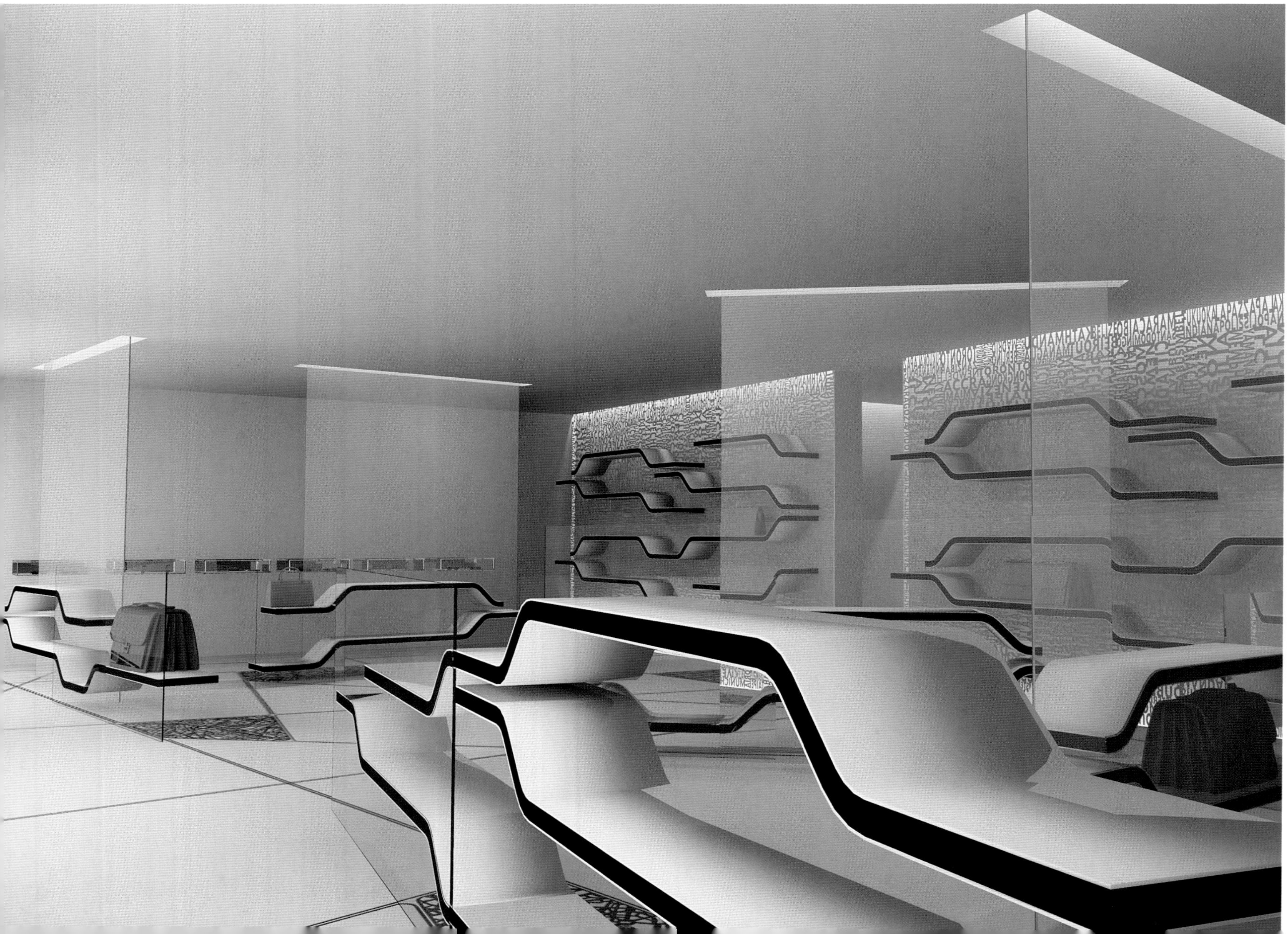

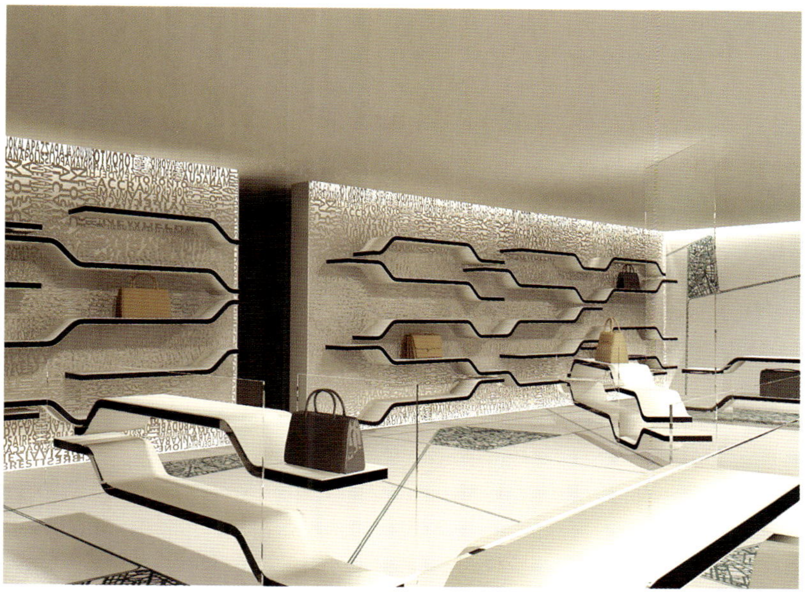

Mandarina Duck – the renown brand since 34 years present and, nowadays, well-established on the market of bags and accessories production – aims at a clientele young and cosmopolitan, open minded, attentive to new products, always on the move, whether that means travelling thousands of miles or simply going from A to B in the city.
And on the idea of evoking the dream of travelling was in fact based the briefing of this project of renewal of the brand's shops in the world; a renewal that should coincide with the strong increment of production given by Mandarina to its leisure devoted products.
Among all different possible ways of travelling (adventure, cruise, work...) the chosen option was to put on stage the metropolitan kind of journey, which – by plain, train or speedway - takes us around the big cities of the world. And is, in fact, an intricate net of speedways – at the junctions of which are the maps of ideal metropolis of our planet – the feature characterizing the surfaces of the spatial container, engraving them with a bas-relief effect (which – on the floor - will then be filled with a transparent resin) and suggesting the idea of an uninterrupted movement along the roads of the world.
Along such routes are displayed a number of assembling standing shelves, generated by different combinations of three simple modules: a long and a short shelf, and a double curve. Thanks to the two parallel and reciprocally slid sheets of glass supporting each shelf, they appear like they're freely fluctuating in the space, so giving an impression of great lightness and granting a maximum visual transparency. The same shelving system is hung to the walls too, standing out against the background of a white lacquered and backlit aluminum trellis, out of which has been laser cut a mass of interlaced names of big cities of all countries. The particular sandwich-like construction of the shelves (two white faces with an interposed dark core), while emphasizing the graphic character of this project, also strongly evokes the classic iconography of underground public transports maps.
All the above is complemented by a series of glazed and frameless caskets – overhanging out of some alcoves expressly carved into the walls - that allow to display small objects and accessories (spectacles, wallets, scents...) recently included within the brand's range of products.

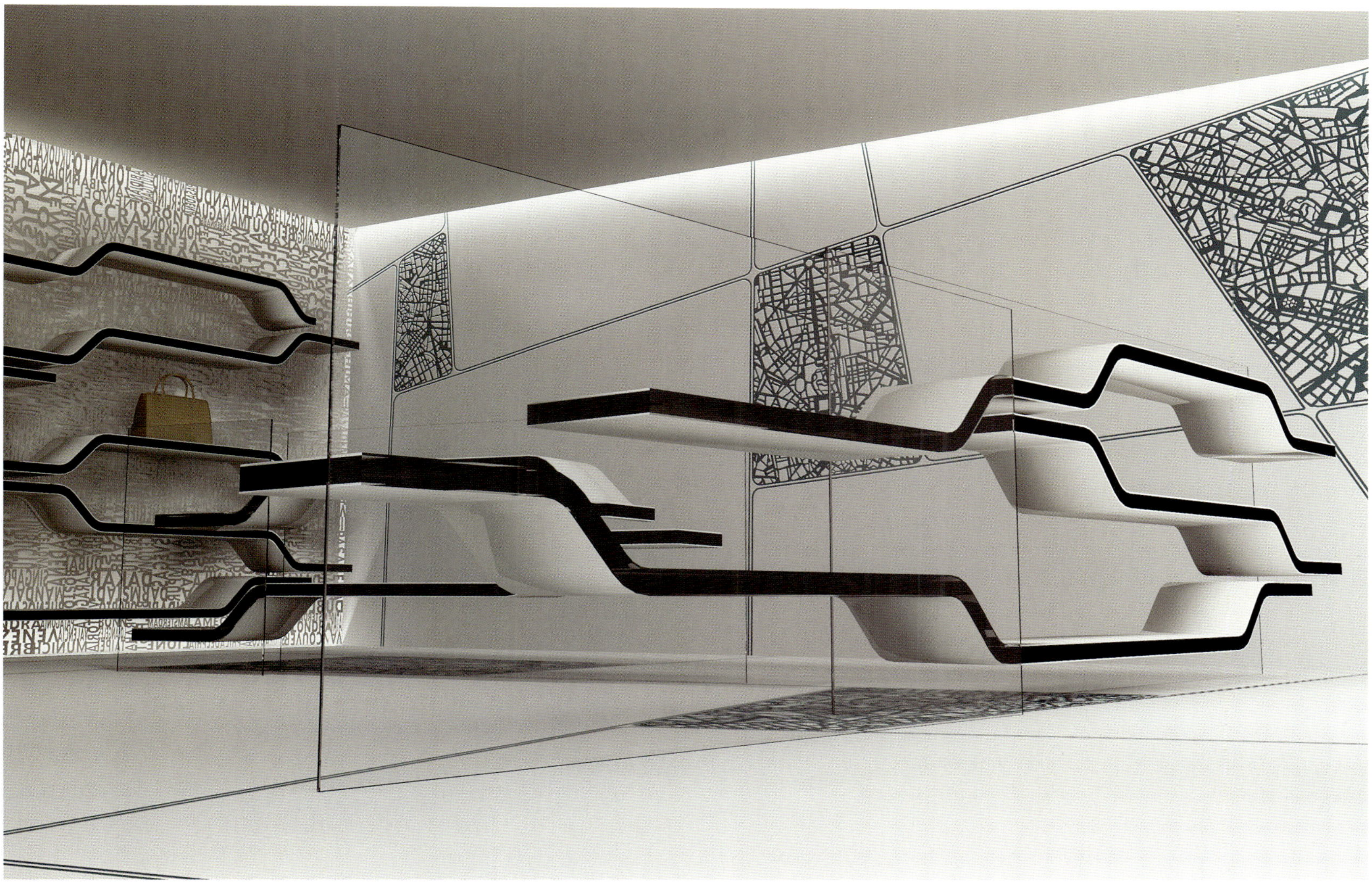

ARCHITECTURE COMPETITION WORKS

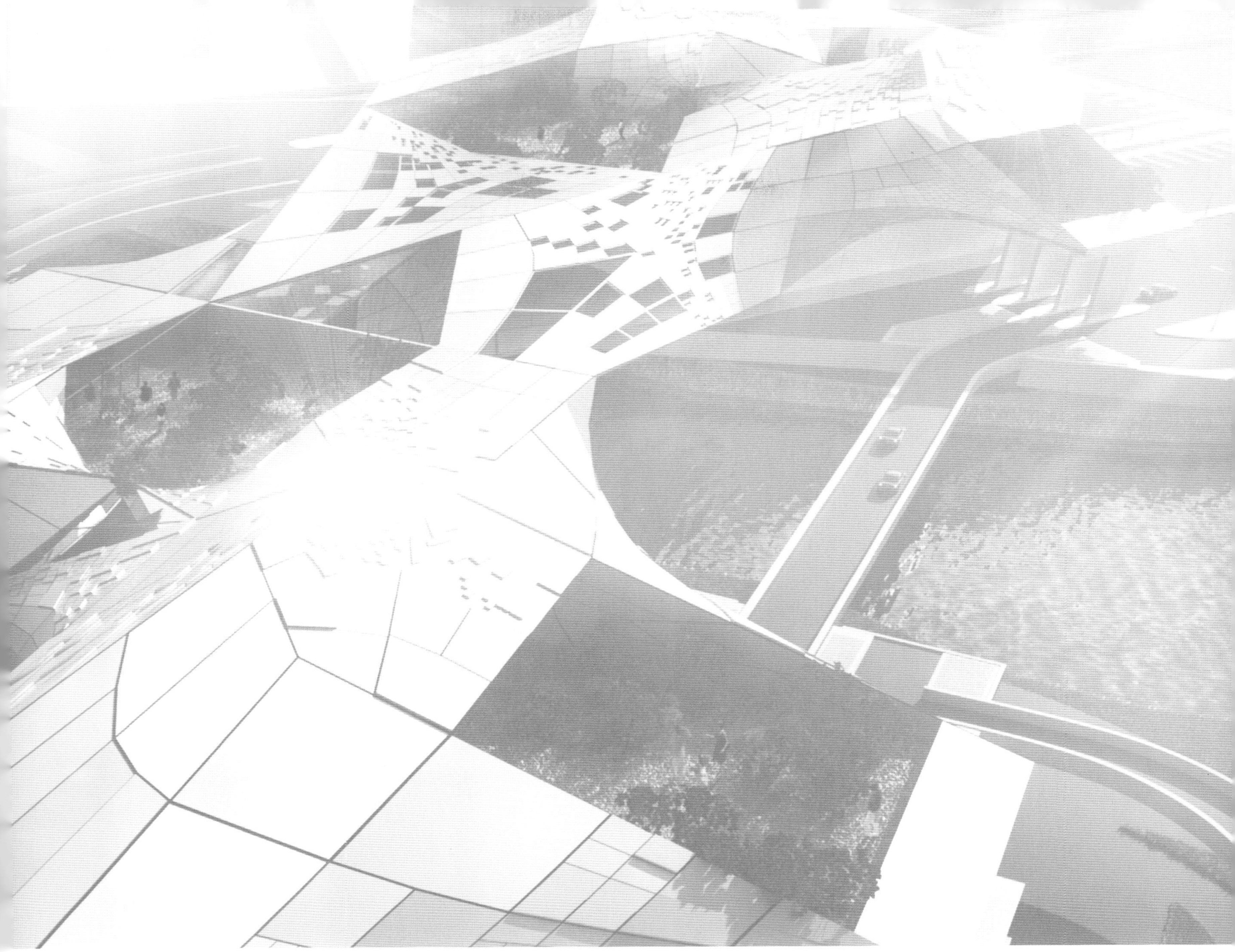

Public Facilities
公共设施

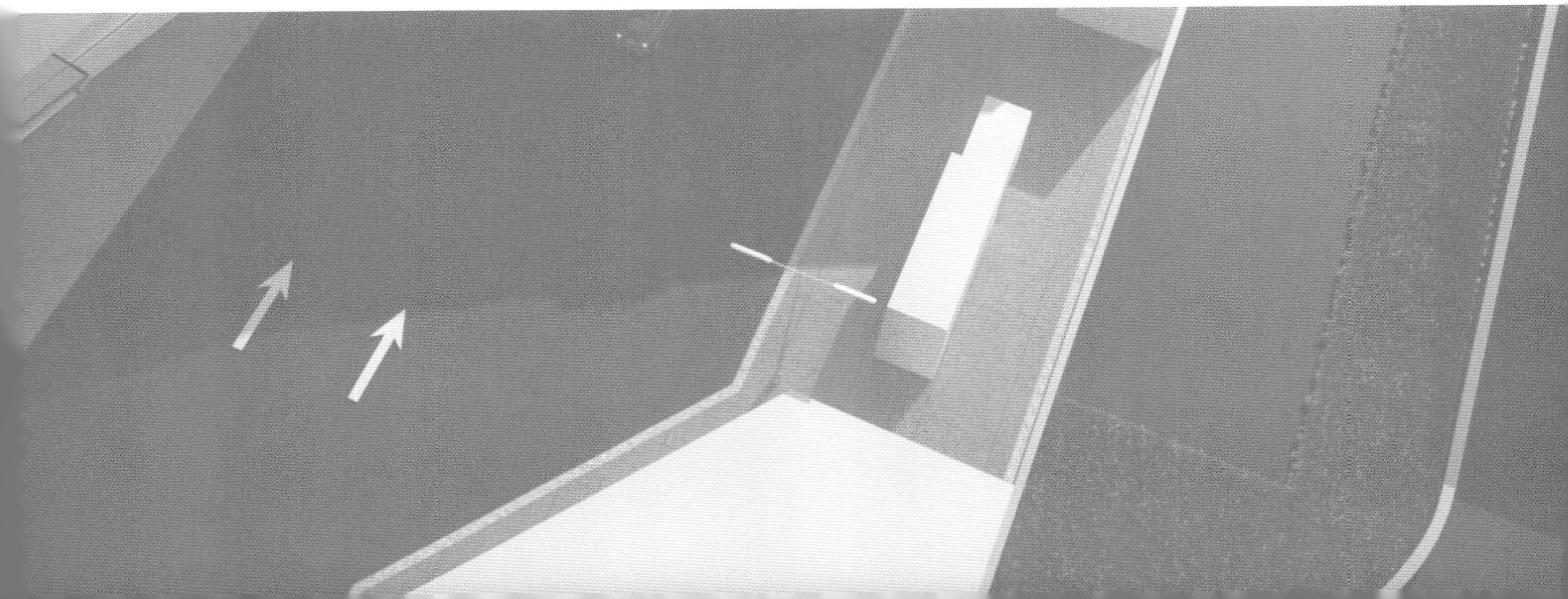

Aarhus University Hothouse in the Botanical Garden – A Sustainable Hothouse

奥尔胡斯大学植物园温室——可持续温室

C. F. Møller Architects

2009 1. Prize in Architectural Competition
2009年建筑竞赛一等奖

Credits
Engineering: Søren Jensen Rådgivende Ingeniører
Architect: C. F. Møller Architects
Landscape: C. F. Møller Architects
Location: Møllevejen, Århus C

Area: 3,300 m² (1,242 m² new tropical hothouse and 2,071 m² renovation and rebuild of existing hothouse)
Year: 2009-2012

C. F. Møller建筑事务所和Søren Jensen Rådgivende Ingeniør公司合作，在为奥尔胡斯大学植物园建造一个新的温室的建筑竞赛中获胜。可持续的设计、新的材料和先进的科技应用到了温室的有机形式中。

奥尔胡斯植物园大学中蜗牛形的温室是温室建筑中的国家级标志。该项目在1969年由C. F. Møller建筑师事务所设计并很好地被周围环境所接受。因此在设计新的温室时参考原来的建筑是相当重要的。

C. F. Møller的建筑师和合作伙伴Tom Danielsen说："这个竞赛旨在寻找一个独立的、与众不同的新温室，但是我们必须保证新的建筑和旧的建筑相互默契配合。"

在这个有机形式和大空间中公众可以在树顶中探测，那里呈现出植物学和不同气候带的景象，使得奥尔胡斯大学新的温室在未来成为欧洲级的温室建筑。

能量设计

新温室的设计是基于在材料知识、室内气候和科技等领域实现能源节约。

建筑师和工程师使用先进的计算器来优化建筑的结构，保证形式与能量消耗的相互作用达到最佳状态并最合理地利用太阳光。建筑半圆形的形状和具有指南针似的定位功能，使它在最小的面积下创造出最大的空间，并且在冬天获得最多的阳光，夏天则相反。

植物知识中心

整个项目还包括对旧温室的修复，那里会改造成一个新的向公众开放的植物知识中心，也作为新的热带温室的延伸。这个项目预计于2012年完工。

VEKSTHUS

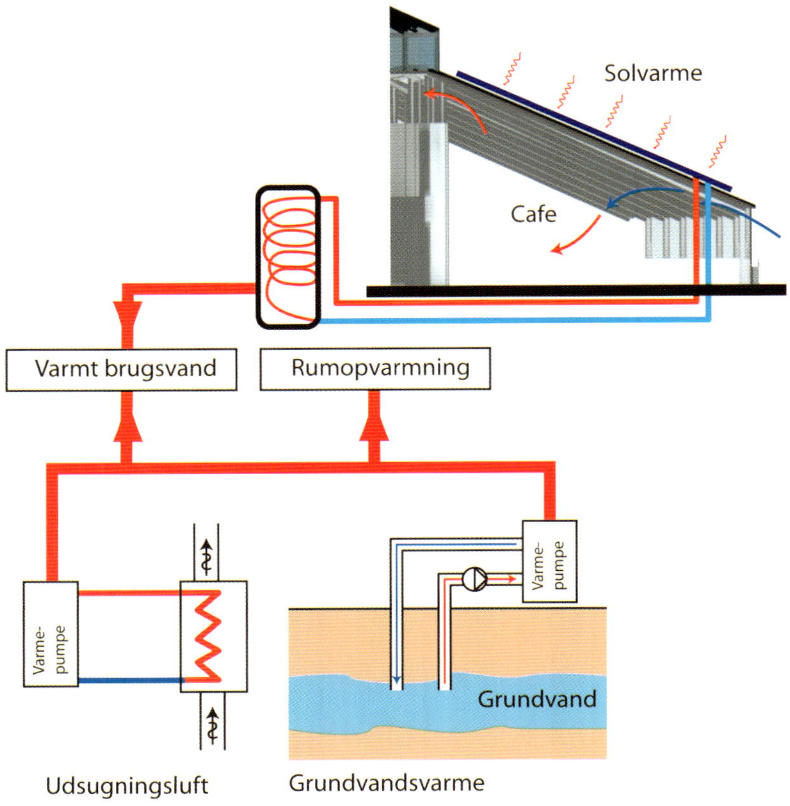

Solvarme
Cafe
Varmt brugsvand
Rumopvarmning
Varme-
pumpe
Varme-
pumpe
Grundvand
Udsugningsluft
Grundvandsvarme

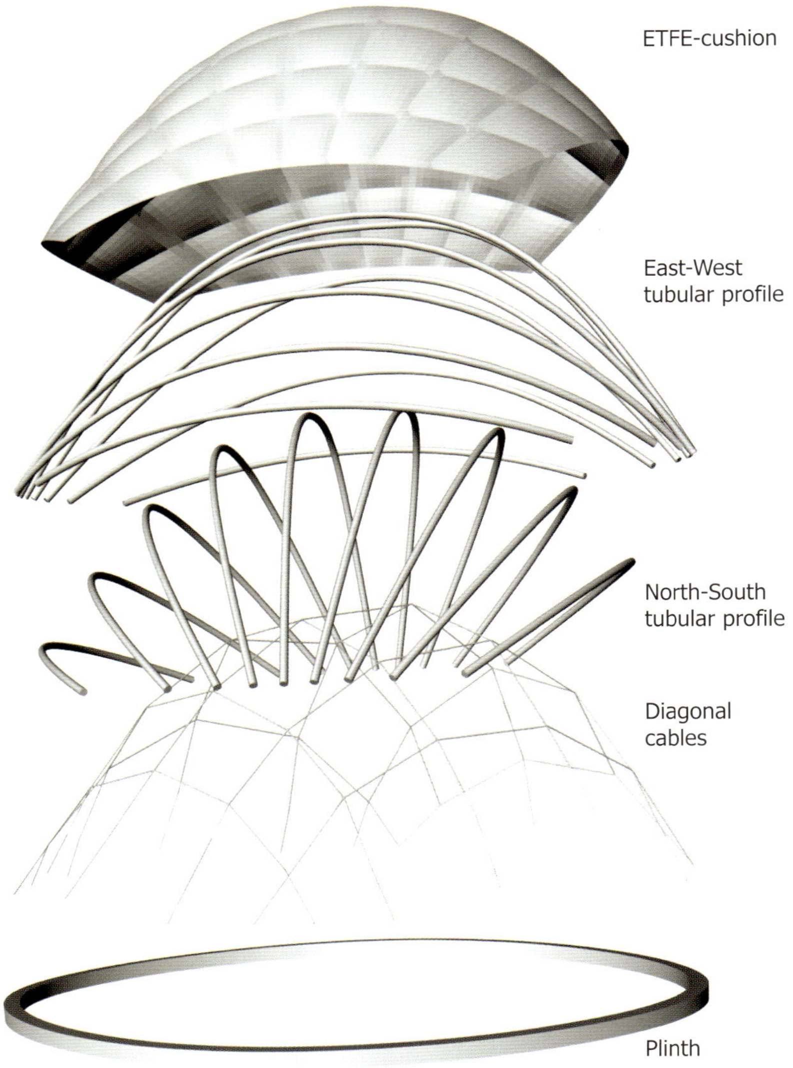

ETFE-cushion
East-West
tubular profile
North-South
tubular profile
Diagonal
cables
Plinth

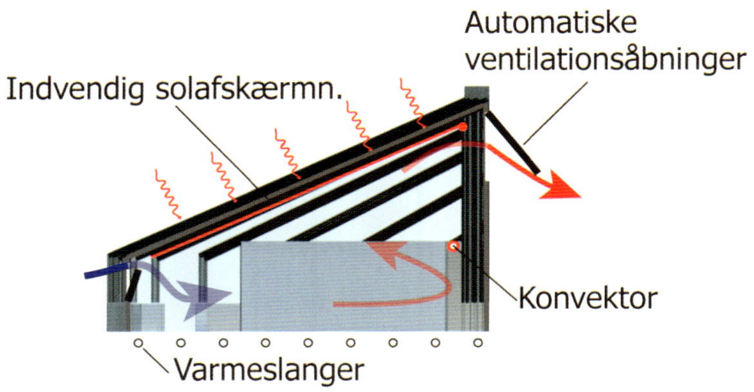

Automatiske
ventilationsåbninger
Indvendig solafskærmn.
Konvektor
Varmeslanger

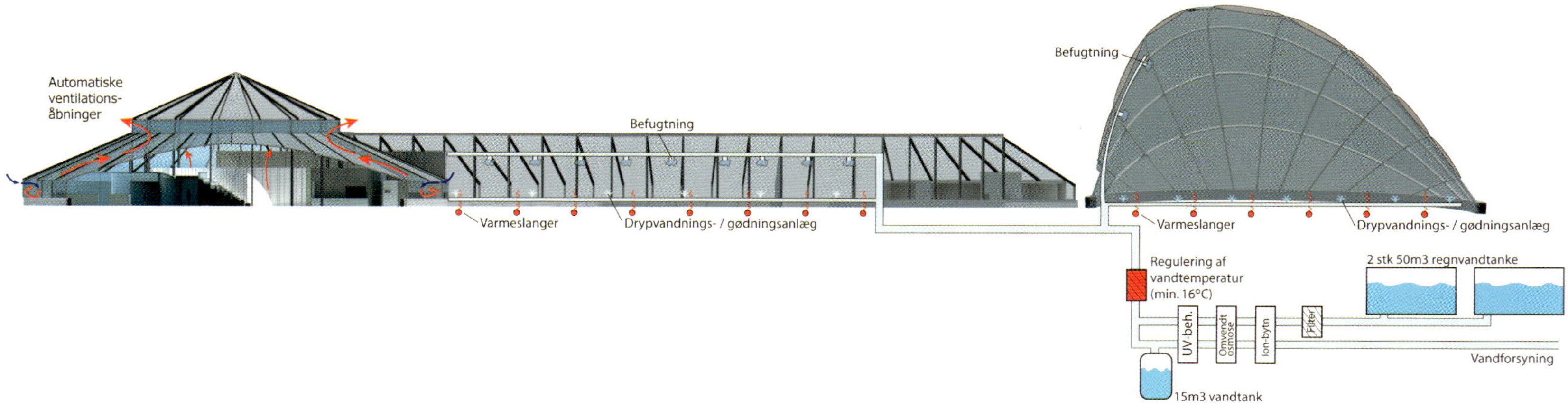

Automatiske
ventilations-
åbninger
Befugtning
Befugtning
Varmeslanger
Drypvandnings- / gødningsanlæg
Varmeslanger
Drypvandnings- / gødningsanlæg
Regulering af
vandtemperatur
(min. 16°C)
2 stk 50m3 regnvandtanke
UV-beh.
Omvendt
osmose
Ion-bytn
Filter
Vandforsyning
15m3 vandtank

Teaching
conference
Workshop
Internal
plaza:
exhibition/shop
plant
Workshop
Mediterranian zone
foyer/cafe
main entrance
Courtyard cafe
Desert zone
Palm house
Mountain Forest zone
Parking
Offices

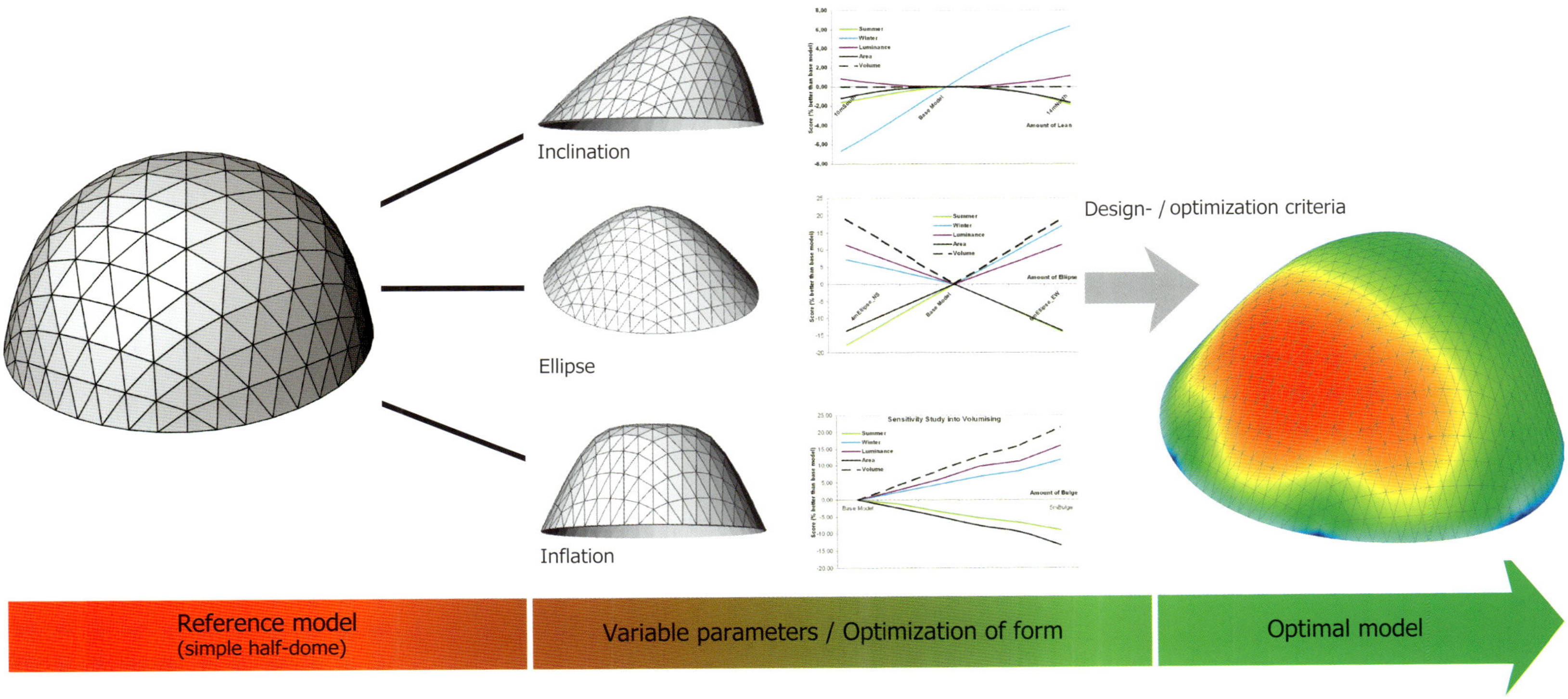
Inclination
Ellipse
Inflation
Design- / optimization criteria
Reference model
(simple half-dome)
Variable parameters / Optimization of form
Optimal model

Kørestolsbrugere / Gangbesværede

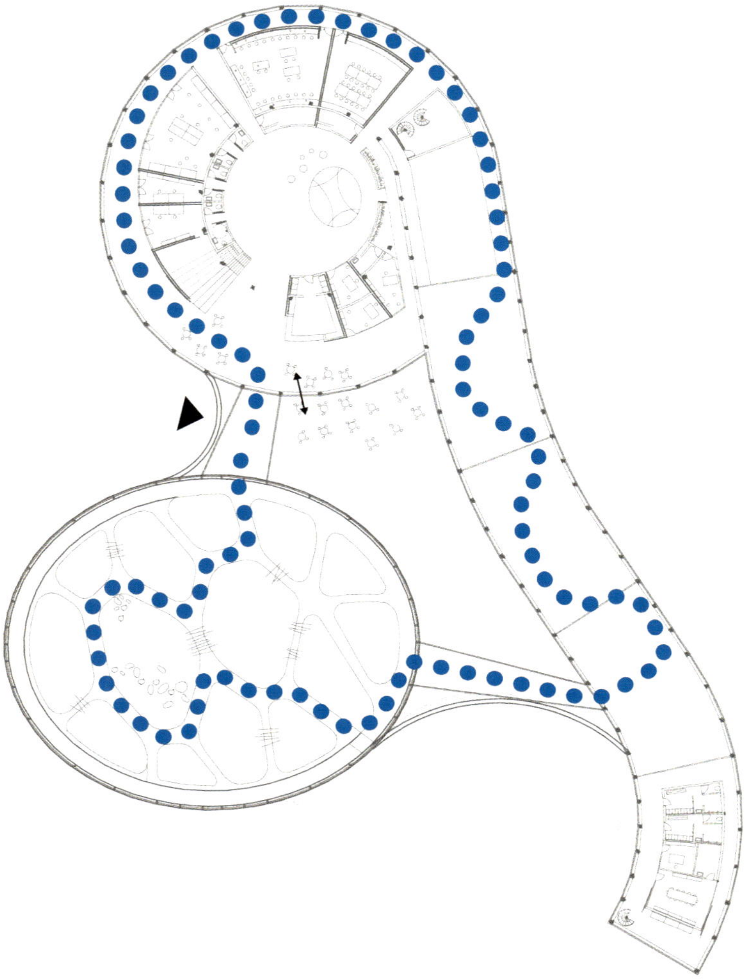

Pædagogisk track

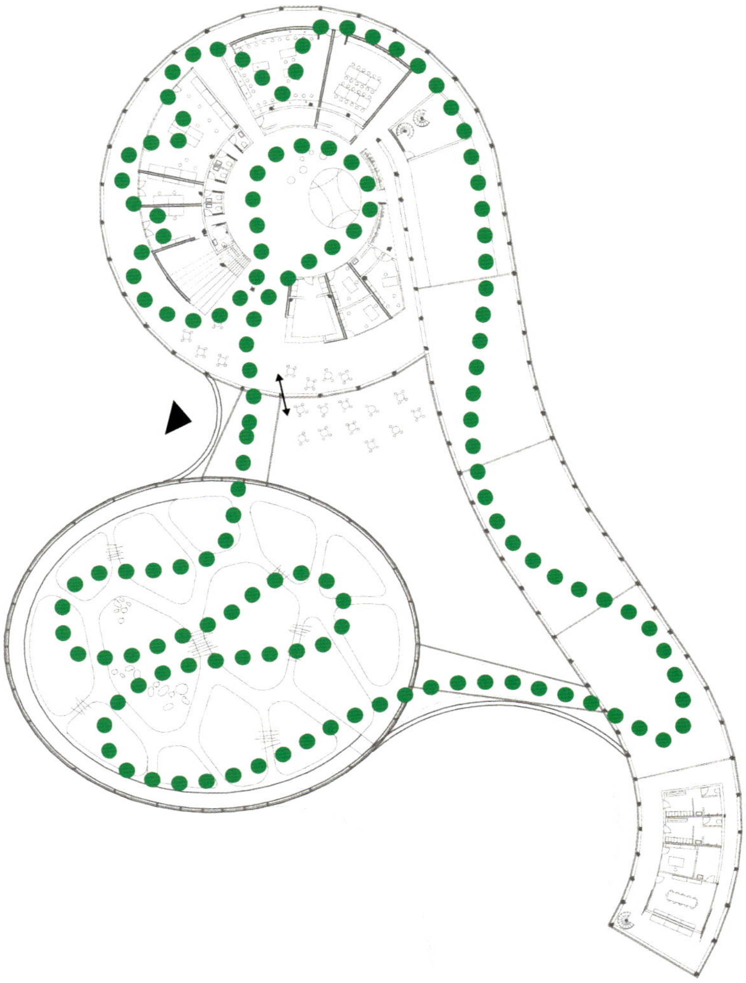

"Fast" track

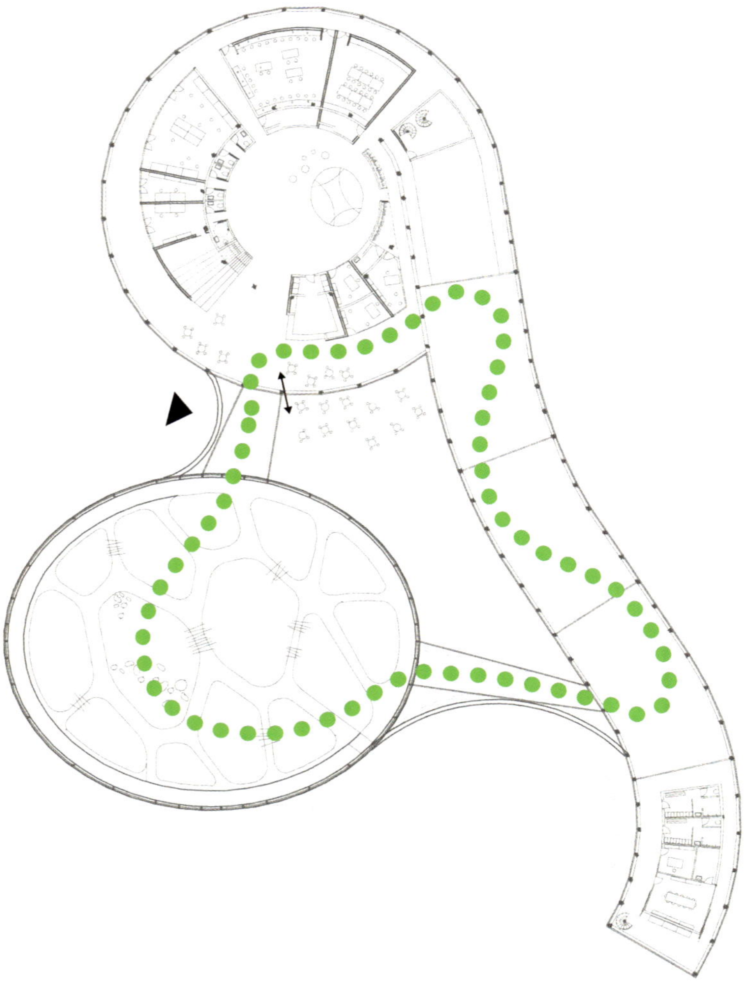

Service track

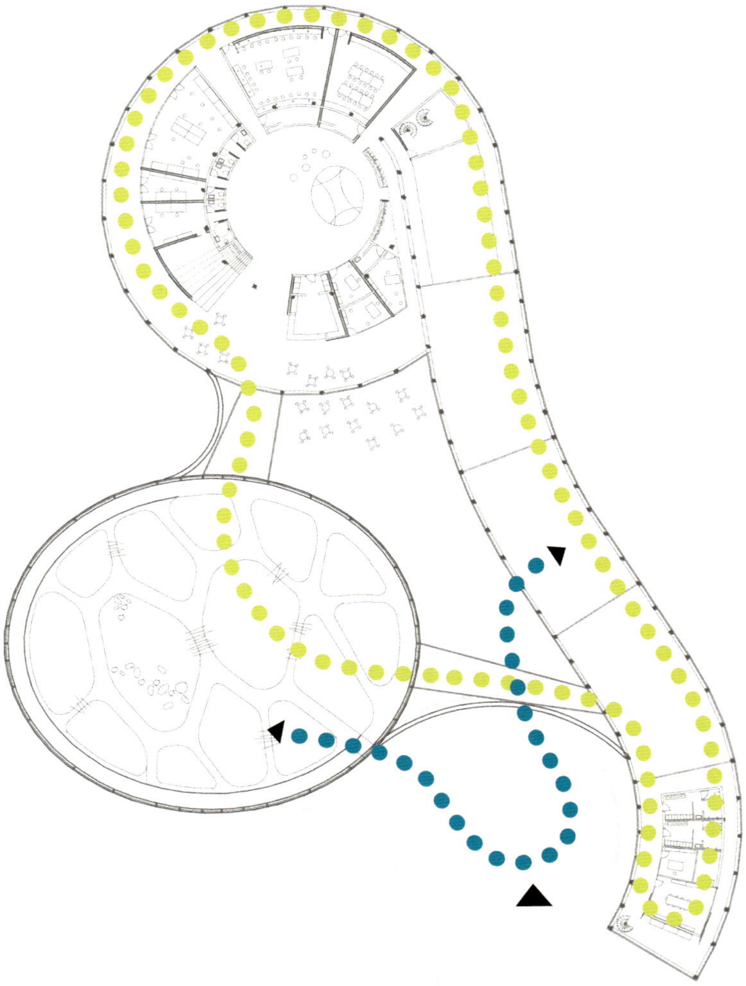

C. F. Møller Architects, in cooperation with Søren Jensen Rådgivende Ingeniør firma, has won the architectural competition for a new hothouse in Aarhus Botanic Garden. Sustainable design, new materials and advanced computer technology went into the creation of the hothouse's organic form.

The snail-shaped hothouse in the Botanic Garden in Aarhus is a national icon in hothouse architecture. It was designed in 1969 by C. F. Møller Architects, and is well adapted to its surroundings. Accordingly, it was important to bear the existing architectural values in mind when designing the new hothouse.

"The competition sought an independent and distinctive new palm house, but it was essential for us to ensure that the new building would function well in interplay with the old one," says Tom Danielsen, architect and partner with C. F. Møller Architects.

The organic form and the large volume, in which the public can go exploring among the tree-tops, present botany and a journey through the different climate zones in a way which will make the new hothouse in Aarhus a future attraction in a pan-European class in hothouse architecture.

Energy Design

The design of the new hothouse is based on energy-conserving design solutions and on a knowledge of materials, indoor climate and technology.

Using advanced calculations, the architects and engineers have optimised their way to the building's structure, ensuring that its form and energy consumption interact in the best possible manner and make optimal use of sunlight. The domed shape and the building's orientation in relation to the points of the compass have been chosen because this precise format gives the smallest surface area coupled with the largest volume, as well as the best possible sunlight incidence in winter, and the least possible in summer.

Botanical Knowledge Center

The total project also includes a comprehensive restoration of the old hothouse in which the palm house will become a new botanical knowledge center aimed at the general public, at the same time as the complex is extended with the new tropical hothouse. The project is expected to be completed in 2012.

New University Hospital in Aarhus

新奥尔胡斯大学医院

C. F. Møller Architects　Cubo Arkitekter A/S

2007 First Prize in International Architectural Competition
2007年国际建筑竞赛一等奖

Credits

Client: Region Mid-Jutland
Client Consultant: Niras A/S
Architects: C. F. Møller Architects in collaboration with Cubo Arkitekter A/S
Landscape Architects: Schønherr Landskab A/S, Tegnestuen Havestuen
Engineering: Rambøll Group A/S, Alectia A/S, Søren Jensen Rådgivende Ingeniørfirma
Other Collaborators: Nosyko AS, Lohfert & Lohfert AS, Capgemini Danmark A/S
Size : 250,000 m², Total Floor Area: Approx. 400,000 m², total Plottage: Approx. 970,000 m²

Year : 2007-2020
Competition Year: 2007
Preliminary Price Estimate: DKK 5.5 – 6.0 billion (excluding equipment)
Staff: Approx. 9,000
Beds: Approx. 1,300
Admissions: Approx. 100,000 Annually
Out-patient Visits: Approx. 600,000 Annually
Total Floor Area: Approx. 400,000 m²
Total Plottage: Approx. 970,000 m²
The Task Will Extend Over the Next 12-15 Years.

新奥尔胡斯大学医院是丹麦历史上最大的医疗建设项目。它在原奥尔胡斯大学医院基础上，将形成一个医院综合体，其面积相当于一个城镇那么大。

这所医院既是大学的附属医院，同时也是该地居民基础医疗与活动中心。

它以城镇的模式建造，社区、街道与广场等的划分，形成多样、动感与绿色的城市化区域。

医院灵活的设计将满足未来医疗技术、诊断方式与工作设施的要求。同时，新的设计也使病人的就诊环境与医院职工的工作环境得到一个大大的提升。

The biggest hospital construction project in Danish history, the New University Hospital in Aarhus, will be built onto the existing Aarhus University Hospital, Skejby, to form a combined hospital complex. The New University Hospital in Aarhus will be the size of a Danish provincial town, and will also be the largest workplace in the city of Aarhus.

The hospital is intended to function both as a university hospital, regional center and basic hospital for citizens in the region.

The large hospital complex will be organized like a town, with a hierarchy of neighborhoods, streets and squares providing the basis for a diverse, dynamic and green urban area.

The hospital has been designed to flexibly accommodate future requirements with regard to technology, forms of treatment and working practices, and it will also bring about a considerable qualitative lift in both the experiences of patients and the working conditions of the staff.

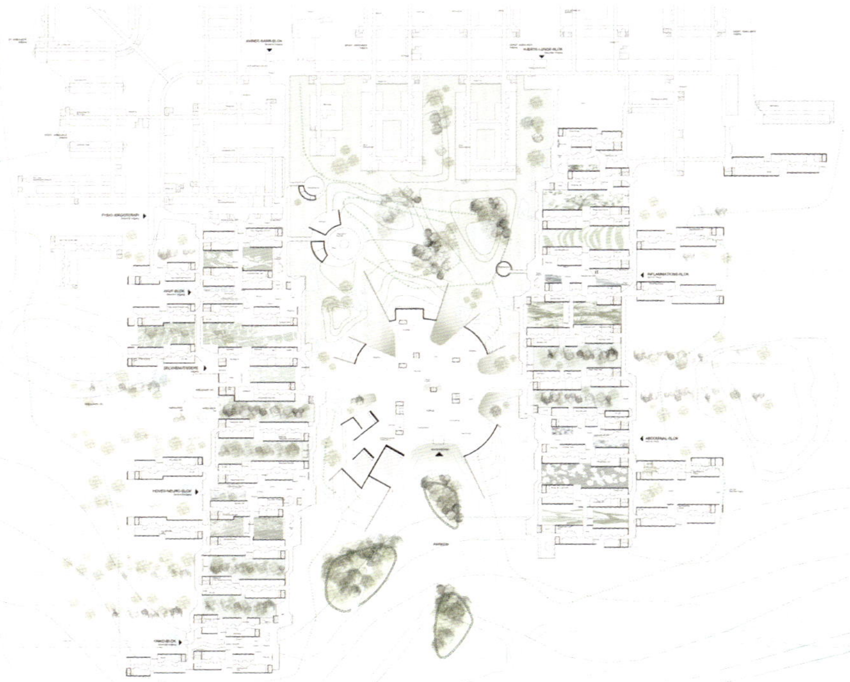

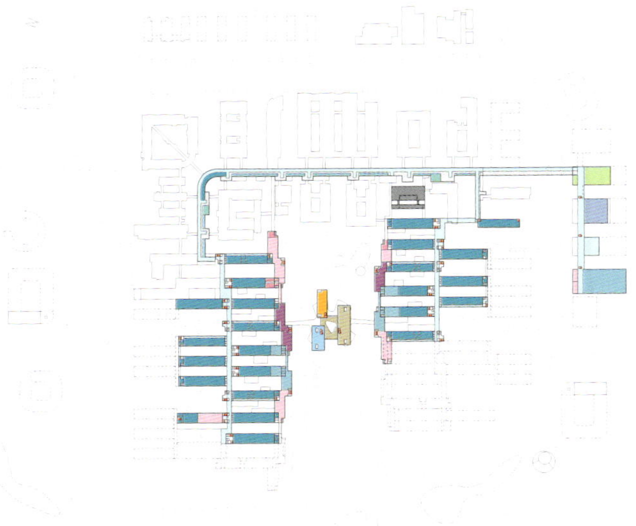

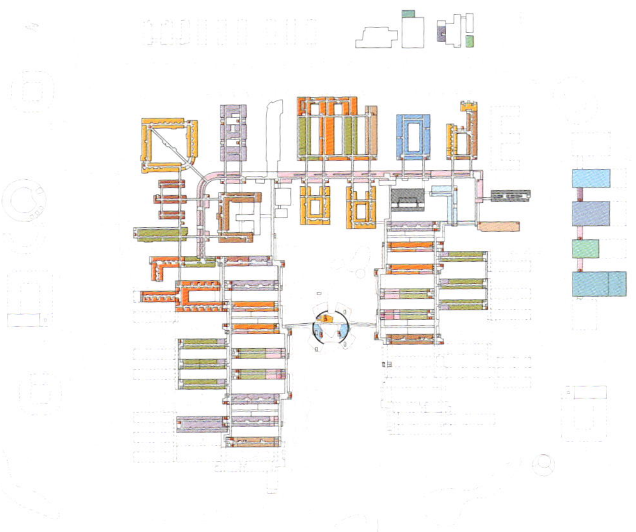

AKUT
onkologi
hoved/neuro

SKEJVIEW

Kryptopyrrol

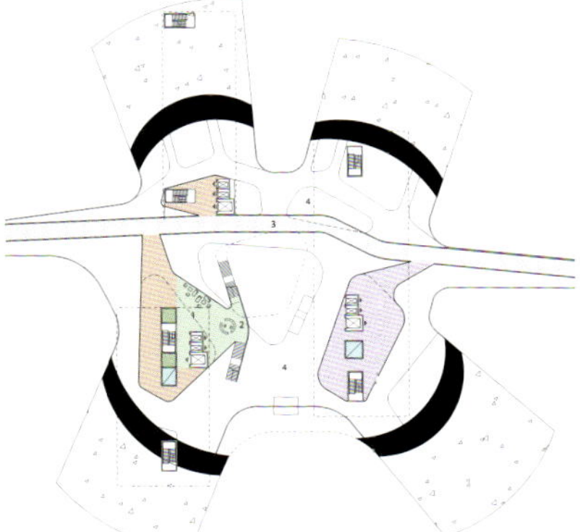

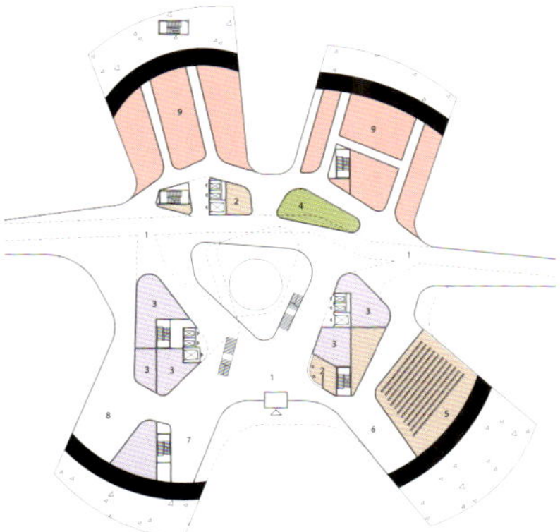

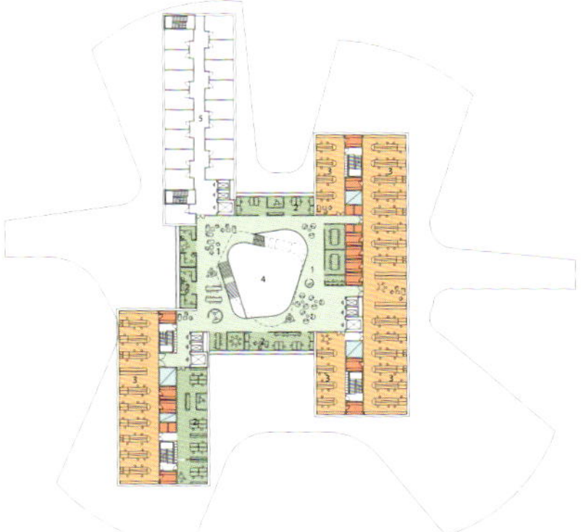

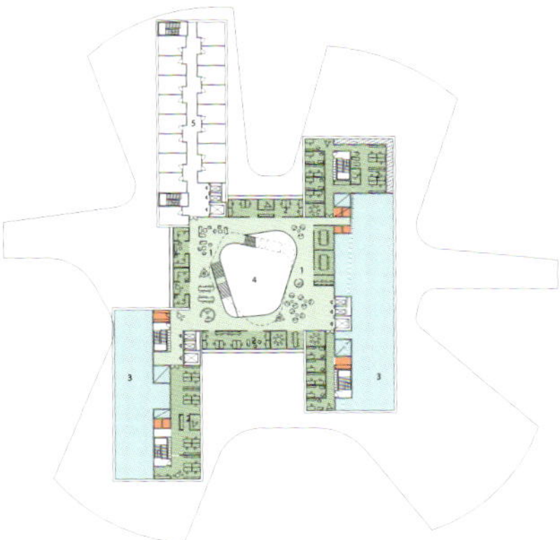

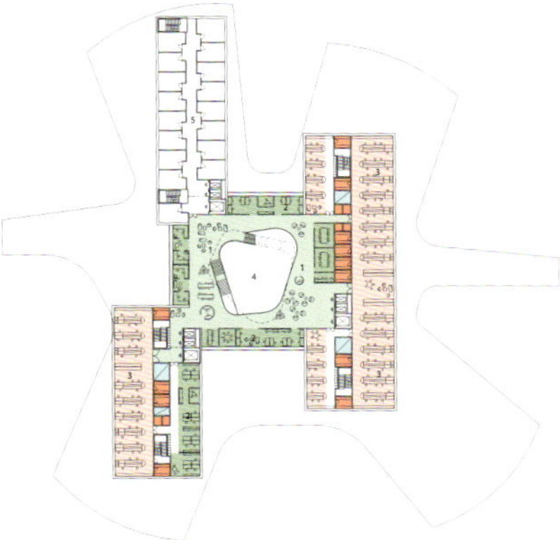

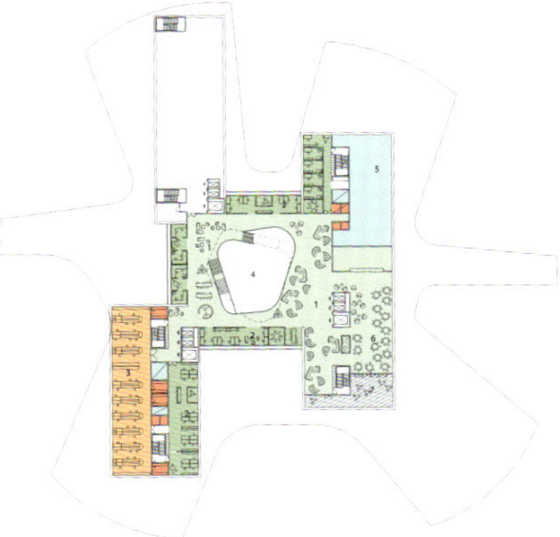

LTHYW Crossing Point

莲塘/香园围渡口 PATTERNS

International Competition, 2011
2011年国际设计方案赛参赛方案

Credits
Project Type: Crossing Point and Passenger Terminal
Location: Liantang, Shenzhen, China and Heung Yuen Wai, Hong Kong
Principals in Charge: Marcelo Spina and Georgina Huljich
Project Architect: James Vincent
Project Assistants: Yeonsu Sue Kim, Reece Satava, Naoko Miyamoto

社会方面

该设计将通过建立综合都市空间——一个制造偶遇机会的社交场所，使人与人之间不再冷漠。设计的最终目的是单纯地用建筑来表达两个城市之间的文化与技术交流。

建筑方面

具体来讲，是要利用一个循环通道连接两个不同的建筑。考虑到设计要求，该设计将呈现一个展示在大型建筑与两个没有关联但又相互依存的建筑之间的和谐与平衡。

设计背景

尽管周围都是道路与设施，但香港与深圳周边景观如亚热带雨林，物种丰富，景色美丽。因此，我们的设计要平衡人工设施的千篇一律与自然环境的个性鲜明。这就使我们在设计这个过境建筑的同时，考虑到周边景致。建筑物表面用附近植物的颜色设计了鞍状装饰图案，不免让人怀恋大自然。

建筑外立面采用了碧绿、翠绿与浅绿色的组合，制造出一种人工合成的自然效果，同时又给周边设施增添了感官上的层次感。

可持续性与建筑系统

从功能上来讲，中庭可以模糊地看到建筑内外各种繁复的活动动态，强化了常被忽略的离港、过境与到港过程。它连通上下，在内部向各个方向凹进去，以保证建筑内部良好的采光与通风。

建筑顶层设有几个观景平台，平台上设有通道。每一栋大楼在顶端景观平台延伸处都设有车辆检查站。户外空间设在中庭不同层次的极端，旅客或者等候船舶的乘客可以一眼看到城市、景观以及周边。这些设施形成一个平台网络，这些平台可以栽种植物，在美化环境的同时净化了空气。

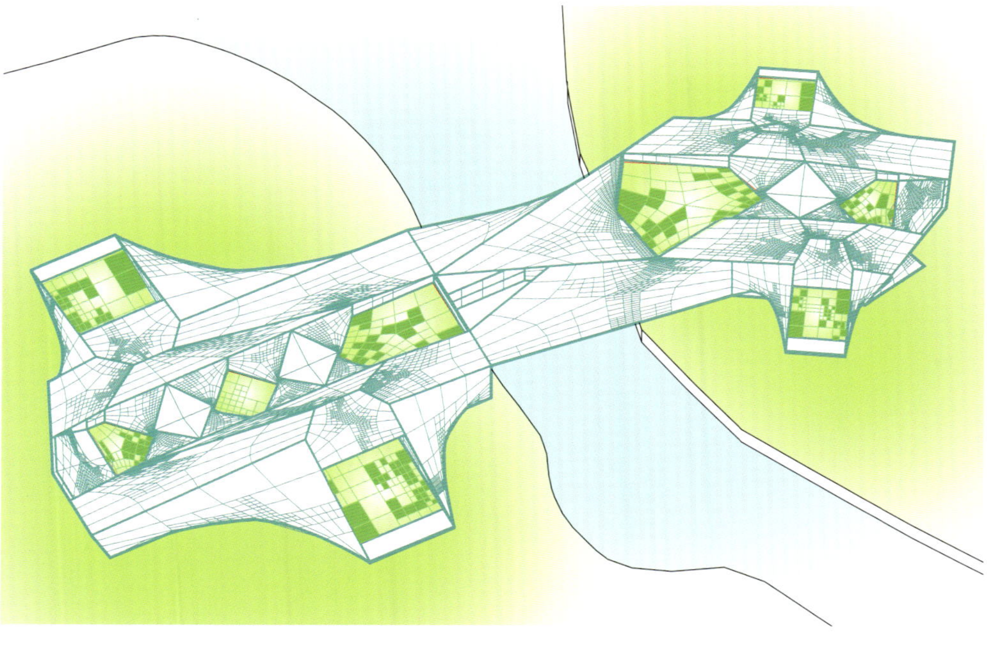

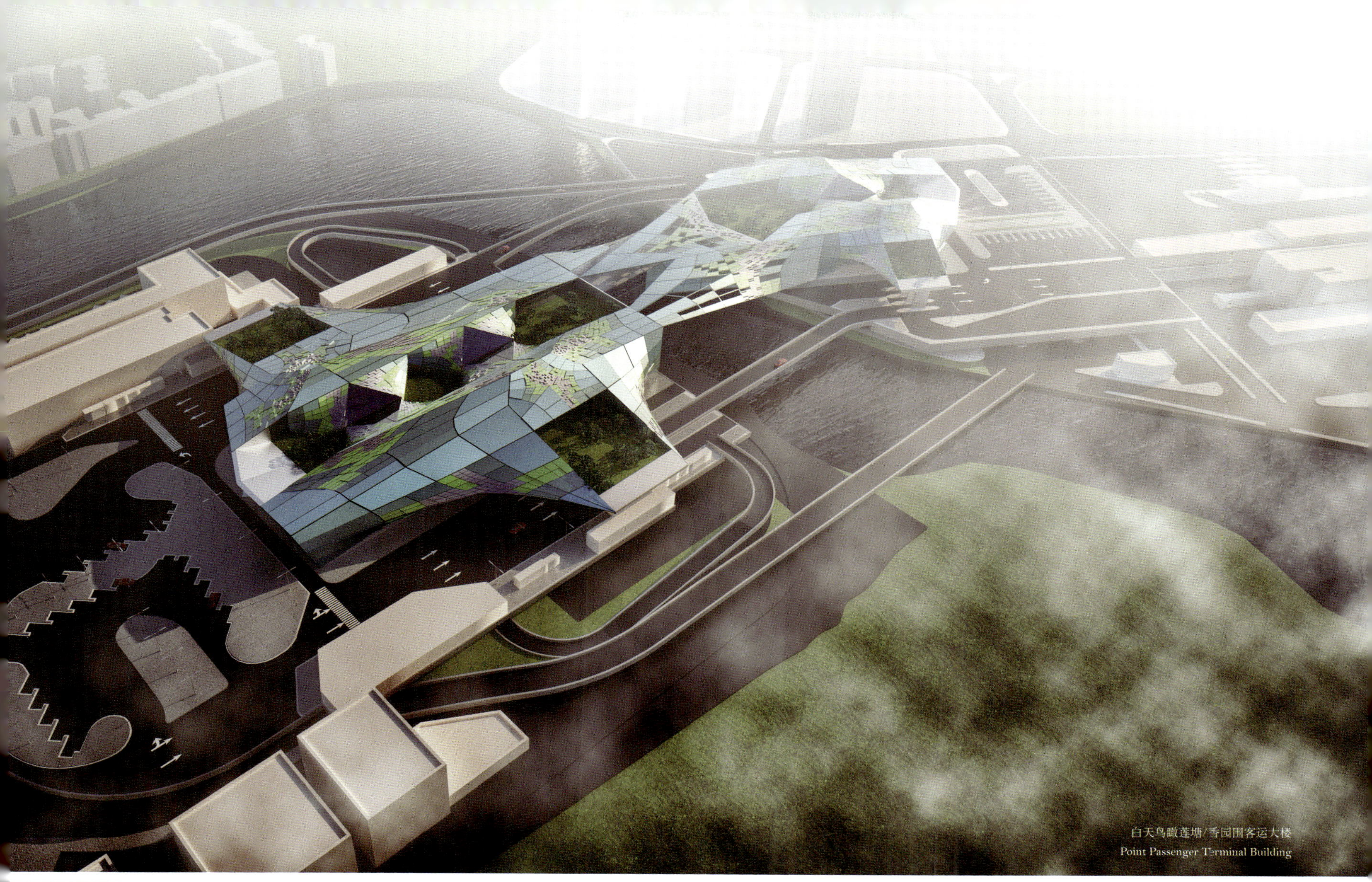

白天鸟瞰莲塘/香园围客运大楼
Point Passenger Terminal Building

夜晚从莲塘向渡桥望去
Evening View towards crossing bridge from Liantang Side

过境中心夜景鸟瞰图
Night bird's eye view of crossing

Contemporary Problem / Security and Segregation do not Engender Public Space

Our project attempts to overcome the tendency of indifferent segregation by generating a complex metropolitan space; a social space of encounter. Hence, the building's ultimate aim is to physically embody the cultural and technological aspirations of two cities within a piece of civic architecture.

Building Problem / When the Whole is not Larger than the Sum of the Parts

Specifically, our proposal intends to resolve the architectural design conundrum of connecting two different buildings with a circulation path. Given the overall determinism of the project brief, our proposal intends to articulate a new whole, producing a harmonious balance between a large manifold monolithic mass and two discrete and interdependent buildings.

Local Context / Global Infrastructure

While the immediate context of the building is surrounded by roads and infrastructure, the mounded landscape around both Honk Kong and Shenzhen constitute an example of subtropical jungle with rich with diverse species. Therefore, our proposal intends to conceptually and physically mediate between opposing problems: the generic artificiality of infrastructure and the natural distinctiveness of the site. Consequently, our project deals with the required instrumentality of a crossing border building by surrounding them with saddle, ornamental patterned surfaces with colors that reminiscent of the foliage of the adjacent landscape.

Turquoise, emerald and light green are combined in the envelope of the project producing synthetic effects and adding a sensorial level to the infrastructure around it. Allowing for multiple openings and vistas, the incrementally tessellated envelope act as a second skin and rain screen, allowing the building to shade itself from the excessive and direct heat and humidity and enjoy shadows and cool air throughout the long summer season.

Sustainability and Building Systems / Nature, Artificiality and the Sublime

Functionally, the atrium entices visitors with oblique and dynamic views of the frenetic activity occurring within and beyond the building, hence re-imagining the often-neglected ceremony of departure, crossing and arrival. The Atrium operates as a connecting tube, a manifold cavity that dynamically extends in all directions of the site allowing natural light [and air during cool months of the year] to pour inside and along the building.

The upper level of the building consists of several associated terraces with access. The wings of each building cover the vehicular checking stations while generating a landscaped terrace expansion in their top. The outdoor spaces are placed at the extremities of the atrium in different levels, and allow for direct vistas towards the city and the landscape and nearby from visitors or people waiting for arrivals or departures. This creates a network of accessible terraces which allow vegetation to become an active source of clean air but also an important form of expression.

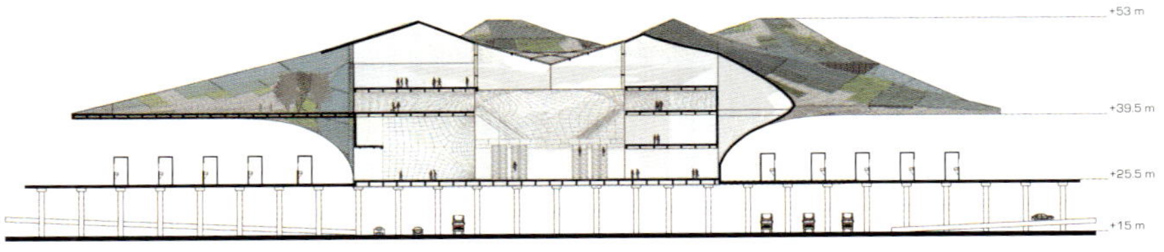
+53 m
+39.5 m
+25.5 m
+15 m

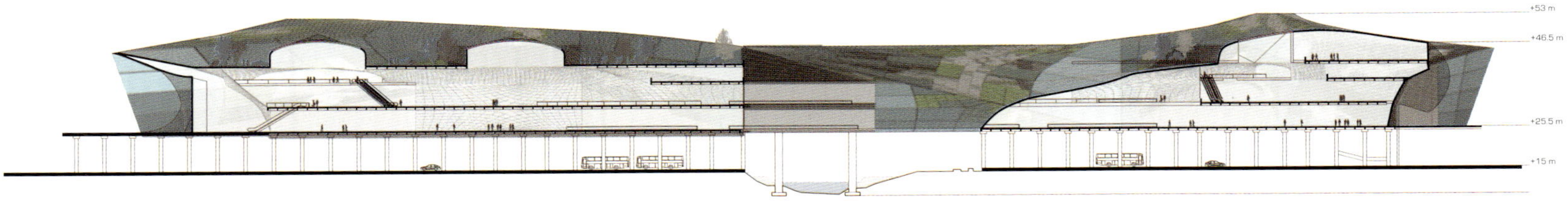
+53 m
+46.5 m
+25.5 m
+15 m

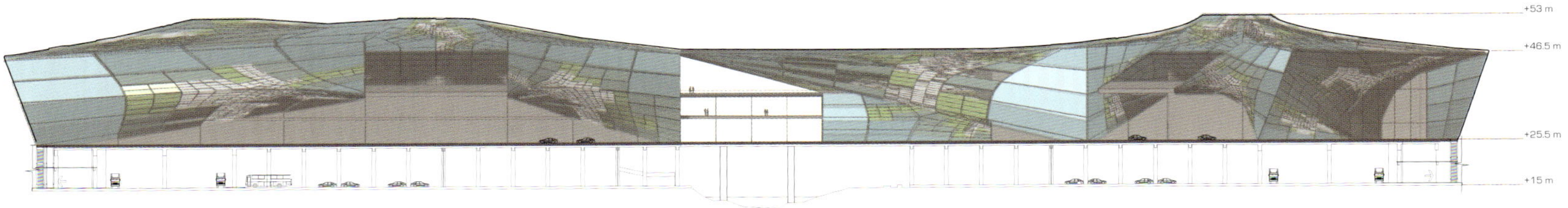
+53 m
+46.5 m
+25.5 m
+15 m

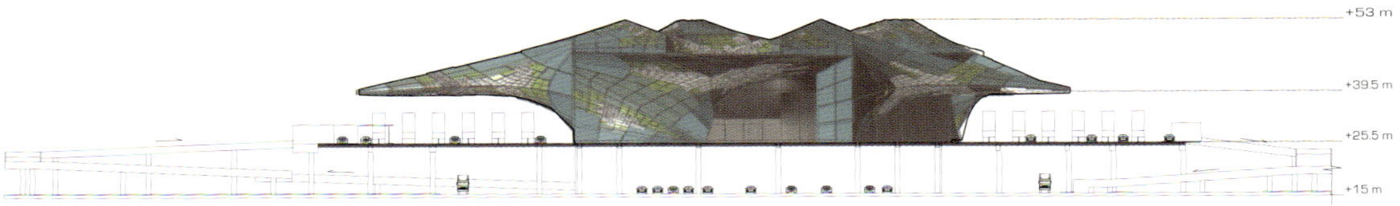
+53 m
+39.5 m
+25.5 m
+15 m

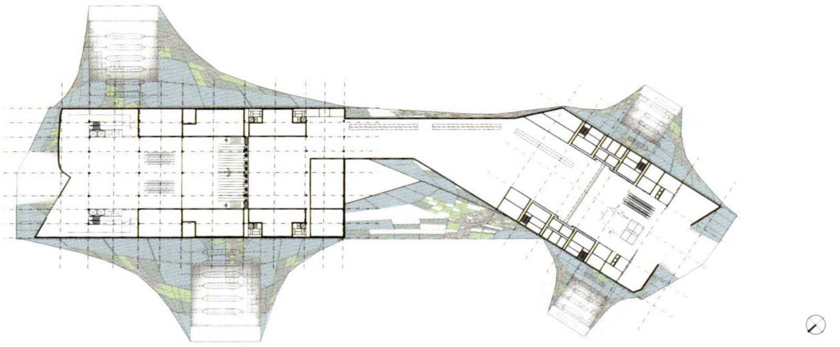

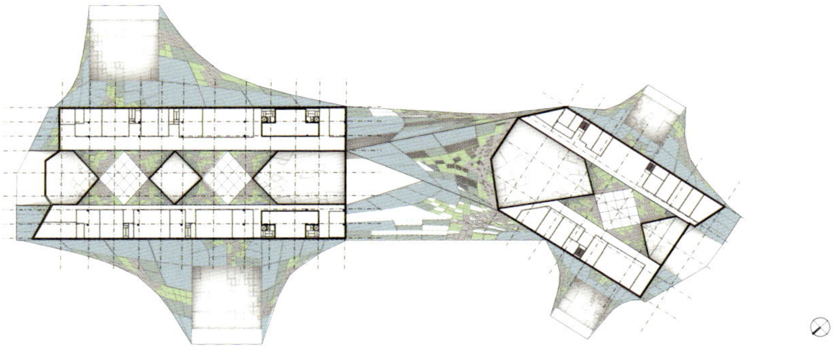

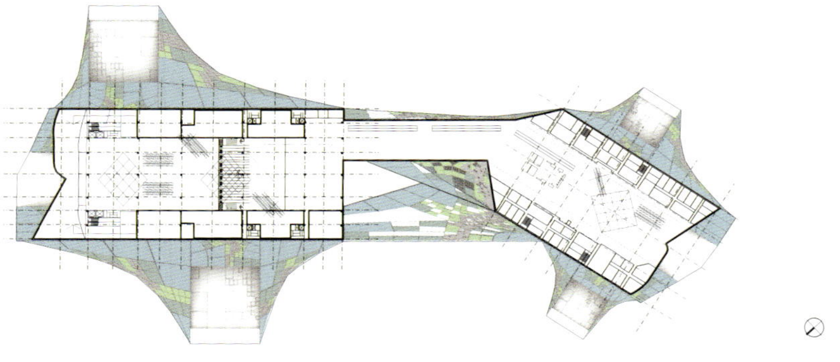

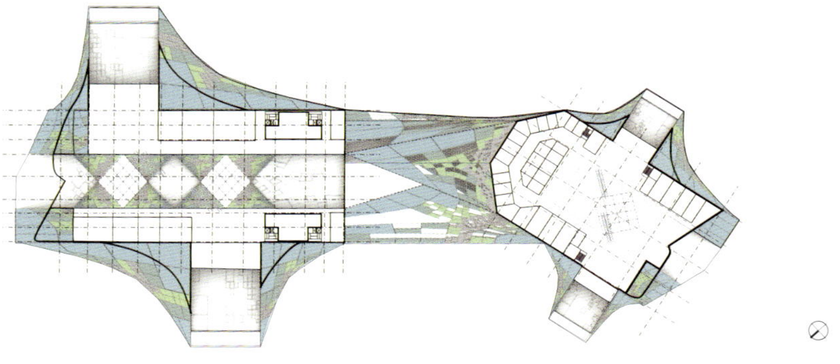

Envelope Articulation /
Fenestration and Panelization
Not To Scale
信封清楚的发音/开窗法和Panelization

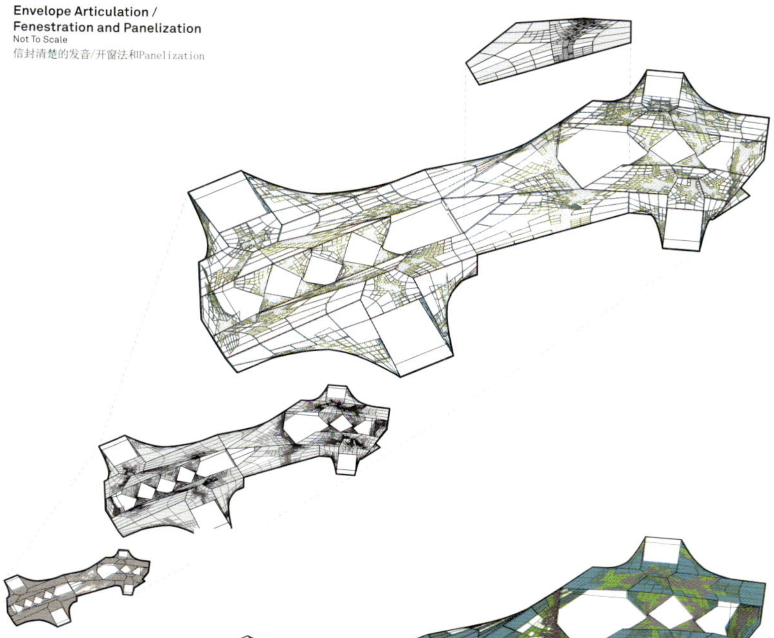

Color Gradation Diagram
Not To Scale
颜色渐进性图

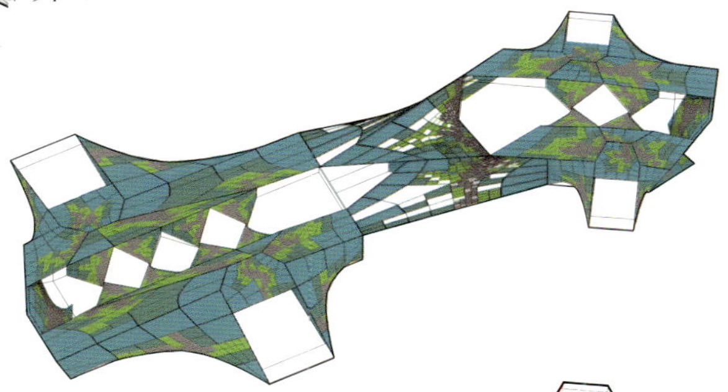

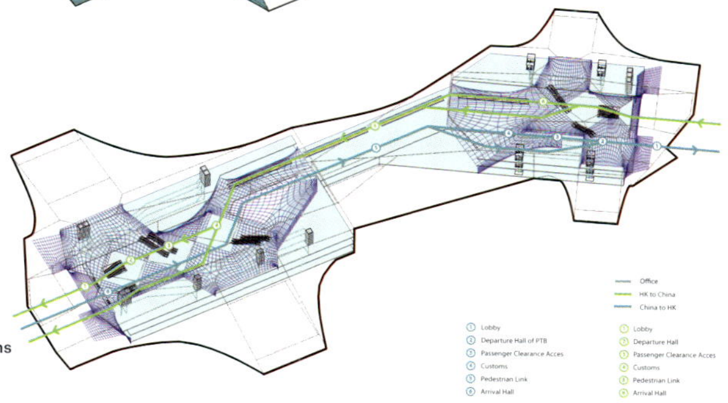

Cores and Movement Systems
Not To Scale
核心和运动系统

Conceptual Diagram / Terraces and Green Areas
Not To Scale
概念性图/大阳台和绿色区域

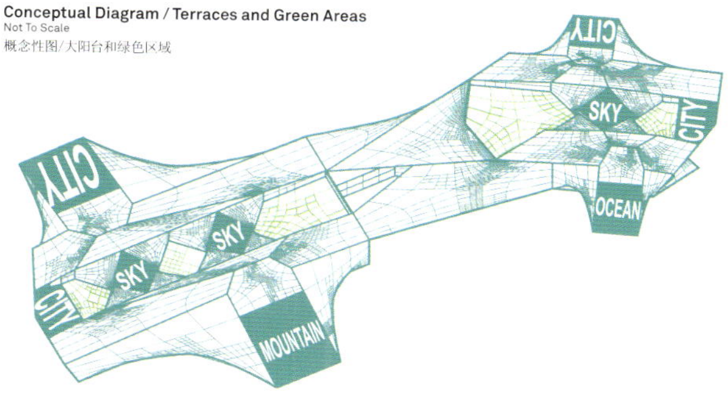

Inner Atrium
Not To Scale
内在心房

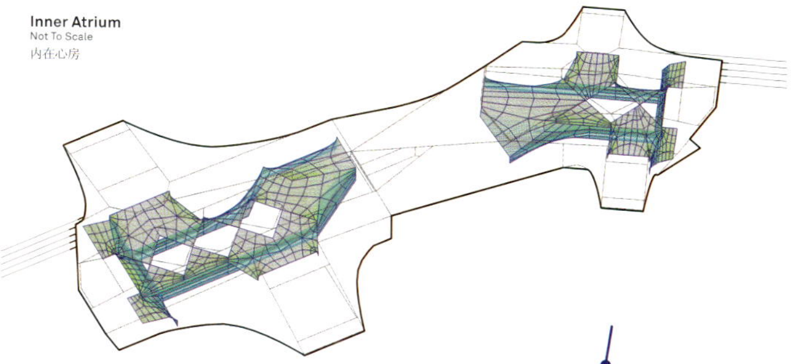

Air Flow Manifold
Not To Scale
气流多头管

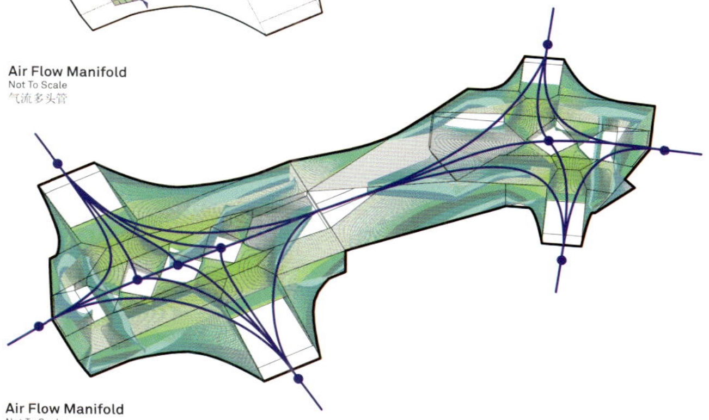

Air Flow Manifold
Not To Scale
气流多头管

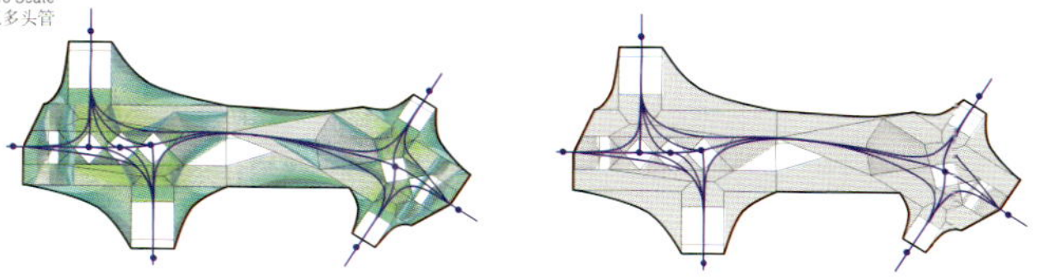

▼向上方人行通道望去
View Looking up towards Pedestrian Bridge

Kaohsiung Port and Cruise Service Center
高雄港口油轮服务中心 PATTERNS

International Competition, 2010
2010年国际设计方案赛参赛方案

Credits
Type: Port and Cruise Terminal
Location: Kaohsiung, Taiwan
Client: Kaohsiung Harbor Bureau, Ministry of Transportation and Communications, Taiwan
Principals in Charge: Marcelo Spina and Georgina Huljich
Project Designers: Bobby Frederick and Yeonsu Kim
Project Assistants: Vince Posick, David Liu, Daniele Profeta

城市结晶体／丰富多样性

为了将其建造成为综合型大楼，设计师把外形设计成块状结晶体。一个包含三层停车场的空间位于三个小高层内部，作为港口中心的底座，顶端与一个观景平台相连。这个底座顶部是一个公共广场，可以鸟瞰整个高雄市以及港湾。这个多晶体在海岸边更加清晰；钻石般的楼体旋转着摆向角落。对称轴增强了从海洋方向望去的认知感，形成内港方向无可比拟的远景图。

纲领性策略

竞赛方案要求将一个中转码头与港务局连通，这本就使设计进退维谷。为了解决这个问题，设计将项目的两个主要组成部分在保持各自功能性独立的前提下，互相紧密连接。因此，项目通过视觉上与形式上融合显著但却相互独立的空间，强调了安全性。

广场

这个公共广场连通建筑物内外。外面设有与高雄最高建筑物平行的大型城市楼梯系统，它前面绵延着连通建筑内部的人行步道。这样，可以将一部分城市户外活动自然推向广场与码头平台。这个大型楼梯系统包括一个感应式自动扶梯，可快速通往广场一层。

广场顶部设有半掩式防雨设施，同时可以在正午及下午时起到遮阳的作用。另外，这里还有植被；绿色区域延伸至人行道，仿佛连接着城市与码头。

多样化中庭

“将人与大海的深邃、蓝天的轻盈动态连接在一起。”

中庭呈中空状态，由于整体比较深邃，在各个方向都设计了凹进去的空间，这样保证建筑内部，包括入口大厅，良好采光以及通风。

外立面坚固而疏松

这个建筑的外立面看上去十分的庞大。多小面与角度的铰链使港口端部看上去沉静而强大。建造使用玻璃纤维加固混凝土与加固纤维聚合物，码头表面的大部分都是无缝连接，把重点放在建筑物上面的大块晶体上。深灰与黑色无光塑料表层使建筑外观看上去异常沉稳，与白色亮泽的军舰形成鲜明对比。

▼港口两岸景色
View across Harbor

东南立面透视图
South-East Perspective

从港口透望码头
Perspective View toward the Terminal from the Harbor

Circulation Diagram of Overall Building

整體動線分析

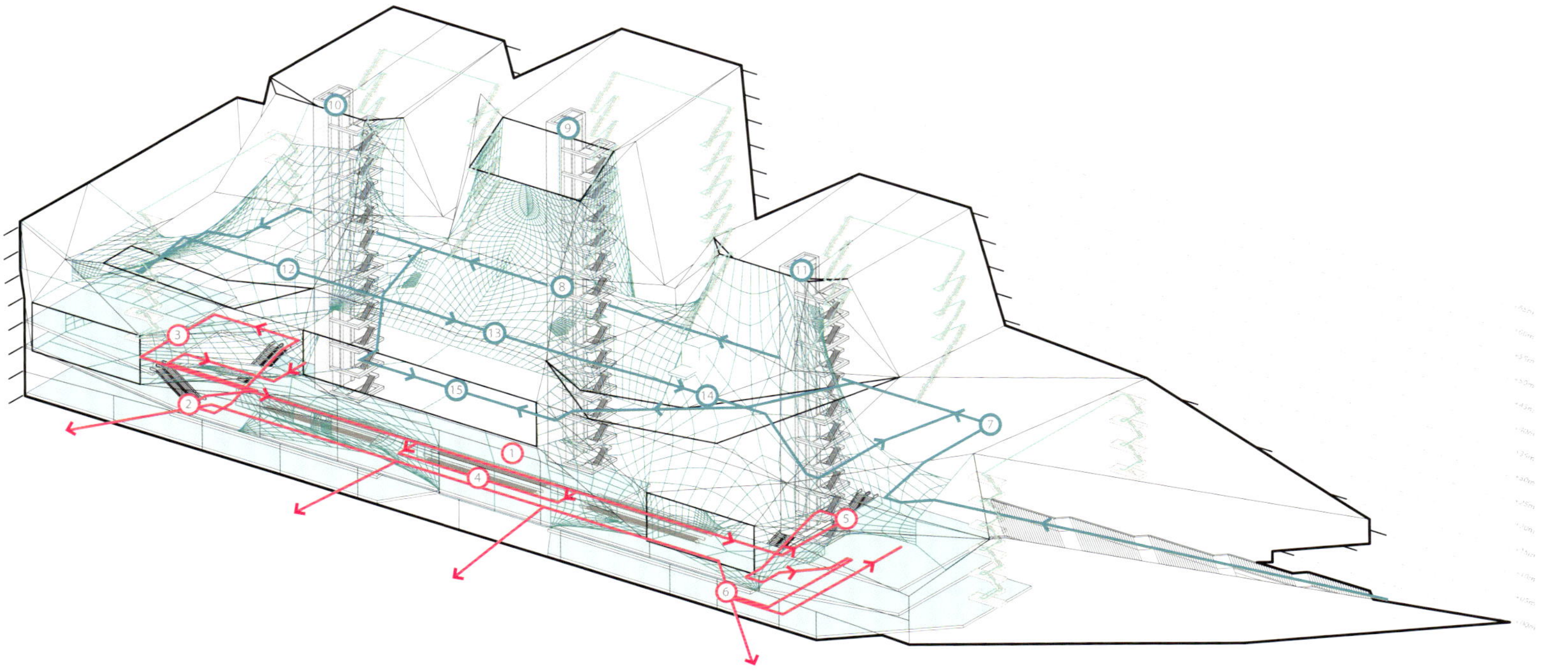

Public Circulation	公共流線
Terminal Circulation	郵輪站台流線
Emergency Circulation	緊急流線

Lobby	入口大廳
Check-in	報到櫃檯及行李託運
Departure Hall	出境大廳
Boarding Concourse	登船廊道
Arrival Hall	入境大廳—管制區
Baggage Claim	行李處理區
Lobby	入口大廳
Meeting Room	型會議室
Office	辦公空間
Vessel Traffic Manage Center	船舶交通管--理第二中心
Conference Hall	梯地
Library	圖書館
Commercial	商業服務區
Dining / Lounge	用餐空間
Terrace	汽車停車空間

Programmatic Distribution Diagram

Not To Scale

空間配置分析

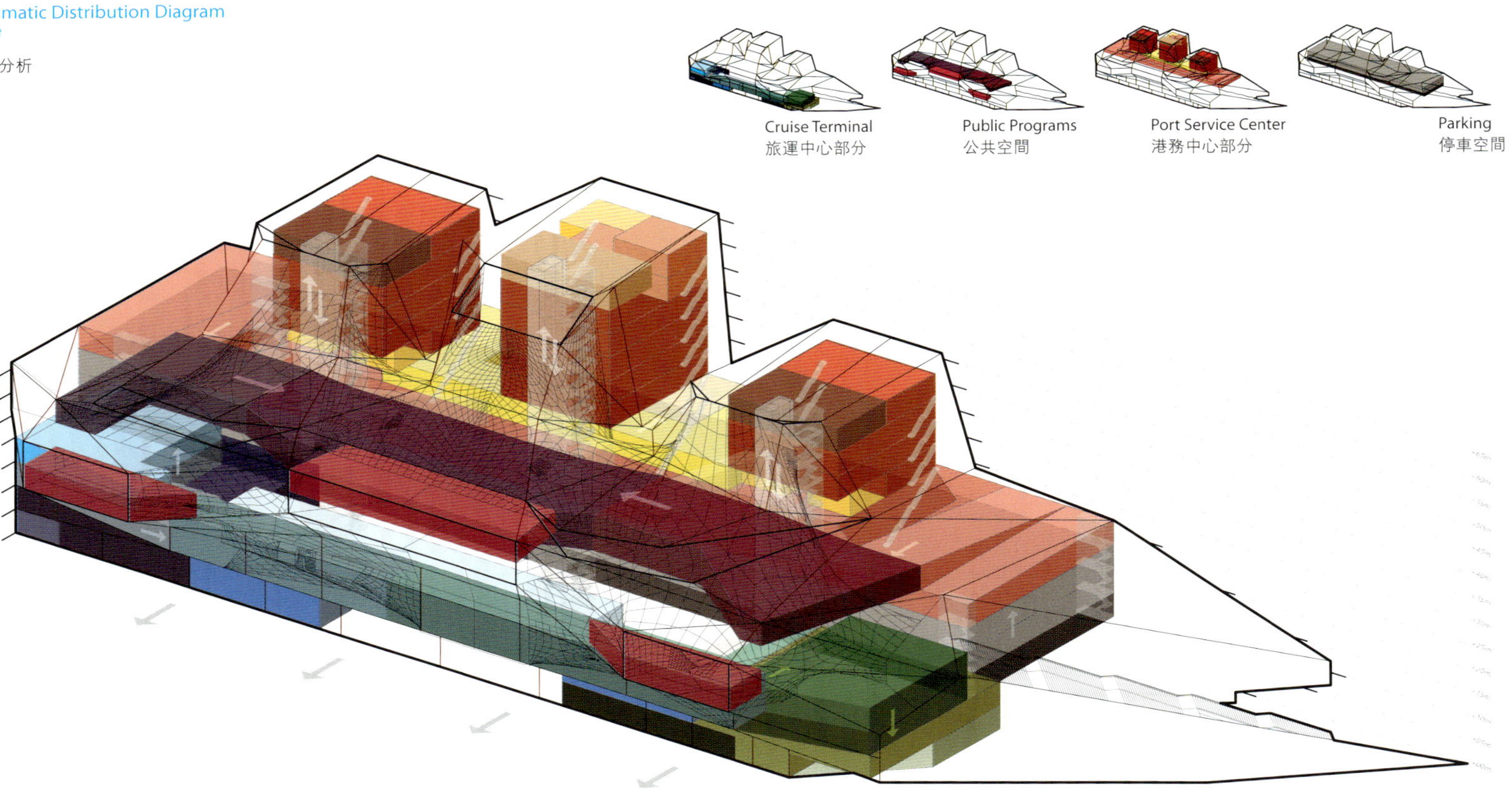

Terrace	梯地
Library	圖書館
Commercial	商業服務區
Dining / Lounge	用餐空間
Cruise Terminal Parking	郵輪碼頭停車場
Port Service Parking	港口服務停車場
Departure Lobby	旅運大廳
Check-in	報道櫃臺及行李托運
Departure Hall	出境大廳
Domestic Lounge	侯船室—國內綫
International Lounge	侯船室—國際綫
Moving Walkways	電動步道
Commercial	商業服務區
Domestic Boarding Concourse	登船廊道—國內綫客輪專用
International Boarding Concourse	國際／國內綫登船廊道
Customs Office	出入境辦公室
Arrival Hall	入境大廳—管理區
Baggage Claim	行李處理區
Traveler's Center	旅客服務中心
Arrival Lobby	旅客大廳
Public Lobby	入口大廳
Port Service Lobby	入口大廳
Conference Hall	國際會議廳
Office	辦公空間
Medium Meeting Room	中型會議室
Large Meeting Room	大型會議室
Lounge	餐廳／休閑
Vessel Traffic Manage Center	船舶交通管理第二中心

▲码头远景图
Perspective View toward the Terminal

Elevations

剖面

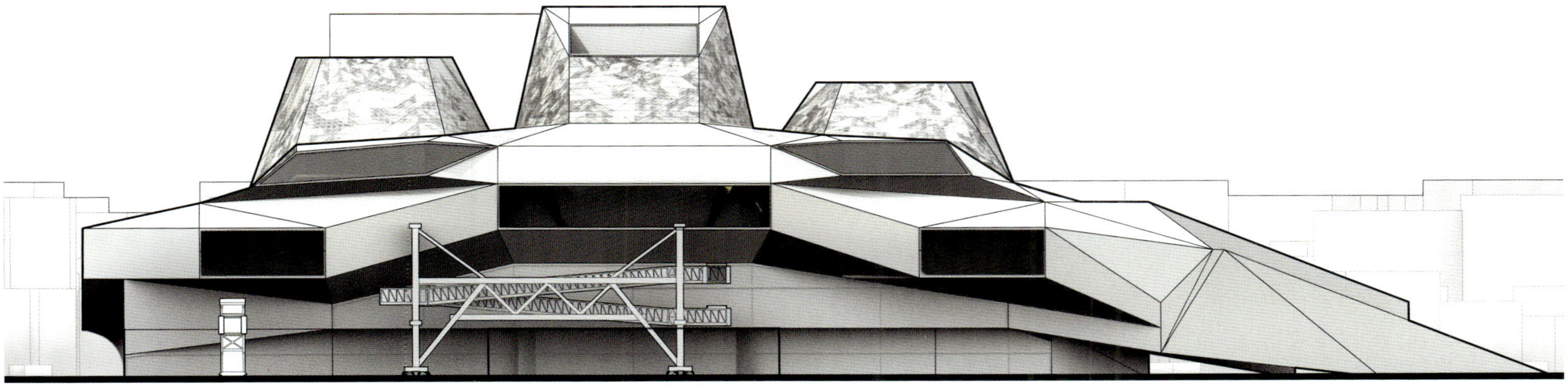

Longitude Section through Atrium

横向剖面

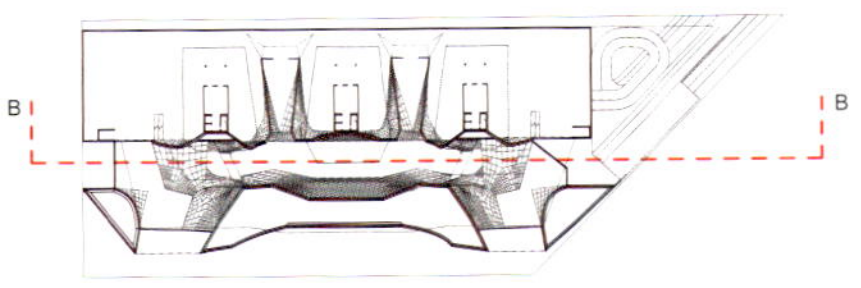

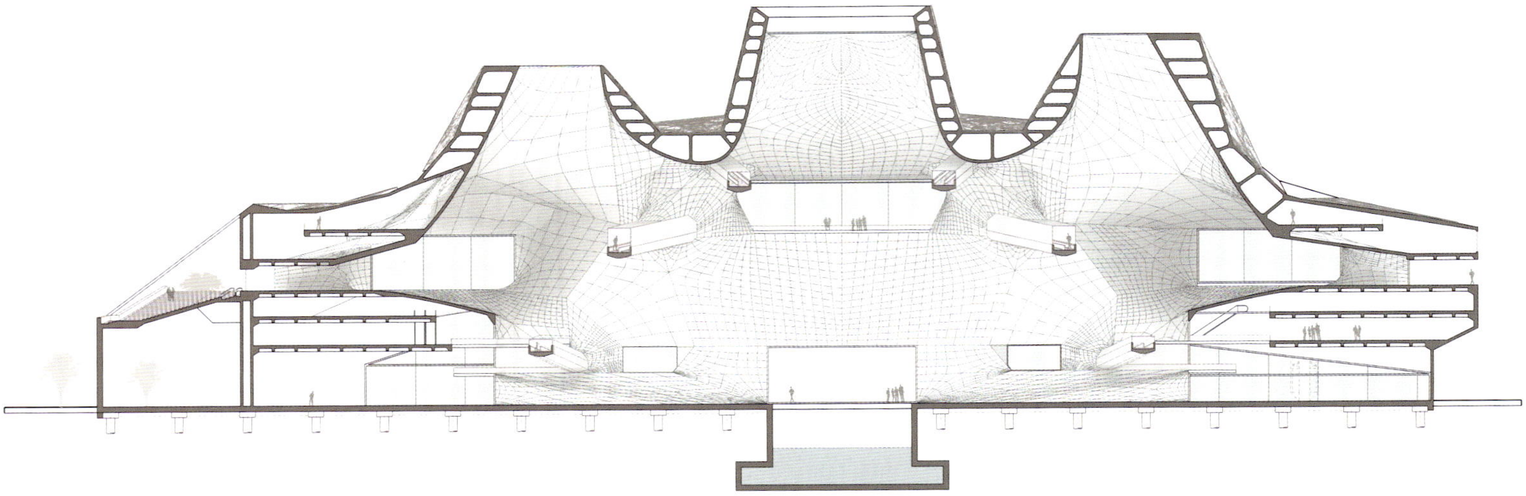

Transversal Section

縦向剖面

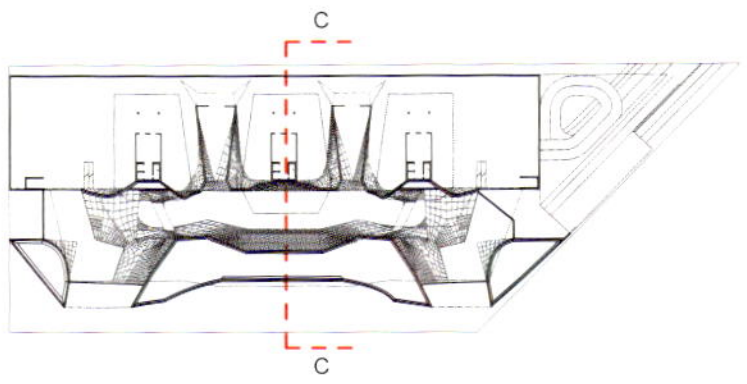

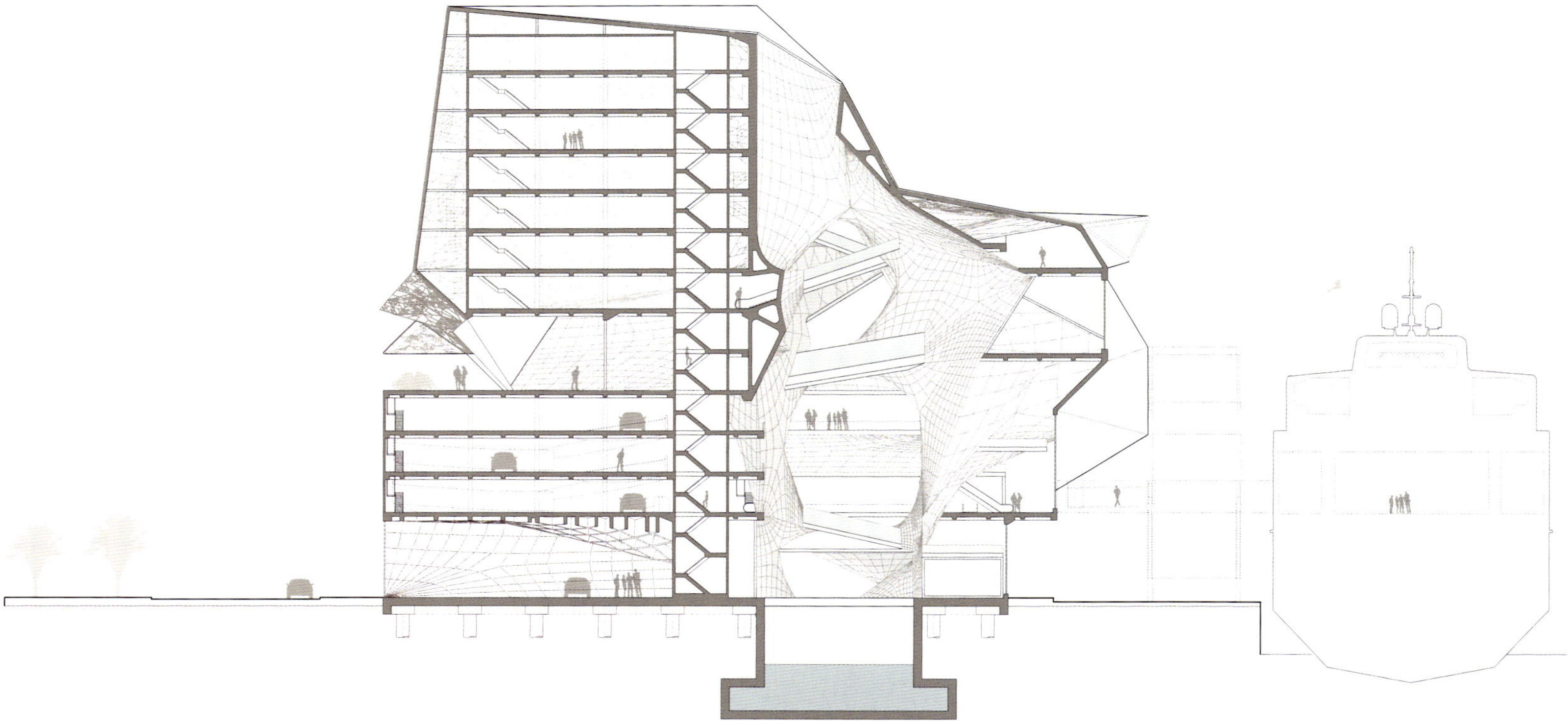

Ground Floor Plan Level 0,00

壹層平面

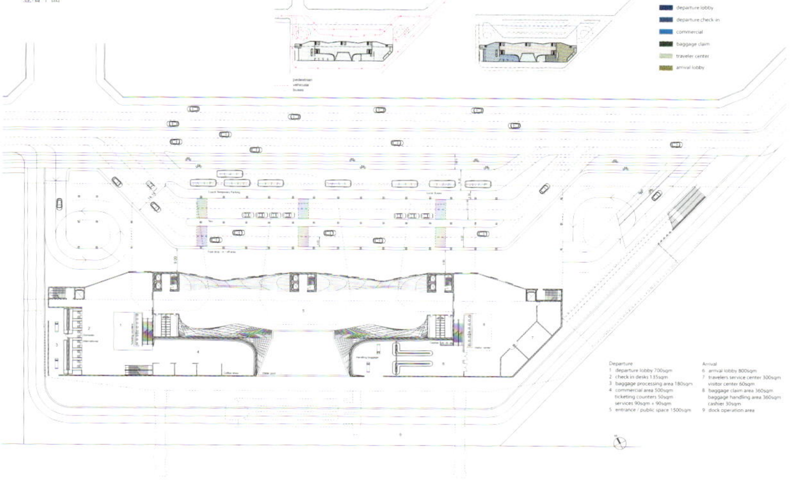

Second Floor Plan Level 10,00

二層平面

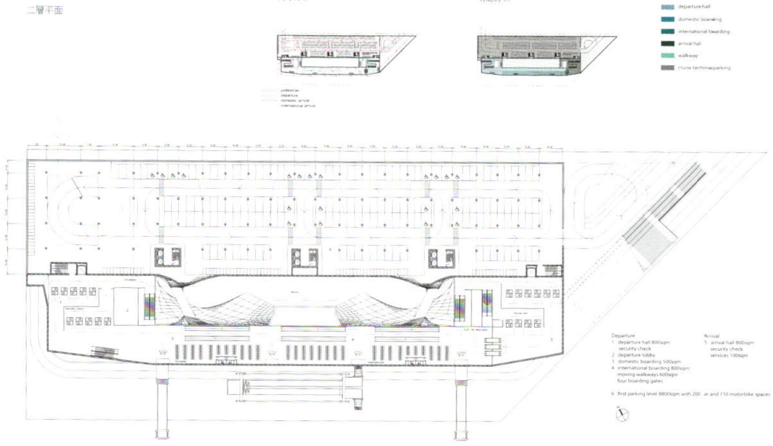

Third Floor Plan Level 20,00

三層平面

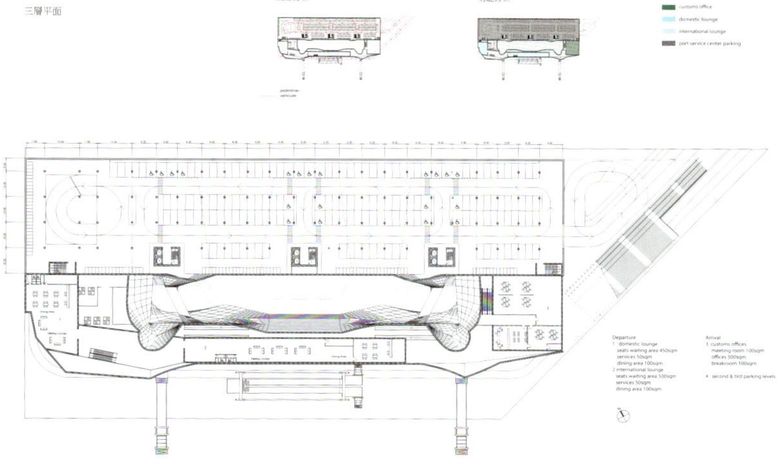

Fourth Floor Plan Level 25,00

四層平面

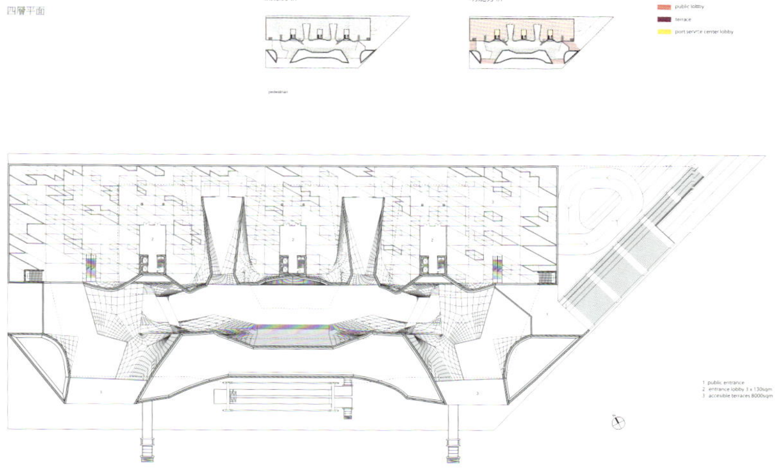

Fifth Floor Plan Level 30,00

五層平面

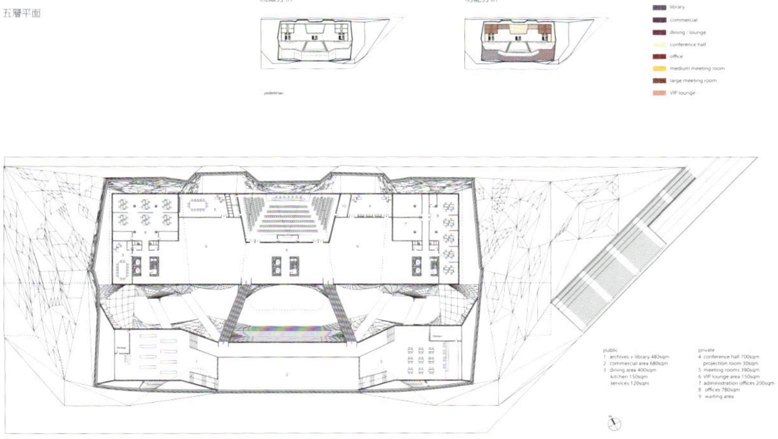

Sixth Floor Plan Level 35,00 - 60,00

六層平面

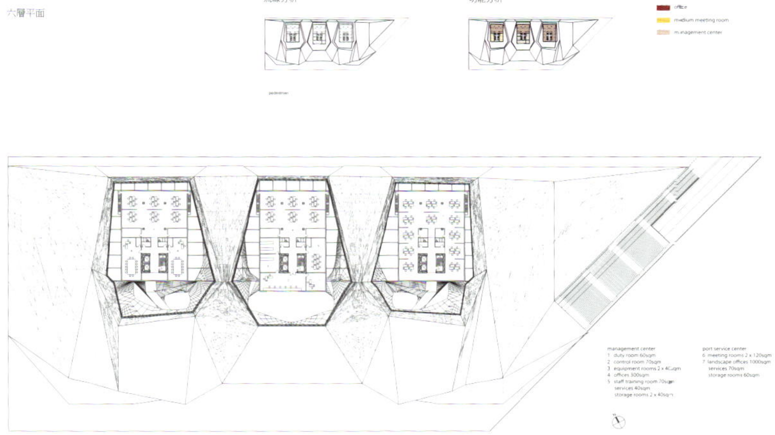

Site Plan

爱爍獶猖亮玶

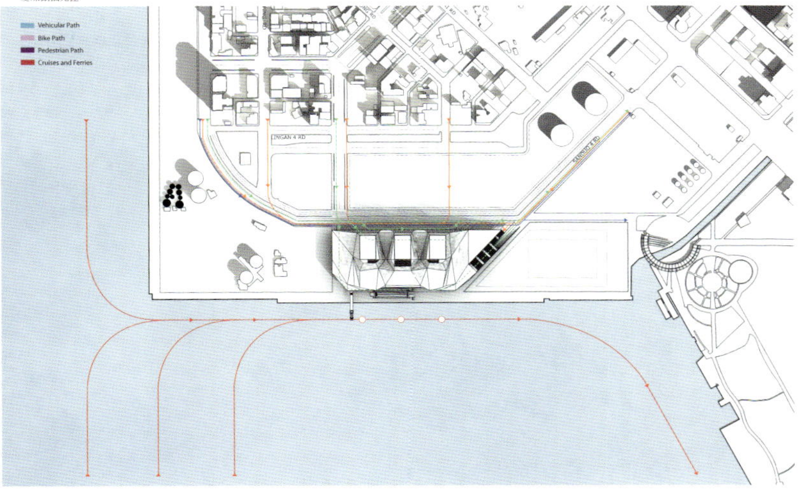

▼外立面细节图

Envelope Detail

▲深庭
Deep Atrium

Urban Mass I A Robust Diversity

Our massing strategy aims to a hybrid type. A volume containing 3 levels of parking serves as a podium for the Port Center located in three small towers connected by a concourse at the top. The roof of the podium allows for an elevated public space with views towards downtown Kaohsiung and the harbor. In the ocean side, the mass is more articulate; its diamond like shape swerves and cantilevers towards the corners. Its axial symmetry allows a strong sense of recognition when seen from the ocean, fostering iconic vistas from the inner harbor.

Programmatic Strategy

The inherent dilemma of the competition brief was to reconcile a space of transition [terminal] with that of a destination [port authority]. To address this problem, the building strategy places the two main and quiet distinct programmatic components of the project in close proximity to each other while allowing their main functions to remain separated. The project therefore addressed security by visually and physically integrating distinct and separated programs.

Podium Plaza

The public plaza is accessible from inside and outside the building. From outside, a large civic stair is aligned with Kaohsiung tallest building, extending the promenade in front of it into the building and connecting with the podium plaza. Hence allowing for a stream of pedestrian activity in downtown to naturally flow up towards the podium plaza and into the terminal terraces. The large stair contains a user activated mechanical escalator which allows quick access to the plaza level.

Partially covered by a porous rain screen system that allows for shading during hot noon and afternoon hours, the plaza includes pockets of vegetation. The green area also extends towards the promenade, generating continuity between the city and the terminal.

Manifold Atrium

"Dynamically connecting people with the Deep Ocean and the Lofty Sky"

The Deep Atrium functions as a connecting void, a deep manifold cavity that dynamically extends in all directions of the site allowing natural light [and air during cool months of the year] to pour deep inside the building reaching

▲油轮码头前厅

Cruise Terminal Lobby

▲广场一角

Plaza Level

all the way down to the entry hall / lobby.

Envelope Outer Shell I from Solid to Porous

The building outer shell is decisively massive. Its faceted and angled articulation aims to produce a mute and yet strong icon at the edge of the harbor. Using composite construction such as GFRC (Glass Fibber Reinforced Concrete) and FRP [Fiber Reinforced Polymers], most of the surfaces in the terminal are seamless, emphasizing mass over building detail and tectonic assembly. Using dark grey and black tones with plastic matte finishes allow the building a controlled coarse finish and a resolutely mute appearance in its exterior, in contrast with the sleek and super articulate white surfaces of the cruise ships.

New Terminal for Stockholm's Permanent Ferry Connections to Finland and the Baltics

新码头——斯德哥尔摩通往芬兰和波罗的海

C. F. Møller Architects

Credits

Location: Värtahamnen, Stockholm, Sverige

Area: 16,500 m^2 and a new customs area of 1,100 m^2

Year: 2009-2010

斯德哥尔摩通往芬兰和波罗的海的新码头建筑无论在结构上还是在环境上，都将是斯德哥尔摩中部皇家港口城市开发中的一座新地标。这座码头建筑的一个侧面由延伸的网状结构覆盖，让人想起前进中的游艇，而建筑还带有大型的吊车和仓库，拥有过去码头的典型特征。同时，设计师还赋予码头建筑可持续发展的设计，体现出整个规划的特点。

设计意在建立斯德哥尔摩中心城区与新城区之间的自然联系，从而与码头连接，使城市生活也渐渐渗透到该区域。所以码头也将成为与城市同等水平的区域，方便行人与车辆的流通。同时，码头建筑的顶部还被设计成一个多变的绿色景观，带有楼梯、斜路、壁龛以及舒适的角落，供斯德哥尔摩市民及过往者散步休息，同时欣赏码头、群岛和城市的美景。

设计的目的是实现该码头能源的自给，为公共设施建筑树立一个环保的榜样。所以该码头建筑集中利用太阳能和风能，例如，顶部的梯田式景观将太阳能电池板和机械设备有机结合。方案通过使用中心配置的电视屏幕向这里的人们传达合理利用资源、可持续发展的理念，提高人们可持续建设的意识。

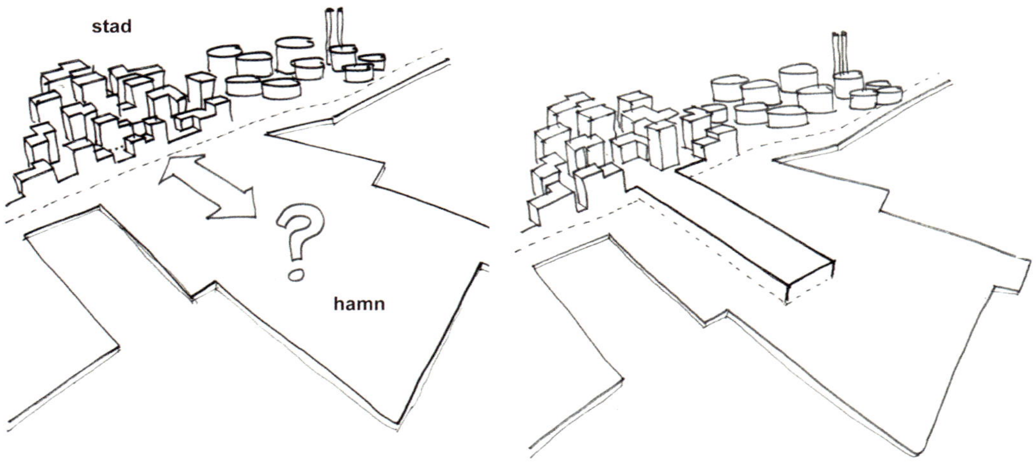

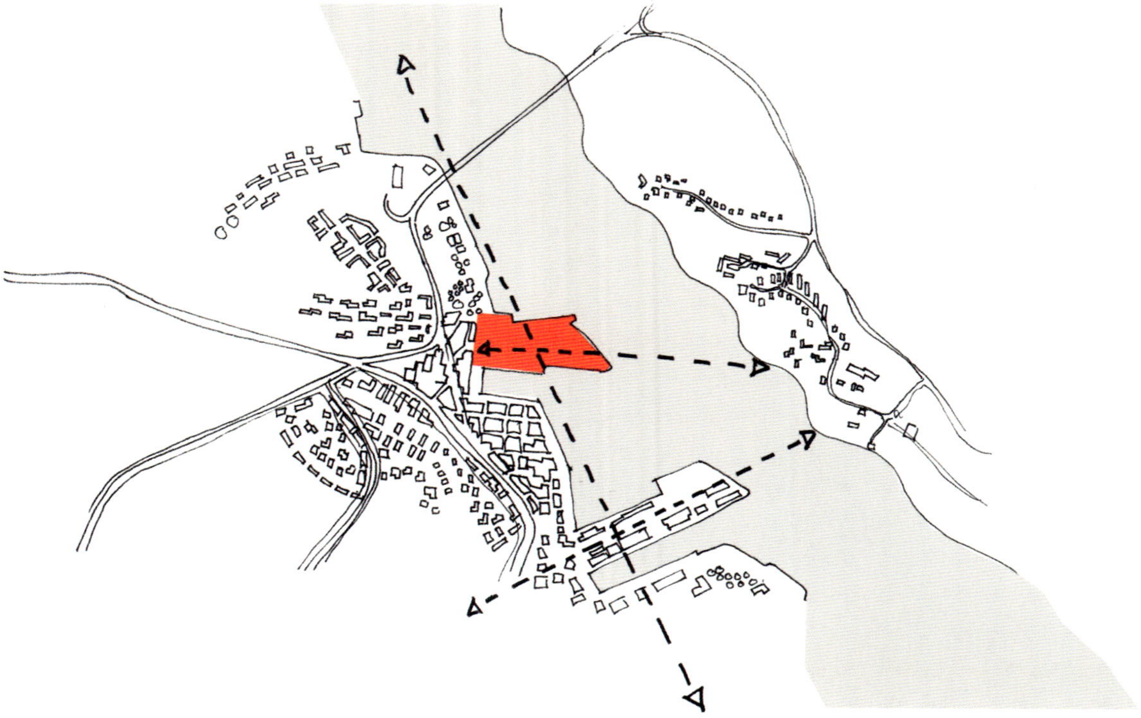

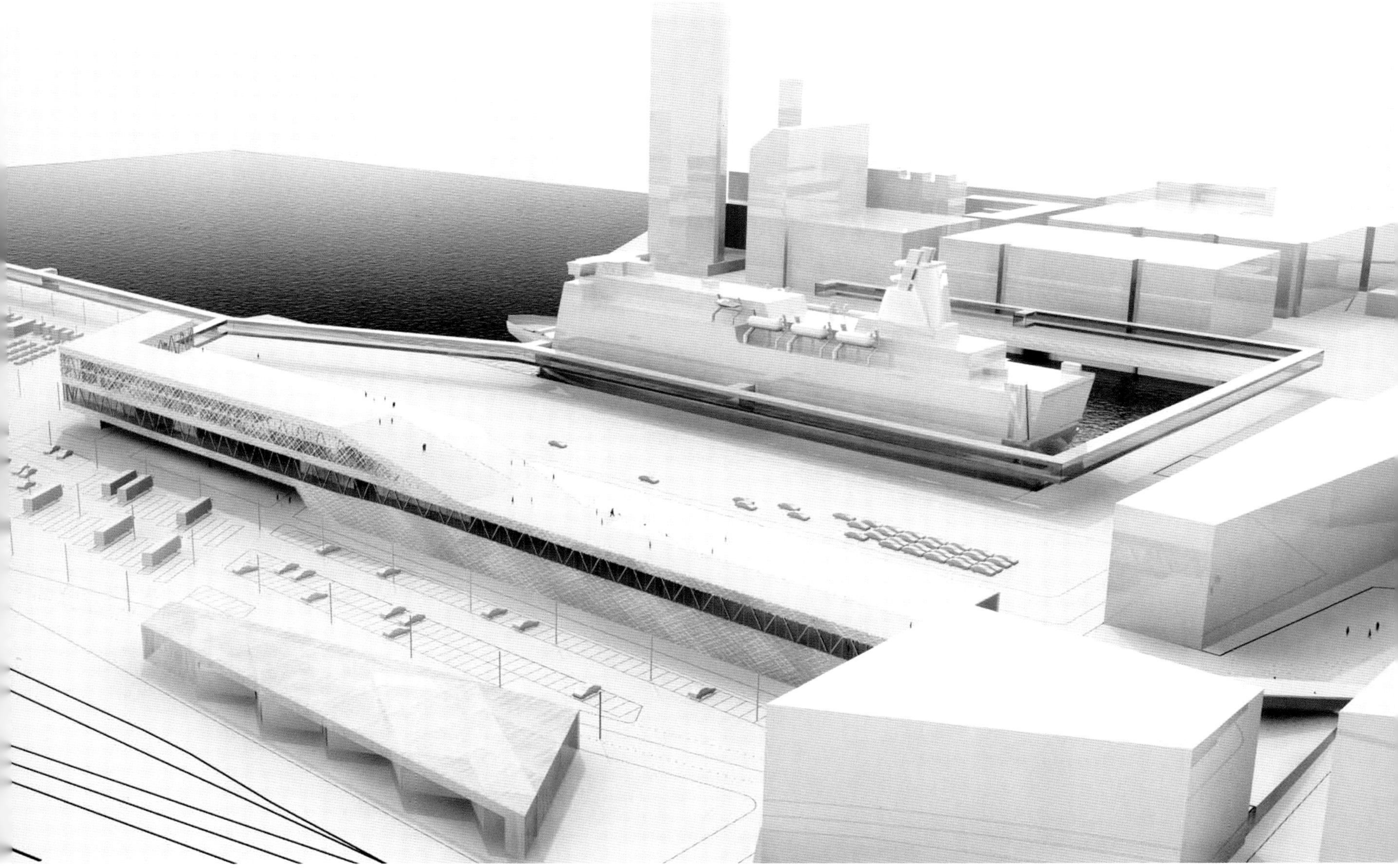
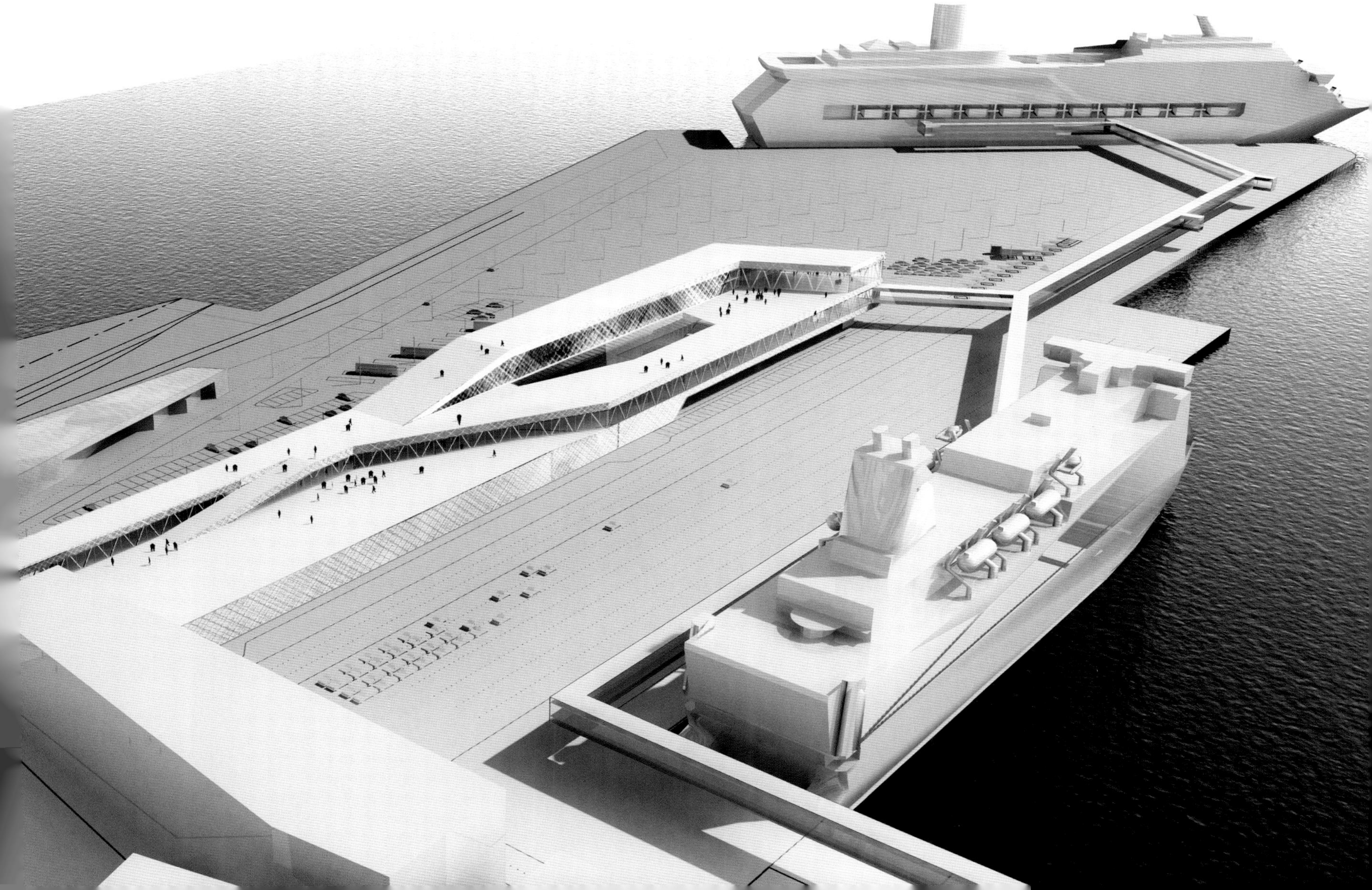

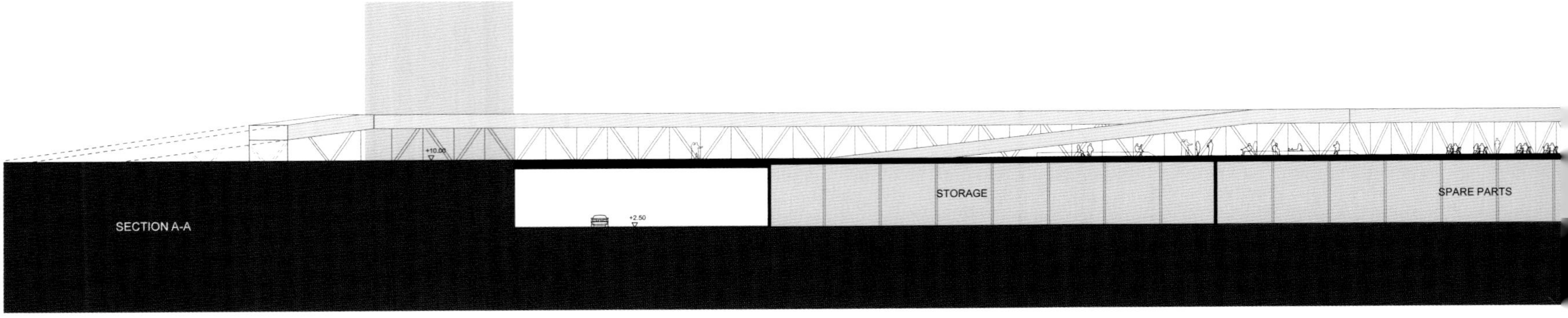
+10.00
SECTION A-A
+2.50
STORAGE
SPARE PARTS

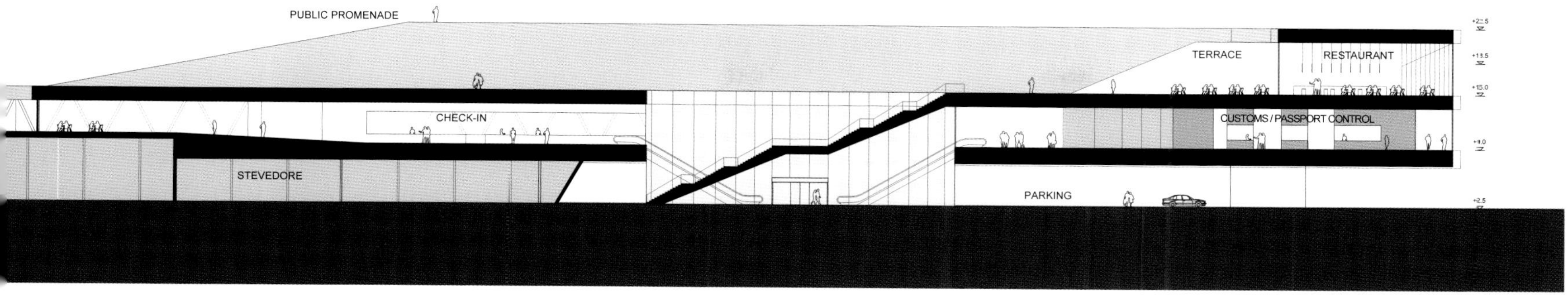
PUBLIC PROMENADE
TERRACE
RESTAURANT
CHECK-IN
CUSTOMS / PASSPORT CONTROL
STEVEDORE
PARKING
+15.0
+2.5

SILJA SYMPHONY

CANAL+

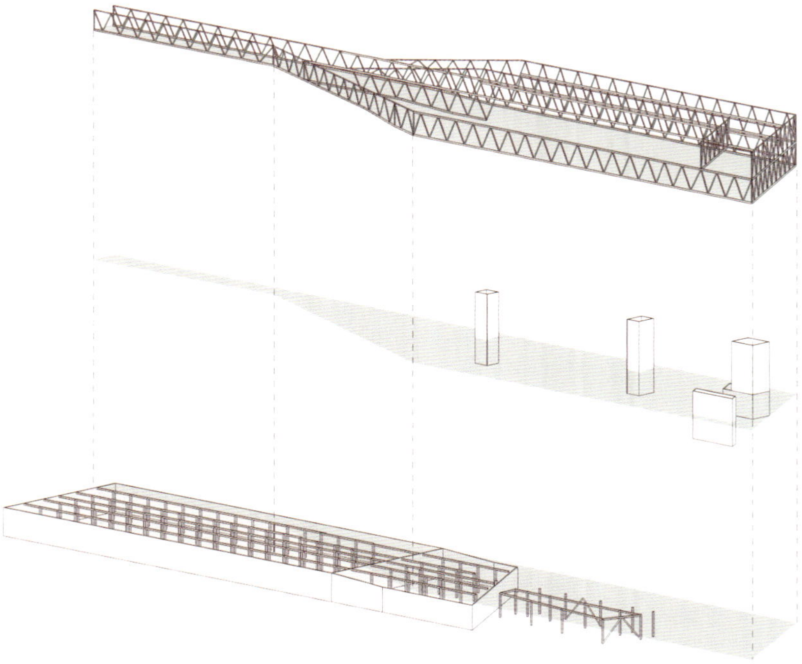

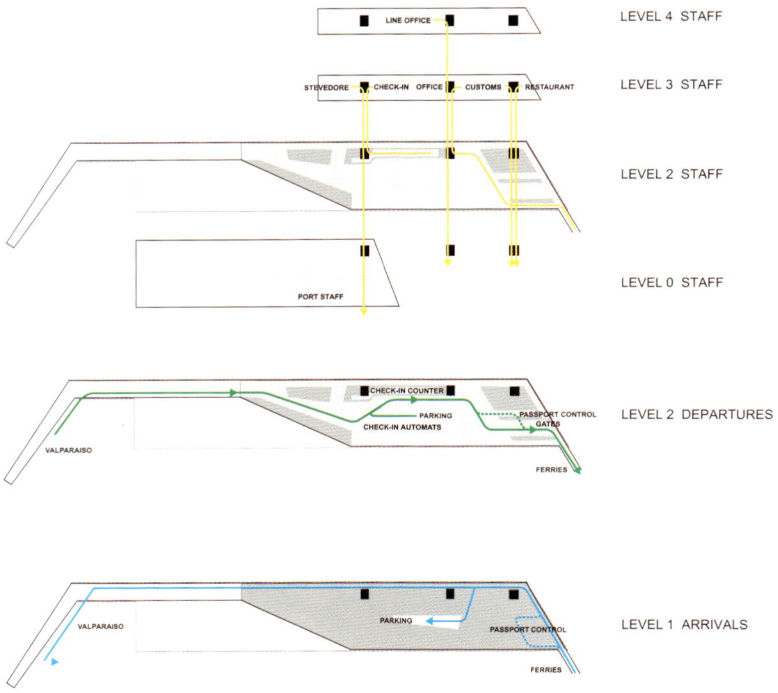
LINE OFFICE
LEVEL 4 STAFF
STEVEDORE
CHECK-IN OFFICE
CUSTOMS
RESTAURANT
LEVEL 3 STAFF
LEVEL 2 STAFF
PORT STAFF
LEVEL 0 STAFF
CHECK-IN COUNTER
PARKING
CHECK-IN AUTOMATS
PASSPORT CONTROL
GATES
VALPARAISO
FERRIES
LEVEL 2 DEPARTURES
VALPARAISO
PARKING
PASSPORT CONTROL
FERRIES
LEVEL 1 ARRIVALS

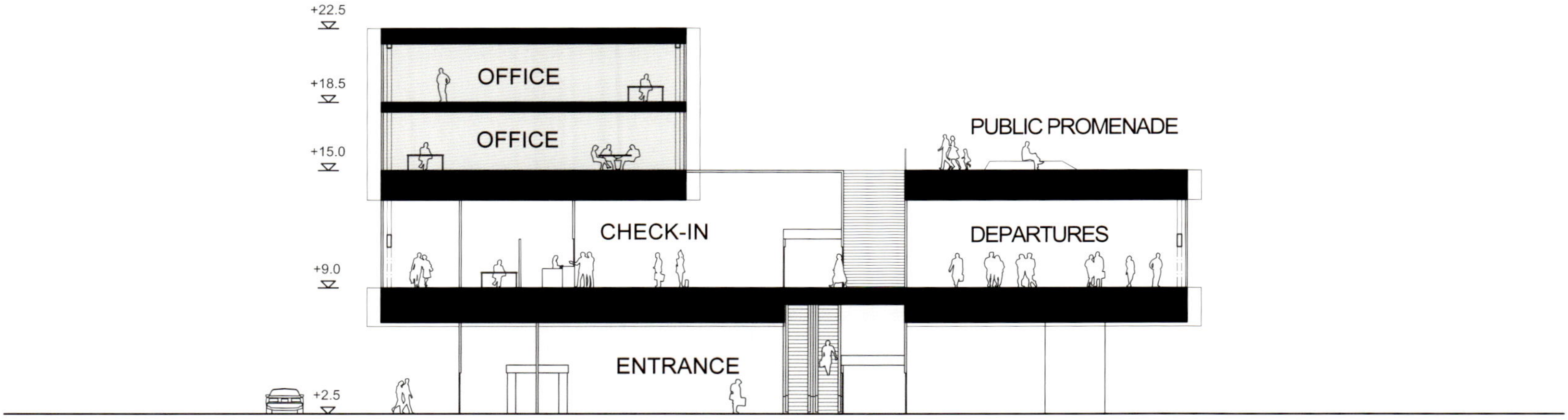
+22.5
+18.5
+15.0
+9.0
+2.5
OFFICE
OFFICE
PUBLIC PROMENADE
CHECK-IN
DEPARTURES
ENTRANCE

TALLINK

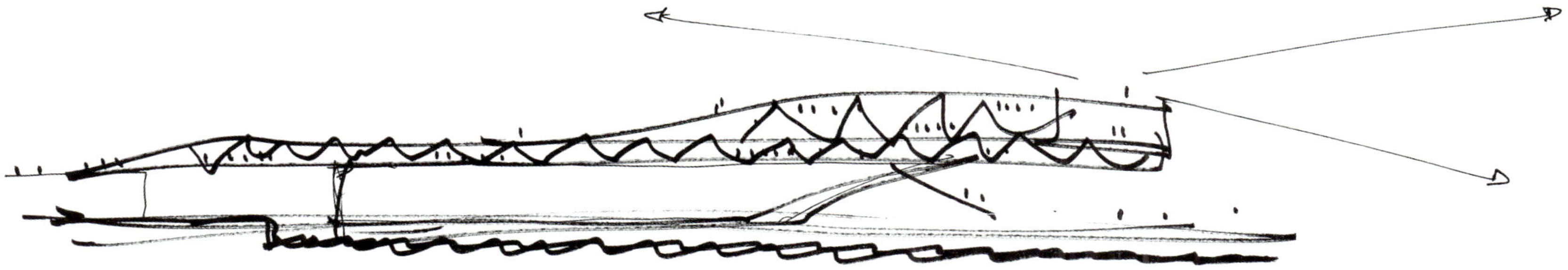

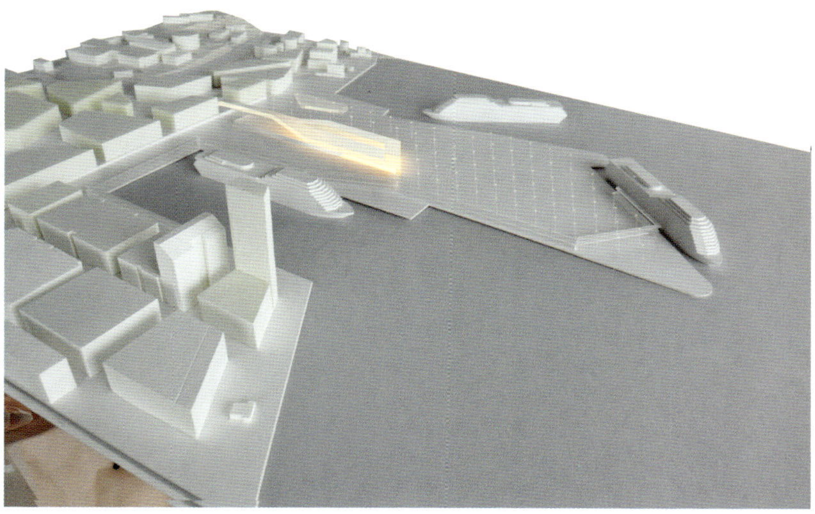

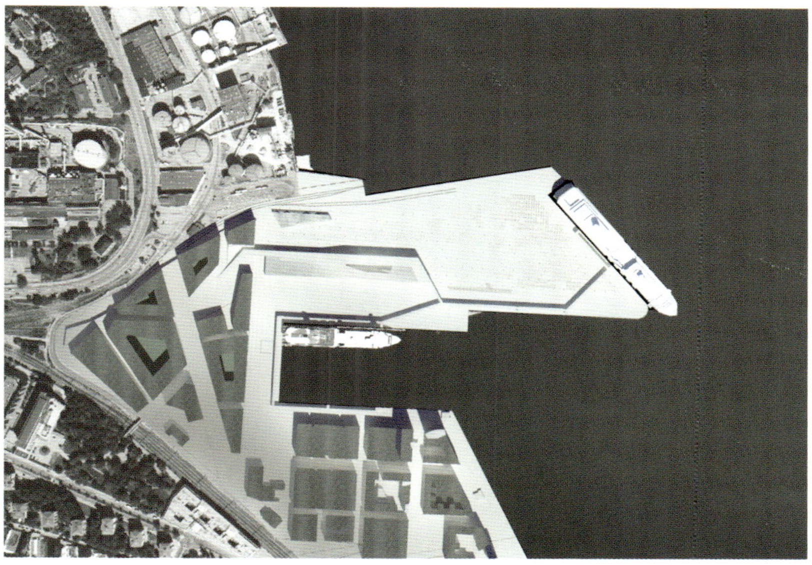

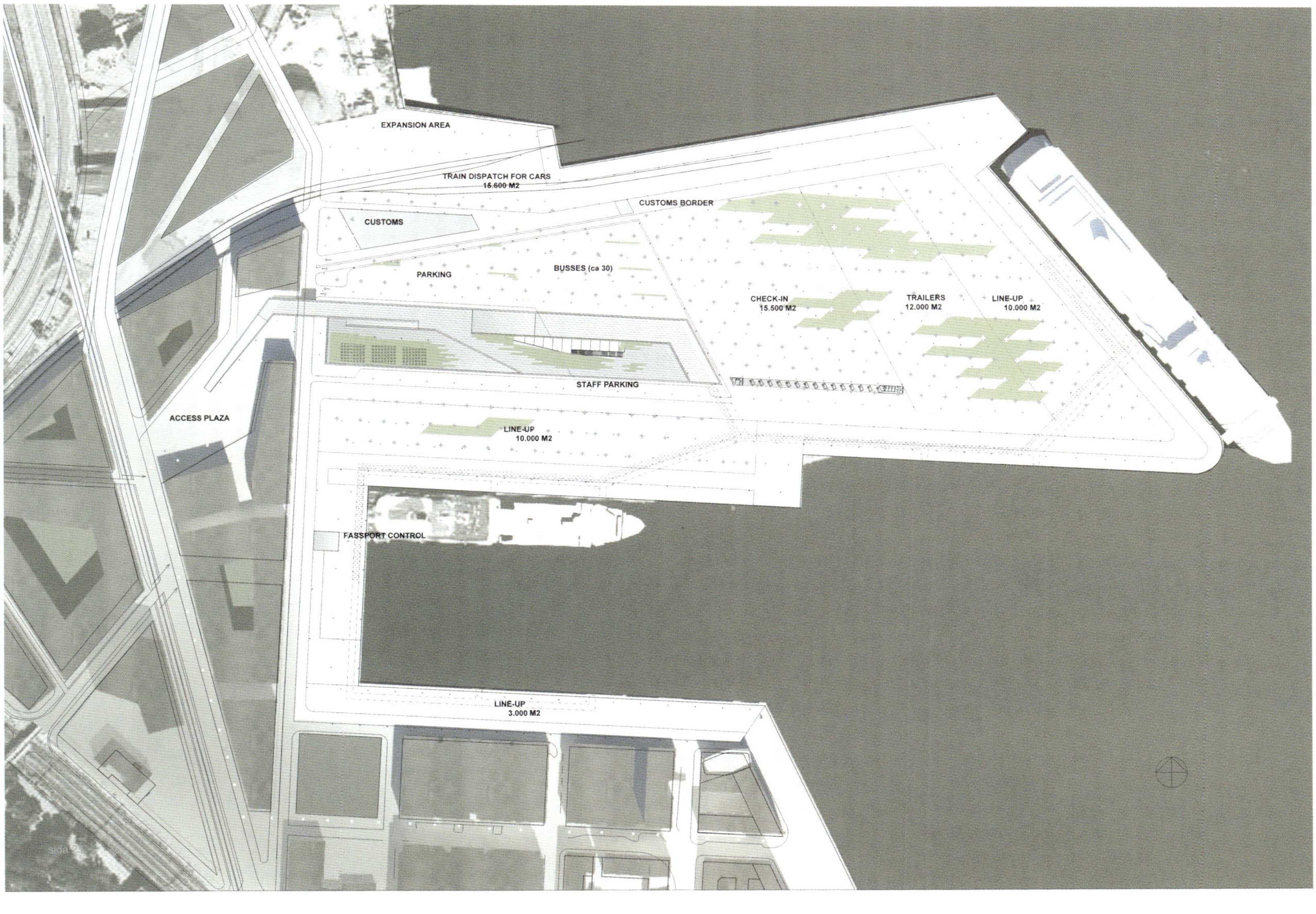
EXPANSION AREA
TRAIN DISPATCH FOR CARS
15.600 M2
CUSTOMS BORDER
CUSTOMS
PARKING
BUSSES (ca 30)
CHECK-IN
15.500 M2
TRAILERS
12.000 M2
LINE-UP
10.000 M2
STAFF PARKING
ACCESS PLAZA
LINE-UP
10.000 M2
FASSPORT CONTROL
LINE-UP
3.000 M2

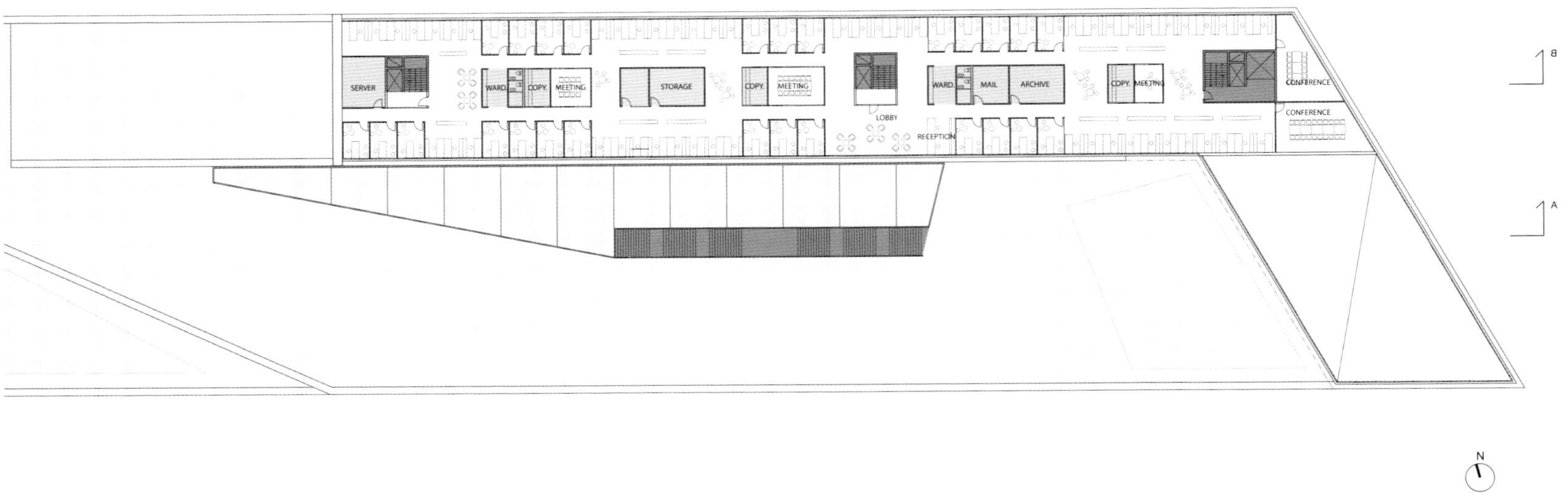
SERVER
WARD.
COPY.
MEETING
STORAGE
COPY.
MEETING
LOBBY
RECEPTION
WARD.
MAIL
ARCHIVE
COPY.
MEETING
CONFERENCE
CONFERENCE
B
A
N

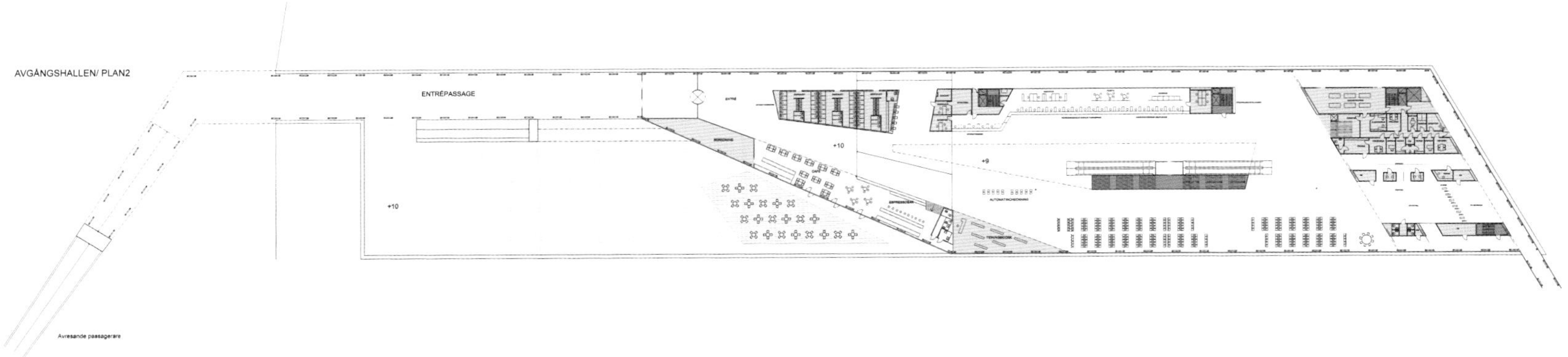
AVGÅNGSHALLEN/ PLAN2
ENTRÉPASSAGE
+10
+10
+9
Avresande passagerare

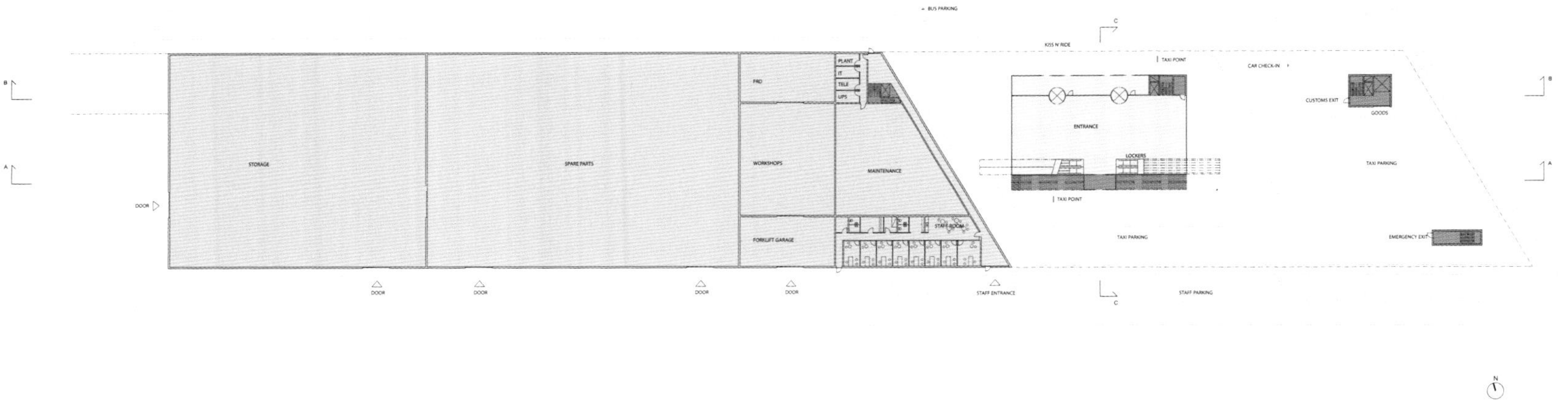
BUS PARKING
KISS N' RIDE
C
TAXI POINT
CAR CHECK-IN
B
A
STORAGE
SPARE PARTS
PLANT
IT
TELE
UPS
WORKSHOPS
MAINTENANCE
FORKLIFT GARAGE
ENTRANCE
LOCKERS
TAXI POINT
CUSTOMS EXIT
GOODS
TAXI PARKING
TAXI PARKING
EMERGENCY EXIT
DOOR
DOOR
DOOR
DOOR
DOOR
STAFF ENTRANCE
STAFF PARKING
N

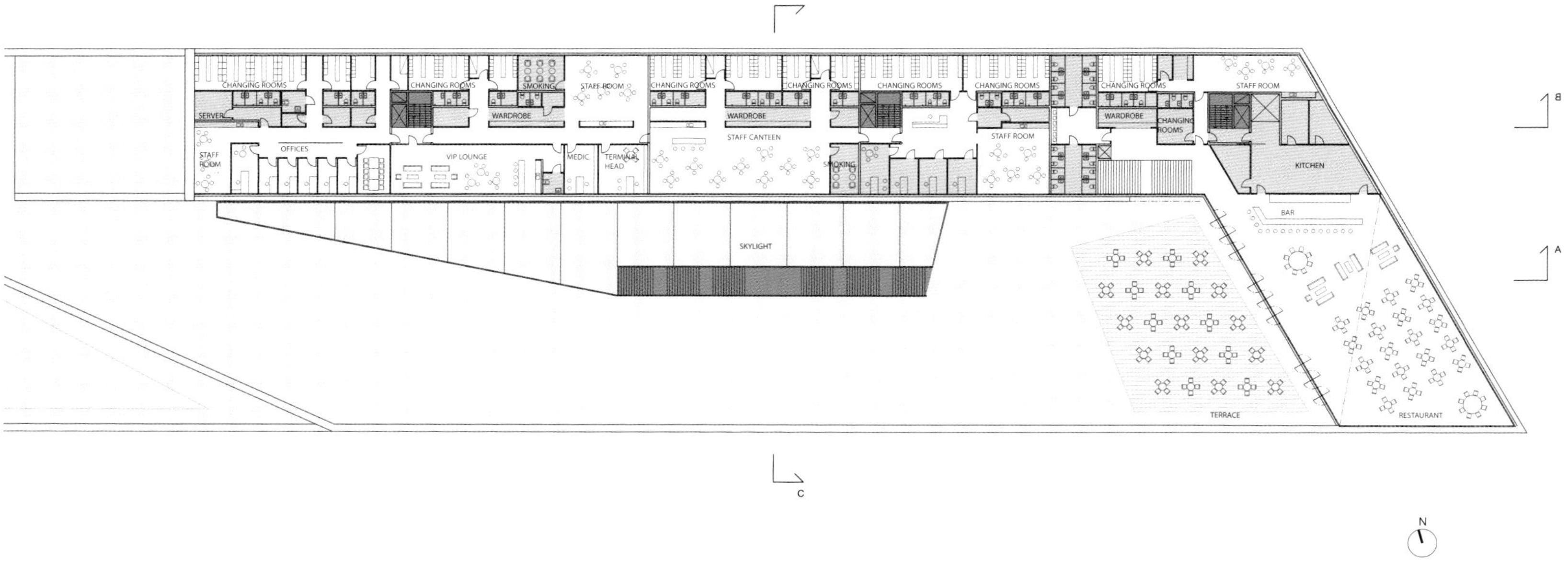
C
CHANGING ROOMS
SERVER
STAFF ROOM
OFFICES
CHANGING ROOMS
SMOKING
WARDROBE
VIP LOUNGE
STAFF ROOM
MEDIC
TERMINAL HEAD
CHANGING ROOMS
WARDROBE
STAFF CANTEEN
CHANGING ROOMS
SMOKING
CHANGING ROOMS
CHANGING ROOMS
STAFF ROOM
CHANGING ROOMS
WARDROBE
CHANGING ROOMS
STAFF ROOM
KITCHEN
BAR
SKYLIGHT
TERRACE
RESTAURANT
B
A
C
N

The new terminal for Stockholm's permanent ferry connections to Finland and the Baltics will be a landmark for the new urban development Norra Djursgårdsstaden - both architecturally and environmentally. The terminal, which will have a facade covered with expanded mesh, recalls the shape of a moving vessel and the architecture - with large cranes and warehouses - that previously characterized the ports. At the same time, the terminal has an ambitious sustainable profile, characteristic of the entire development.

The main idea has been to create natural links between central Stockholm and the new urban area in connection with the terminal, so that city life will naturally flow into the area. Therefore the terminal is raised to be at level with the urban zone, so it is easy for both pedestrians and traffic to access.

At the same time the roof of the terminal building is designed as a varied green landscape with stairs, ramps, niches, and cosy corners, inv ting both Stockholmers and passengers for a stroll or relaxing moments, while enjoying the view of the ferries, the archipelago, and the city skyline.

The aim is that the ferry terminal will be predominantly self-sufficient in energy and thus stand as an environmental model for public construction. Therefore the architecture of the terminal will integrate i.e. solar and wind power, for example the terraced landscape on the roof will integrate beds of solar cells along with the planting. The plan is to communicate the sustainable efforts to the people in the building by using i.e. centrally placed television screens, helping to raise awareness of the potential of sustainable construction.

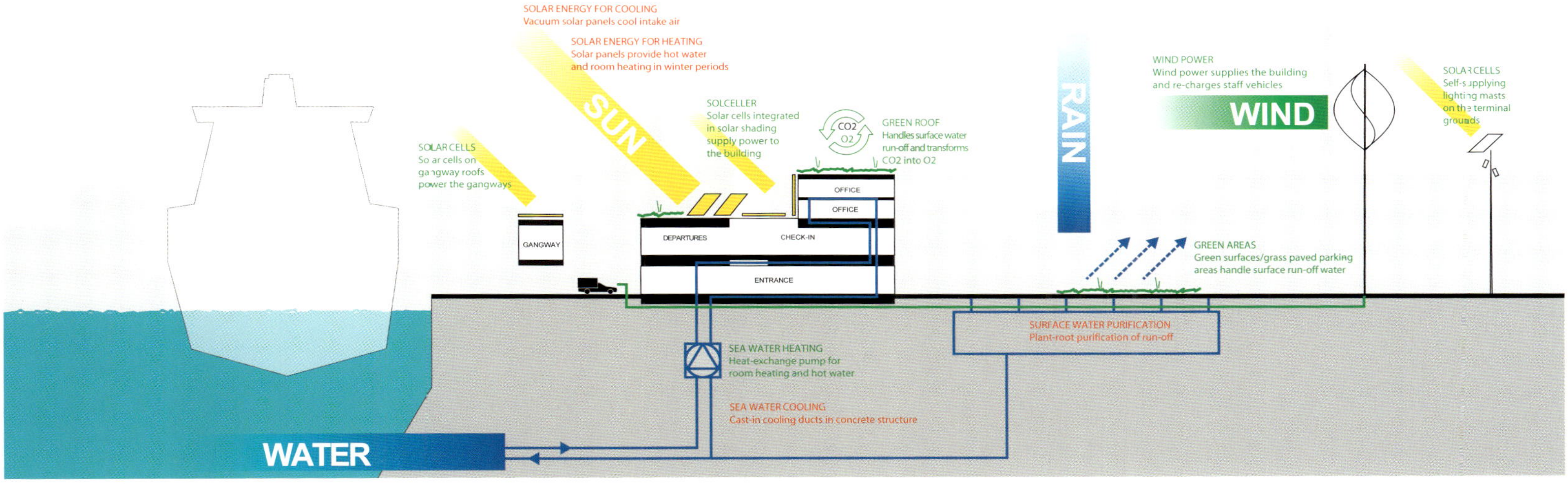

Passive energy strategies

Airbaltic Passenger Terminal

波罗的海航空客运大楼

MSB Arquitecto/ahad Arquitectos

Architectural Ideas Contest Riga. Latvia
拉脱维亚里加建筑创意赛参赛方案

这项工作开启了MSB建筑事务所与伦敦ahad建筑事务所合作的序幕，如今两者已建立了战略伙伴关系。

设计师计划将大楼建筑设置于两个码头中间，以展示出与它们之间强烈的关联。形式、材质以及表达的相互关联塑造了建筑内部与外部的形状。而且，拟定的设计方案目的是将其解读为一个对象，一个在其内部两个码头都发挥共谋作用的实体。

这反映出空中波罗的海的抱负，强化了连接天与地、东欧与西欧的理念。

建筑位于固定的网格之上，生成自拉脱维亚文化中找到的几何形状。此网格定义了空间和客运航线，相互作用于建筑主体与两个码头之间。钻石外形的构造与对角网格对建筑开发至关重要，建筑形式是通过一个象征性的过程获得的内部空间的结果。由复杂图案带来平面图和立面图几何形态的强烈视觉冲击，仿佛飞机场从图案中成型且作为大型雕塑脱颖而出一般。忠于这样的设计理念，内部和外部的熔覆面板都依照一种独特的模型剪切出来，使得柔和的光线弥漫机场内部。

三角形的入口在游客抵达或出发时都呈现出戏剧化的效果。它采用了动态图像，反映了飞行的真正活力。

多孔镶板薄膜与充满活力的形式相结合，产生大量富有表现力的光线，使该建筑成为地标。

设计师的目标是设计出一座机场航站楼，反映社会的价值观，并超越客户的期望。该设计成功结合了航空公司愿望以及技术/运营要求。

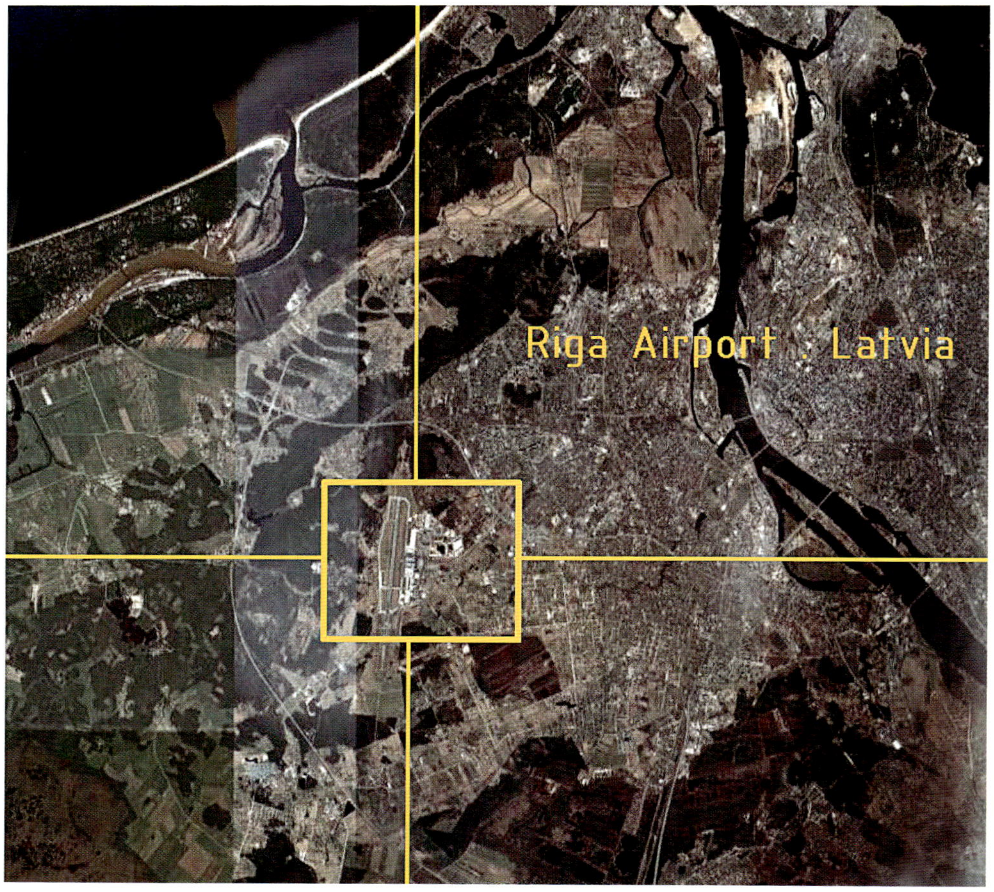

airBaltic
Baltic

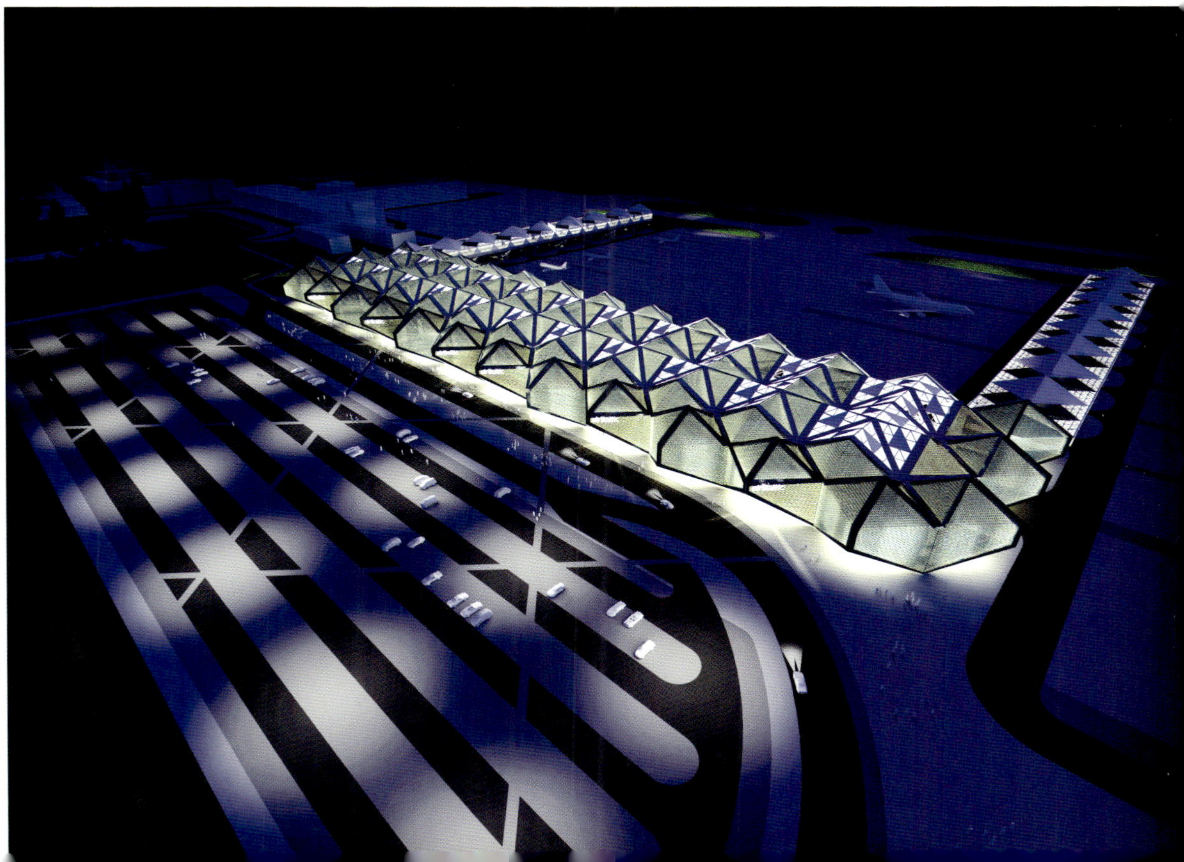

Terminal Floor Plan
Level 01 (+ 5.50)

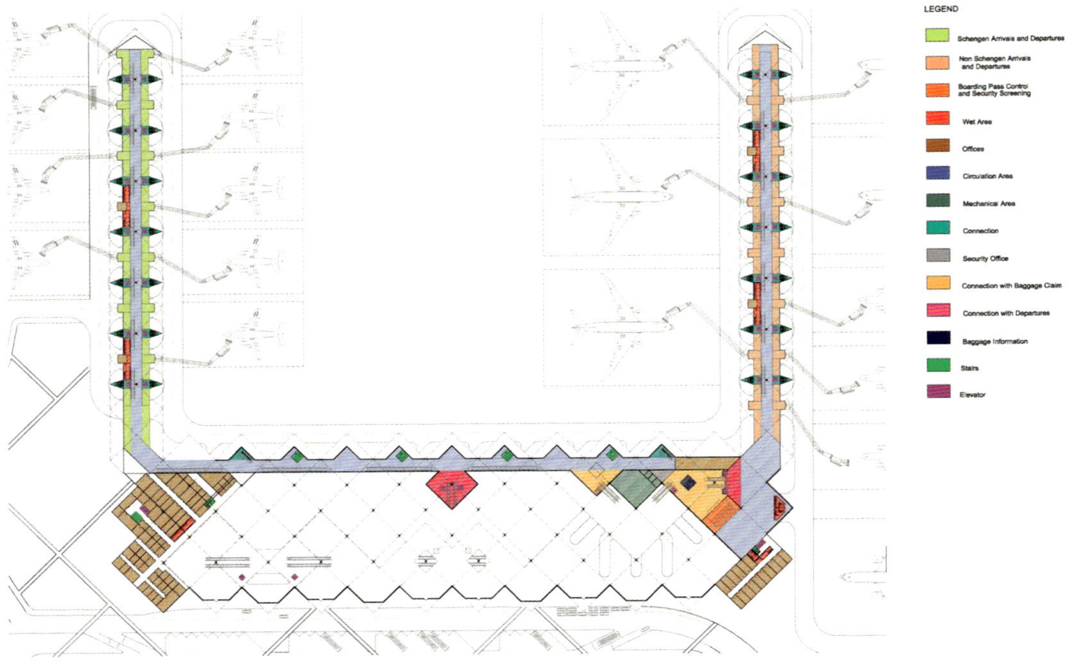

Terminal Floor Plan
Level 00 (+ 0.00)

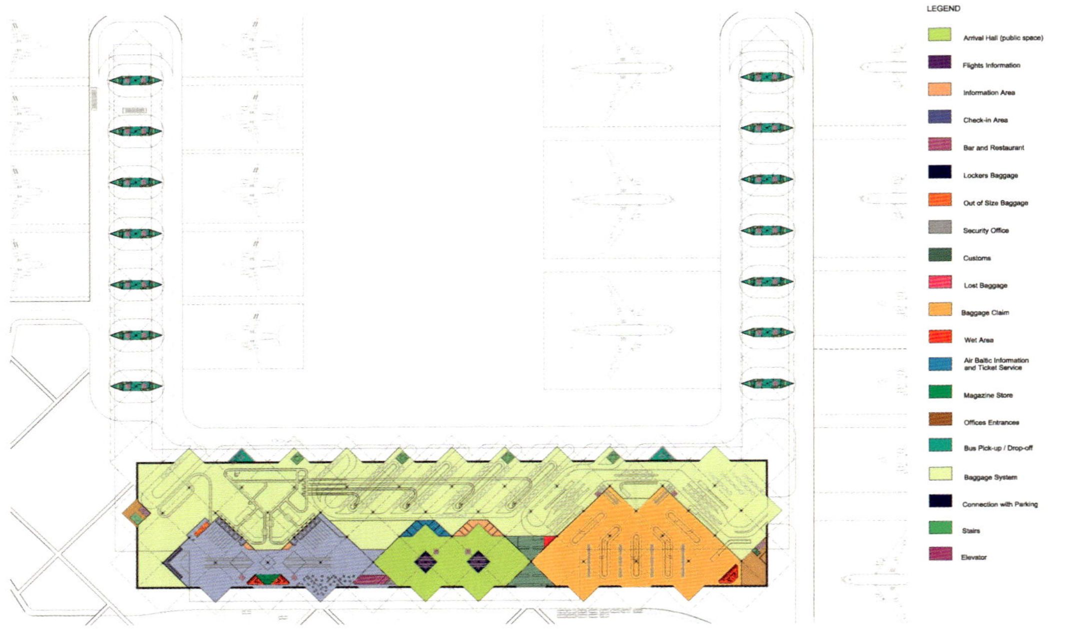

Terminal Floor Plan
Level 02 (+ 10.40)

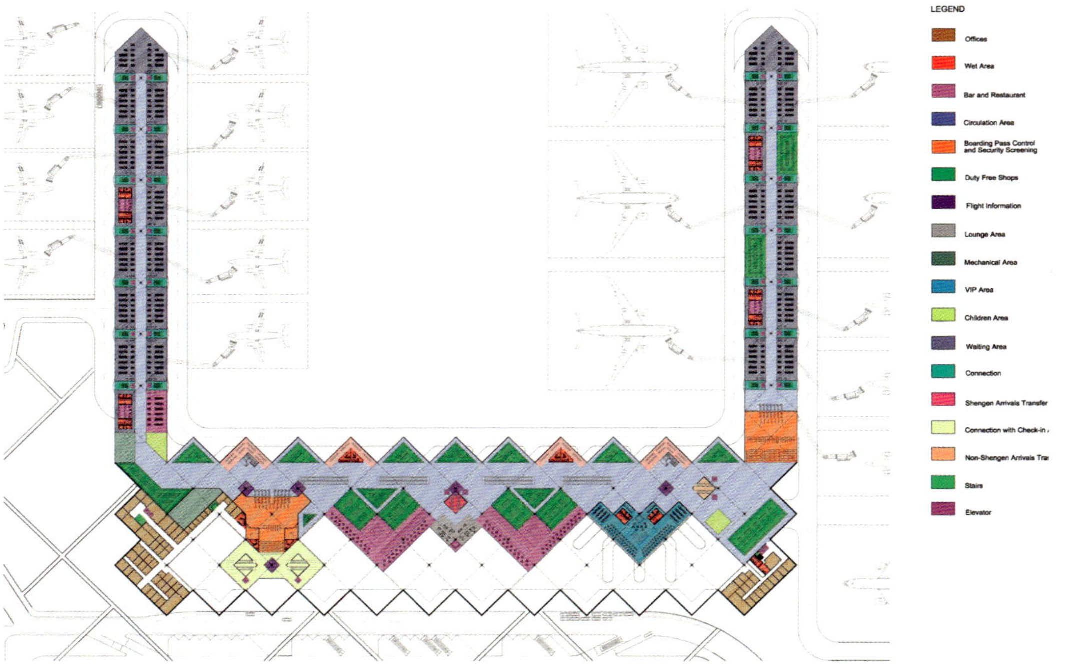

This work opens our participation with ahad - architects, based in London, a studio which we now have a partnership.
Our scheme sets the terminal building in the middle of the two piers, demonstrating a strong connection with both of them. The continuous language of form, materials and expression shapes both interior and exterior of the terminal. Further more, the proposed design is intended to be read as one object; one entity where both piers play an inner complicity.
This reflects Airbaltic's aspiration, reinforcing the idea of linking earth and sky, Eastern and Western Europe.
The building lies on a fixed grid generated from the geometries found in Latvia's culture. This grid defines spaces and passenger routes that inter-play between the main body of the terminal and both piers. The diamond like configuration and the diagonal grid are fundamental for the development of the terminal, where form is the result of the interior space obtained through a symbolic process. The studies of the intricate patterns result in a strong geometry both in plan and elevation as if the airport would take shape from the patterns and emerge itself as a large scale sculpture. Faithful to this idea, both interior and external cladding panels have been cut out with a distinctive pattern that allows the smooth light to flood the interior of the airport.
The main facade with the angular entrances has a dramatic effect, whether arriving or departing. It adopts a dynamic image, which reflects the true dynamism of flying.
The perforated paneled membrane is combined with the vibrant form generating light flooded and expressive volumes that make of this terminal a landmark building.
Our aim was to deliver an airport terminal that reflects the community's values, and exceed customer expectations. We believe that our design successfully combines airline aspirations, national charisma and its development, passenger flows and technical / operational requirements.

Departures

Baltic
Atvykimas
Arrival

Memorial Bridge

纪念之桥 Studio 3LHD

纪念之桥通过对空间的介入，成为一项标志性的城市公共设施，可以让人们感知。这座步行桥既具有实用性的功能，又是一个纪念参加克罗地亚独立战争场所，一个充满了回忆与虔诚的社交聚会和社交活动的地方。里耶卡市中心的河道将历史城市中心和过去的港口分开，纪念之桥就坐落在这条河道上。对于未来城市建设，这是一个非常重要的区域（目前作为停车场之用，未来会改建成为公园）。

纪念之桥如同一个厚约65厘米的薄板，以特有的L形状将河道两岸连接起来，并延伸至一端的纪念碑。仅仅通过纪念之桥本身的结构就实现了公共空间的定义，而纪念之桥也被公认为是一座具有纪念意义的建筑。该项目是通过长达四年的研究才得以建设，尤其是项目的设计与建造。这个150吨重的钢结构桥梁是当地一个造船厂制造的，每一部分都是工业化设计与建造的成品，通过特制的驳船经水路运抵项目现场组装成为现在的建筑。由于纪念之桥的结构部件要经由两座桥才能抵达项目现场，所以驳船要经过个性化的设计，使其在水位最低时和经过两座桥时可以尽量下沉，运载纪念之桥的结构部件安全通过。

纪念之桥已经通过其材料和精神意义改变了里耶卡市的愿景，也实现了其作为纪念之桥的目标。

The Memorial Bridge is a spatial intervention and an urban public object, a construction with tactile properties and a symbolic object. It is a pedestrian bridge that embodies the utilitarian function and a memory of the Croatian war of Independence veterans, a place of memory and piety, but also a social gathering and socializing spot. It is located in the center of Rijeka, on the channel which separates the historic city center from the former port, at one of the most potent sites for future construction (it is currently occupied by a parking lot, in the future it will become a park).

The bridge is designed as an extremely thin plate (65cm) connecting the two banks of the channel in a characteristic L-shape, it extends across the channel and ends with a monumental vertical accent. The definition of the public space was achieved only through the structure of the bridge itself, which at the same time must be recognized as a memorial object.

A four year long research resulted in the construction specially designed and manufactured exclusively for this project. Each part of the bridge was an industrially designed and manufactured product that was assembled on site. The 150 tons heavy steel bridge structure was produced in a local shipyard, and carried by sea in one piece on a specially designed barge to the location. As the Memorial Bridge had to pass under two existing bridges on its way to the site, the barge was designed so that it can sink at the time of lowest water levels and go under the existing bridges with its burden.

This urban public hybrid has changed the vista of the city (in the material as well as mental sense) and has achieved its goal of being a Memorial Bridge.

1 Section
2 Plan

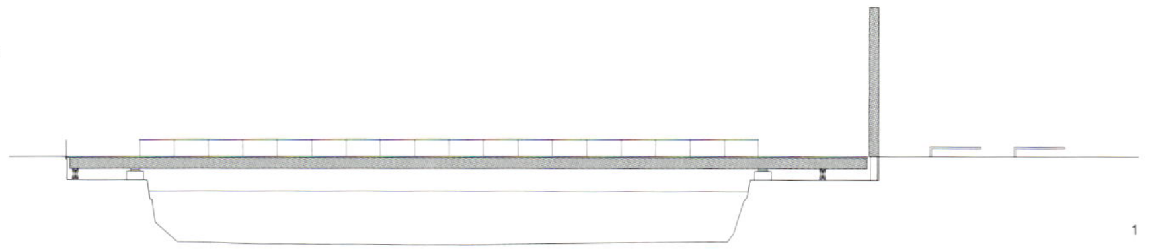

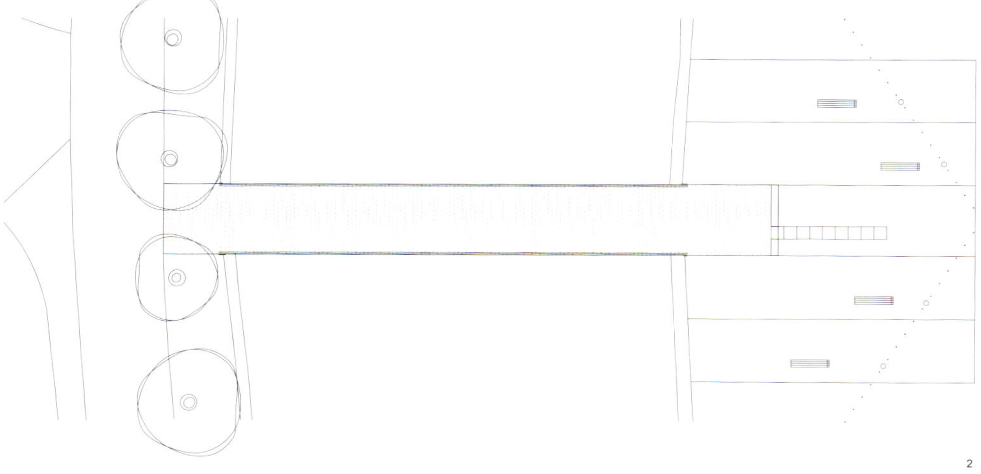

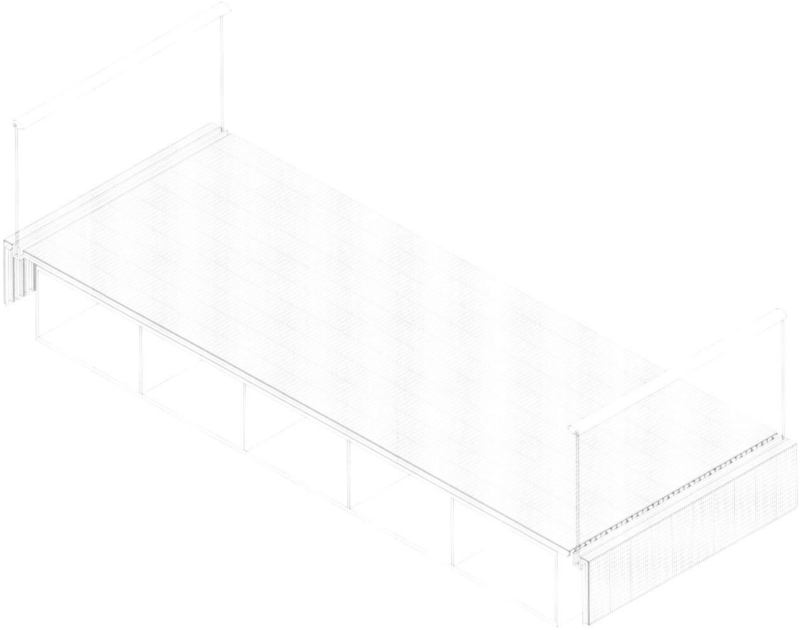

1 UNP6.5
2 Al 6063T6
3 DIN 912A2
4 Rubber 70 shora + PUtackdrive
5 Seal 50 shora
6 Nytril rubber
7 Al 6063T52
8 DIN 933A2
9 AISI 316TI
10 DIN 912A2
11 DIN St 37-3N
12 Sarnafil T6
13 Tar epoxy
14 Al 6063T6
15 DIN 933A2
16 Al 6063T52
17 Al 6063T52
18 Alcaten tube
19 Planibel linea azura
20 Rubber 50 shora
21 LED LM01A-W1
22 Al 6063T6
23 Teak wood

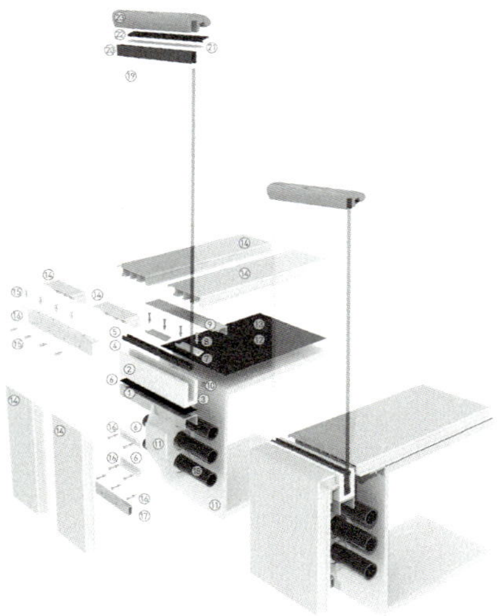

1 Aluminium tie
2 AlMg0.7Si (6063 T6) anodized 20 μ
3 Kerf
4 L - profile 80x80mm

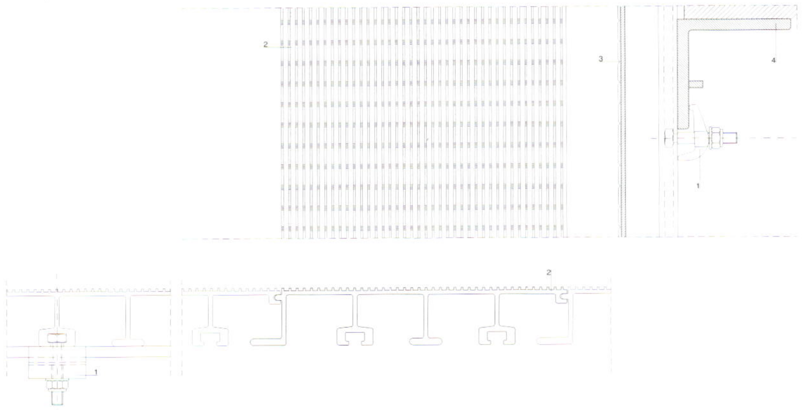

1 Prefabricated RC wall 3,15 x x12,15 x 0,4m
2 Prefabricated RC wall 1,15 x x12,15 x 0,4m
3 Symmetrical continual girder over three spans (3.5,36.9,3.5) with extensions (0.35,2.6m)
4 Transversal section of steel hollow square girder 500 x 65cm DIN RST 37-3N top and bottom chord 34mm 6 x chord 15mm
5 6 x tension bearings set
6 Reinforced concrete abutment 1mm
7 2 x fixed pressure bearings set DIN-U4141
8 6 x RC drilled pilots Ø 17m 80cm

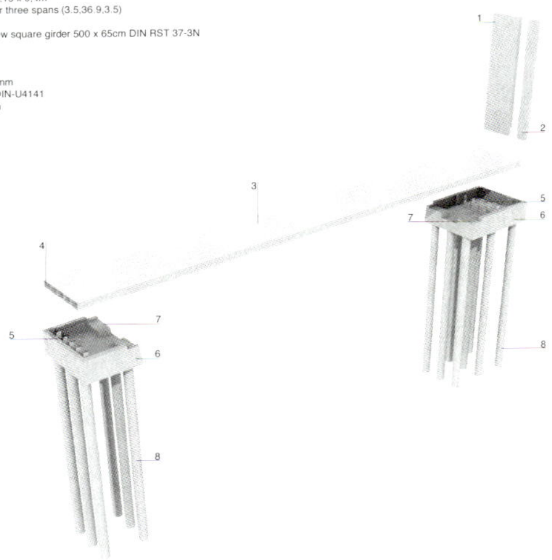

Redevelopment of Plaza de la Encarnacion, Seville, Spain

塞维利亚恩卡纳西翁广场改建项目

J. Mayer H. Architects

Permanent Collection of Museum of Modern Art, NY and Staatliche Museen zu Berlin, Preussischer Kulturbesitz, Berlin
Permanent Collection of DAM, Deutsches Architekturmuseum Frankfurt, Germany
纽约现代艺术博物馆与柏林普鲁士文化遗产——柏林州立博物馆永久收藏品
德国法兰克福德意志建筑博物馆永久收藏品
2005年豪西姆奖，即欧洲可持续建筑铜奖

Credits
Client: Ayuntamiento de Sevilla und SACYR
Architects: J. MAYER H. Architects
Project Architect: Jürgen Mayer H., Andre Santer, Marta Ramírez Iglesias
Project Team: Ana Alonso de la Varga, Jan-Christoph Stockebrand, Marcus Blum, Paul Angelier, Hans Schneider, Thorsten Blatter, Wilko Hoffmann, Claudia Marcinowski, Sebastian Finckh, Alessandra Raponi, Olivier Jacques, Nai Huei Wang, Dirk Blomeyer (Management Consultant 1st Phase)
Competition Team: Jürgen Mayer H, Dominik Schwarzer, Wilko Hoffmann, Ingmar Schmidt, Jan-Christoph Stockebrand, Julia Neitzel, Klaus Küppers, Georg Schmidthals, Andre Santer, Daria Trovato
Technical Support for Plants – Competition 2nd Phase only: Coqui-Malachowska-Coqui with Thomas Waldau
Technical Consultant and Multidisciplinary Engineers for Realization: Arup
Manager: Jose de la Peña, Jan-Peter Koppitz
Team Statics: Romain Buffat, Kayin Dawoodi, Steffen Janitz, Andres Garzón, Enrique González, Carlos Merino, Estrella Morato, Victor Rodríguez Izquierdo, Volker Schmid
Team Building Services: Salvador Castilla, Alborada Delgado, Marta Figueruelo Calvo
Team Fire Protection: George Faller, Benjamin Barry-Otsoa, Jimmy Jonsson
Timber Engineering and Detail Design: Finnforest, Aichach
Photographers: David Franck, Fernando Alda

由J. Mayer H. 建筑事务所翻新设计的塞维利亚恩卡纳西翁广场——“大都会梧桐”成为了塞维利亚的新象征，该广场是一个有认同感的场所，同时阐明塞维利亚作为世界上最具魅力的文化城市之一的角色定位。“大都会梧桐”探索了恩卡纳西翁广场成为新的当代城市中心的可能性。其置身于中世纪塞维利亚市致密结构内的独特城市空间，将各种各样的活动诸如纪念、休闲以及商业考虑了进来。一项高度发达的基础设施激活了广场，使其成为观光客和许多本地居民的旅游目的地。

“大都会梧桐”将令人印象深刻的木结构大量运用于考古博物馆、农夫市场、低空广场、多个酒吧和梧桐内外的餐厅，以及梧桐最顶端的全景露台等建筑中。作为规模最大、最具创新性的，以聚氨酯涂层黏合的木质结构之一，梧桐从考古挖掘现场破土而出，成为一座当代的里程碑，它定义了历史与当代城市的独特关系。“大都会梧桐”的多用途特征为塞维利亚心脏内外的文化和商业激发出一种动态的发展。

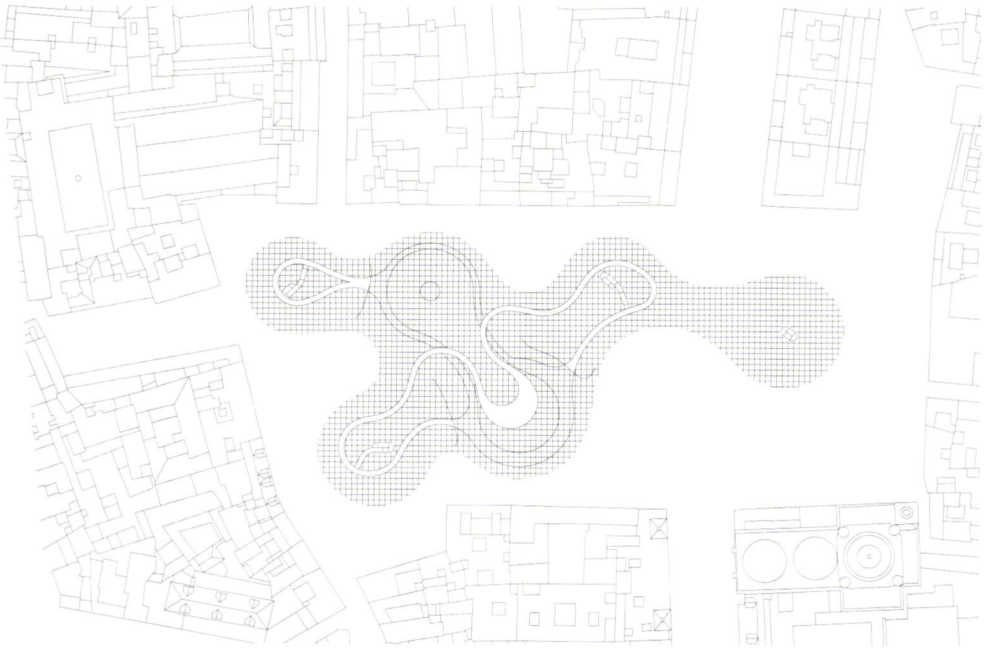

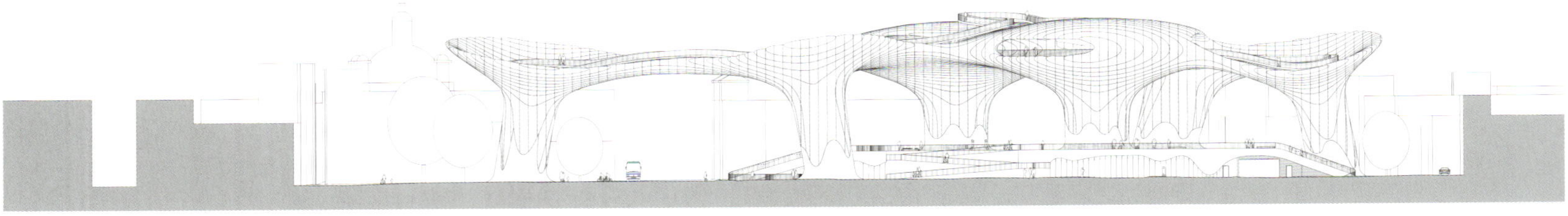

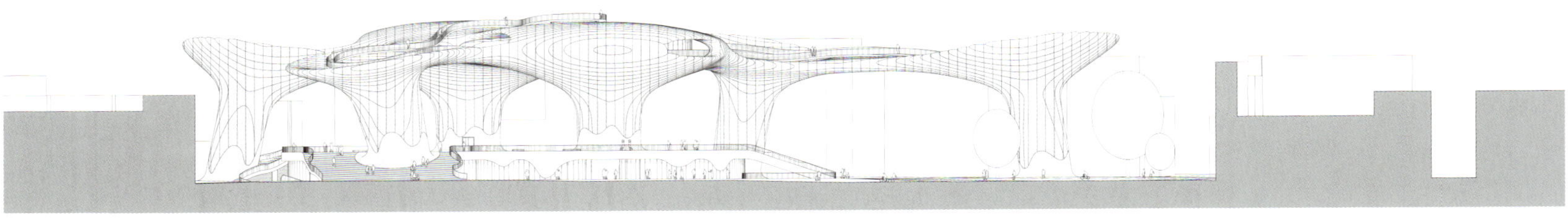

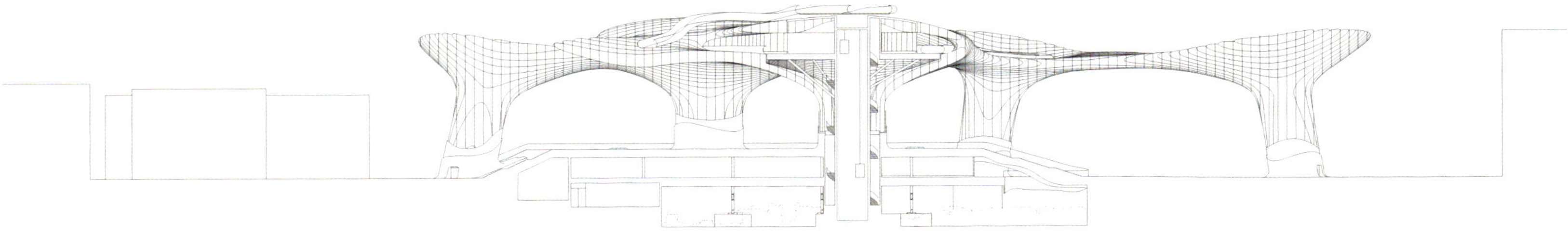

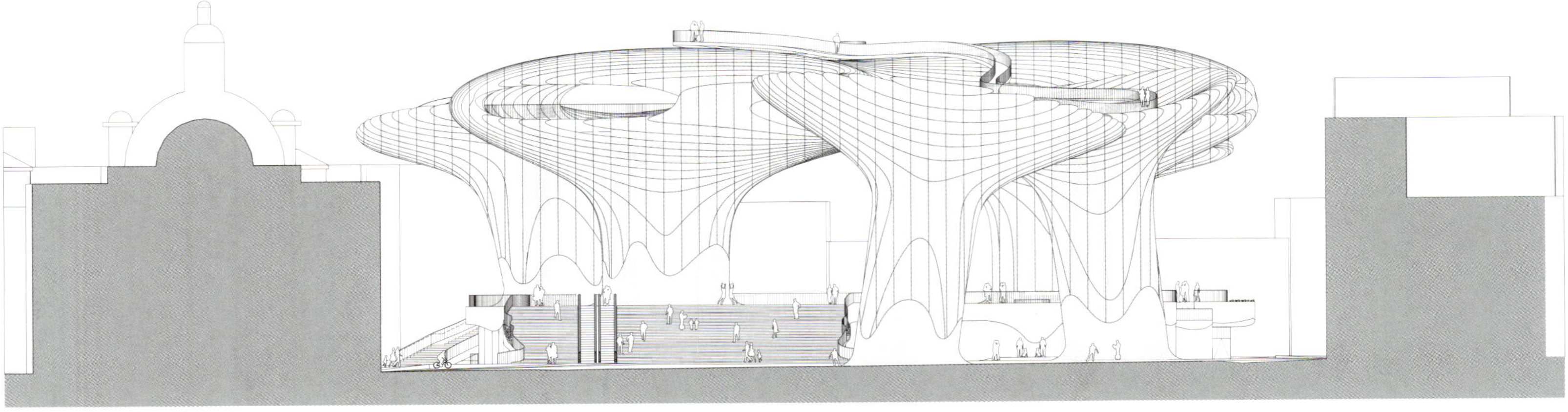

"Metropol Parasol", the Redevelopment of the Plaza de la Encarnacion in Seville, designed by J. Mayer H. architects, becomes the new icon for Seville, - a place of identification and to articulate Seville's role as one of the world´s most fascinating cultural destinations. "Metropol Parasol" explores the potential of the Plaza de la Encarnacion to become the new contemporary urban centre. Its role as a unique urban space within the dense fabric of the medieval inner city of Seville allows for a great variety of activities such as memory, leisure and commerce. A highly developed infrastructure helps to activate the square, making it an attractive destination for tourists and locals alike.

The "Metropol Parasol" scheme with its impressive timber structures offers an archaeological museum, a farmers market, an elevated plaza, multiple bars and restaurants underneath and inside the parasols, as well as a panorama terrace on the very top of the parasols. Realized as one of the largest and most innovative bonded timber-constructions with a polyurethane coating, the parasols grow out of the archaeological excavation site into a contemporary landmark, defining a unique relationship between the historical and the contemporary city. "Metropol Parasols" mix-used character initiates a dynamic development for culture and commerce in the heart of Seville and beyond.

RYD D.Q. Celebration Hall

利雅得外交区庆典大厅 Labics

International Competition, Third Prize
国际竞赛三等奖

Credits
Client: Kingdom of Saudi Arabia, High Commission for the Development of Ar Riyadh
Location: Riyadh, Saudi Arabia
Program: Celebration Hall, Parking
Time of Design: 2011
Total Area: 10,580 m²
Cost: 108,500,000 euro
Design: Labics (Maria Claudia Clemente, Francesco Isidori)
Architectural team: Leonardo Consolazione, Billy Erhard, Elisa Frisetti, Piergiorgio Loconte, Dominique Rethans
Model: Modelab
Renderings: Federico Pitzalis
Video: Piero Perilli
Structural Concept, Services and Cost Estimation Consultant: Favero & Milan ingegneria

该方案的目标与设计原则源于阿拉伯的传统文化，旨在为利雅得打造一个展望未来的新形象。为了实现该目标，设计师提出了两项明确的原则：一是打造一个典型的清真寺式建筑，层次分明，井然有序；二是要形成一个帐篷式的外围空间，灵活而具有吸引力。综合两项原则，设计师呈现的是一个庄严而亲切的建筑空间，将公共属性与私密性融合在一起。

该设计突出的几何线条将结构、空间和装饰性元素统一在一起。这样一个严密的形态能够清楚地显示出其各个组成部分，从大的屋顶到细节性的装饰物，无不清晰可见。所有设计元素组合在一起可以控制和改善日光的射入。

当用户从外面进入到建筑内部时，会发现建筑结构的密度在不断改变，主厅更开放与空旷；而在会议区和餐厅，为保证私密性，结构的密度有所增加。

优化自然光的效果是该设计的一个最基本的目标。设计不仅要减少太阳光的直接照射，同时要改善建筑内部空间的质量，达到舒适的效果。阳光从屋顶透过建筑外层时，外层的各种材料会导致其强弱发生不同程度的改变，这样一来就在室内形成了一种温暖、柔和而舒适的氛围。

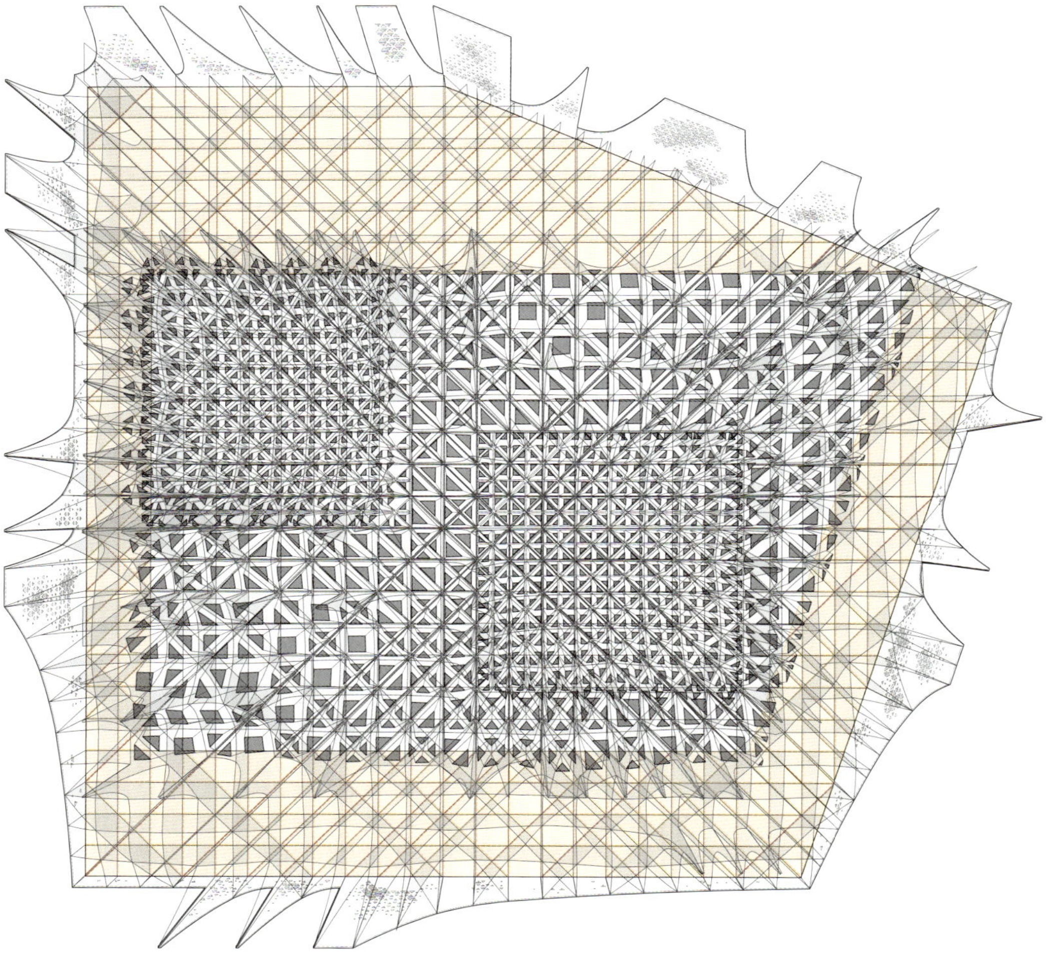

The proposal is based upon objectives and design principles that originate from Arabic traditions, with the aim to provide a new image for Riyadh, an image that looks towards the future. In order to achieve this ambition, the design is based on two clear and recognizable principles: The first design principle refers to the hierarchically structured, geometric and well-organized space of a typical Mosque; the second principal refers to the inviting, enveloping and flexible space of the typical tent. The union of these two principles defines a solemn space, simultaneously capable of providing hospitality, user friendliness and protection; effectively combining an inviting public character with a more intimate and private dimension.

The design, as in the best Arab tradition, is governed by a strong geometry that unifies structure, space and decorative elements. It is a rigorous pattern that is able to define all the elements that the design is composed of; from the large roof structure to the detail of single decorative elements; all collaborating to control and refine the access of daylight.

As the user moves from the outside into the building, he is accompanied by variations in the density of this structural pattern; from the Main Hall with a more public and open character to the Majlis and Dining hall where the structural pattern has been densified to provide intimacy.

The quality of natural light is a fundamental objective of the design, not only in order to reduce the direct radiation of sunlight, but principally to increase the spatial quality and the comfort inside the building. The light, captured from the roof and filtered through the building skin is modulated by the differences in this envelope and creates a warm, soft and comfortable atmosphere inside the spaces.

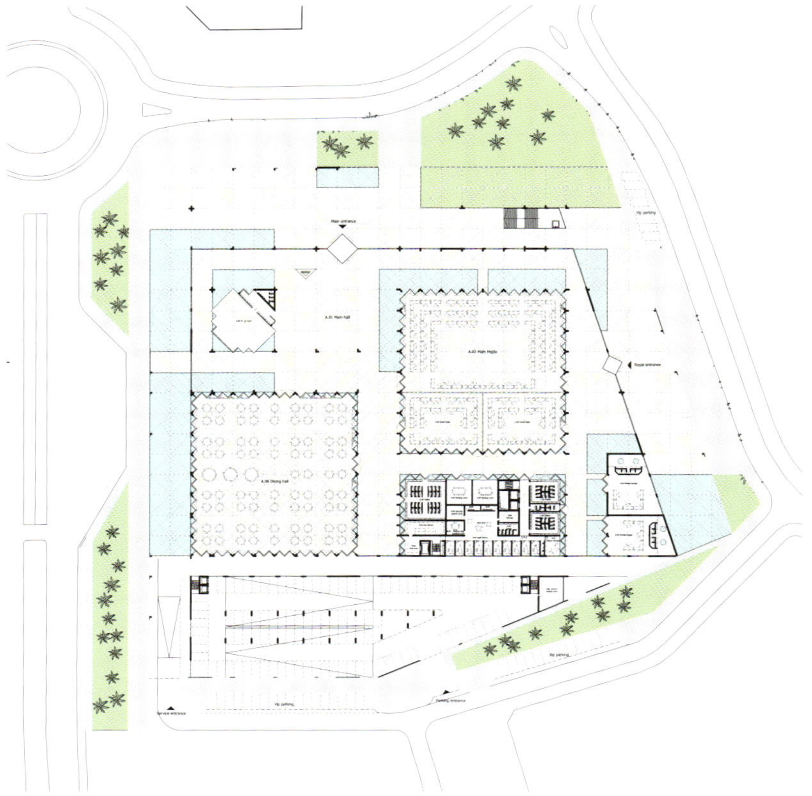

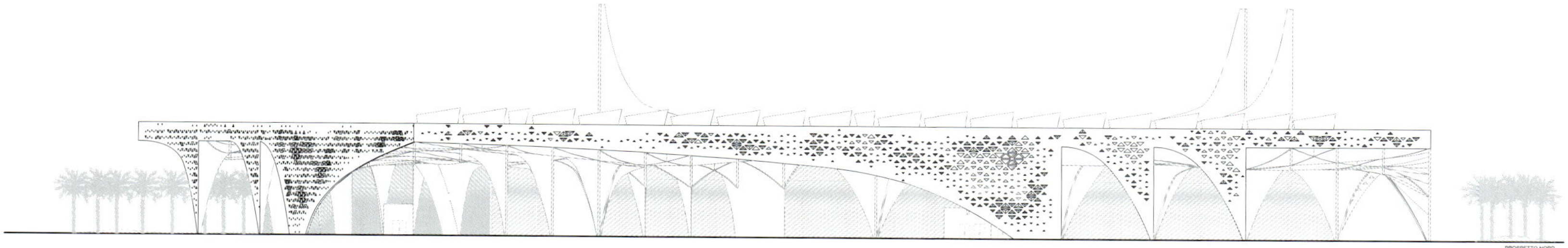

PROSPETTO NORD

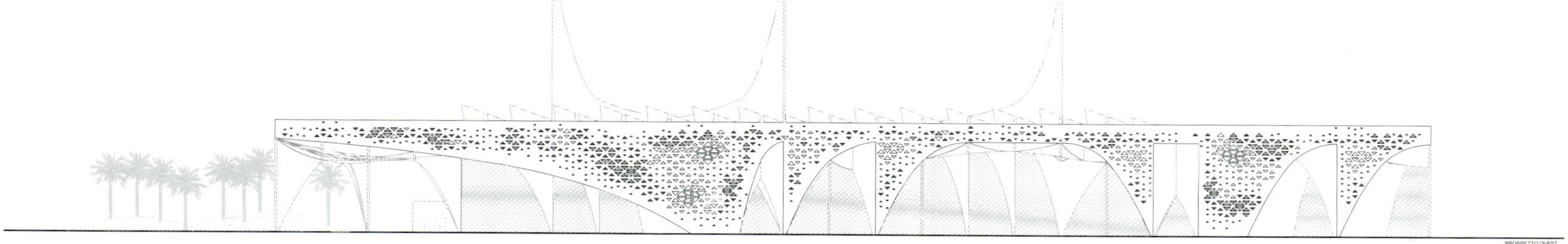

PROSPETTO OVEST

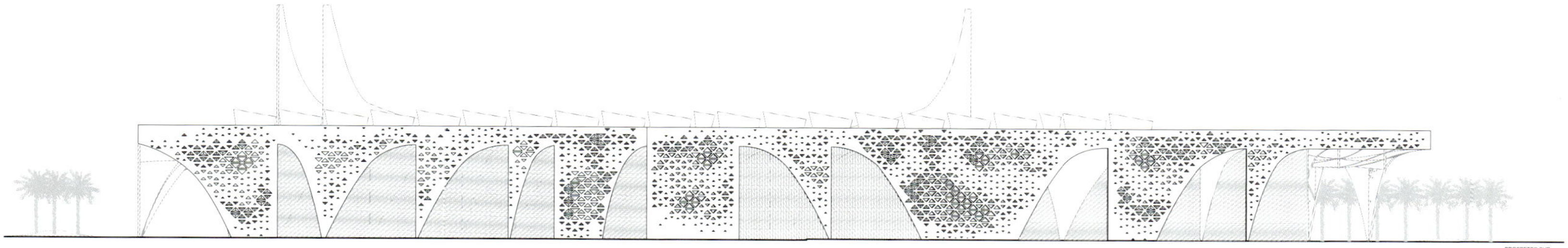

PROSPETTO SUD

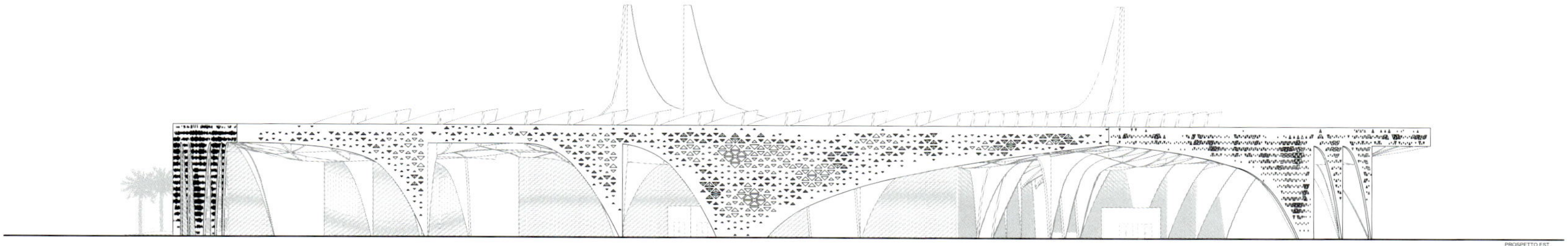

PROSPETTO EST

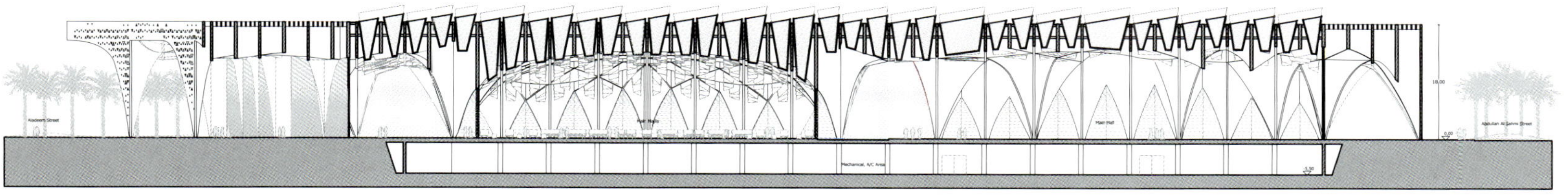

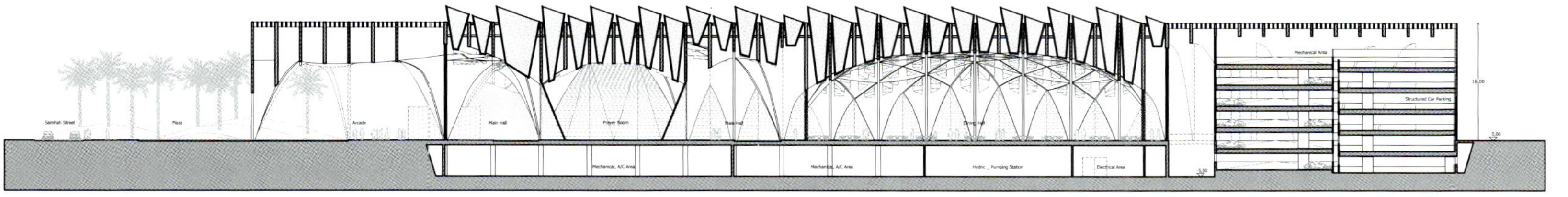

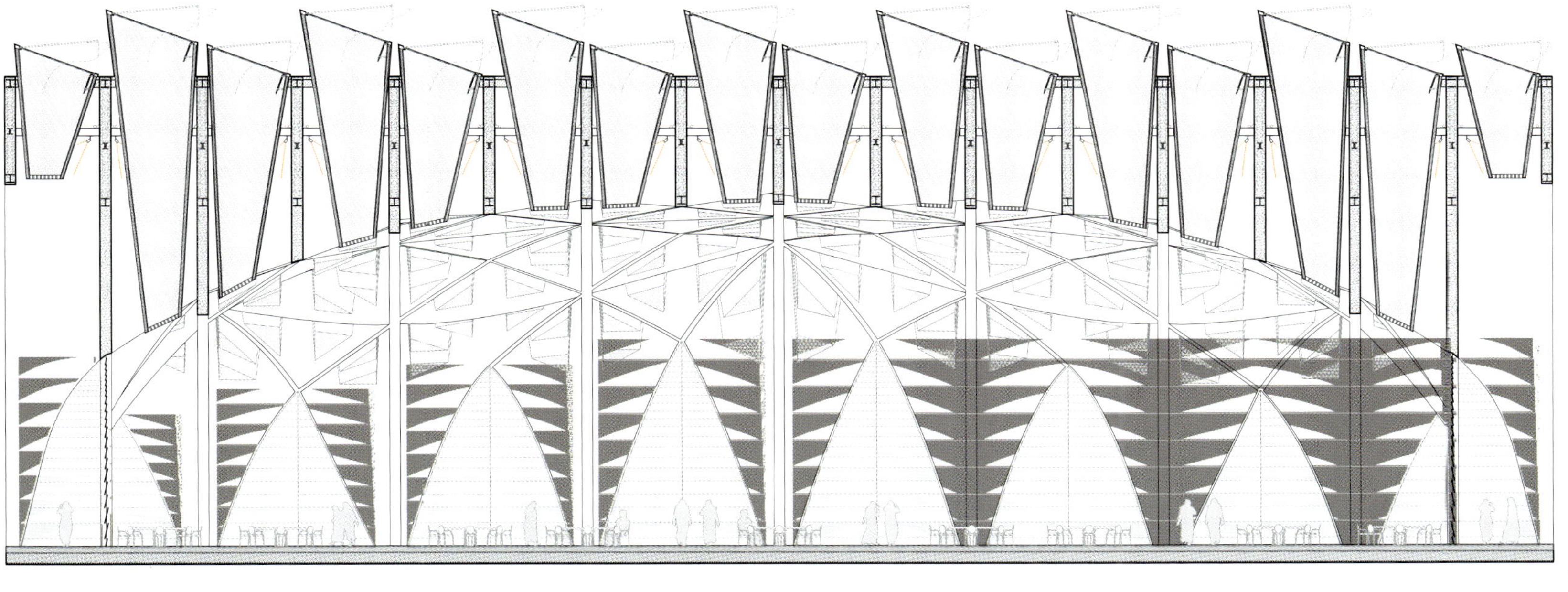

Components

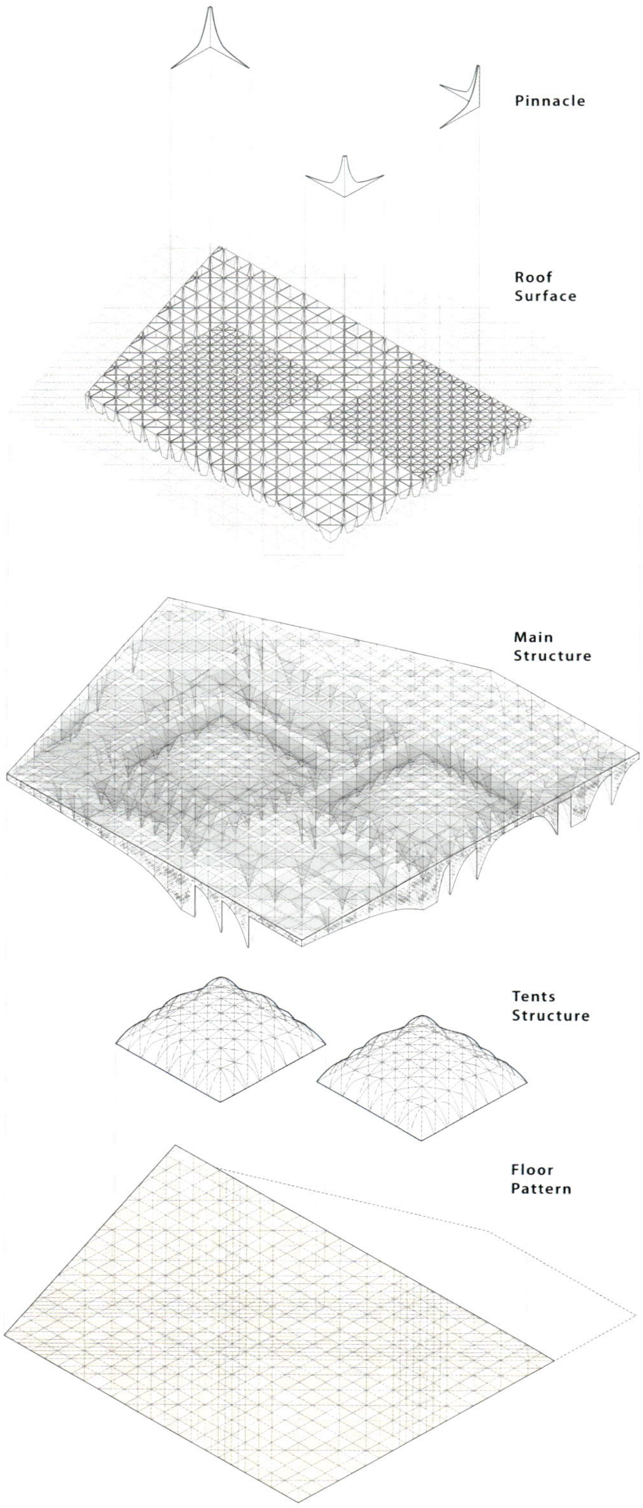

Pinnacle
Roof Surface
Main Structure
Tents Structure
Floor Pattern

Ceremony hall
Parking
Service entrance
Southern entrance of the D.Q.

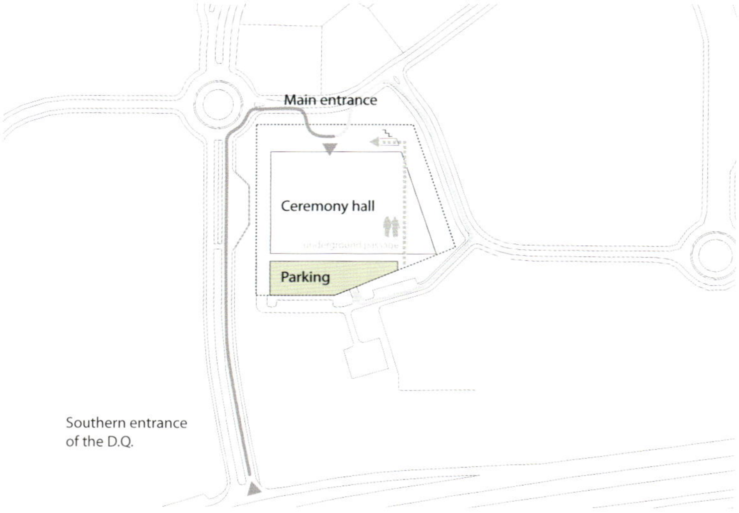

Main entrance
Ceremony hall
Parking
Southern entrance of the D.Q.

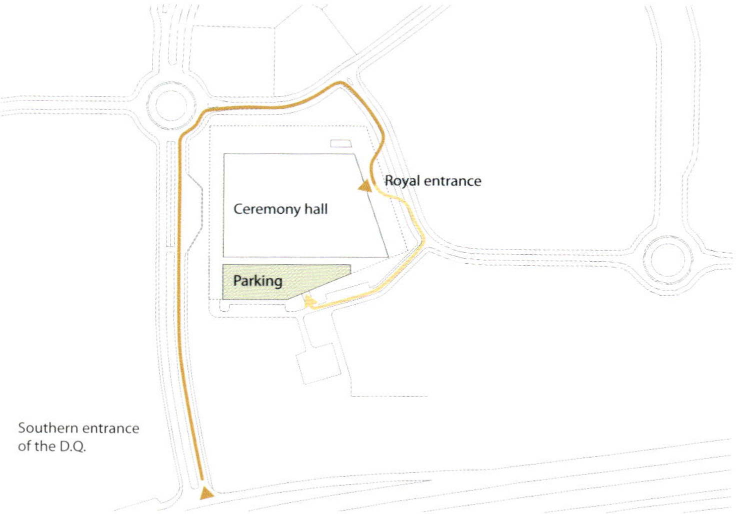

Ceremony hall
Royal entrance
Parking
Southern entrance of the D.Q.

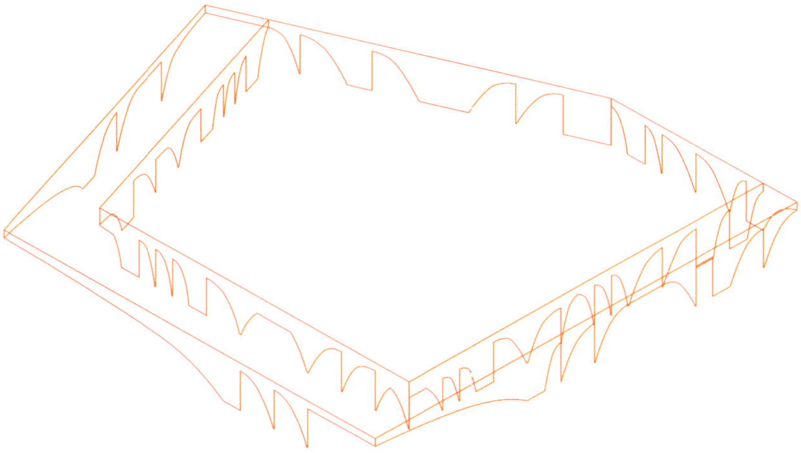

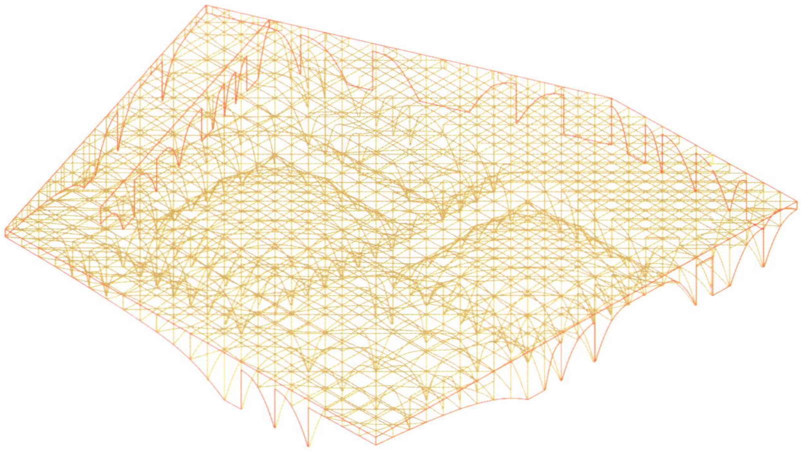

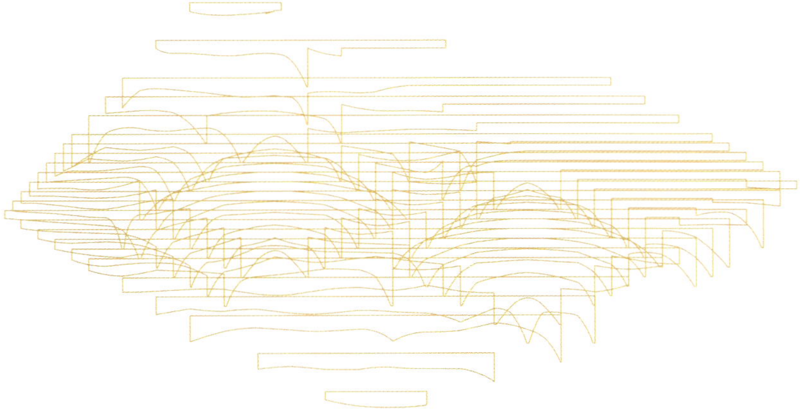

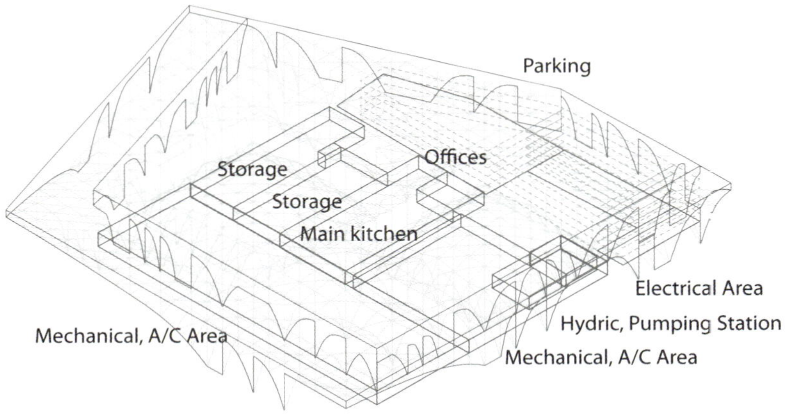
Parking
Storage
Offices
Storage
Main kitchen
Electrical Area
Hydric, Pumping Station
Mechanical, A/C Area
Mechanical, A/C Area

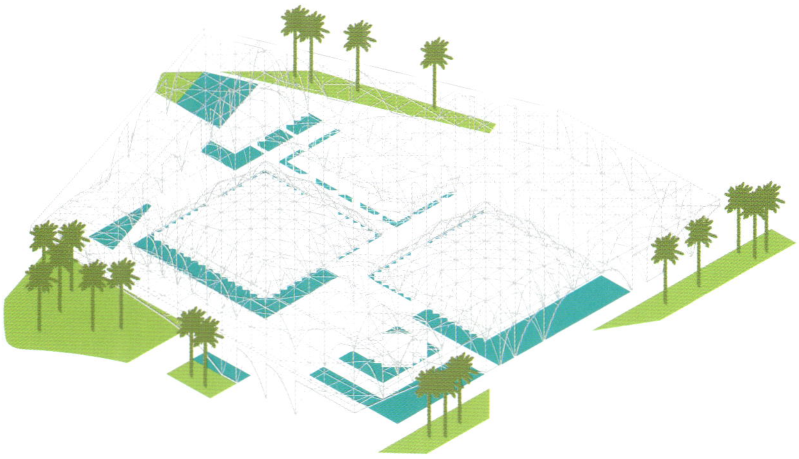

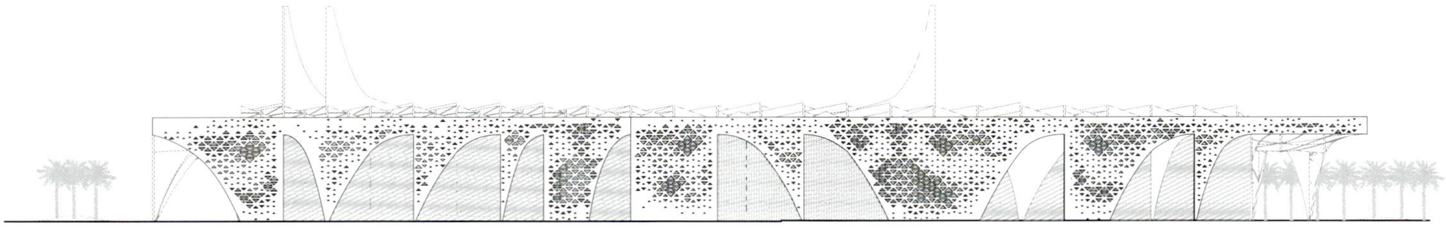

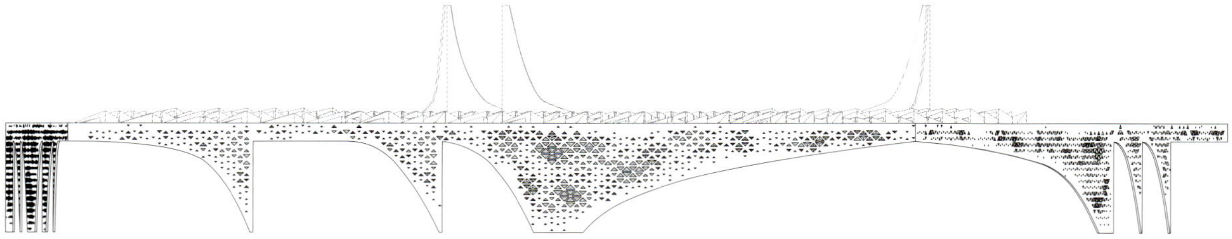

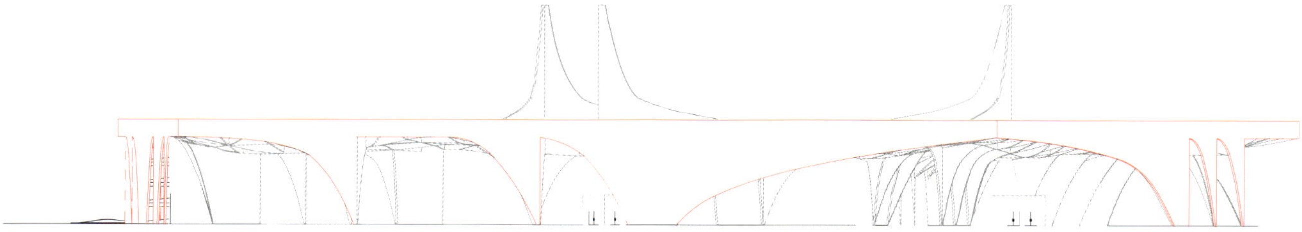

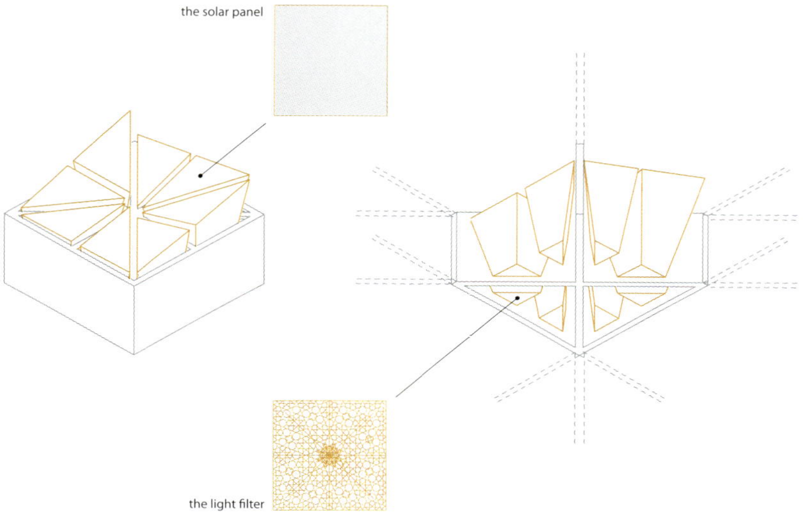
the solar panel
the light filter

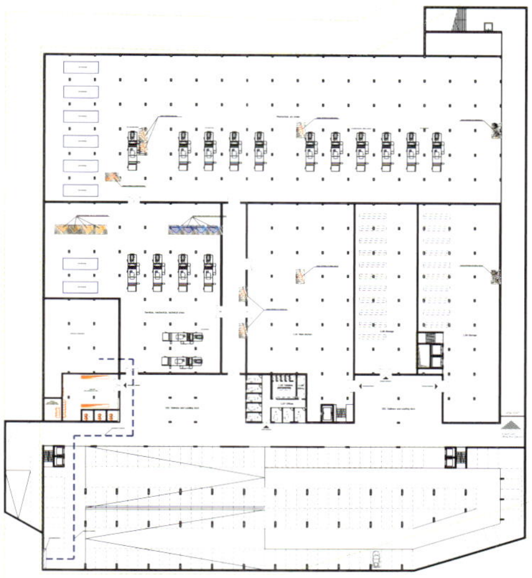

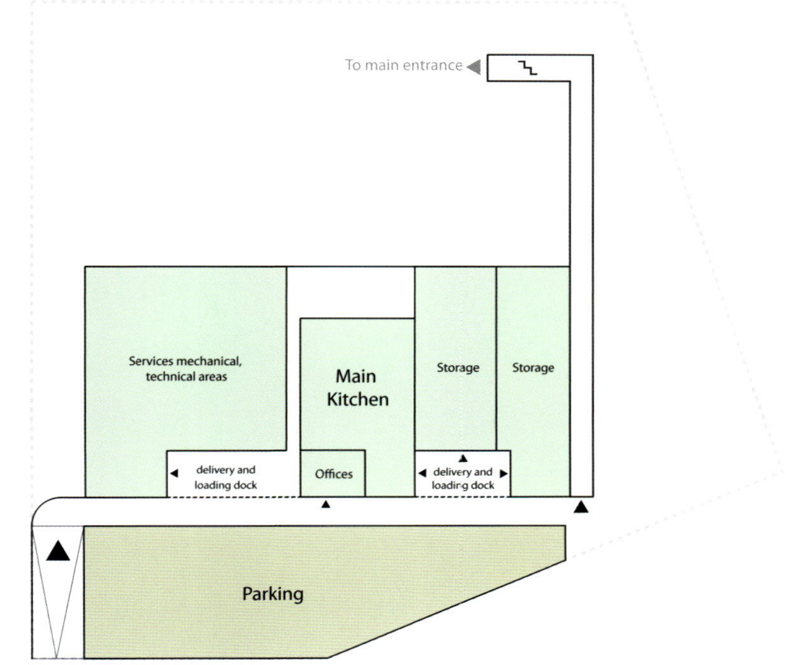
To main entrance
Services mechanical, technical areas
Main Kitchen
Storage
Storage
delivery and loading dock
Offices
delivery and loading dock
Parking
Service entrance

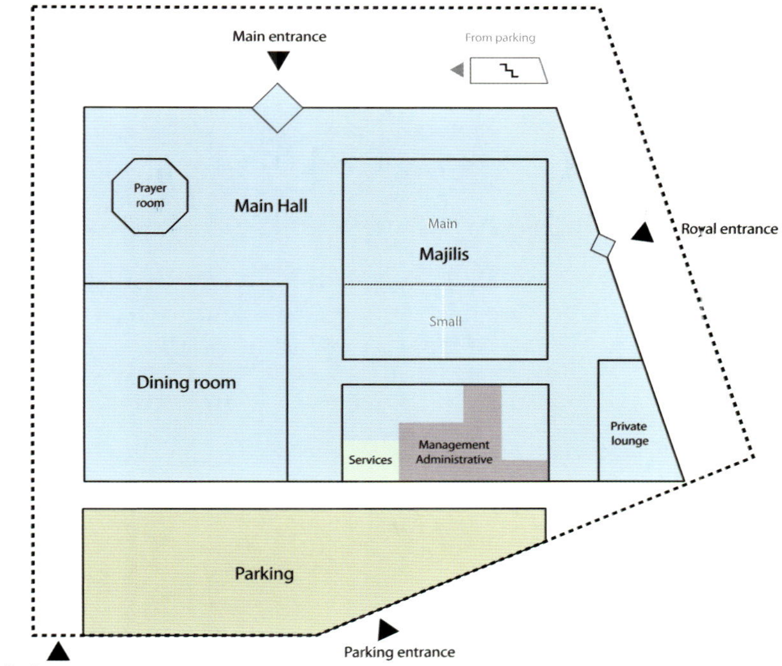
Main entrance
From parking
Prayer room
Main Hall
Main
Majilis
Small
Royal entrance
Dining room
Services
Management Administrative
Private lounge
Parking
Parking entrance
Service entrance

Roman Square Bouquet: Defining a "Square"

罗马广场之绿色定义

University of Arkansas Community Design Center

Revitalization of the Roman Square
罗马广场重建竞赛参赛方案

Credits
Location: Brno, Czech Republic
Design Year: Summer 2010
Architect: University of Arkansas Community Design Center

布尔诺举行了一场国际性的设计竞赛，征集罗马广场的整修方案。罗马广场坐落在这个历史性城市的中心，毗邻马萨里克大街——布尔诺主要的零售与散步场所之一，并且与该市主要的两个广场相隔不远。但是目前，罗马广场已被忽视，那里的土地使用、绿化、照明以及建筑都处于很糟糕的状态。

广场目前主要被一个临时的室内市场以及一些酒吧所占据——这种状态主要是多年以来对周围空间与建筑的忽视所造成的。广场上最突出的建筑便是一座13世纪的法兰西斯修道院，目前已经被遗弃。除此之外，还有其他各种类型的建筑，如哥特式商人旅馆改造的酒吧、20世纪20年代的鞋店、哥特式酒窖综合体。酒窖比地平面要稍微低几米，由于其所依附的房子已经被拆除，现在已经无法进入酒窖之中。最具讽刺的是，该地原来只是一条狭窄而蜿蜒的街道，直到19世纪50年代的时候，由于两栋中央建筑的拆除，才成为了一个广场。

阿肯色大学社区设计中心为广场的整修提交的方案之一——"罗马广场之绿色定义"，主要是借助两种城市空间来定义该广场，一是依附于中央火车站、自由广场和农贸市场的文化轴；二是环绕这个中世纪城市的绿化带，与街区繁茂的小庭院以及公园相融合。罗马广场的设计重点并不在于与布尔诺的两个主要广场相比拼，而是将其转化成一个通透的空间和通道，在广场中央种上一些树木，使其成为建筑群中的焦点。

在过去的100年间，广场周围建筑的变更缺乏对广场环境的考虑，该方案新增两栋新的综合性大楼，通过人行拱廊促成空间的视觉联系，并设置通往哥特式酒窖的通道。两座建筑简约而现代，作为现代建筑设计融入历史环境的代表，势必给人以深刻的印象。

1858 2010 proposed

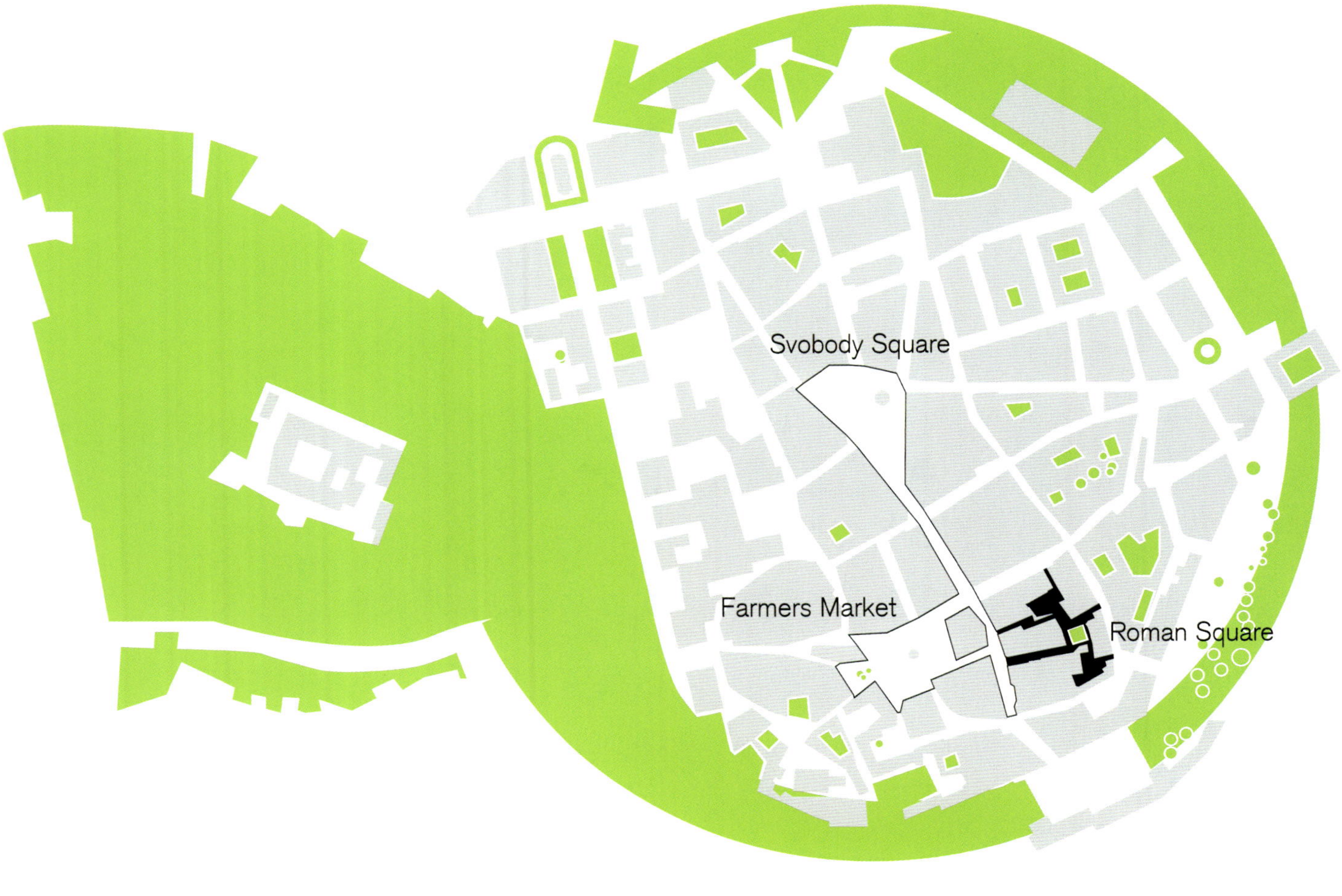

The city of Brno held an international design competition to renovate Roman Square, located in the heart of the historic city center. Roman Square is prominently located adjacent Masaryk Street--one of the main retail and entertainment promenades--and in close proximity to the city of Brno's two main squares. However, it is currently underutilized and neglected. The square suffers from poor land use and urban nuisances such as unkept vegetation, poor lighting, and decaying buildings.

The issues of the Roman square are primarily a result of decades old neglect of the surrounding spaces and buildings. Roman Square is underwhelming–occupied mainly by a makeshift market hall and pubs. The most prominent building defining the square is a 13th century Franciscan Monastery, which is currently abandoned and in decay. Other buildings include a highly varied collection of building types including gothic merchant houses converted to bars, a 1920s' functionalist shoe store, and a complex of gothic wine cellars, which are several meters under the surface and currently inaccessible, since the houses they belonged to have long been demolished. The most ironic aspect of the site is that it is not a square by design – what used to be a narrow, winding street until the 1850s' became a 'square' as a result of demolition of two central buildings, which were never rebuilt.

One of the schemes presented by the University of Arkansas Community Design Center-Roman Square Bouquer is based on two major urban structures that define the city center; a cultural axis that is anchored by the central train station, Svobody Square, and the Farmers Market, and a green beltway that surrounds the medieval city, pixelating into small lush courtyards and parks at the scale of the block. Instead of competing with the Brno's two main squares, Roman Square is transformed into a system of porous spaces and passages, with a bouquet of trees at the square's center, redefining it as the focal point of the block.

The edges of the square, currently unresolved as a result of insensitive additions and subtractions of buildings in the past 100 years are reestablished by adding two new, mixed-use buildings, with pedestrian arcades on the ground floor to visually reunite the space and extend the commercial program on the ground level as well as provide access to the gothic cellars. The buildings are simple and contemporary and should serve as an example of sensitive modern additions to the historical fabric, a challenge which is often underwhelming in recent construction within the city.

Orlí Street
Josefská Street
Masarykova Street
Roman Square
Masaryk Street
Josefská Street

ČESKÉ FUSEKLE

Roman Square: An Episodic Approach

罗马广场之插曲

University of Arkansas Community Design Center

Revitalization of the Roman Square

罗马广场重建竞赛参赛方案

Credits
Location: Brno, Czech Republic
Design Year: Summer 2010
Architect: University of Arkansas Community Design Center

这是阿肯色大学社区设计中心为罗马广场的整修提供的第二项方案。罗马广场作为布尔诺这一历史性城市的中心，在城市的发展与居民生活中扮演着重要角色。

与罗马广场之绿色定义不同的是，该方案将重点放在空间中一些小插曲的设计上。广场上最具价值的历史建筑应属哥特式酒窖综合体。酒窖作为重要的文化遗产，受到了布尔诺政府的重视，这也增加了对这座古老城市实行新建设的难度。

该方案通过一系列的设计插曲吸引行人的目光，为罗马广场策划了更多新的活动。这些设计包括一顶从农贸市场延伸至广场的顶篷，一面作为密集型城市缩影的绿墙，以及通往住宅区的象征性的大门。广场上的主要视觉元素是西侧的凹陷，将13世纪的酒窖综合体显露在外。这些酒窖看似随处可见，但是很多居民和游人不会一眼就发现它们。这样一种方案将城市酒窖的保存呈现在公众的面前。

该方案并没有采用重建的方式，而仅仅只是增加了一些新的元素，以此将广场上零散的空间综合在了一起。

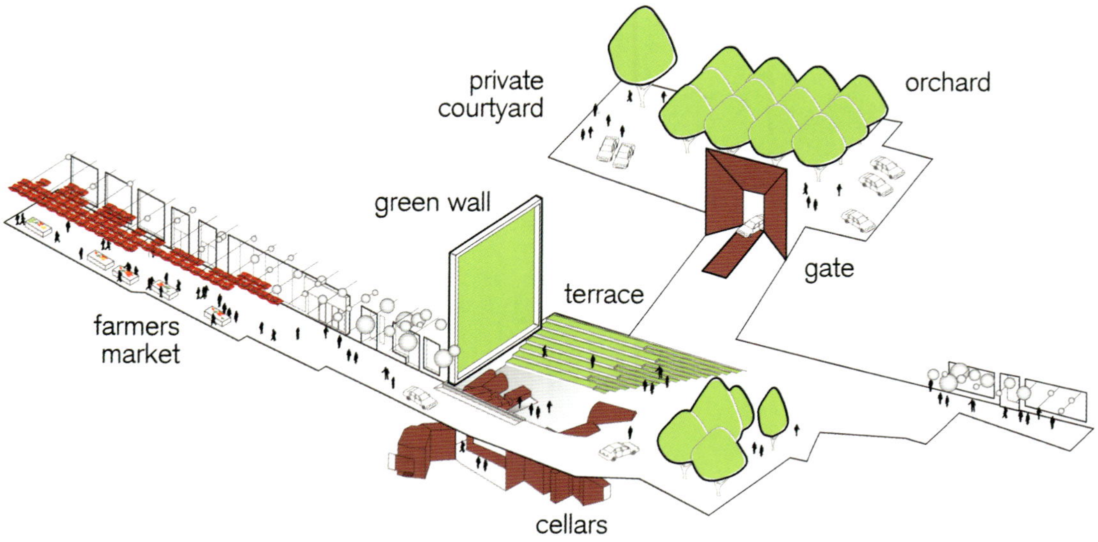

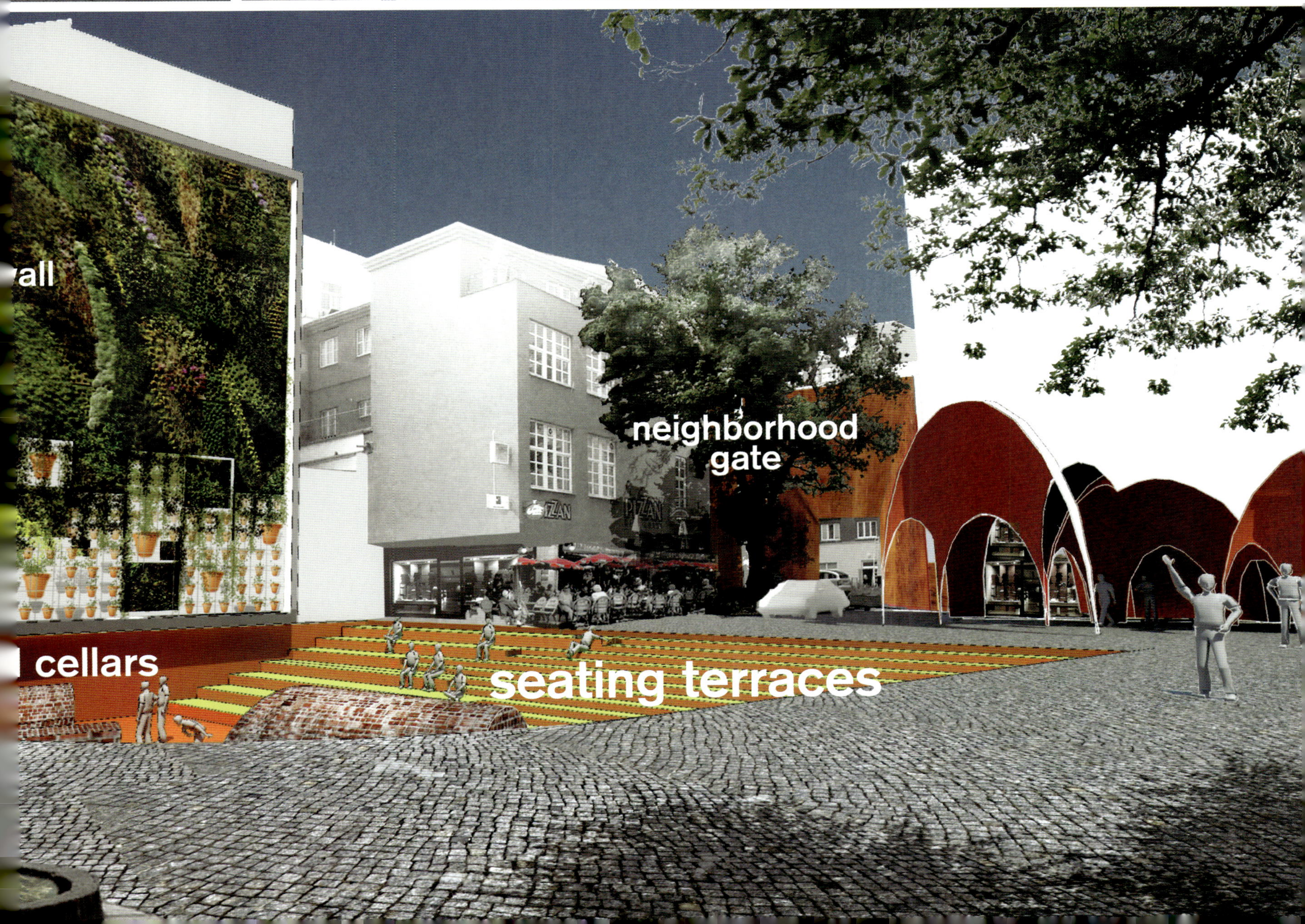
all
neighborhood
gate
cellars
seating terraces

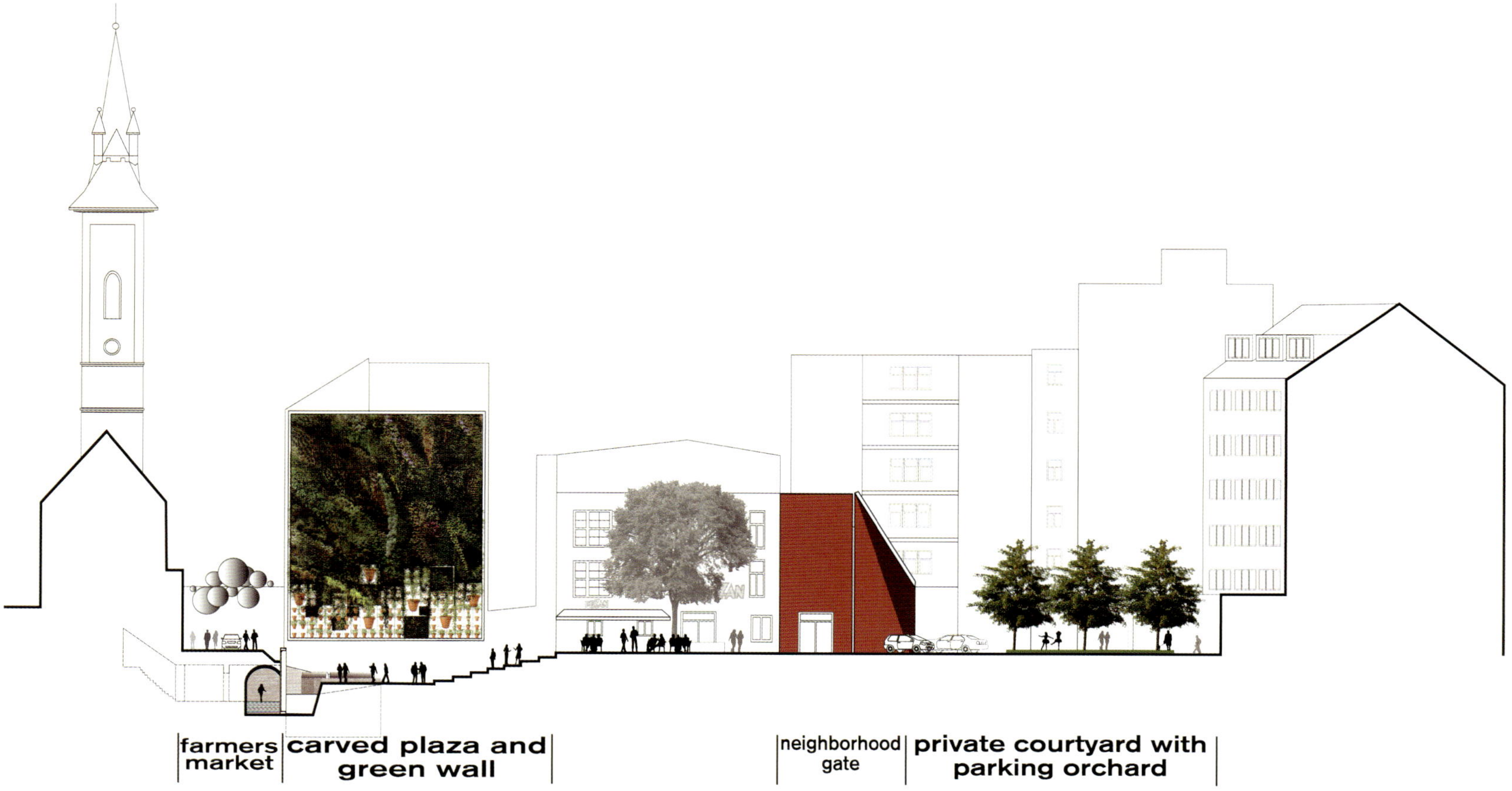

This is another proposal submitted by the university of Arkansas Community Design Center for the Renovation of the Roman Square, which as the center of the historic city Brno plays an important role in the CIty's development and residents' life.

Different from Roman Square Bouquet,this proposal focuses on the small design interventions. The mast valuable aspect on the square should be the gothic wine cellars.While these cellars are an important cultural heritage, the Brno city government treats them as a liability that complicates new construction in the old city.

The scheme consists of several small-scale design interventions which creates a sequence of events that visually draw the pedestrian through the site, while reprogramming Roman Square with new activities. These episodic moments include a hanging canopy that extends the farmers market into Roman square, a green wall a surprising microcosm in a dense city, and a figural gate to the residential inner block. The main visual element of the square is a topographic excavation of the western edge, revealing the 13th century cellars. These cellars are ubiquitous in the city center; however, their presence is hidden from view of most residents and visitors. This proposal provides a unique opportunity to bring the preservation of the city's cellars to the forefront.

Instead of new construction, it utilizes an incremental addition of elements in order to reunite the disjointed spaces of the square.

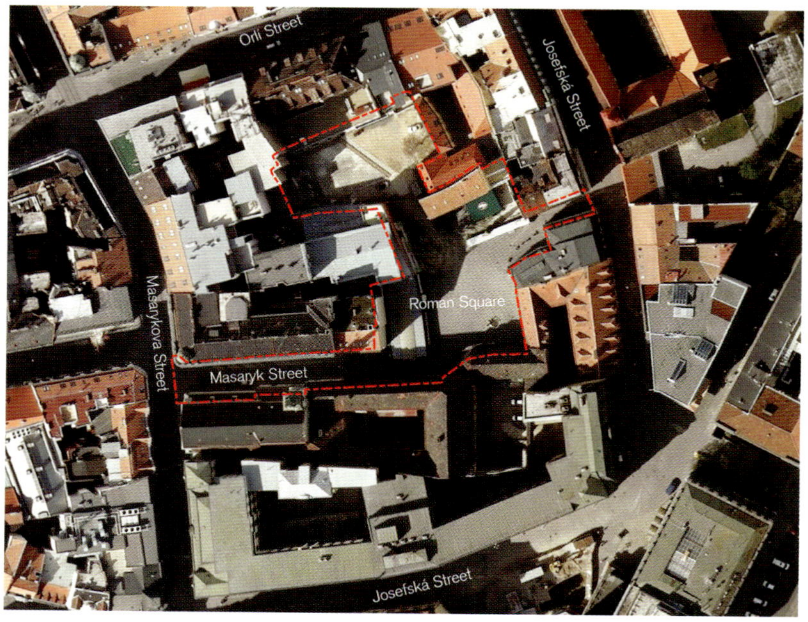

ARCHITECTURE COMPETITION WORKS

Residence
住宅

Sonnenhof

森内霍夫住宅综合体

J. Mayer H. Architects

Credits

Team: Juergen Mayer H., Jan-Christoph Stockebrand, Christoph Emenlauer, Jens Seiffert, Max Reinhardt, Christian Pälmke
Client: Wohnungsgenossenschaft "Carl Zeiss" eG, Jena
Total Floor Area: approx. 10,000 m^2
Start: 2008
Expected Completion: 2012
Project and Construction Management: Kappes Partner IPG, Berlin
Structural Engineering/Structural Fire Protection: Ingenieurbüro Dr. Krämer GmbH, Weimar
Building Services:Scholze Ingenieurgesellschaft mbH, Dresden
Consultant Building Physics:Ingenieurbüro Santer Bauphysik, Duisburg
Consultant Ventilation: Ingenieurbüro Rau, Heilbronn
Consultant Traffic Engineering: GRI Ingenieure,Berlin
Consultant Outside Facilities: Ingenieurbüro Abraham, Berlin
Lighting Engineers: Lichttransfer, Berlin
Model Making: Werk 5, Berlin

森内霍夫建筑由4幢办公楼和居住楼群组成。它位于一部分合并过的小型场地之上——德国耶纳历史上的中心位置，其分隔的建筑结构纳入了通过地面连接的自由出入的通道。建筑场地外侧边缘的位置界定了小规模户外空间，与中世纪的城市结构重合。其户外设施延伸了建筑的整体设计理念，超出了场地的边界。对商业、住宅、办公用途的按计划合并开启了一种小规模、分段式、灵活化的使用模式，同时从概念上将建筑自身集成到周边环境之中。

Sonnenhof consists of four new buildings with office and residential spaces. Located on a consolidated number of smaller lots in the historical center of Jena, Germany, the separate structures allow for free access through the grounds. Their placement on the outer edges of the plot defines a small-scale outdoor space congruent with the medieval city structure. Its outdoor facilities continue the building's overall design concept past the edges of the lot. The planned incorporation of commerce, residence and office enables a small-sectioned and flexible pattern of use that also integrates itself conceptually into the surroundings.

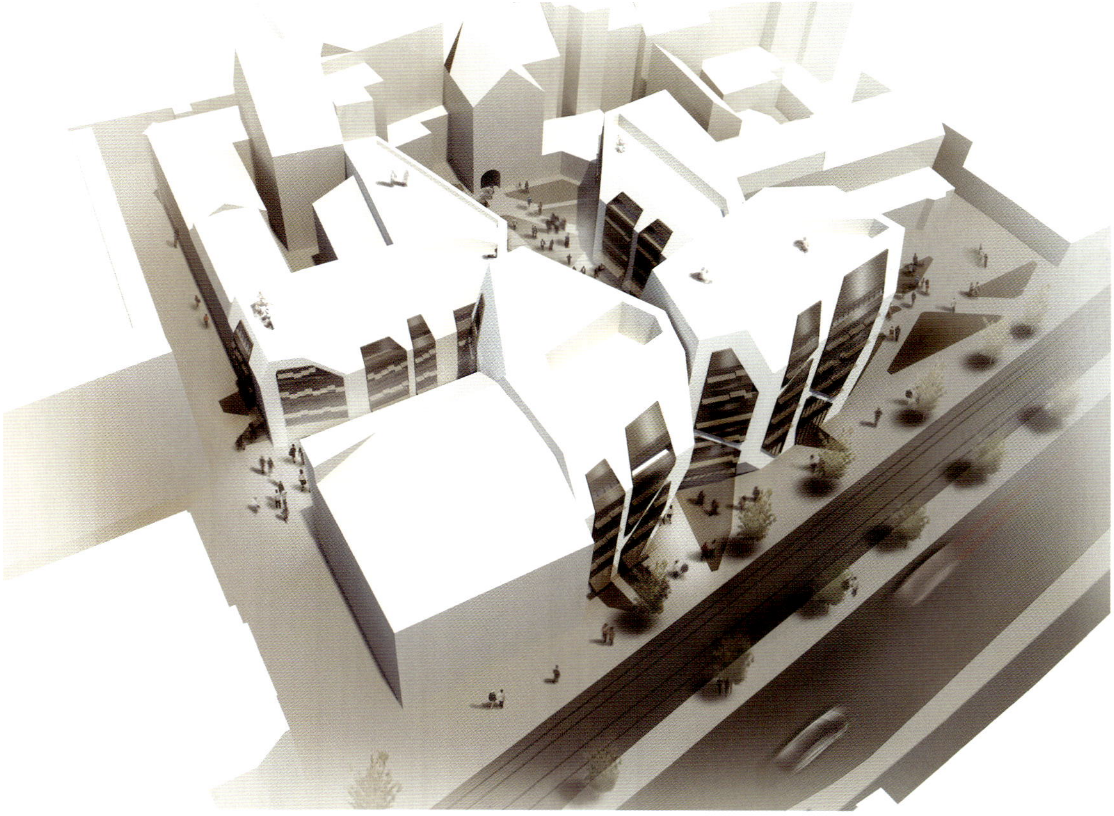

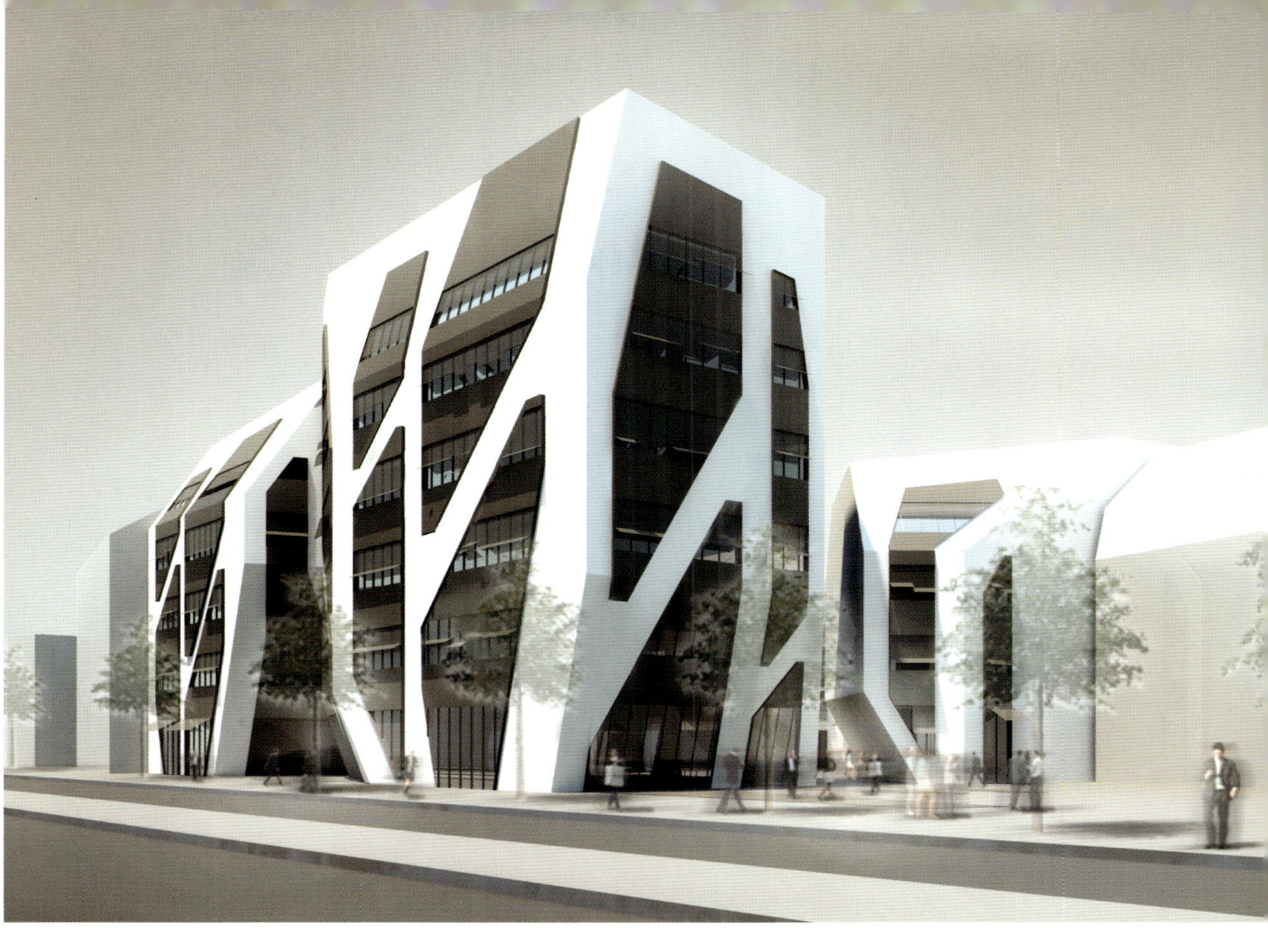

JOH 3 – Apartmenthouse Johannisstraße 3, Berlin

柏林JOH 3公寓 J. Mayer H. Architects

Invited Competition 2008, First Prize
2008年设计方案邀请赛一等奖

Credits
Team: Juergen Mayer H., Hans Schneider, Wilko Hoffmann, Filipa Frois Almeida
Competition Team: Juergen Mayer H., Thorsten Blatter, Marcus Blum
Project: 2008-2011
Completion: Spring 2011
Client: Euroboden Berlin GmbH, München
Architect on Site: Architekturbuero Wiesler, Stuttgart with Thomas Quinten Projektmanagement, Berlin
Structural Engineers: EiSat GmbH, Berlin
Building Services: Ingenieurgesellschaft Striewisch mbH
Model: Werk5, Berlin
Model Photographer: Ludger Paffrath, Berlin
Visualization: Buenck + Fehse, Berlin

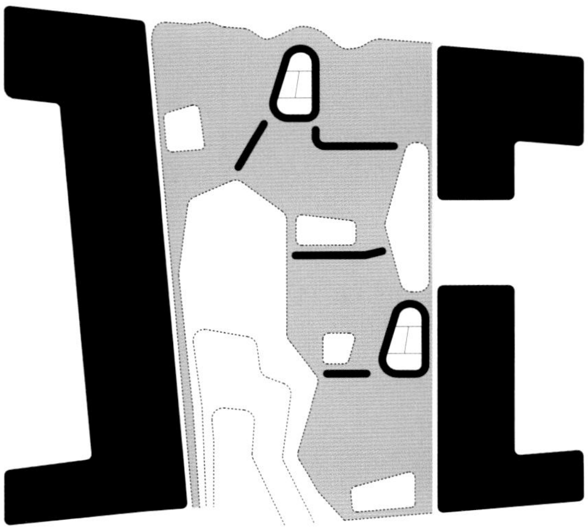

房地产开发商欧柏登（Euroboden）在柏林市中心——米特区建造了一座独特的公寓。对于以博物馆岛和弗雷德里克大街为邻的大楼，J. Mayer H. 建筑事务所的设计，以其多单元的结构与绿色的内部庭院，重新诠释了经典柏林房屋的内涵。大楼正面的悬浮式帘片的雕塑设计吸收了城市中的景观概念，一种置于渐变的庭院花园和大楼的轮廓和布局之中的可视化品质。大楼底层面向街道的方案中，同样包括一大批商业空间。占地面积庞大的公寓将会是东南朝向，面向一派恬静的、精心设计的庭院花园的景色。宽敞活泼的过渡空间和外部区域，各层楼面的不同高度，提供了一种在城市中心的开放式居住体验，同时提供了一种有意义的房间演替。各单元不同的平面规划与分布使用户具有了房屋类型选择的可能，公寓被设计成工薪族带有私家花园的联排别墅，或是可以尽享老弗雷德里希大街景色的经典公寓或是顶层复式住宅。集成的设计理念包括，从建筑外立面到楼梯，从电梯到公寓室内装饰的一切，承诺一种着眼于空间和生活体验的高端独享。

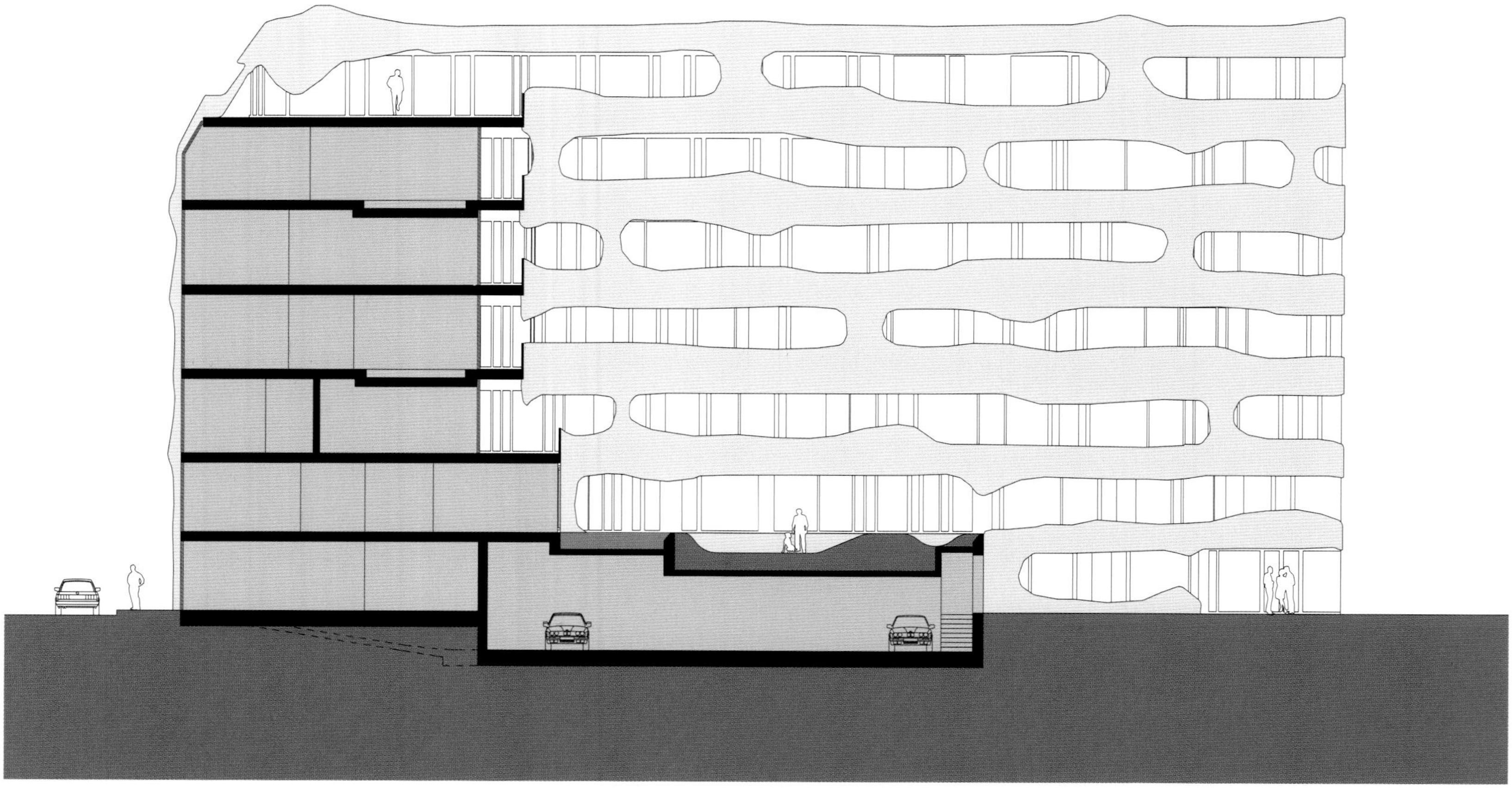

Property development group Euroboden is building a unique apartment house at Johannisstraße in Mitte, Berlin's downtown district. J. MAYER H. architects' design for the building, which will soon neighbor both Museum Island and Friedrichstrasse, reinterprets the classic Berliner Wohnhaus with its multi-unit structure and green interior courtyard. The sculptural design of the suspended slat facade draws on the notion of landscape in the city, a quality visible in the graduated courtyard garden and the building's silhouette and layout. Plans for the ground floor facing the street also include a number of commercial spaces. The generously sized apartments will face south-west, opening themselves to a view of the calm, carefully designed courtyard garden. Spacious, breezy transitions to the outside create an open residential experience in the middle of the city that, thanks to the variable heights of the different building levels, also offers an interesting succession of rooms. The units' varying floorplans and layouts indicate a number of housing options; condominiums are organized into townhouses with private gardens, classic apartments or penthouses with a spectacular view of the old Friedrichstadt. The integrated design concept, which incorporates everything from façade to stairwells, elevators to apartment interiors, promises a unique spatial and living experience with an eye to high design.

Standard Floor

1st Floor

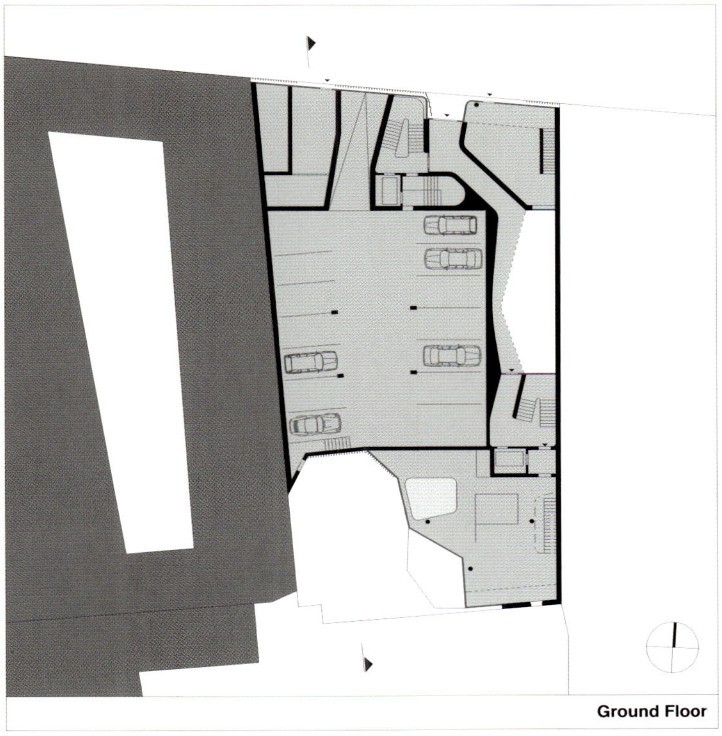
Ground Floor

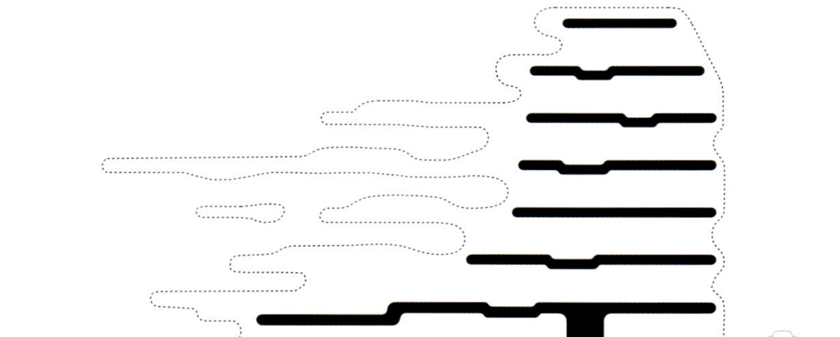

Tango Towers

“探戈楼”——住宅综合体

C. F. Møller Architects

First Prize in Architectural Competition
建筑设计方案一等奖

Credits

Client: Holstebro Municipality & Nordicom Properties
Architect: C. F. Møller Architects
Landscape Architect: C. F. Møller Architects
Engineers: Søren Jensen engineers a/s

Area: 9,000 m²
Location: Holstebro, Denmark
Year: 2007

该项目包括一个有两栋独立建筑的街区，将占据约2,500 平方米作为其扩建空间的荷史迪堡的市政厅，大约6,500平方米的办公场所与公寓。

该综合体力图成为城市的一个新地标，为公共服务设施进驻到城市中心地带提供机会，同时也为医科诊所、办公及公寓等其他功能性建筑提供了空间。

因为两栋建筑像极了一对跳着探戈的舞者，所以被命名为“探戈楼”。两栋纤细修长的建筑如雕塑一般矗立在城市中，随着你脚步的移动，两栋建筑表现出来的形式和相互关系也会在视觉中逐渐改变。如果从远处观望，它们则又是城市景致中的新篇章。

单一线状的建筑水平地分布在建筑区域内，逐渐弯曲然后汇合形成现在的高层，汇合点经过折叠方式处理形成一个具有斜面覆盖的主入口。

两栋建筑由一个玻璃面的垂直体连接，中间有通道可以到达两栋建筑的不同楼层。实际上，外观呈现出来的两栋建筑是一体的，只是以两栋的形式呈现更显出它们外形的纤细与精致。建筑的外立面是呈现不同色泽的光板，表现了其高雅而连续的特征。

作为建筑物的外观艺术处理的一部分，连接两栋建筑的两个立面突出了较强的配色方案。Holstebro在公共空间包括城市空间、市政厅、学校建筑等的艺术设计方面很有建树，所以该项目继承了这种艺术与建筑相结合的传统。

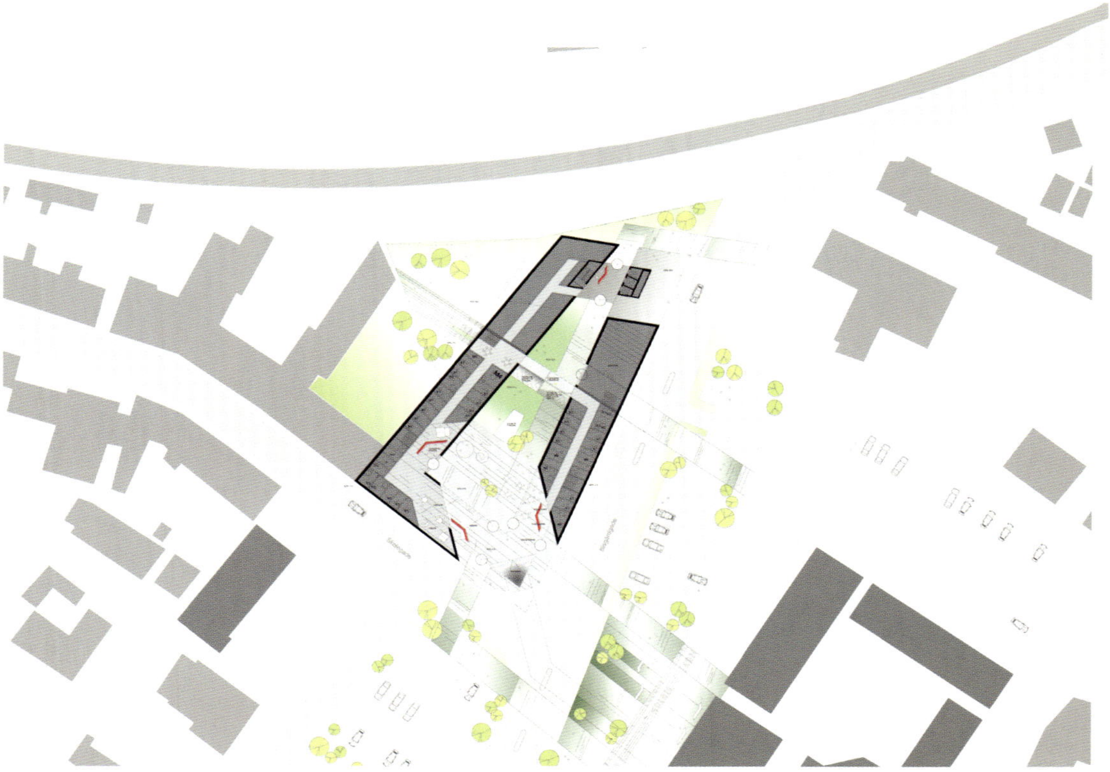

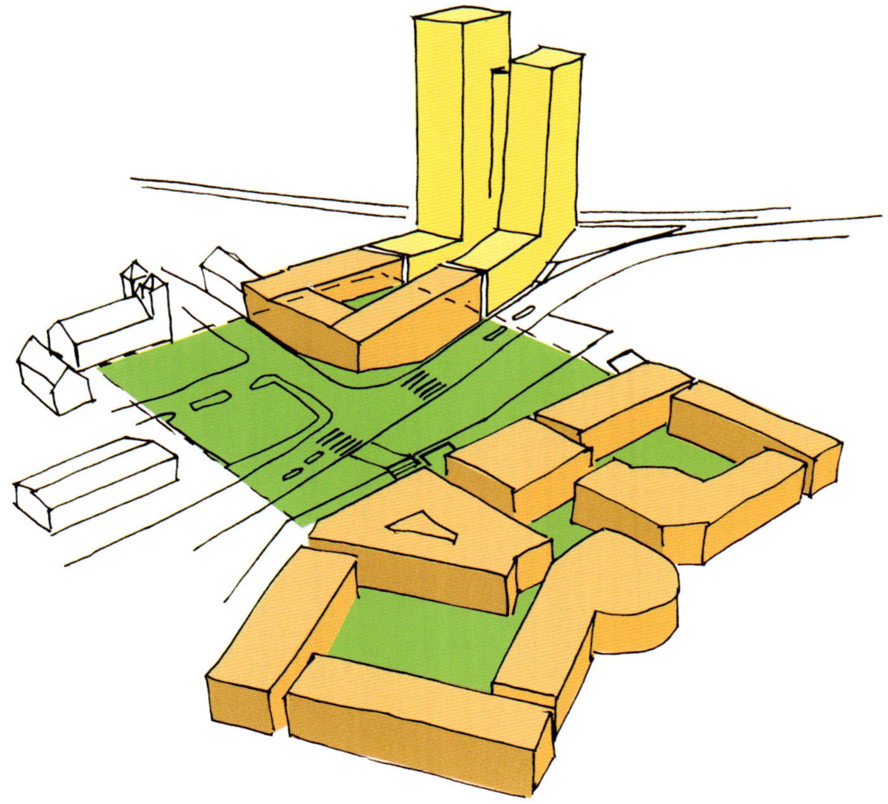

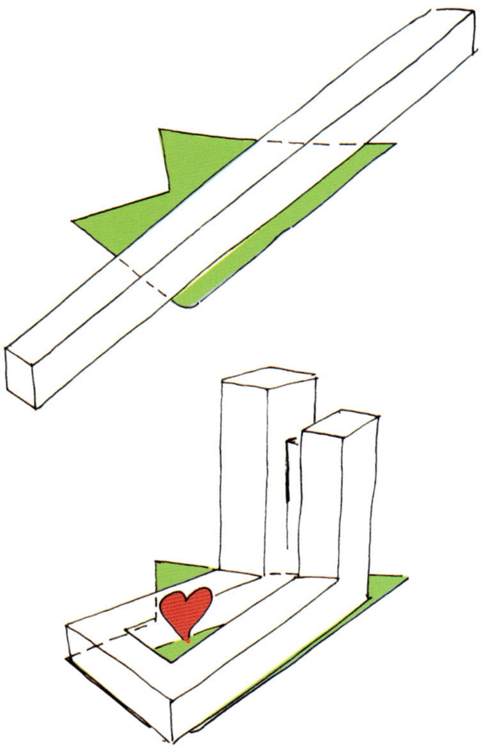

The symbolism has been drawn from the world of dance in the lines of the two 70-meter tall towers to be built in connection with Holstebro Town Hall. The development, which encompasses a split tower block of up to 20 storeys and 70 meters in height, will provide room for approximately 9,000 m² of offices and apartments, of which Holstebro Municipality will use approximately 2,500 m² as an extension of the town hall.

The complex is aimed to become a new landmark of the city. It also provides the opportunity to draw together public service functions in one central location, and further provide spaces for added uses such as a health clinic, offices and new quality housing units.

The working title of the development is Tango Towers, as the architecture is reminiscent of a dancing couple. The complex forms a sculptural shape with two slender towers, and the expression and mutual relationship of the towers will change continually as you walk around the complex - or view them from a distance as part of the town's new profile.

The shape of the complex is the result of the manipulation of a single 'strip' of building volume, which runs horizontally around the site perimeter, doubling up at both ends to become the towers, and at one point folding upwards to create a covered main access.

The two towers are linked by a glazed, set back vertical core, which gives access to the different levels. They are in fact one building, though they appear as two, which makes for a much more slender and vertical appearance. The facades are clad in light coloured panels, as part of the continuity and elegant understatement of the exterior.

As part of the artistic treatment of the building's appearance, the interior facades of the gap between the towers are highlighted by a stronger colour scheme. Holstebro is well known for its art in the public space – including urban spaces, town hall, school buildings etc. As a result, the project continues this tradition of integrating art and architecture.

Jobnet-Café
DIFKO
Product Manager
Ydelses kontor
Job - Annoncer
JKS
Job - Annoncer
GRV Group

Absolute Green Urban Tree

纯生态树形住宅 Geotectura Studio

Honorable Mention in the International Lisbon Ideas Challenge Competition, 2006
2006年里斯本国际创意挑战赛荣誉奖

Credits
Architecture Team: Dr. Joseph Cory (Geotectura Studio)

PV4All是应可持续发展环境的需要而对光伏技术的原始应用示例。Geotectura建筑事务所在此设计上给大家呈现了这项技术。设计的目的是要将生态、成本效益融入建筑设计当中。PV4All将口头的概念推广到全球认知与全民责任，它不仅仅是一个单独的产物或者模型，而是一个完整的设计理念。这项技术的施行应当提升全球认知与意识。

设计主要以空中立方体的形式呈现，以最小化对地面环境的污染。每一个立方体都可以居住，其中设有温室平台。生物气象结构舱体使空间的利用以及功能性更加灵活。每一个模块的规模与功能设计的灵活自主也使每一个立方体都能观赏到外界风景，并保证最佳的空气流通。连同多边形绿色树形住宅楼以及天井，这个多样化自给自足的舱体实质上反映了个人责任与义务的问题——即便是在高空中。这两个螺旋面预制结构的全新设计理念如同都市里一棵生长的大树，不断提升空中生活人群的生活品质。运用了光伏技术的能量体与水塔使大楼能源独立化。这些能量立方体将大自然与生物群落带回密集的城市。

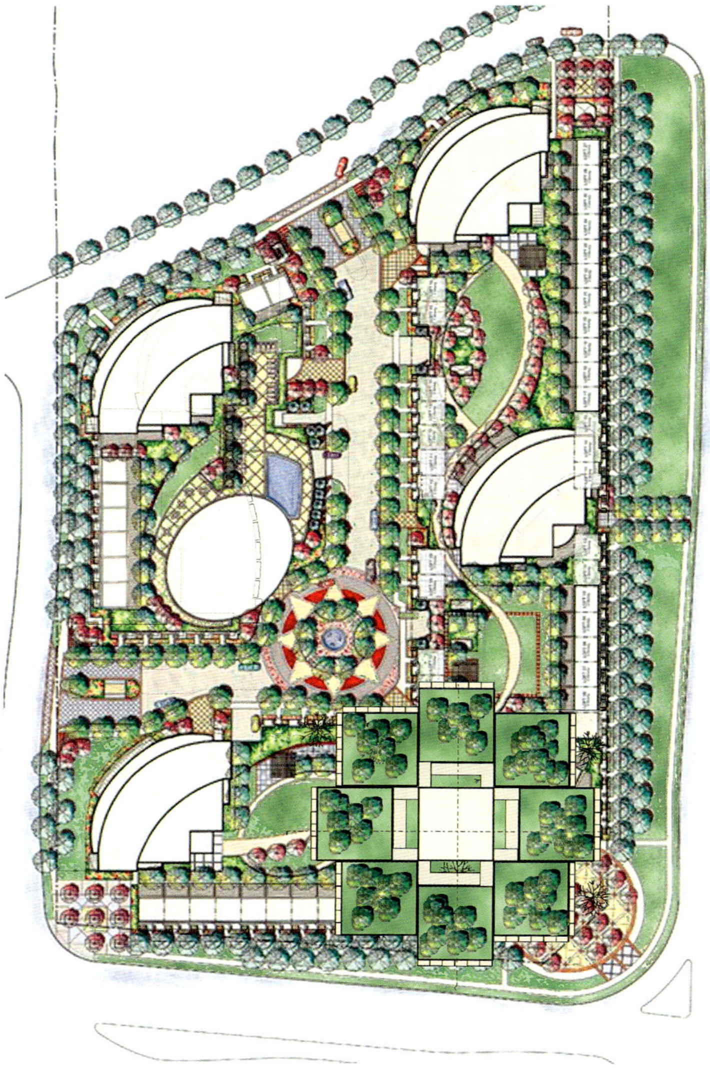

PV4All is a planning paradigm that explores original use of photovoltaic technologies in a growing need for a sustainable environment. Whatever we design and no matter what the size of our product is we must overlay it with PV in the manner you can see in all of GEOTECTURA's work. The main motivation of the PV4All design is the need to merge architecture together with ecological values and cost effective technological design. PV4All promotes global awareness and communal responsibility and sees beyond a single short term solutions. It is a complete design philosophy and not just a single product or prototype. This technological implementation concept should be for the people, increasing awareness and willingness around the globe.

This urban scale structure contains greenhouse platforms on each of the dwelling floating cubes in the sky while keeping a minimal footprint on the ground. The bioclimatic structure capsule approach enables space and function flexibility. Freedom & Individuality in design of the size and function of each unit are preserved within each cube giving panoramic viewpoints and optimal air flow. Together with the multi dimensional absolute green environment and the terrace sky courts this versatile self sufficient capsules are reflecting a personal responsibility and commitment even in a higher ground. This new concept of two helicoids prefabricated infrastructure is like a growing urban tree trying to improve the dwellers' life quality while living in the sky. The energy cubes incorporate PV technology and water tank facilities making this building energy independent. These mega cubes are weaving nature and communities back into the dense city.

Northern Style Housing Complex in Aomori

青森市北国风格集合住宅

Katsuhiro Miyamoto & Associates

Jury' s Special Award
评审团特殊奖

Credits
Location: Aomori-city, Aomori, Japan
Principal Use: Apartment house
Total Floor Area: 28,420 m²
Structure: steel, reinforced concrete, steel reinforced concrete, 1 basement and 4 stories
Design: Katsuhiro Miyamoto & Associates

超级屋顶

设计师意欲将集合住宅打造成一个巨大的屋顶，最高部分设三层楼，而一般的只设两层楼，对于各部分之间的空间，上部设计成"夏日花园"，而下部设计成"冬日花园"。

"超级屋顶"底层架空柱之间的空间设有社区美发店、商店和日间托儿所等，布置井然有序。这一区域有可能成为疏散空间，在人形灯笼节（日本具有代表性的节日之一）时，这里也可能成为睡魔车游行的起点。

夏日花园与冬日花园

夏日花园——夏日花园呈碗状，在辽阔的天空之下，掩映于绿树之中，但可以被风直接吹到。冬天的时候，地平面会由于积雪而上升。

冬日花园——日光通过顶部的开口射入内部，底层架空柱再将阳光反射出去。冬日花园不仅没有积雪，还能一直保持暖和的状态。这一部分有些像希腊雅典女神的神殿。

沿着起伏的地面走过，人们可以体验到夏日花园与冬日花园交替出现的感觉，并且它们在冬天与夏天所呈现的是不同的景色。

积雪地带的新通道

"超级屋顶"提供了一个新的通道，代替日本传统的带顶柱廊，这样一来，人们可以享受到厚厚的积雪所带来的乐趣。该通道属于东西向的后巷的一部分，后巷经过中央城市区域、本町公园、通道所在部分以及文化中心，与四号国道平行。

雪障调节住宅与周围环境的关系

住宅外墙以及边缘空地被多孔折板环绕起来，形成雪障，与内部的温暖通道形成鲜明的对照，雪障主要是在冬天的时候阻挡暴风雪的袭击，它形成了住宅楼的第二个外层，通过穿孔率的改变与场地周围的办公楼联系起来。

第五立面

从住宅楼四周的楼层俯视该楼，各种花园与私人庭院就如同从"超级屋顶"上凿出的一个个小洞，形成第五立面。尤其是在晚上的时候，住宅楼如同一艘母舰或者说是躺下的高层公寓楼。整栋楼就如同一辆睡魔车载着花灯停在雪障之中。

留下记忆的地方

该地曾经坐落着一所小学，有120年的历史，为了延续这一记忆，现有的树木将被移植到夏日花园中，而娱乐设备将被转至冬日花园中。

Hyper Roof

We intend that the residential building itself will become a large roof (named 'hyper roof'), of which the maximum thickness is for three stories and the average one is for about two stories, by making the void named 'garden for summer' in the massive volume from above and 'garden for winter' from below.

In the space of the pilotis with the diverse aspects under the 'hyper roof', each facility (community salon, shops, day nursery and so on) is placed properly without being loaded from the mass above, and this area may become a evacuation space and the starting point of 'Aomori NEBUTA festival' which is one of the Japanese representative festivals.

Garden for Summer and Garden for Winter

In the garden for summer ~ the sky expands largely. A wind goes through. Trees make a shade. In winter, the ground level rises according to the deep snow cover. The space is like a bowl.

In the garden for winter - The beam of sunlight streams through the opening at the top. The reflected light from the pilotis spreads out. The space is not snowed. A warm air keeps still. The section is like Parthenon.

In the undulating ground, we can experience 'garden for summer' and 'garden for winter' alternately. So the different landscapes appear in summer and winter.

New 'Passage' for a Snowy Region

The 'hyper roof' gives this area new 'passage' that replaces a traditional covered portico named 'Gangi', so that people take pleasure in heavy snow. This passage plays a part of a network of the back lane which runs from west to east (central urban area ~ 'Honmachi' park ~ this site ~ cultural

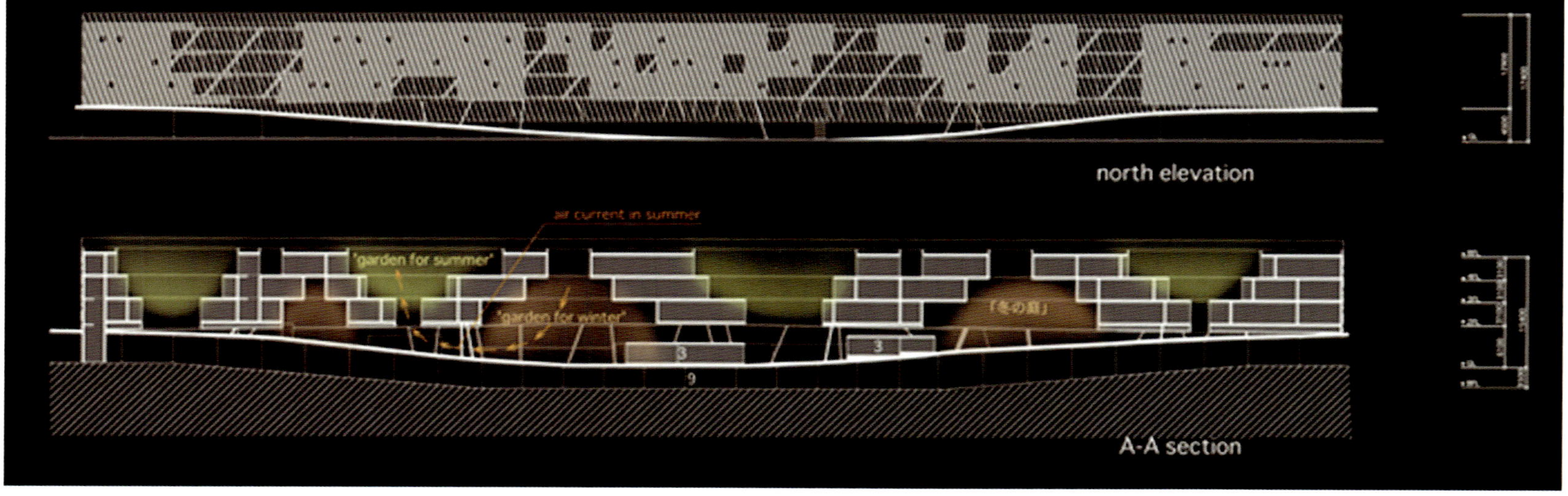

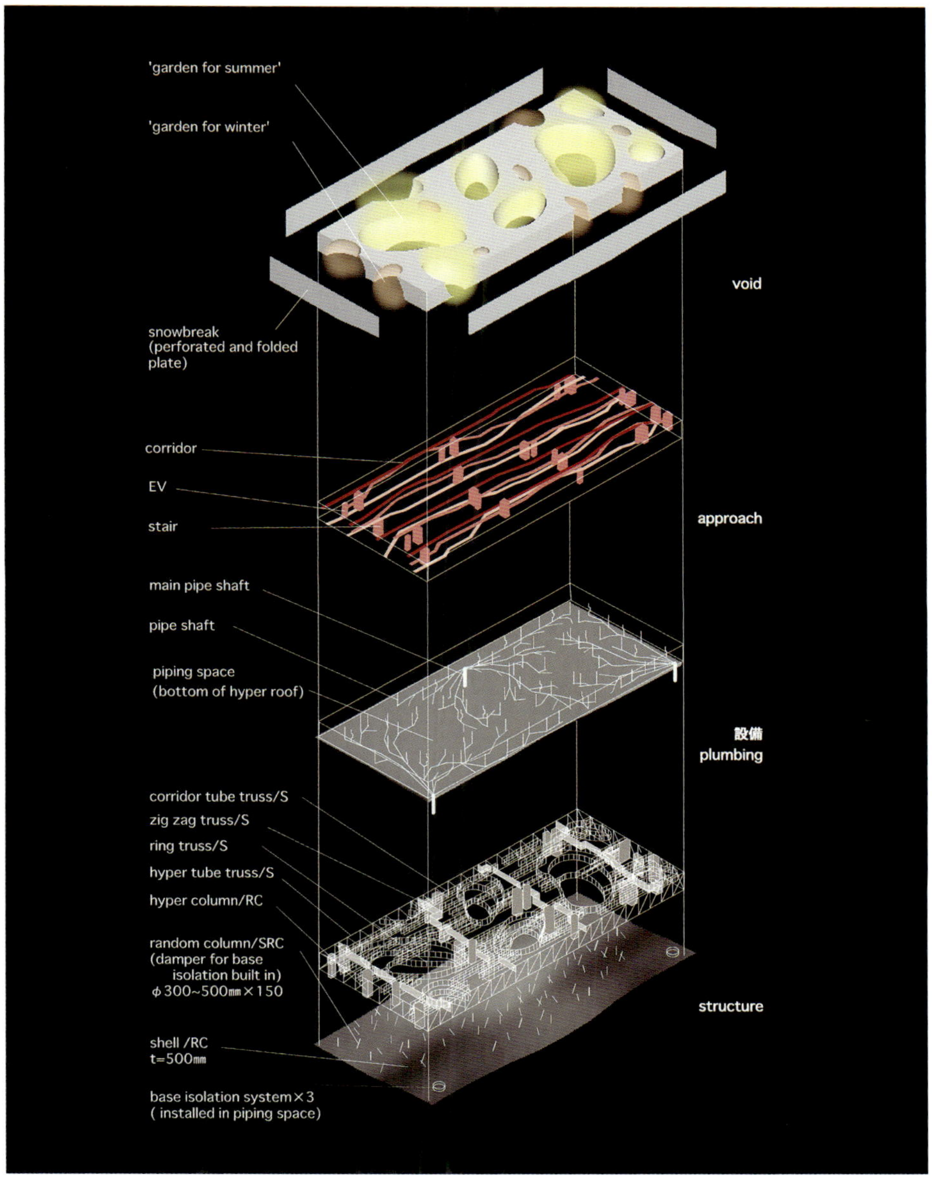

hall), parallel to the National Highway 4.

'Snowbreak' Adjusting the Relation with the Surroundings

The outside wall and the void along the boundary line are surrounded with the perforated and folded plates as 'snowbreak.' The snowbreak, which gives a sharp impression in contrast with the something warm in the 'passage' space, is a shield that protects the daily life of the residents from the snowstorm in winter, and, at the same time, become a 'double skins' with a function adjusting the relation between the residential spaces and the office buildings around this site by altering the rate of the perforation.

The Fifth Facade

In looking downward this residential building from the buildings around it, the figure that the hole of various gardens and private courts are bored in the flat 'hyper roof' can be regarded as 'the fifth facade.' Especially in the night view, it looks like a mother ship or a highrise condominium building which lies down. The whole building appears as the volume of 'luminous field' like a NEBUTA with the light through the snowbreak.

Memory of this Place

The existing trees will be transplanted to the garden for summer and the existing play equipments will be replaced to the garden for winter, so as to hand down the memory about the elementary school that had been in this site for about 120 years.

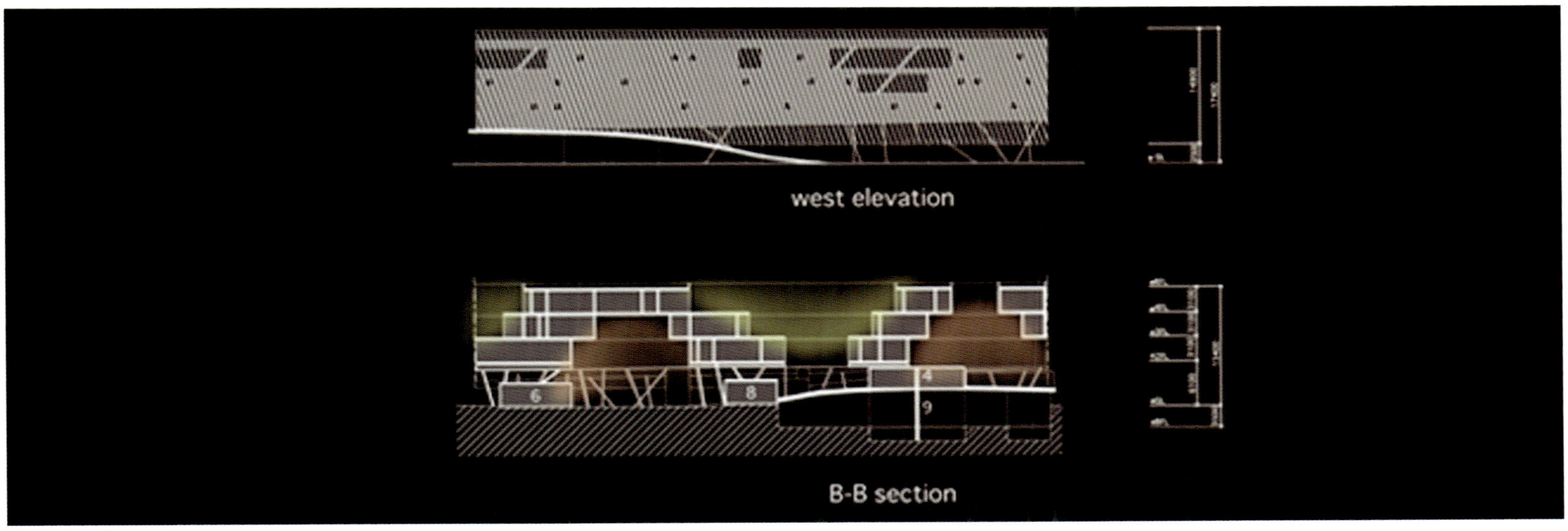

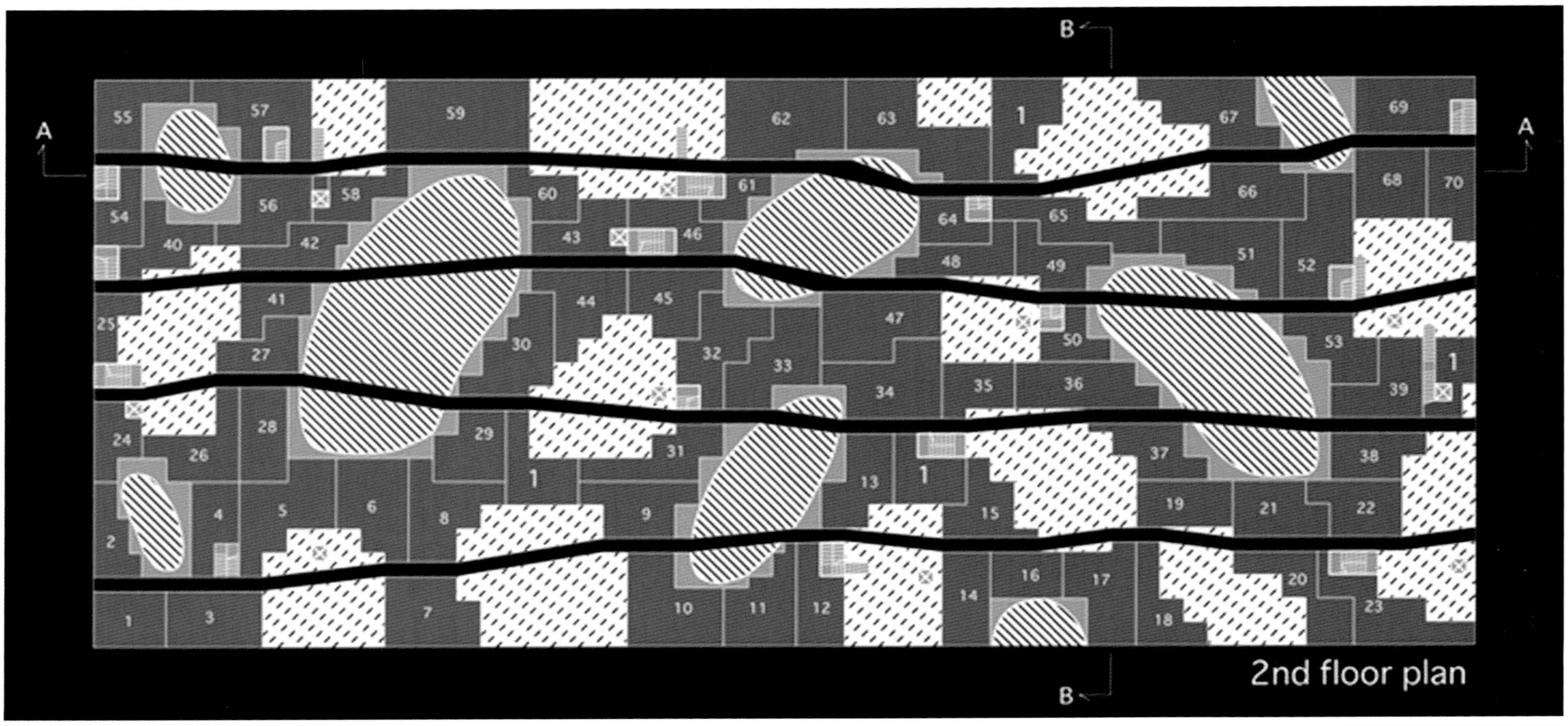

2nd floor plan

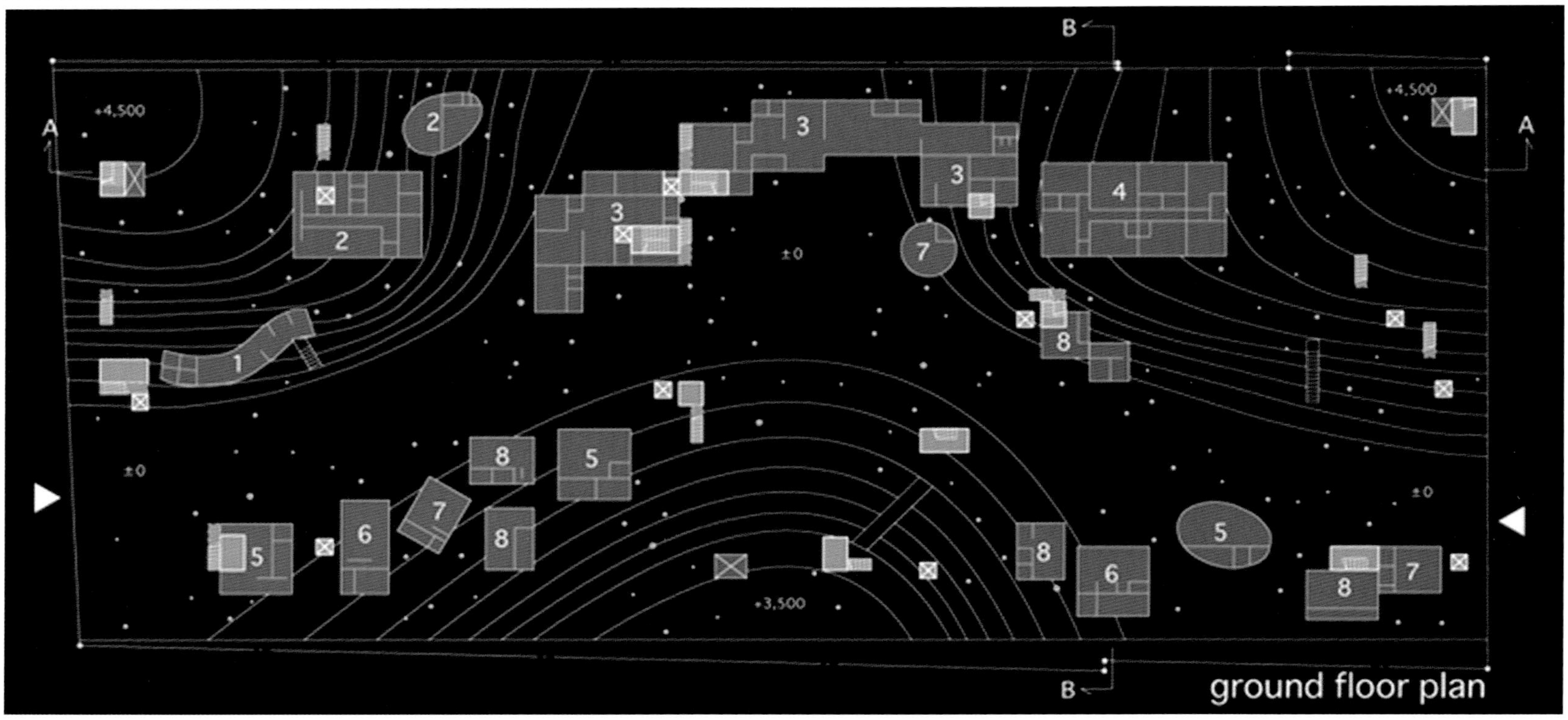

ground floor plan

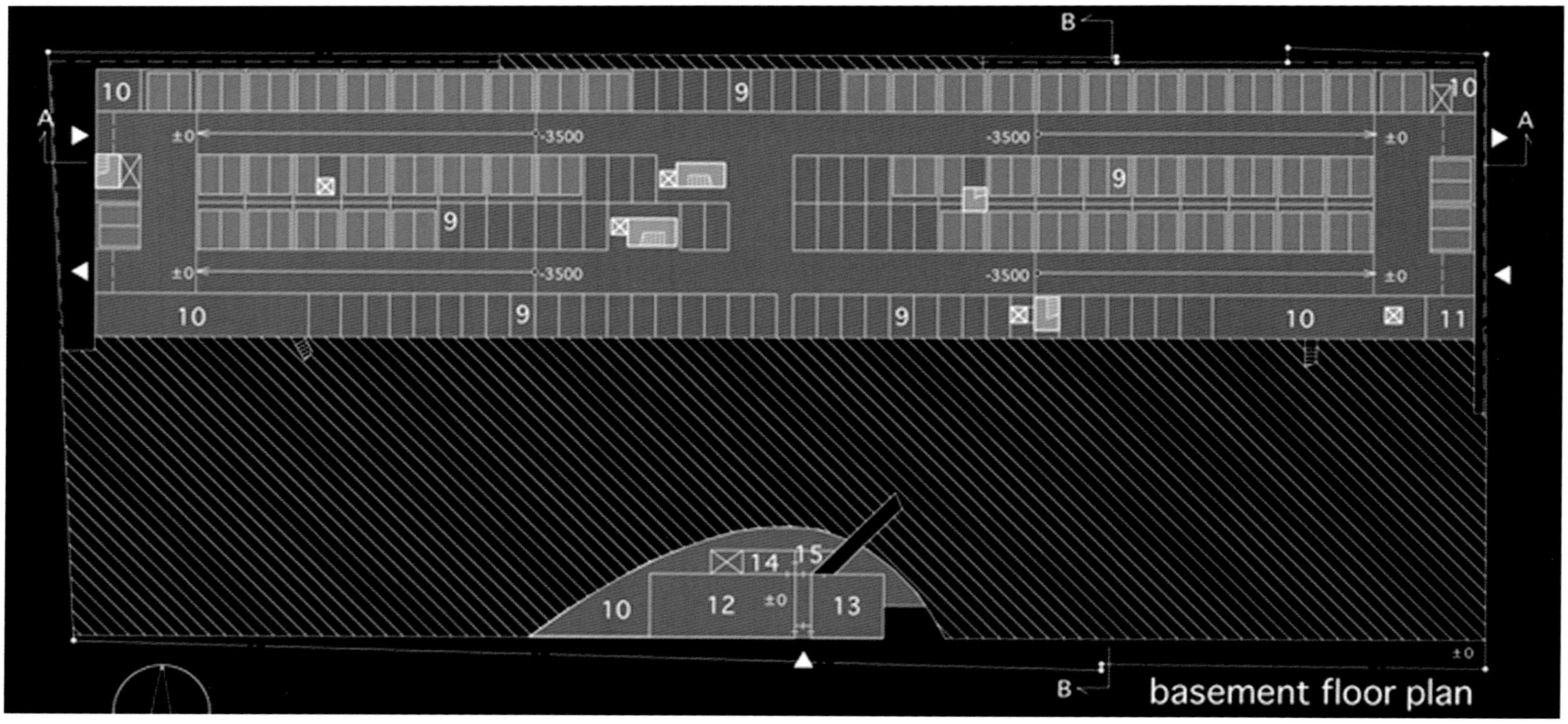

basement floor plan

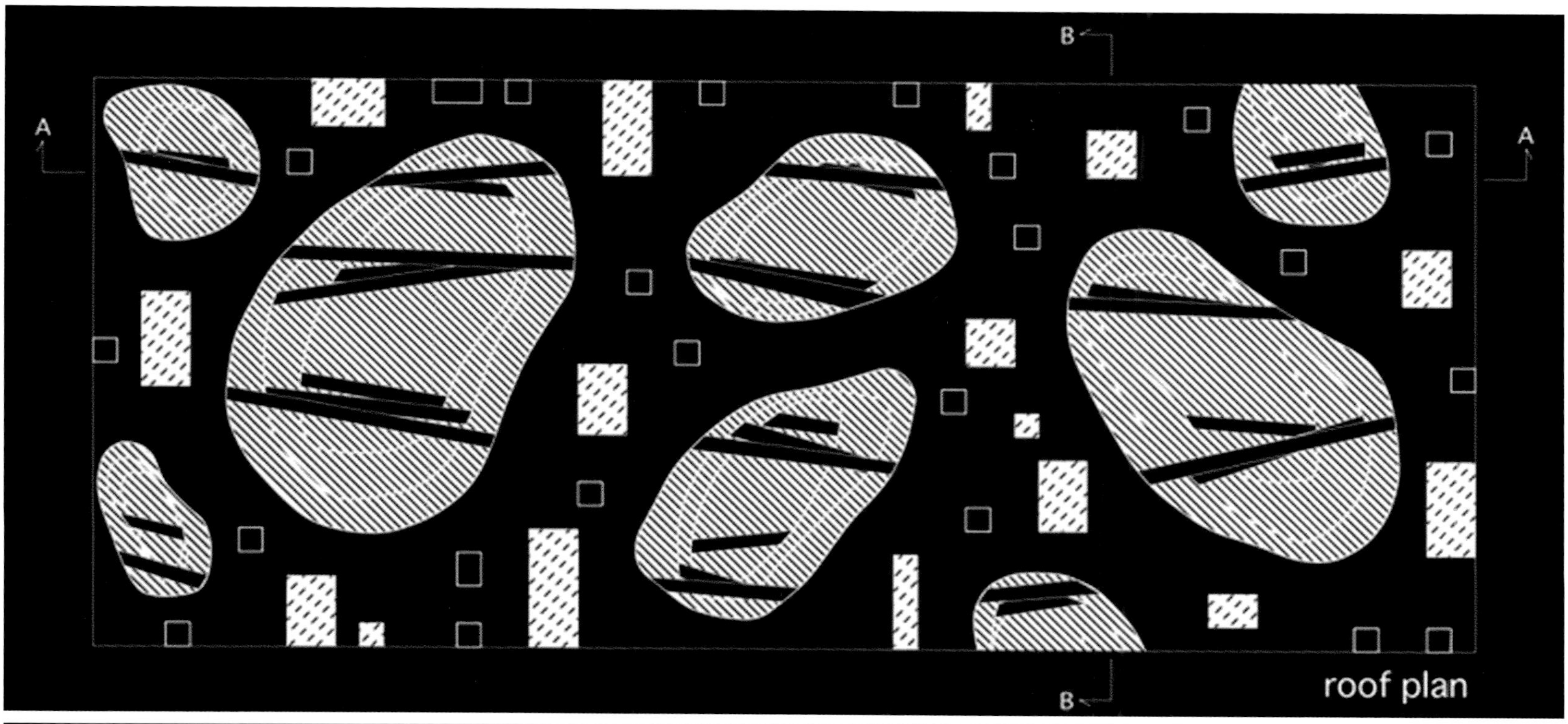

roof plan

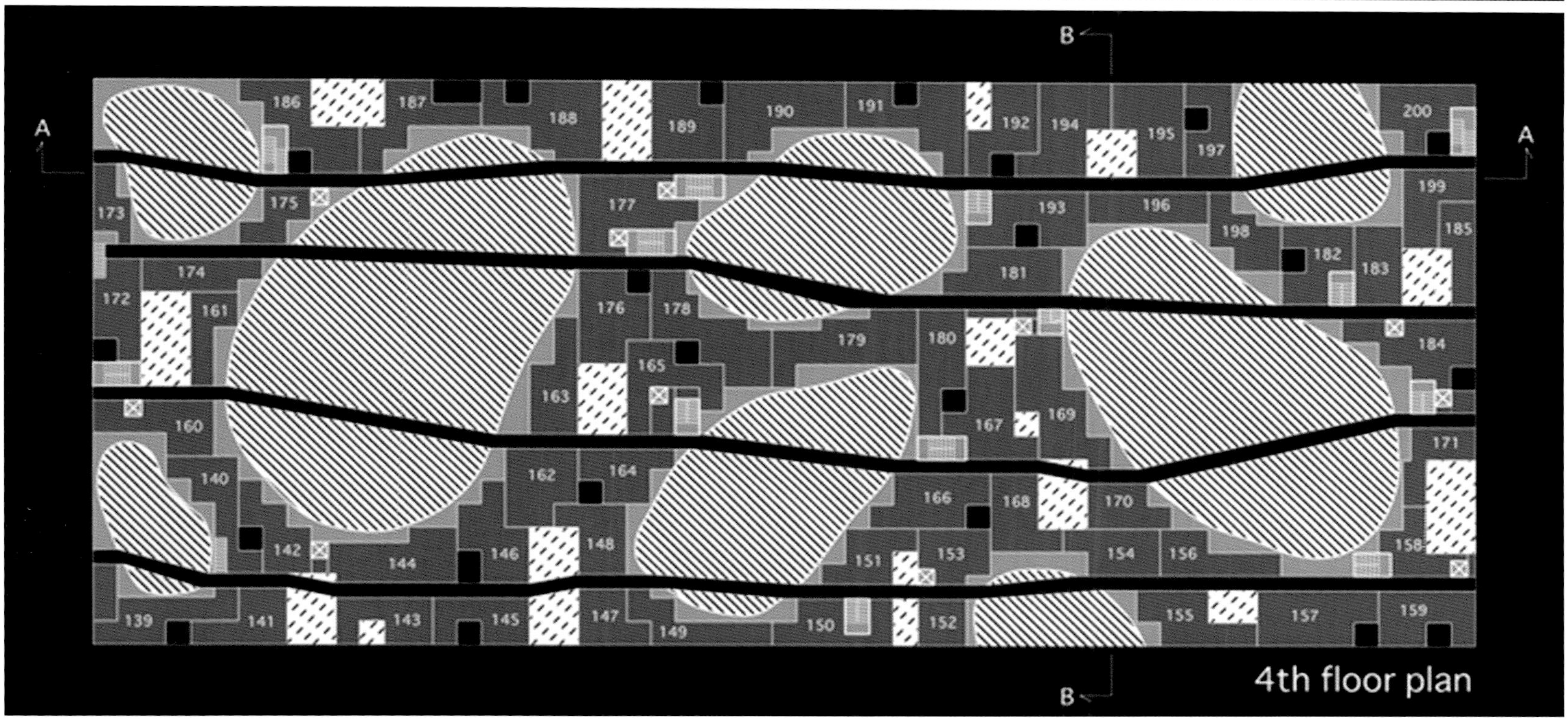

4th floor plan

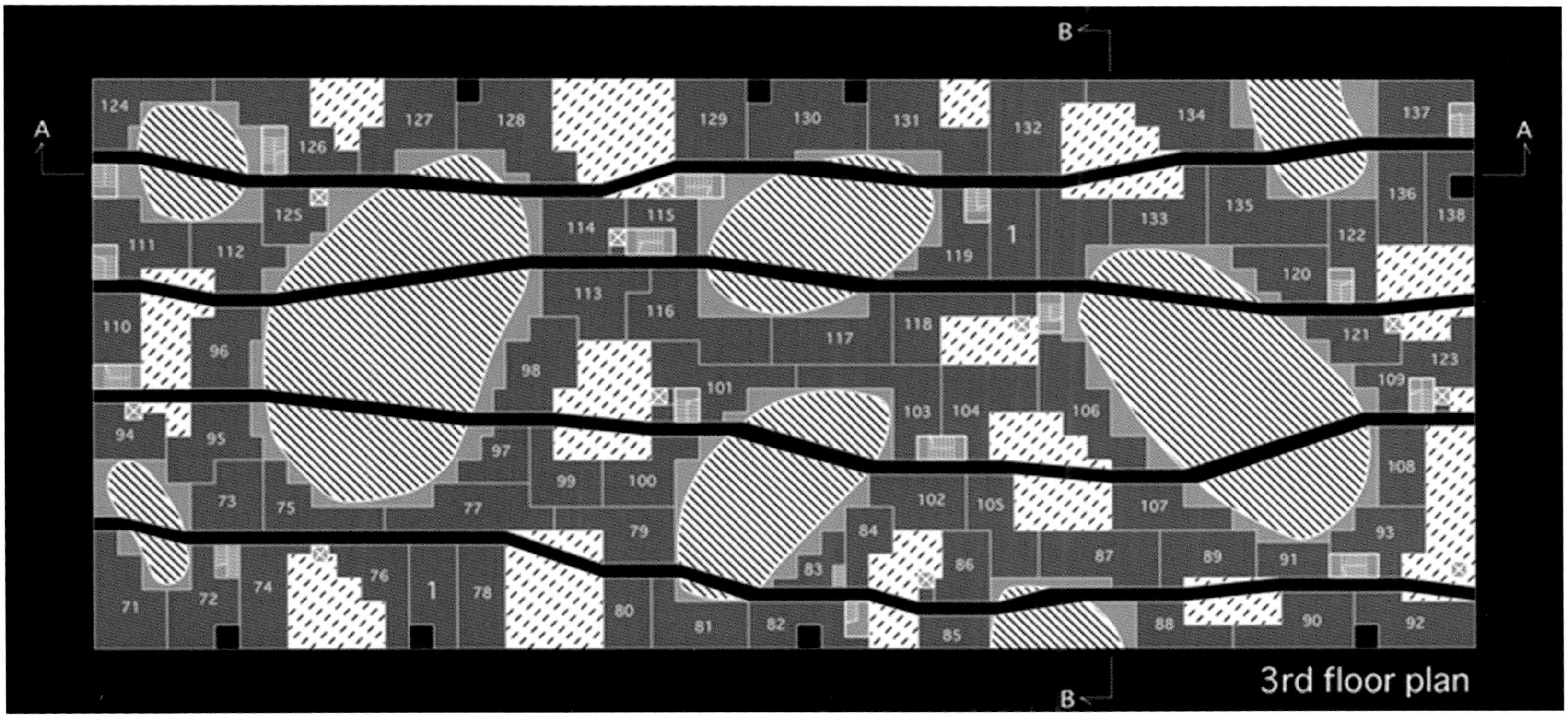

3rd floor plan

Alvik Tower

阿尔维克雕塑住宅综合体 C. F. Møller Architects

2008 First Prize in Competition
2008年竞赛一等奖

Credits
Location: Alviks Strand, Stockholm
Functions: Housing, Commercial, Restaurant, Parking, Leisure
Area: (Floor Area): 20,000 m²
Competition Year: 2008
Completion (Expected): 2012

阿尔维克塔是一个雕塑住宅综合体，其设计意图赋予其所在区域以及斯德哥尔摩一种特性。项目面朝群岛，在城市和景观间形成了一个过渡的区域，这也激发了建筑设计的灵感。项目分为一个约为9层的折形街区建筑与一个21层的地标性塔楼建筑两部分，这样的设计从倾斜的角度描绘出塔楼的形状和线条以及与梅拉伦湖接壤的山区的结构。

该设计还包括商店，餐厅和绿色空间，为阿尔维克提供新的功能与身份，在一个连续的、极具吸引力的，活跃的城市环境中创造一个现代的综合居住社区。

该塔楼将成为设计中的地标，而建筑的形式也会保证每一栋住宅最好的采光度和最广阔的视野。同时，设计方案还提供一些自由的场地，可以作为广场或者开放空间，例如水边的观景台，朝向城区的中心广场。开放的空间通过透明而面向公众的街道级的区域与城市生活紧密相连。

整个综合体计划通过卓越的热性能和现场太阳能发电来节约资源，该设计还包括一个具有200个停车位的多层停车场。

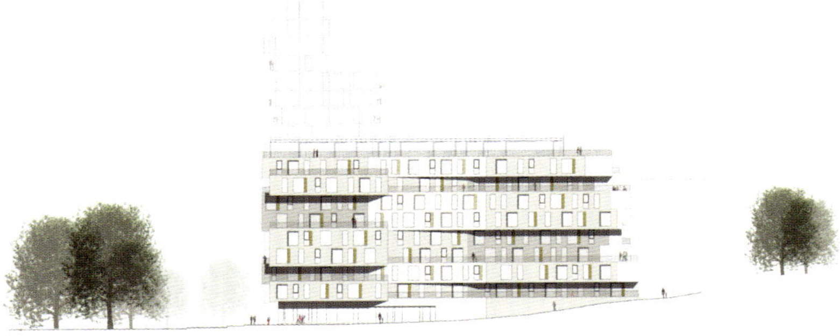

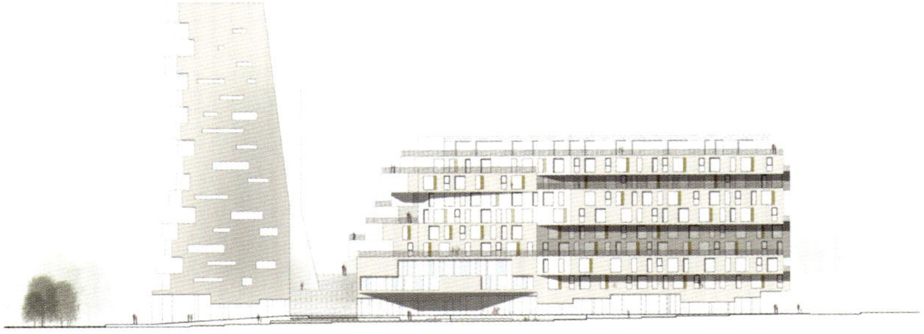

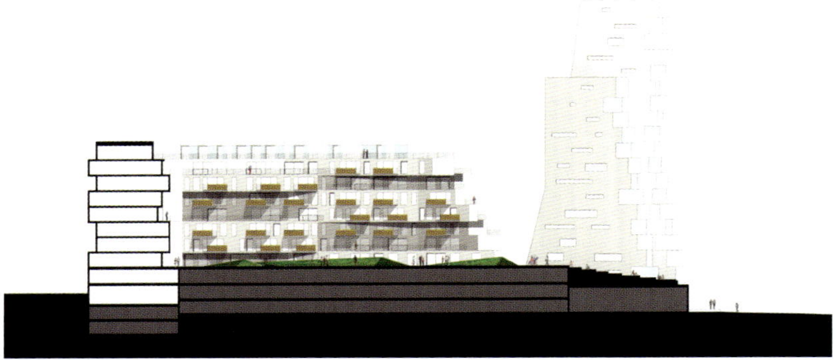

PLAN 03

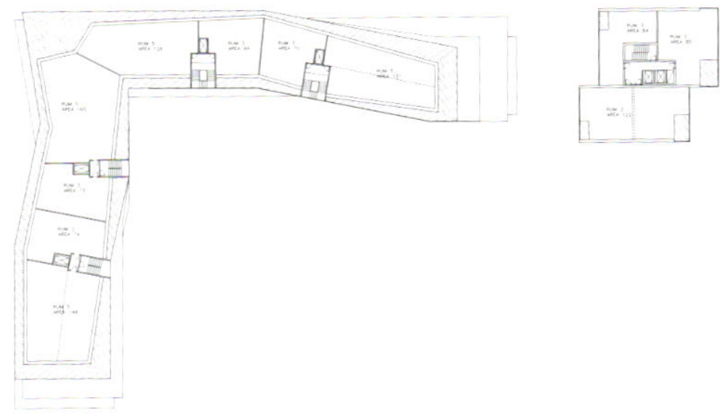

PLAN 09

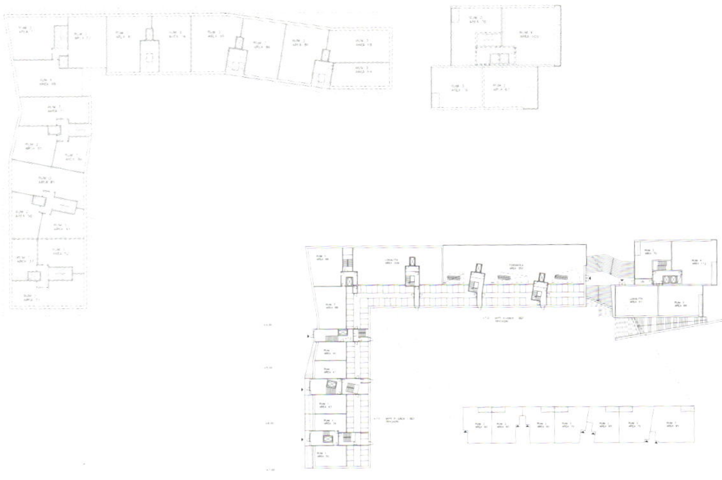

PLAN 01

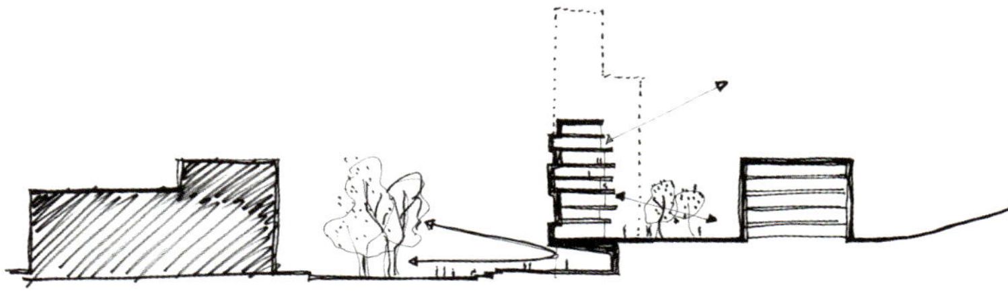

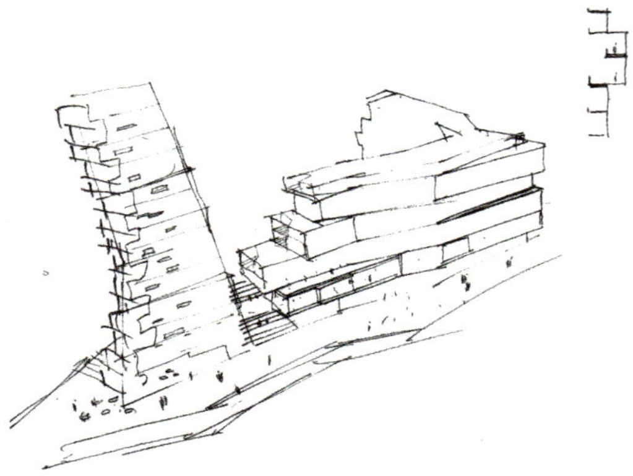

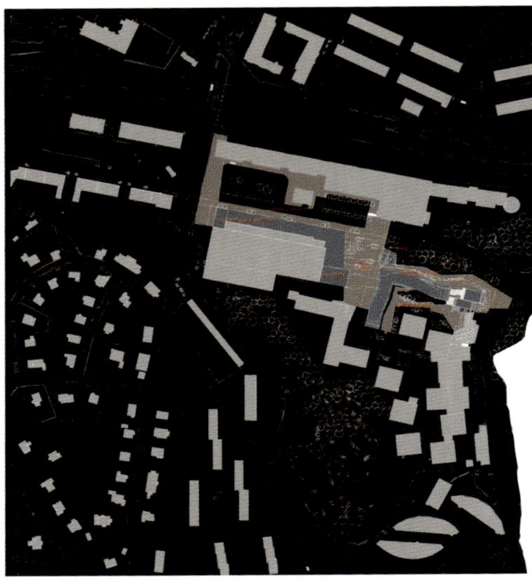
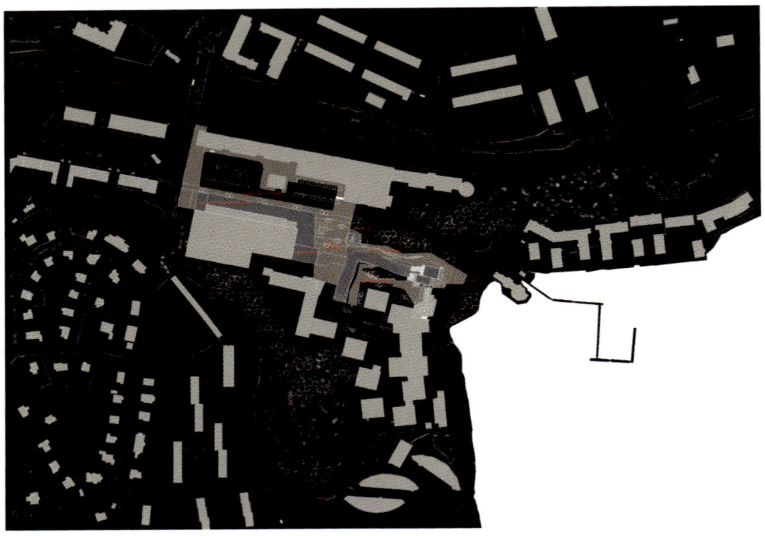
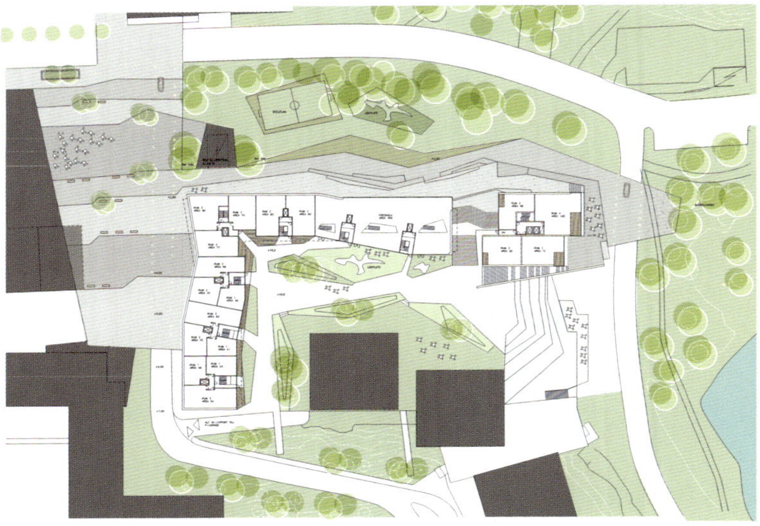
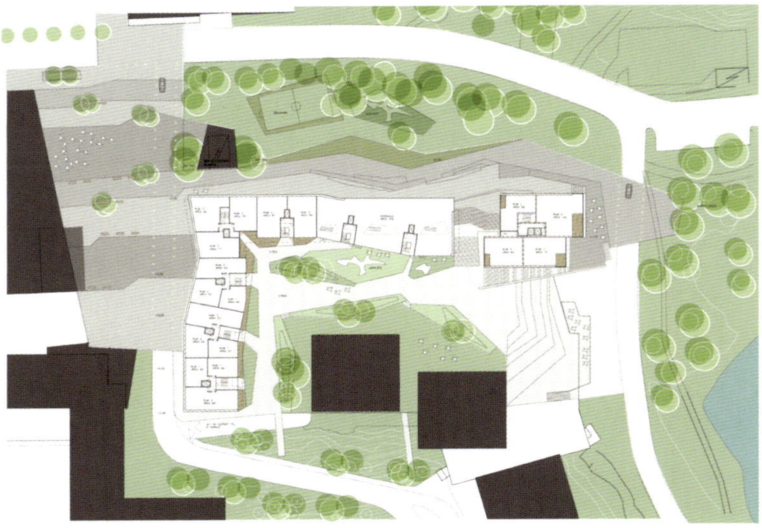
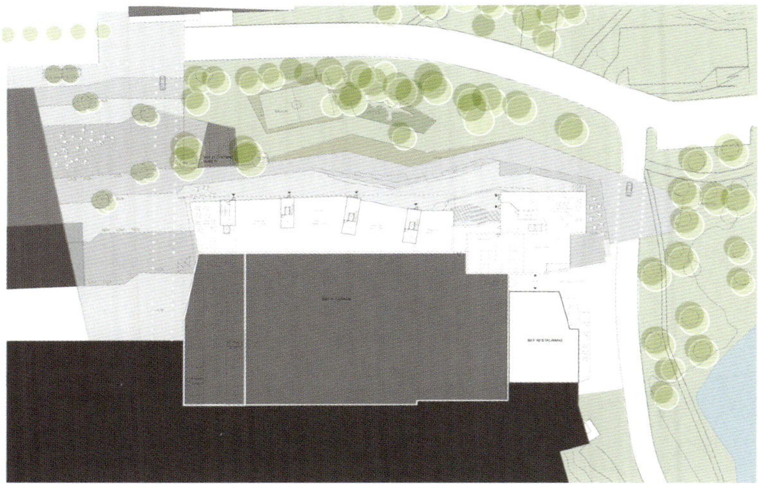
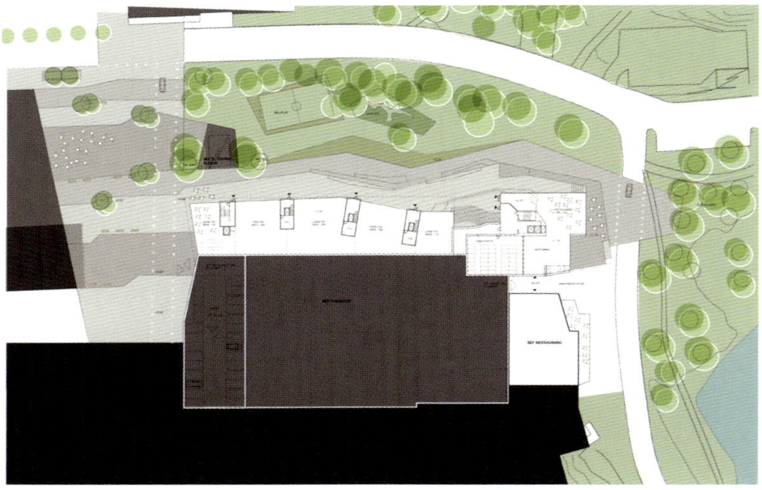
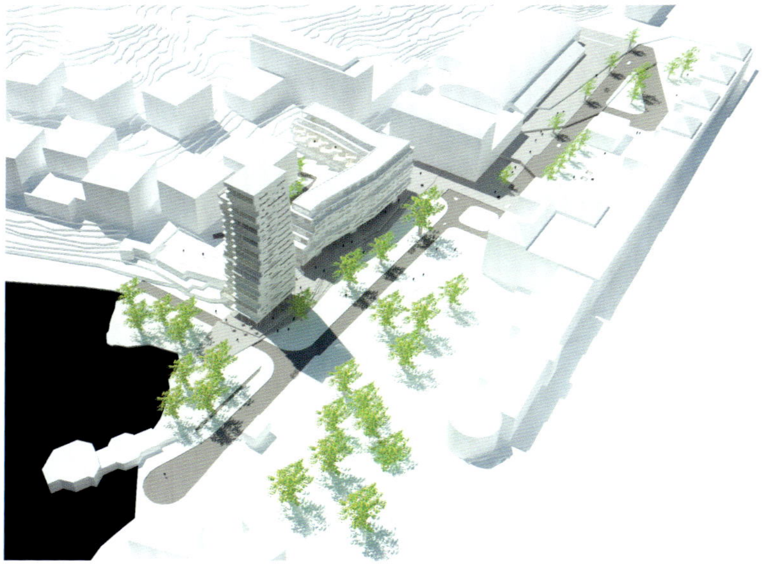

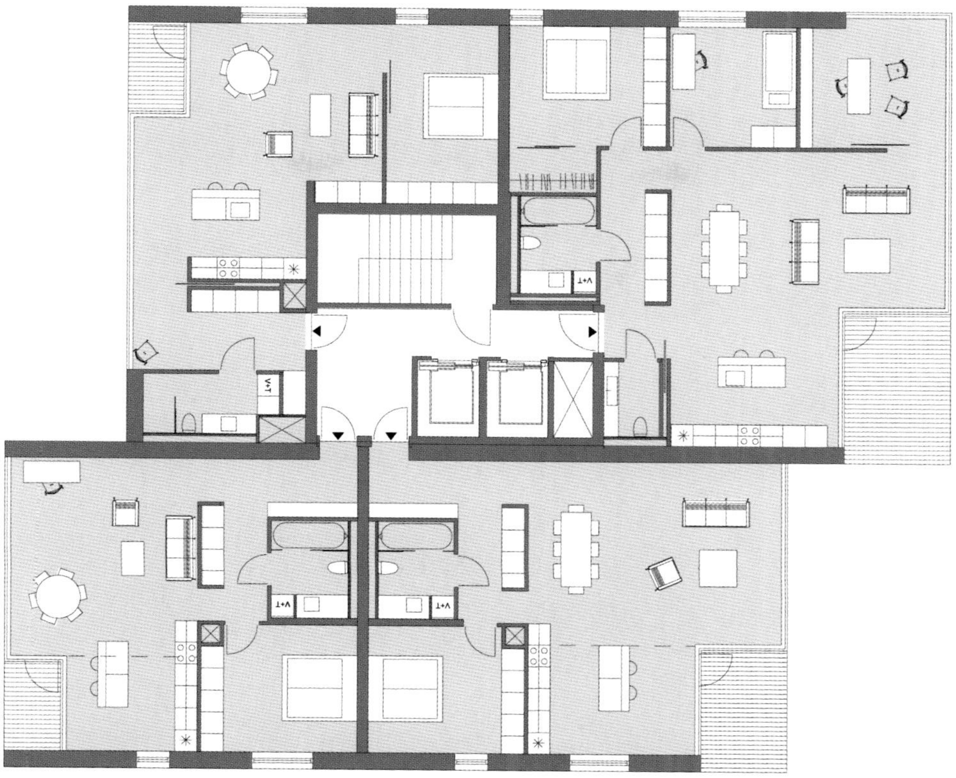

TYPE A - 2 værelser

TYPE A1 - med gesims mod syd
3 værelser

TYPE A2 - med gesims mod nord + altan mod syd
3 værelser

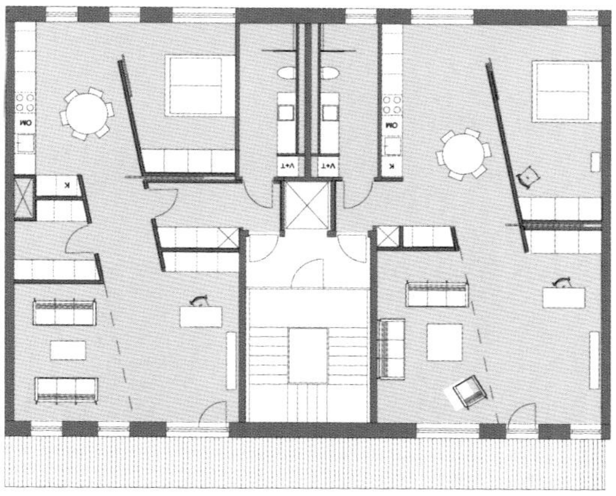

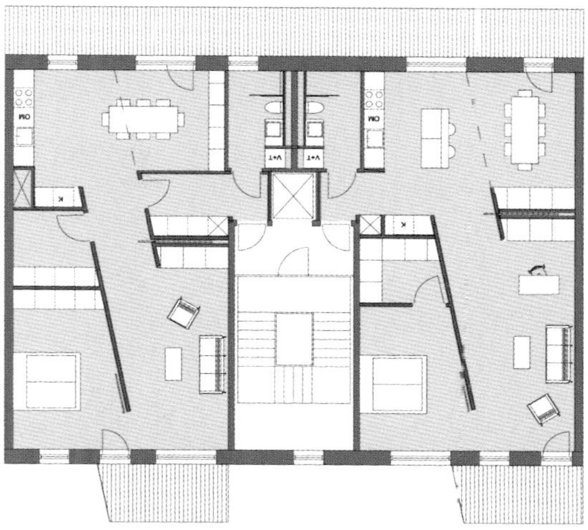

TYPE A - 3 værelser

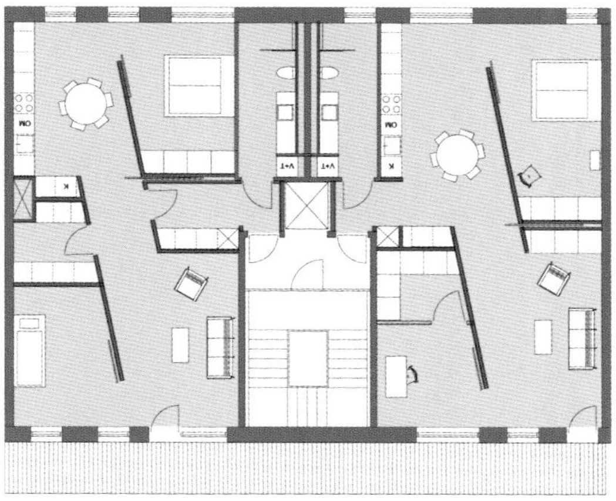

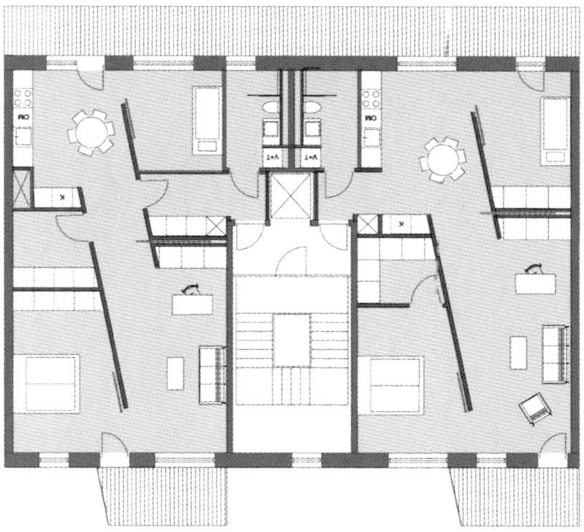

TYPE A1 - med gesims mod syd
3 værelser

TYPE A2 - med gesims mod nord + altan mod syd
3 værelser

Alviks tower is a sculptural housing complex designed to bring identity to the district - and thereby, to Stockholm. The site lies in a prominent location towards the archipelago, and forms a transition zone between the city and the landscape, which has inspired the architecture. The development - a fractured block of approximately nine storeys and a landmark 21-storey tower divided into two parts - draws its shape and lines from the steepness and structure of the mountains bordering the lake Mälaren.

The development also contains shops, a restaurant and green spaces to provide the Alvik area with new functions and identity, and create a modern living compound in a continuous, attractive and active urban environment.

The tower will become the landmark of the development, and the buildings' forms secures each residence a maximum of daylight and view. At the same time, the solution provides free areas which can be used as squares or open spaces – such as a viewing level near the water, and a central square oriented towards the city. The open spaces unite the life of the city with transparent, public-oriented areas at street level.

The entire complex is planned to conserve resources, through excellent thermal performance and on-site solar energy generation. Integrated into the development is a 200 spaces multi-storey car-park.

Sustainable TownHouses –Norra Djurgaardsstade, Stockholm

斯德哥尔摩可持续的小镇住宅

C. F. Møller Architects　Berg Arkitektkontor

First Prize in Competition
竞赛一等奖

Credits
Client: Byggnadsfirman Erik Wallin AB
Location: Norra Djurgaardsstaden, Stockholm, Sweden
Year of Competition: 2010
Construction Period: 2010
Architect: C. F. Møller Architects | Berg Arkitektkontor

位于斯德哥尔摩的Hammarby Sjöstad因为之前的工业港口改造为可持续发展的城市规划区的示范性案例而享誉全世界。现在，斯德哥尔摩正在策划另一份高端环境区开发方案，位于港口前沿，称作Norra Djurgaardsstaden。这里，C. F. Møller Architects建筑事务所已经取得了小镇住宅开发计划的设计权。

18幢小镇住宅周边有前为瓦斯站的辖区文化中心，以及流入附近多岛海的Husarviken河，更有斯德哥尔摩国家城市公园，项目与上述三者为邻。建筑灵感来自于前瓦斯站（煤气包）的红色砖墙、简单的几何形状和区域的绿色品质。

五大类不同的居住区将会以分散化的模块形成一种带有充足阳光和壮丽水岸景色的独一无二的居家环境。该设计方案同样为私人外部空间和露台预留了空间。

在可持续发展的基础上，享受现代生活。该项目使之成为可能。

Norra Djurgaardsstaden居住区的目标是与全球气候变化相适应，到2030年该居住区不再使用化石类燃料，因此不增加二氧化碳排放。居住单元的能源消耗不会超过55千瓦时/平方米·年，包括15千瓦时/平方米·年的电力消耗最大值。

能源消耗将会通过使大楼密集化的方式降低到最小程度，比如，小镇住宅错开排列以最大化利用阳光日照，且密集建造。同样地，诸如智能照明、太阳能加热面板和热能复原的解决方案对节能也有益处。

小镇住宅的特征还包括，带有野生幼兽和草本植物的绿色屋顶，以及带有池塘的普通庭院，除了采集雨水、转换二氧化碳，还为生物多样性提供肥沃的土壤。

可持续方法贯穿大楼的整个生命周期，从建设期到运作期再到可能的后期分解期，这也就是常说的摇篮到摇篮设计法。

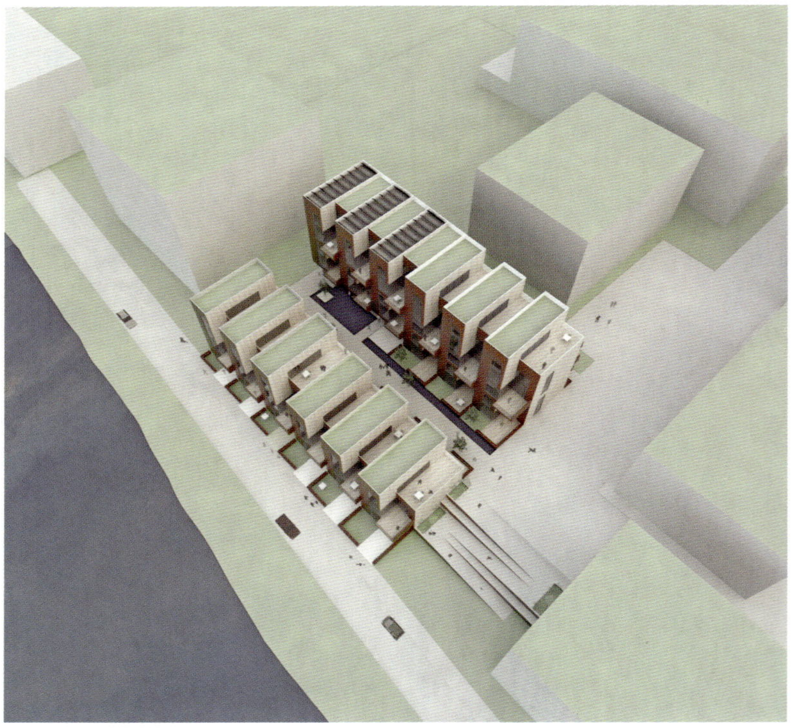

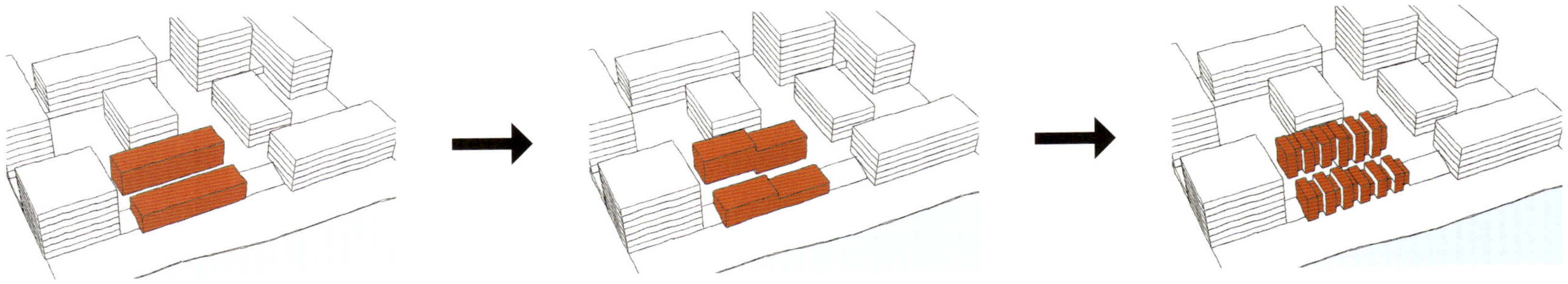

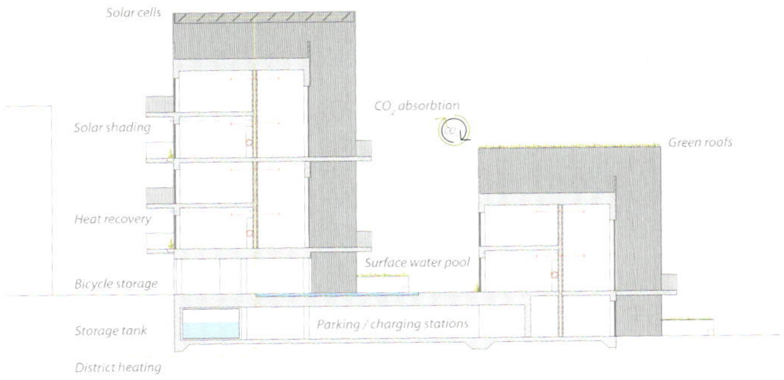
Solar cells
Solar shading
Heat recovery
Bicycle storage
Storage tank
District heating
CO_2 absorbtion
Green roofs
Surface water pool
Parking / charging stations

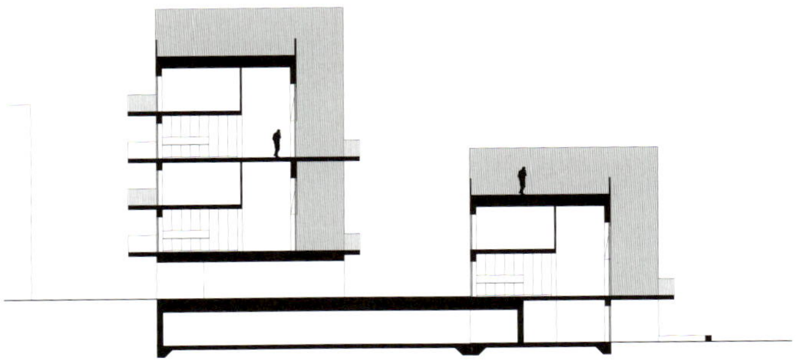

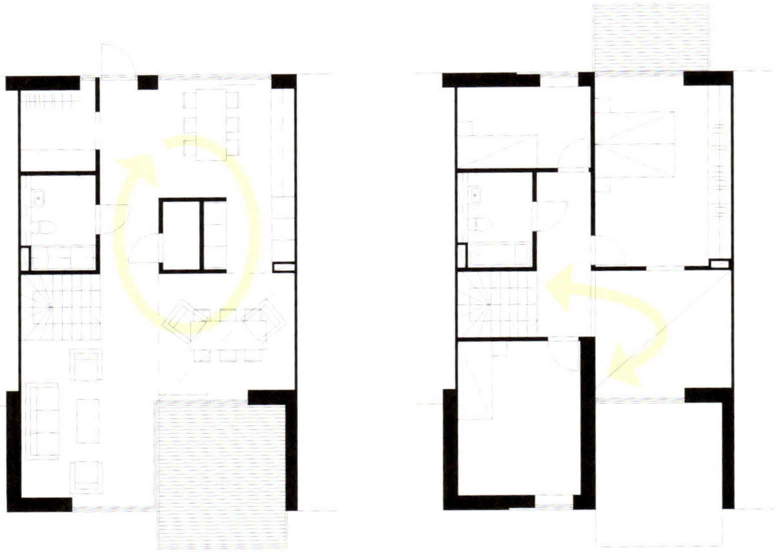

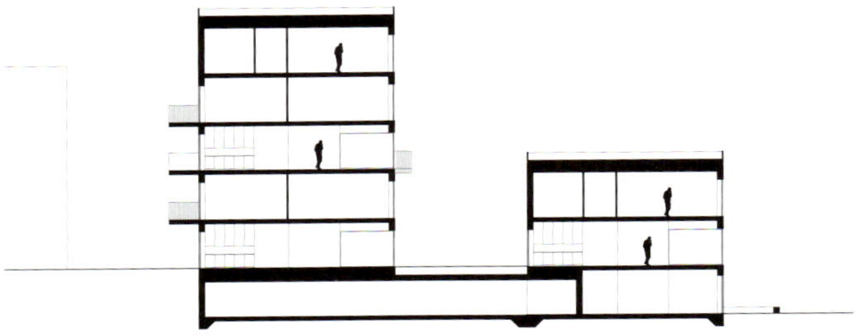

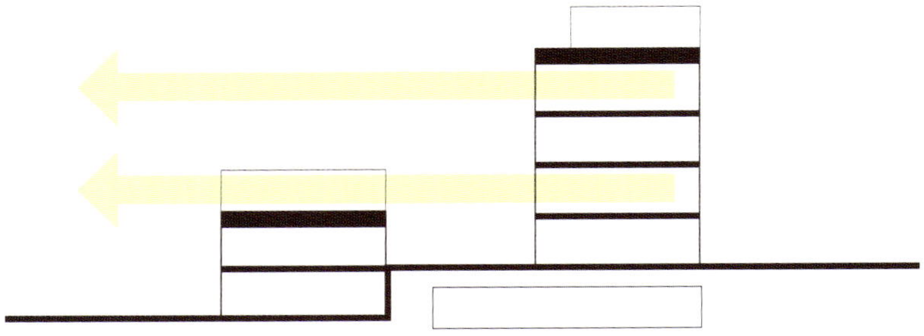

Hammarby Sjöstad in Stockholm is known worldwide as an exemplary example of how a former industrial port area can be transformed into a sustainable urban development. Now Stockholm is planning yet another high-profile environmental area on the harbourfront called Norra Djurgaardsstaden. Here, C. F. Møller Architects has won the competition for a town house development. The town houses, 18 in total, will be neighbouring a former gasworks, which will house the district's cultural center, as well as Husarviken, which flows into the archipelago, and the Stockholm National City Park. The architecture is inspired by the gasworks' red bricks and simple geometry and by the area's green qualities.

The residences, five different types all in all, will be built as staggered modules, a solution creating unique homes with rich daylighting and splendid views of the waterside. The design also gives room to private outdoor spaces and terraces.

The project makes it possible to live a modern life based on sustainable solutions.

The aim of the Norra Djurgaardsstaden housing district is to adapt to global climate changes, so that in 2030 the district will no longer make use of fossil fuels - and thereby not contribute to the emission of CO_2. The energy consumption of the living units will not exceed 55 kWh/m²/year, including a maximum of 15 kWh/ m²/year used for electricity.

The energy consumption will be minimized by means of the massing of the buildings, for example, the town houses are staggered in order to maximize daylight and have dense constructions. Also contributing are solutions such as intelligent lighting, solar panels for heating, and heat recovery.

Characteristic for the town houses will also be green roofs with e.g. wild honeysuckle and herbs, which - along with a landscaped pond in the common yard - will collect rainwater, convert CO_2, and provide a fertile ground for biodiversity.

The sustainable approach is continuous throughout the building life cycle - from construction phase to operational phase and a possible later decomposition phase, i.e. Cradle to Cradle Design.

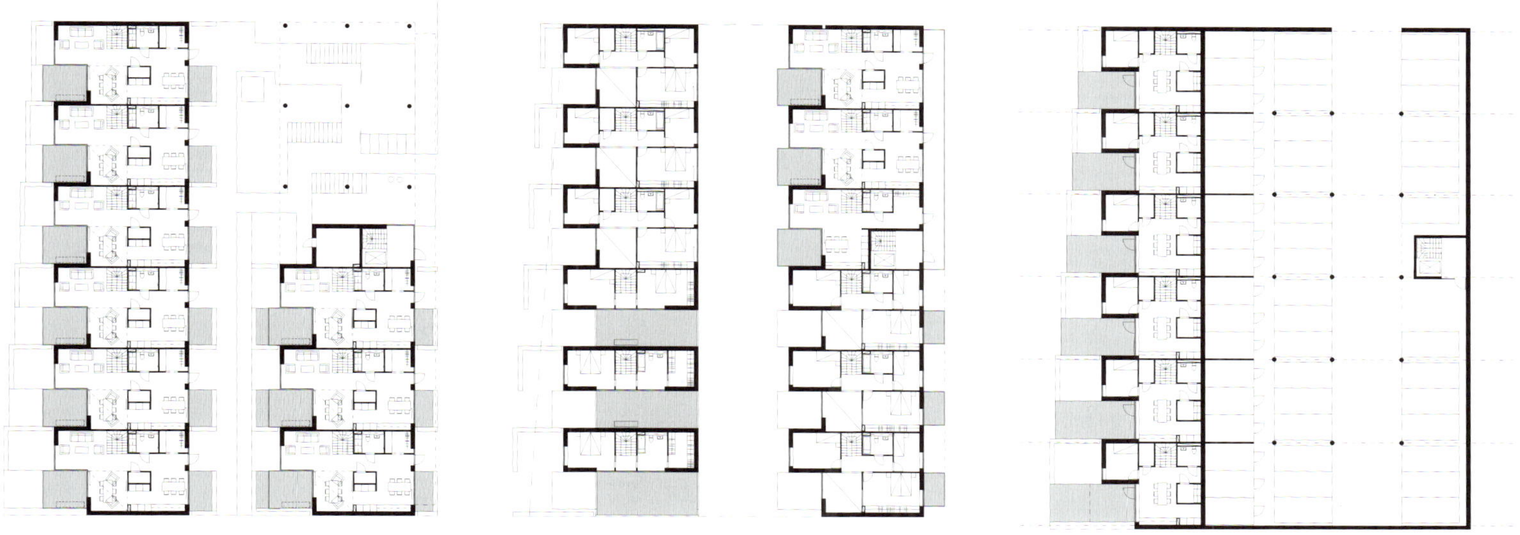

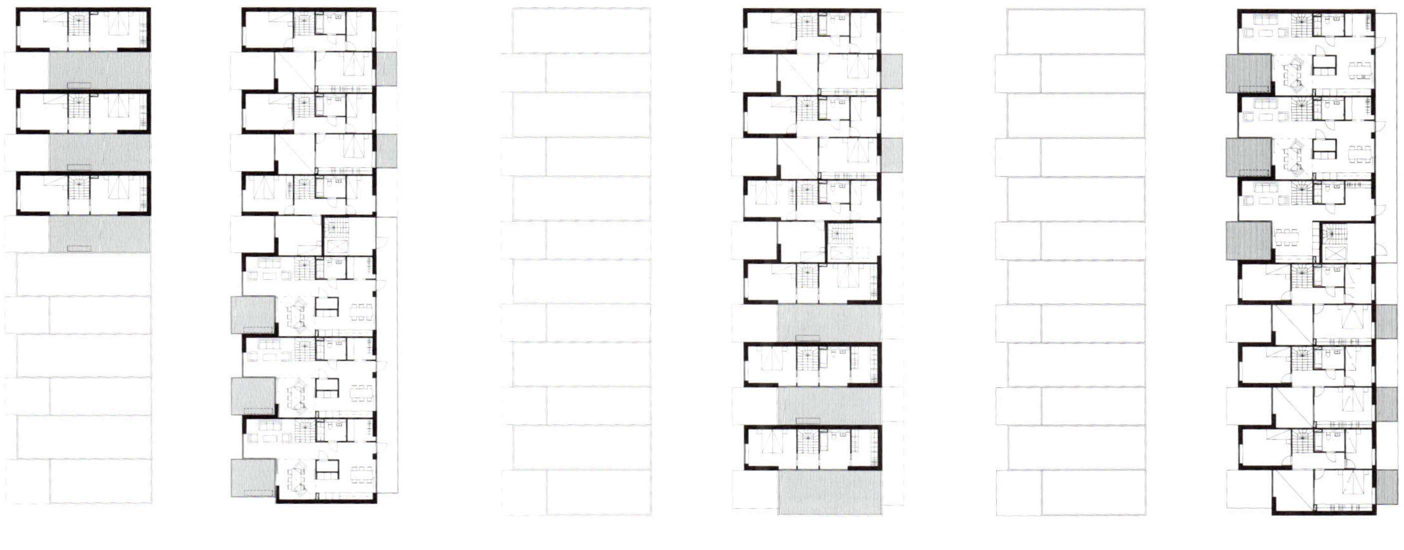

ARCHITECTURE COMPETITION WORKS

City Complex
城市综合体

Samsong Greens
Samsong Bromex 数字媒体公司绿色规划

Sasaki Associates

Credits
Name: Samsong Greens
Location: Goyang City, Seoul, Korea
Completion Date: 2009
Area: 34 hectares

Services: Landscape Architecture, Urban Design, Planning
Project Team: Denis Pieprz, RomilSheth, Isabel Zempel, SarfarazMomin, VeePetchthevee

该项目总体规划是为Samsong Bromex数字媒体公司在韩国首尔高阳市的基地设计的综合开发方案。整体思想是要设计出Samsong Bromex的特色，通过将不同的主题元素与自然体系、可持续发展策略与建筑楼层相融合，打造活力无限的综合社区。规划充分利用了周边自然环境资源（河流、滨水公园与小山），成为这个开发地段，甚至是整个高阳市的"心脏"。

这颗"心脏"以两个公共场所的形式呈现——中央媒体公园与小型城市喷泉广场。配套运动设施有多功能运动场与数字体育馆，鼓励不同人群之间的集体活动。有几个南北走向的线性公园沿着绿色的山脊分布，使每一块地面都有自己的特色。另外，每一条街也因栽种的树木而别有一番风景。这些景观在为大家提供公共娱乐与休息场所的同时，都在不同程度地演绎着"绿色Samsong"的创想，在水域管理与可持续发展战略中占据重要地位。

我们设想要建造一条"生命绿毯"。这条地毯是由两种基本类型组成的：内部带庭院的低层建筑物与高楼。除此之外还零星点缀着一些别具一格的建筑，如数字体育场、零售亭等，其中零售亭还设有媒体和数码相关的互动型娱乐教育设施。低层建筑物4到6层，主要作为城市工厂。楼顶平台将设计成规整的花园，以集中城市农业和花园的形式展现一个高低错落的生产层，组成生活绿毯的一部分。高楼则与绿毯系统有机结合，作为办公场所，CBD写字楼项目以及其他城市工厂的栖息之所。

Samsong Bromex项目山坡的景观设计依据高阳地形，富有当地文化特色。它在山水之间创造出动感活力的现代多媒体区域，见证了城市与自然的结合，让人们在享受大自然赋予的宁静的同时，领略活跃城市生活与自然美景之间的不同。

设计加入了韩国传统景观元素，如满山的松树与樱桃树、河边垂柳、杜鹃花、波斯野菊花、荷花池、桥、芦苇丛、竹与银杏小树林，打造如诗如画的自然景观，与私密的楼顶城市景观形成强烈对比。

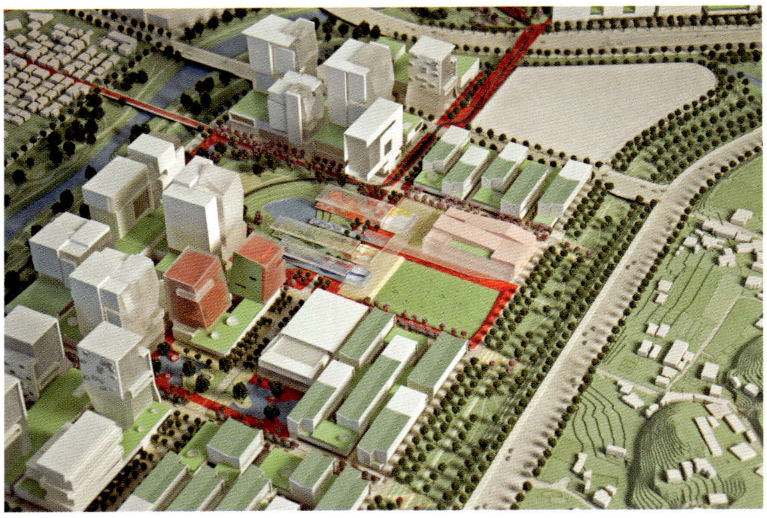

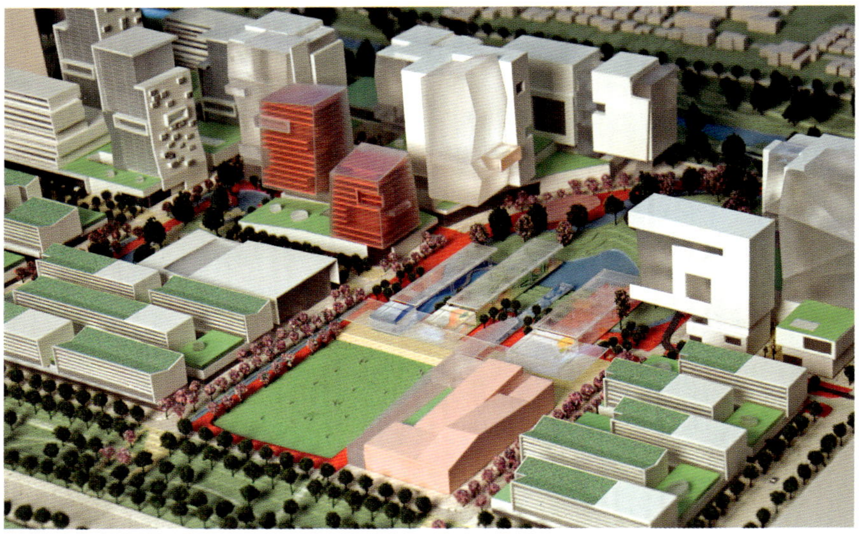

Samsong Landscape Plan

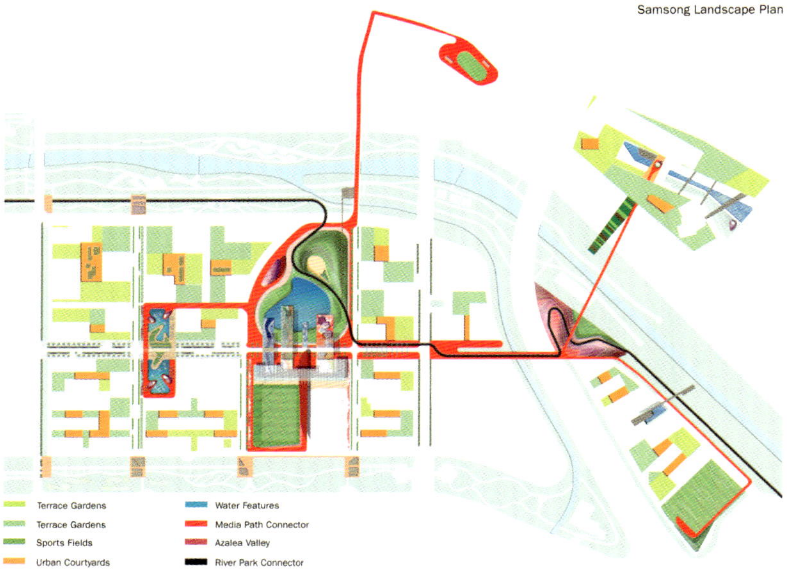

Non-monsoon

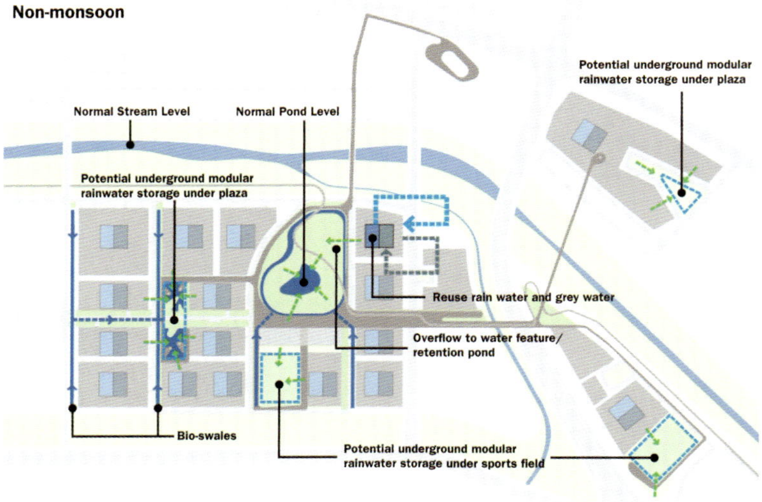

Ground level land use

Monsoon

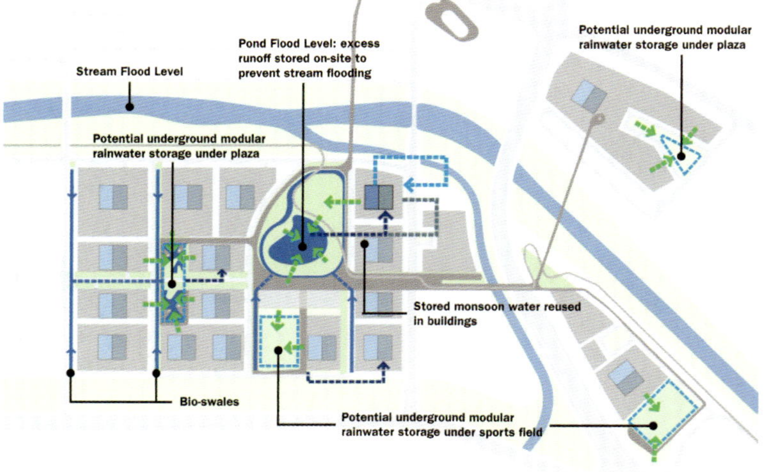

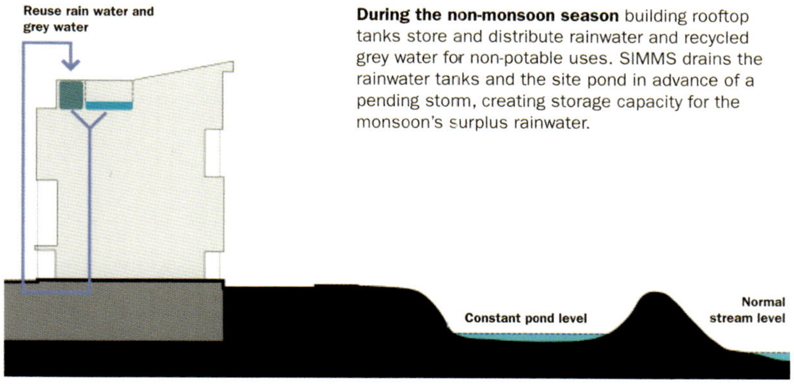

During the non-monsoon season building rooftop tanks store and distribute rainwater and recycled grey water for non-potable uses. SIMMS drains the rainwater tanks and the site pond in advance of a pending storm, creating storage capacity for the monsoon's surplus rainwater.

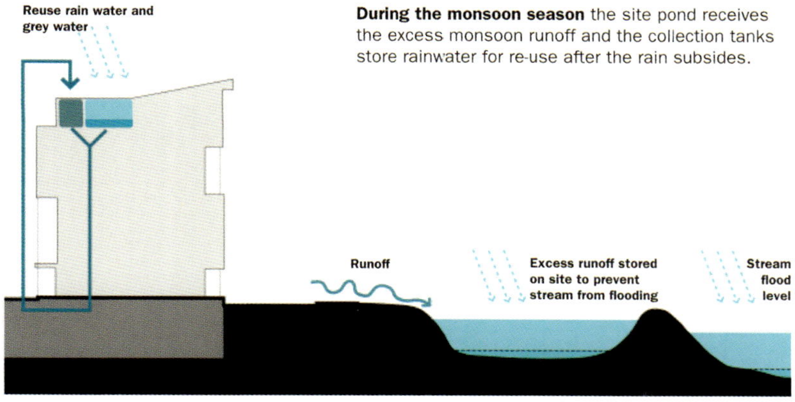

During the monsoon season the site pond receives the excess monsoon runoff and the collection tanks store rainwater for re-use after the rain subsides.

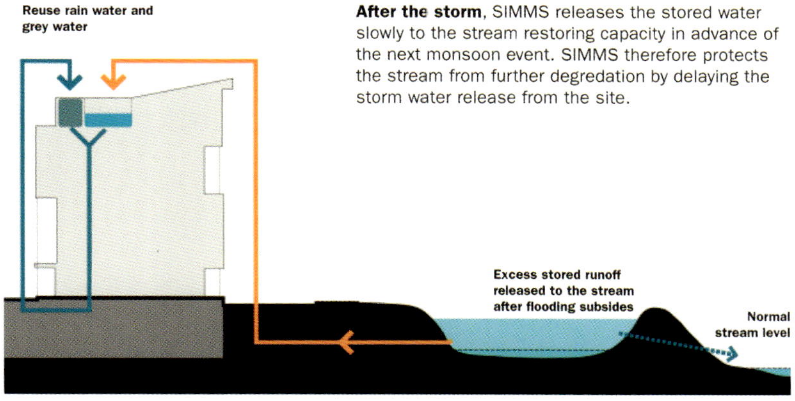

After the storm, SIMMS releases the stored water slowly to the stream restoring capacity in advance of the next monsoon event. SIMMS therefore protects the stream from further degredation by delaying the storm water release from the site.

The SamsongBromex Master Plan establishes a comprehensive development strategy for a growing and dynamic Media Company. The overall vision for the site is to create a distinct identity for Samsong Bromex by integrating the various programmatic elements with natural systems and sustainability strategies at the site and building level to create a vibrant mixed use community. The proposed master plan harnesses the various natural assets of the surrounding context (river, river park, hills) to create a 'heart' both for the development as well as the larger context of Goyang city.

This 'heart' takes the form of two public spaces – the central 'Media Park' and the smaller 'Fountain Plaza' that is more urban in character. These spaces are supplemented by sport facilities in the form of the Multi-playground and stadium (media) that encourage communal activities between the different entities within the development. A series of linear parks that run in the N-S direction give each parcel within the development a unique location and an address along these green spines. This is re-enforced by the tree planting strategy that is designed to give each street a unique character. All these together constitute the larger idea of 'Samsong Greens' as being a series of designed public spaces that provide various spaces for recreation and retreat as well as play a vital role in the water management and sustainability strategies for the development.

The built form is envisaged as a 'living green carpet'. This carpet consists of two primary typologies: low-rise buildings with internal courtyards and a series of towers. Interspersed in this system are a series of distinct buildings such as the stadium (media) and retail pavilions that also house media and digital related interactive recreational and educational programs. The low-rise buildings are 4-6 storey's tall and primarily house Urban Factories. The terraces of these buildings are designed as a series of terrace gardens that form an alternate layer of production in the form of focused urban agriculture and flower gardens that constitute the living carpet. The towers are integrated with this system of the green mat and house Officetel, CBD office programs and additional Urban Factories.

The Landscape Concept for Samsong Bromex Hillside reflects the indigenous values, cultural memories and the local geographic context of Goyang. It creates a vibrant and contemporary Multi Media District within the site setting of the Mountains and the River. It is where nature and city meet and where people enjoy a sense of inner peace from becoming one with nature but also find a highly active, community based park system that offers exciting contrasts between urban city life and natural scenic environments.

Traditional Korean landscape elements such as pine and cherry tree covered mountains, willow trees along the river, azalea, and cosmos wildflowers, lotus ponds, bridges, reed grasses, bamboo and ginkgo tree groves are used to create a picturesque landscape setting which sits in an exciting contrast to the more private urban rooftop landscape.

Creative Gateway-- Urban Regeneration of Sungang and Qingshuihe Areas of Shenzhen

创意门户
——深圳笋岗、清水河片区发展规划

KCAP Architects & Planners

Win the International Competition for the Regeneration of the Sungang and Qingshuihe Areas
中国深圳笋岗-清水河片区发展规划国际咨询竞赛获奖项目

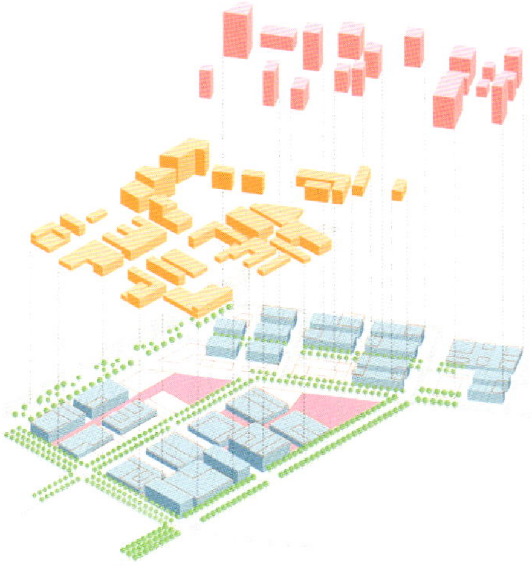

Credits
Client: The People's Government of Luohu District, Shenzhen Urban Planning, Land and Resources Commission of Shenzhen Municipality
Program: 540 ha. strategic masterplan for the regeneration of the areas Sungang and Quingshuihe for creative industries including fashion, industrial and furniture design, media and film, additionally high quality housing public facilities
Year: 2011
Role: Masterplanner

笋岗和清水河片区是罗湖地区的一部分，是深圳的城市中心之一。近年来，由于此区域陈旧的仓库及货栈设施不能再满足现代社会发展的需求，因此逐渐开始衰败。尽管作为物流中心的重要性逐渐消失，但是丰富的城市生活却开始占据这一区域的边缘，并不断渗透蔓延到整个区域。针对这一540公顷的场地，KCAP并没有像中国的传统做法那样，提出一个有限的总体更新规划，而是带来了一个发展框架，这个框架包含一系列的策略性措施。这个框架指明了该区域的发展方向，并且解决了一系列重要的难题。

KCAP设计的出发点是考察及检验城市现状。和传统概念中的直接开始一个全新的大型项目不同，KCAP是针对现状提出了一系列调整措施以及潜在的开发项目，并将它们充分整合到现有的城市肌理中。通过充分考虑现有的建筑肌理、独特的道路脉络和铁路设施、大型的开敞空间以及场地内的地形，整个发展框架将能延续深圳城市发展的历史。

在具体地形研究的基础上，此规划根据不同尺度以及不同转型潜质，将场地进行了分区。在另一个研究中，建筑师提出了一些新规划，这些新项目规划是基于现有的微观经济网络，并将提升它们的层次和经济地位。除了设计、创新、管理以及时尚销售之外，工业设计、家具设计以及多媒体、电影制造业也将是这一区域的主要产业。再通过增加高品质的住宅以及一系列的公共建筑，将最终完善该区域的经济网络。在此规划中，这一区域将不仅仅是罗湖区中心的一个拓展，更将成为连接香港的一个创意门户。

Blue & Green Strategy
蓝色绿色品质

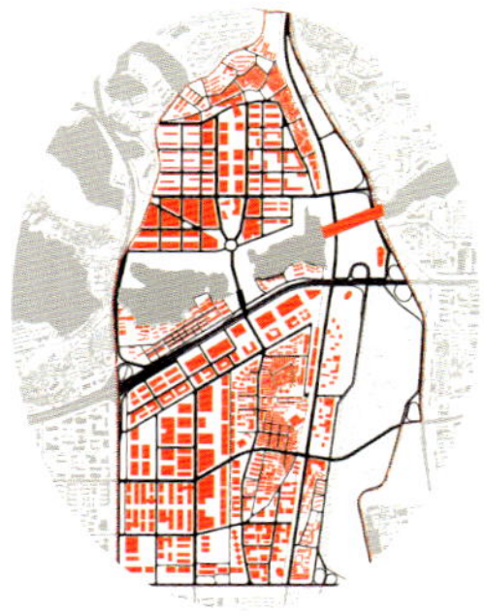

New Urban Quarters
新的都市小区

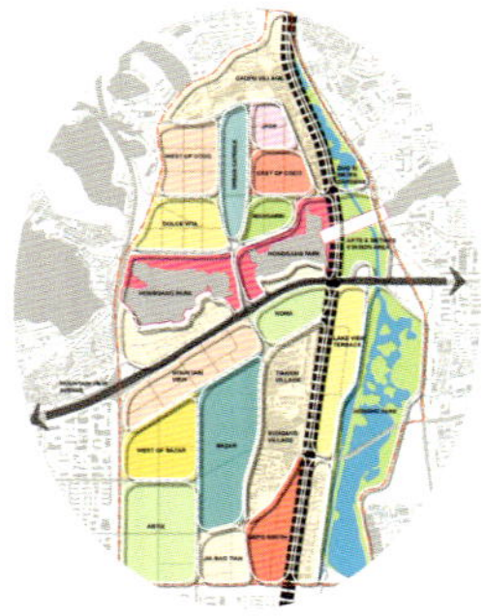

Urban Mosaic
城市马赛克 — 混合功能

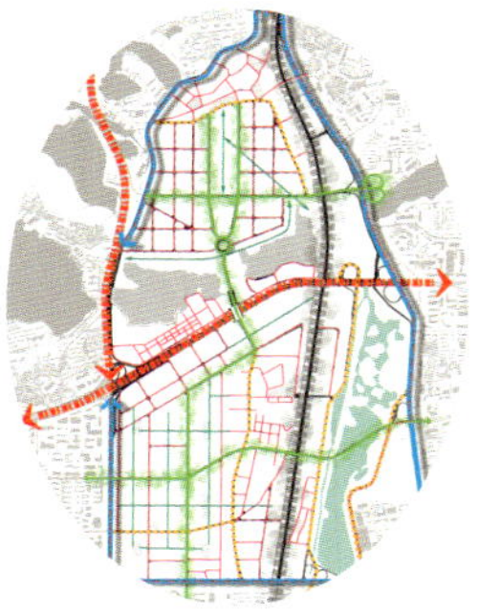

Transport Strategy
交通模式

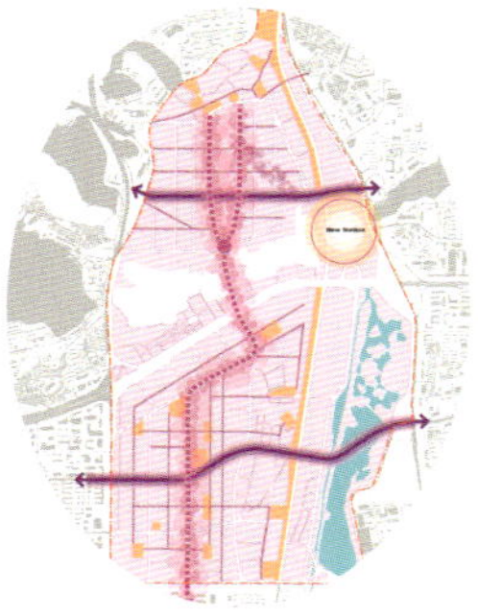

Implementation Strategy
实施策略

Pilot Areas
试点区域

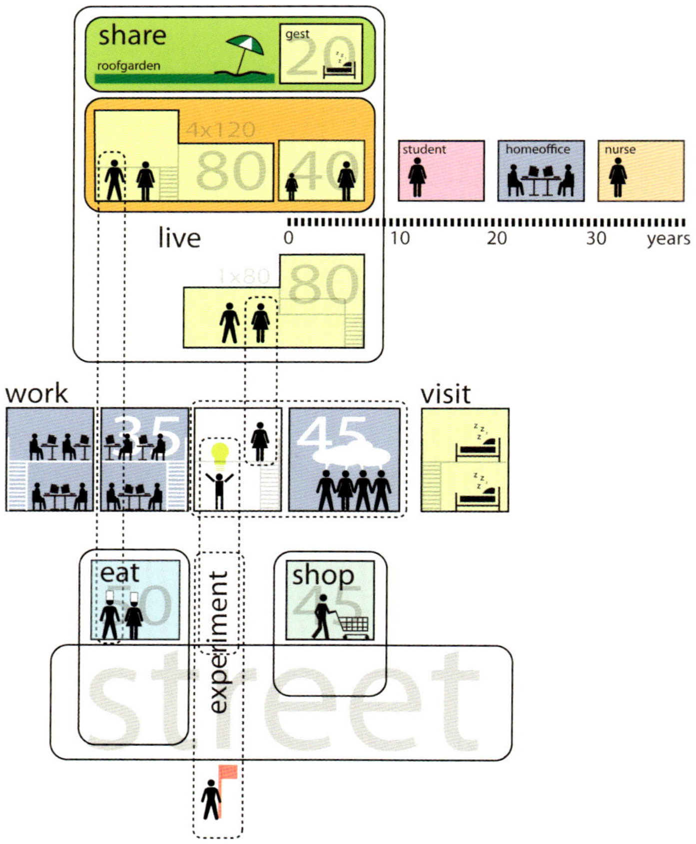

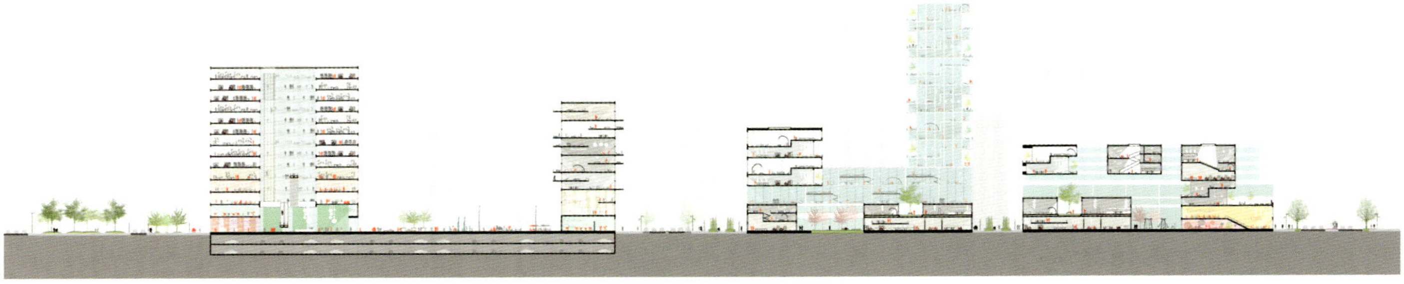

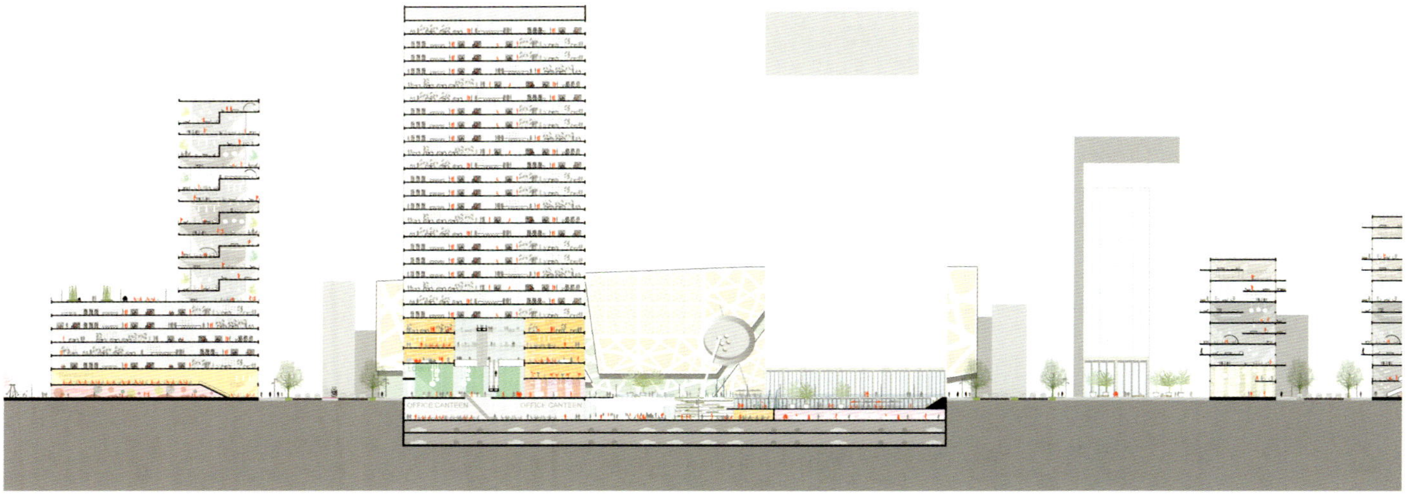

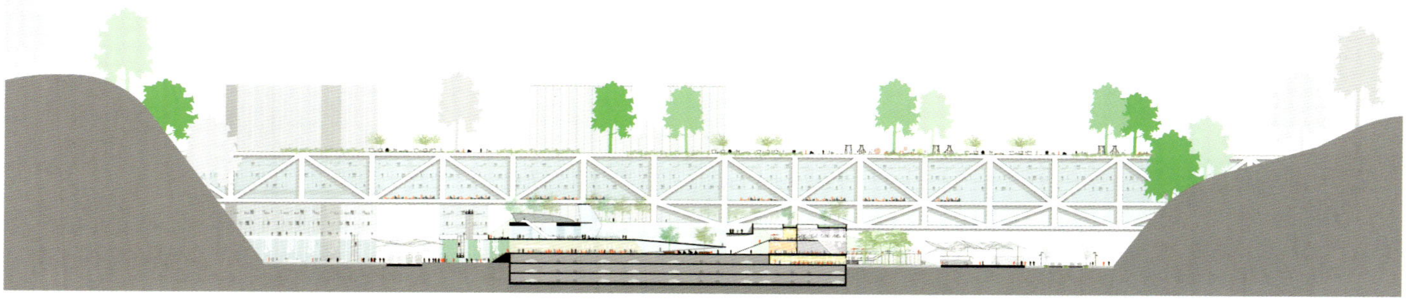

The areas Sungang and Qingshuihe form part of Luohu district, one of the city centers of Shenzhen. Both have become outdated since the warehouse and storage facilities there cannot fulfill contemporary needs anymore. Despite the vanishing importance as logistic center, rich urban life has established itself at its fringes and gradually infiltrates the area. For the urban regeneration of the total 540 ha site, KCAP does not propose a finite regeneration masterplan plan, as is common practice in China, but a framework for development with a set of strategies. They set out the direction of development and resolving the most important problems and questions.

The approach is based on the acceptance of the current urban condition as a starting point. Instead of grand projects, the design introduces a series of measures and projects to be integrated in the existing fabric. With respect for the partly unused built environment, significant road and rail infrastructure, large open spaces and the topography of the site. the framework will continue the urban history of Shenzhen as the result of pragmatism succeeding over monumentality.

The plan defines, based on detailed typological studies, zones of different scale and transformation potential. In a separate study, new programs have been defined, that are based on the already established fine grain of economic networks but bring them to a higher level of profile and economic importance. Next to design, creation, management and sale of fashion, the focus will lie on industrial and furniture design as well as media and film industry. This economic profile will be complemented by high quality housing and a chain of public buildings. In its positioning the area will not only play an important role as the extension of Luohu district center, but also as a creative gateway to and from Hongkong.

Urban Development Plan "FredericiaC"

海洋工业区滨水再生框架 KCAP Architects & Planners

Credits

Client: FredericiaC (partnership of Local Authority of Fredericia and Realdania By (former Realdania Urban Development)

Program: 21 ha masterplan with 126,000 m² residential, 82,000 m² offices/workspaces, 22,000 m² culture and education, 15,000 m² retail, 15,000 m² hotel, restaurants, café

Year: 2011

Role: Masterplanner

Participating parties:Fakton, Ramboll UK, Ramboll

FredericiaC项目是Fredericia海滨的占地21公顷的棕色地块的城市发展战略规划，它将使市中心区的面积扩大25%，它设定了二十六万平方米的城市综合功能空间。此项目将在该地区建立一个新的目的地，在一个有吸引力的海洋环境区创造与城市密切相关的一个独特的内城。它具有可持续性发展和实现碳中和的抱负。Fredericia是在繁华的三角地区的城市之一，在丹麦的中心。它作为一个军事驻防城的历史很完整地呈现在拱顶和文艺复兴的网格布局中。在东边，内城已有一小块带状的有吸引力的海滩，但是在南面，城市被港口从海滨中分离。在很长一段时间里，这片工业区一直是市中心的负担。现在，一个历史性的独特的机会为我们展现了延伸市中心至滨水区，并且塑造一个充满活力的综合城市区域的愿景。尊重城市的规模和历史，发展战略提出了一个坚实的框架和一个具有敏感网络的公共空间。运河网格将现有的海滨丰富化，为城市提供了亲水区。运河区域原本的设计是为表现驻军区的历史感，但是它从来没有被实现。运河在城市、新建筑与水域之间设定了集中的互动场所，并且提供了有吸引力的，海洋性的整体风格。包罗一切的城市规划设计理念将城市与海港特征融合在一个连贯的城市合奏里。其规划原则决定了综合性用途、结构类型和不同的建筑组合。在这些原则里，城市发展高度灵活地适应未来的变化，并且可能会积极影响发展进程。该计划提出了一个灵活的短期策略，以干预措施和临时活动促进该区域的发展。金融策略则担保了工程每个阶段的可行性。

café

kole

Rotterdam, 23 May 2011 - KCAP Architects&Planners, together with advisors Ramboll and Fakton, win the international competition for the urban development plan 'FredericiaC' in the Danish city of Fredericia. The plan for the 21 ha brownfield site at the seafront of Fredericia will expand the existing inner city by 25% and accommodate a mixed urban program of 260,000 m^2. The project will establish a new destination in the region, create new correlations in the city and add a unique inner-city district in an attractive maritime environment. It has high ambitions on sustainability and is planned to be CO_2 neutral.

The multidisciplinary team elaborated the winning proposal in an exceptionally open process, guided by FredericiaC, a partnership between the Local Authority of Fredericia and Realdania By. This form of competition allowed an ongoing dialogue with the organisation, the stakeholders, local citizens, experts and the competing design teams during the design process.

In their decision, the jury concluded: 'Team KCAP's coherent and well-prepared physical and structural concept, springing from historical considerations about Fredericia's urban structure, has created special qualities in the form of a canal structure and a pronounced urbanism as the key means for a robust development strategy for FredericiaC. The proposal is distinct with its urbanism and its very strong concept of bringing great attractiveness and value to all lots.'

The all-embracing urban planning concept combines the best and most characteristic elements of Fredericia - a historic garrison town and a harbor city - in a coherent urban ensemble. The renaissance grid layout of the inner city is continued, extending view lines into the site. The impressive vaults around the city will be connected and made accessible by a new park, and the existing seafront will be enriched by a network of canals, strengthening the cities relation to the water.

'KCAP provides a coherent interdisciplinary proposal that differs from the other proposals by bringing into play more expressive physical elements such as the canals,' states the jury. 'We see Team KCAP's proposal as a very strong unifying physical concept with a regional appeal and intended to increase the area's attractiveness, attention and value.'

The planning principles dictate a varied mix of different uses, urban typologies and architectures. Within these principles, the urban development is highly adaptable to any future changes that might affect the development process. The plan sets out a flexible phasing strategy with short term interventions, pioneers and temporary events to promote the site. A financial strategy guaranties the feasibility of the project in every stage.

The concept takes a sustainable approach to the redevelopment of the brown field area. It includes solutions for the present issues of soil contamination and environmental risks of neighboring harbor activity and prescribes a green building code, waste strategy and sustainable transport strategy. With feasible investments in sustainable energy supply from wind turbines, district heating and sea water cooling, the whole development will become CO_2 neutral.

In an interactive and inclusive process KCAP, Ramboll and Fakton will elaborate the winning proposal towards a definite development plan which is planned to be presented the first half of 2012.

Waterfront Regeneration of the Bay of Pasaia

帕萨亚海湾滨水区改造

KCAP Architects & Planners

Credits

Client: Provincial Council of Guipuzcoa

Program: 70 Ha Masterplan with 840,000 m^2 Built Area for Residential and Commercial Use, Parks

Time: 2009 - Present

Role: Masterplanner

Participating Parties: LKS (Engineering, Sociology and Urban Economics); Hoz-Fontan (Local Architect); ECOFYS (Sustainability Consultancy)

西班牙帕萨亚海湾占地70公顷，现在主要用于开展港口业务，将逐渐改造成新的开发区。此项目对于周围需要更新重建的地区来说，是个绝佳机遇，将给整个区域带来重要影响。

总体规划对帕萨亚海湾丰富多样的群岛海岸线进行了开发，并创造各种条件鼓励当地社区及居民参与到设计中。城市规划方案为当地社区提供了发展空间与推进自身形象发展的可能，并发掘出多样性魅力。区域发展主要围绕公共交通枢纽进行。围绕码头和林荫大道的公共空间网络将不同类型的区域统一起来。在此稳定的结构内，形成灵活的公共空间网络，为现有和未来的居民的亲水娱乐休闲活动创造条件。以海湾为核心，该区域还将建造文化中心和海洋博物馆等新的地标性建筑。大量新型经济活动将为各种行业提供新的就业机会。

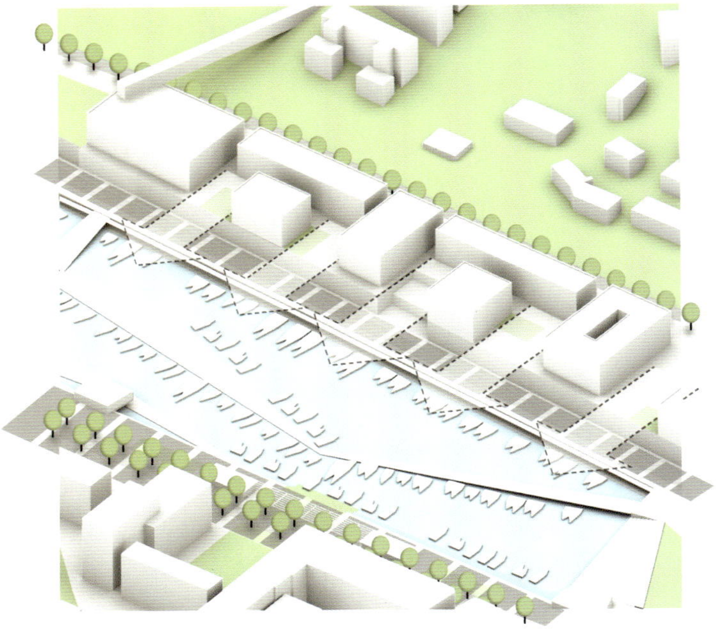
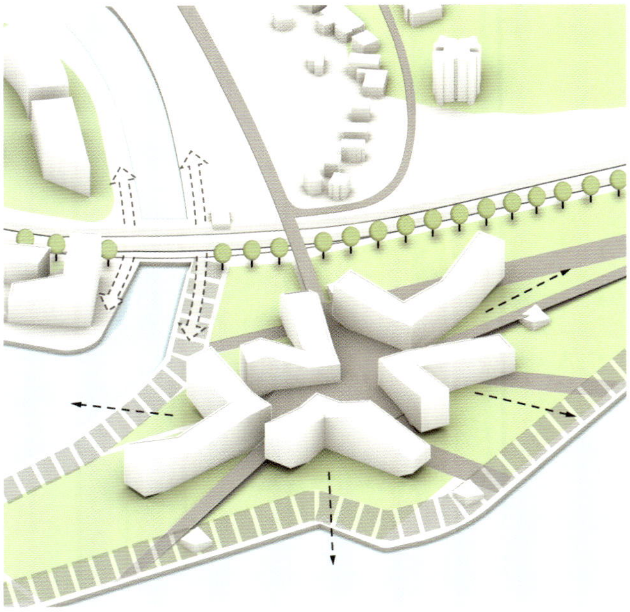
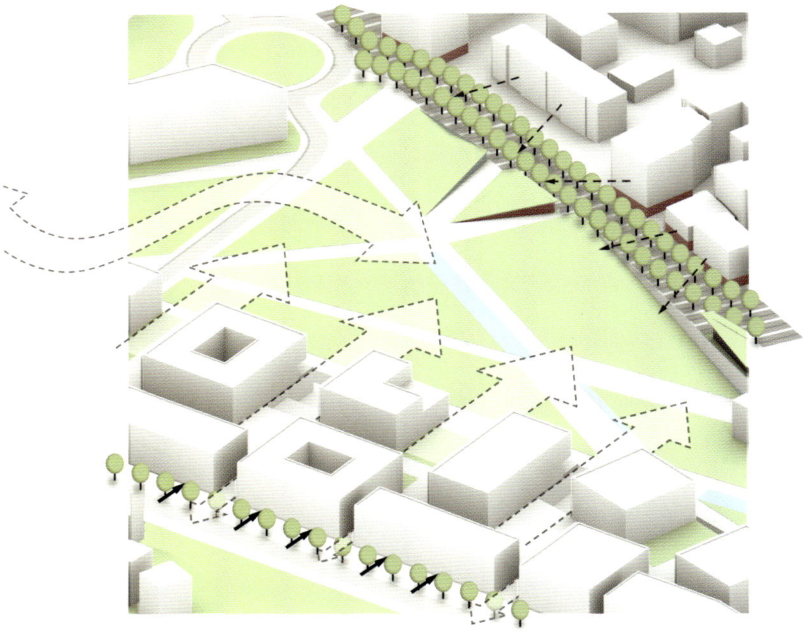
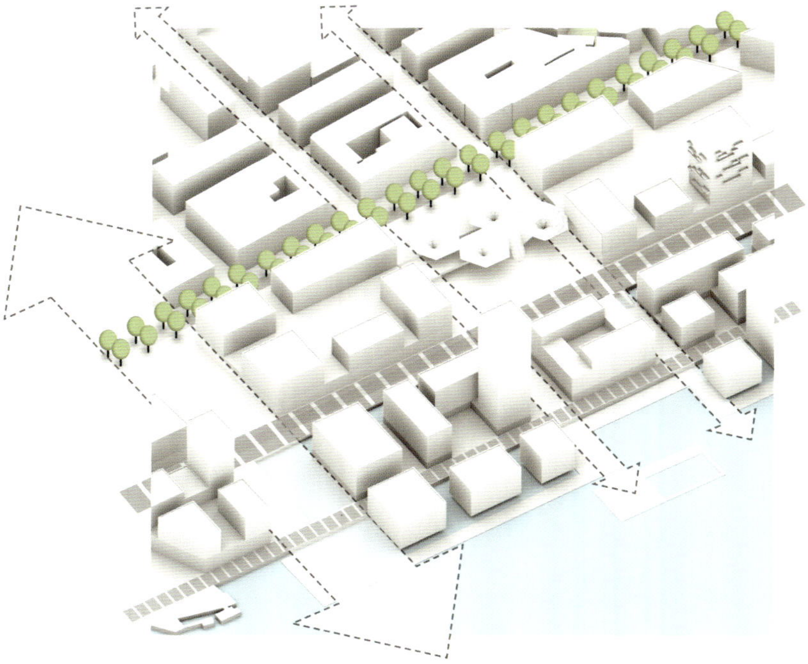

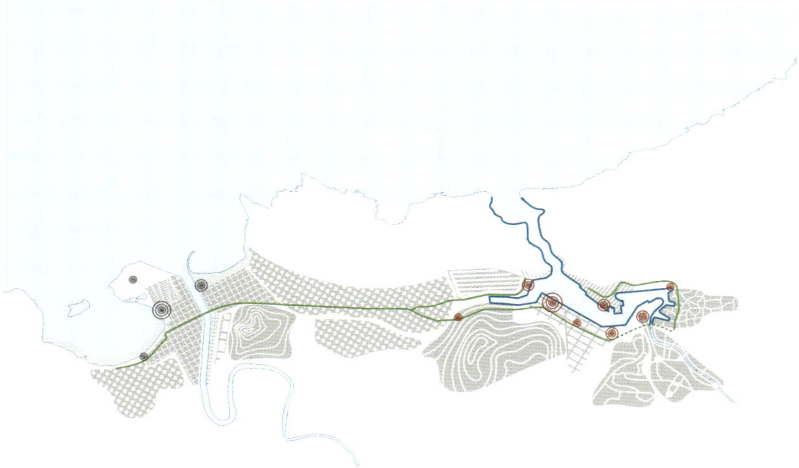

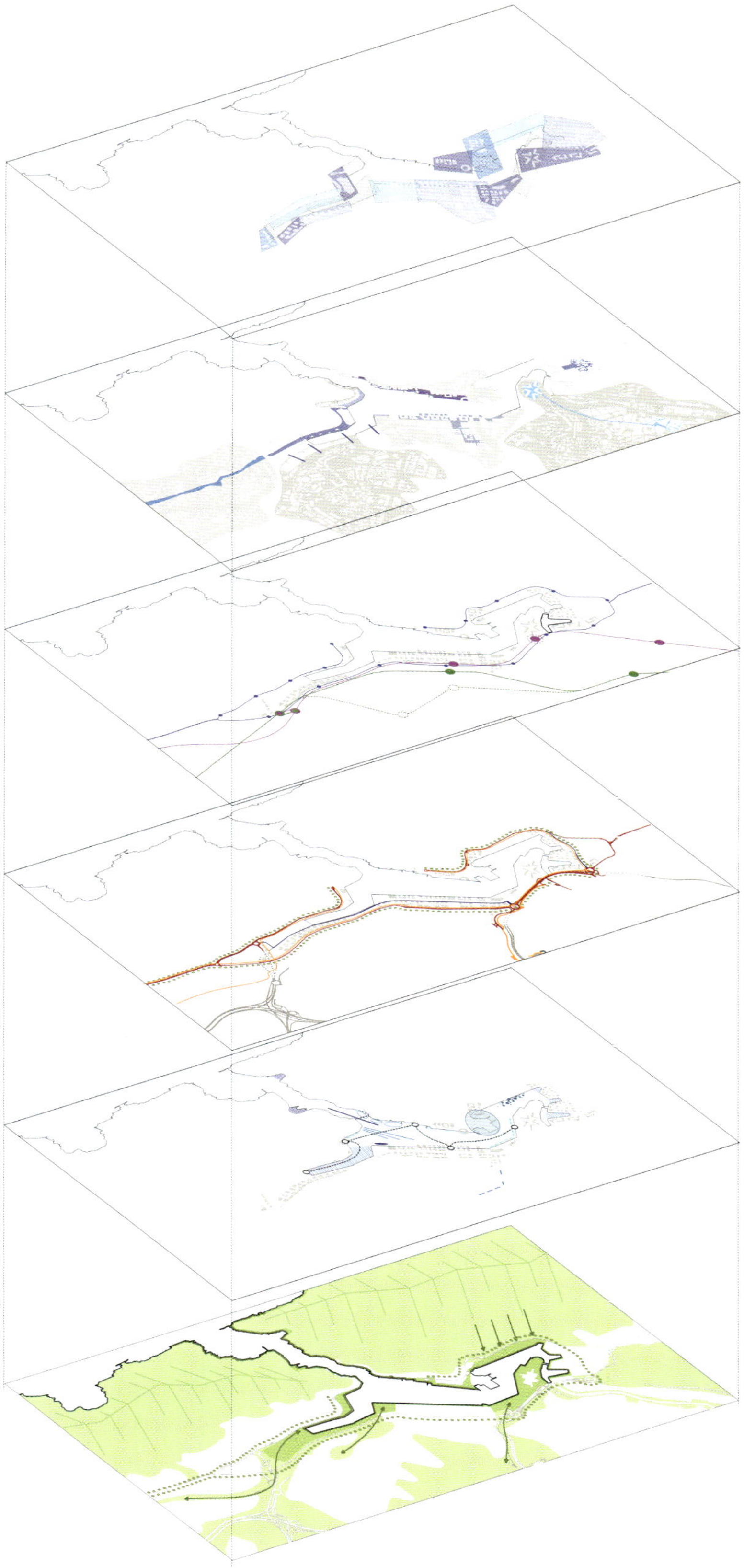

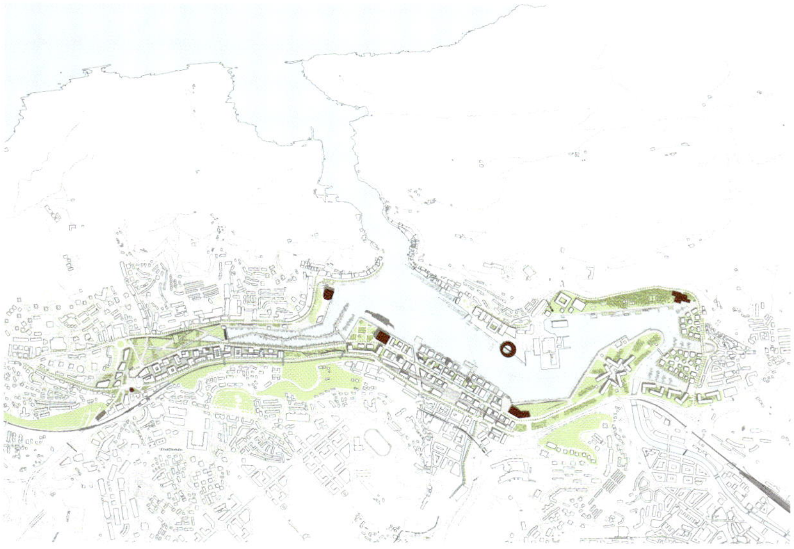

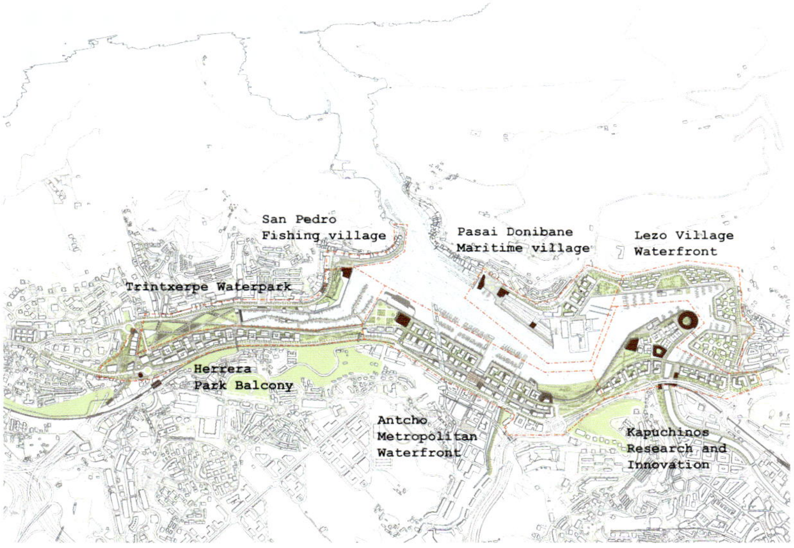

Renteria Belvedere

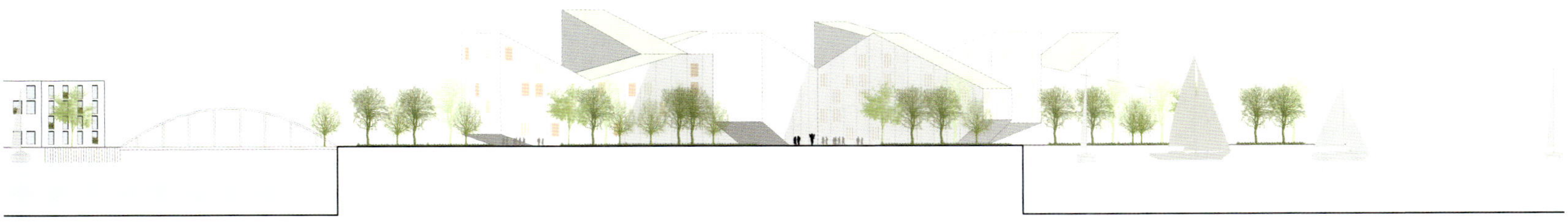

Antxo Metropolitan Waterfront

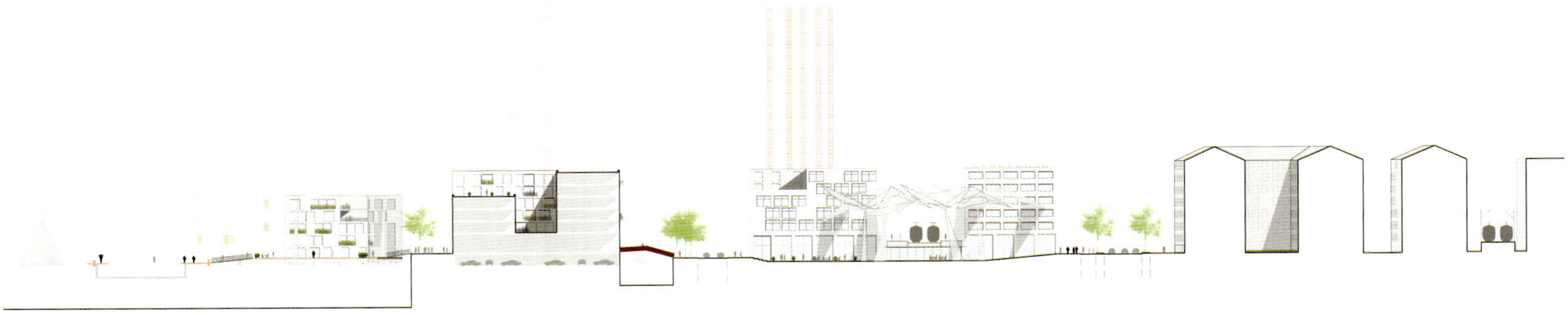

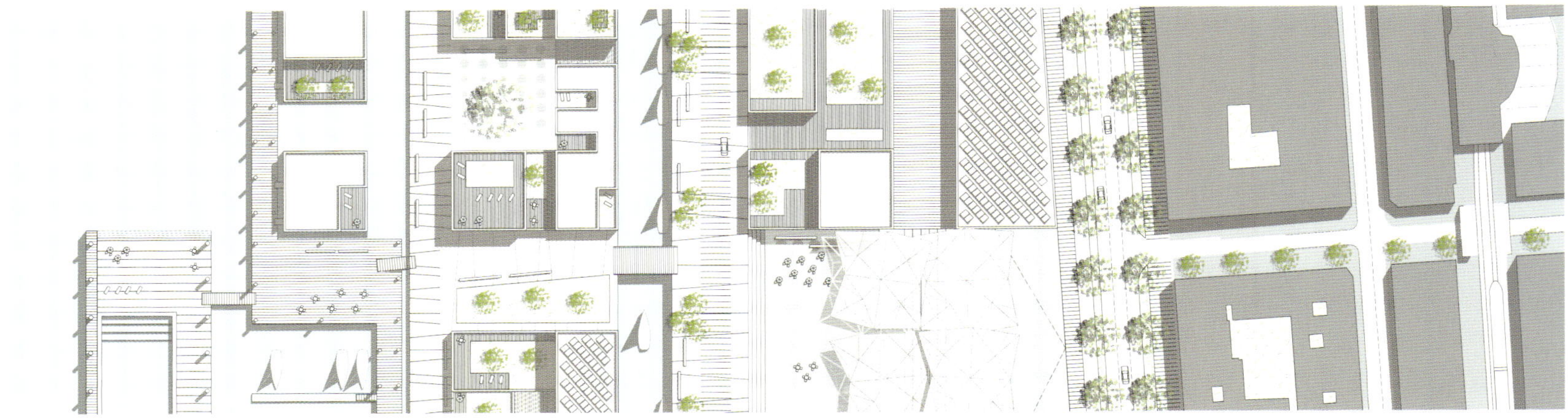

Trintxerpe Waterpark

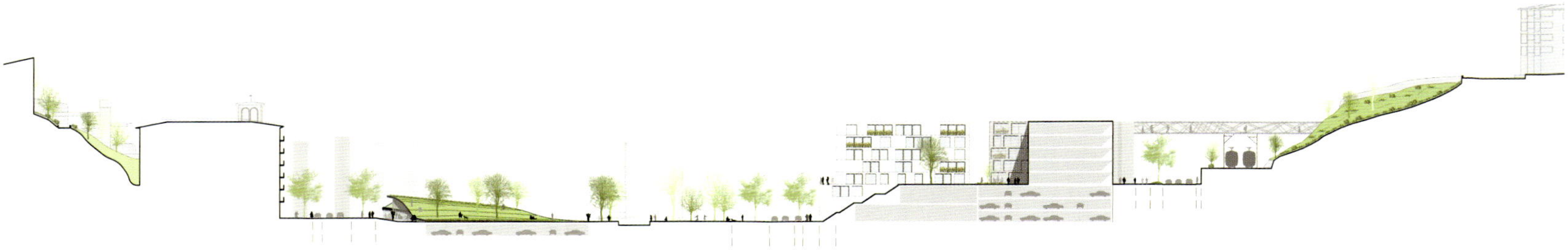

Pasai Donibane Naval Center

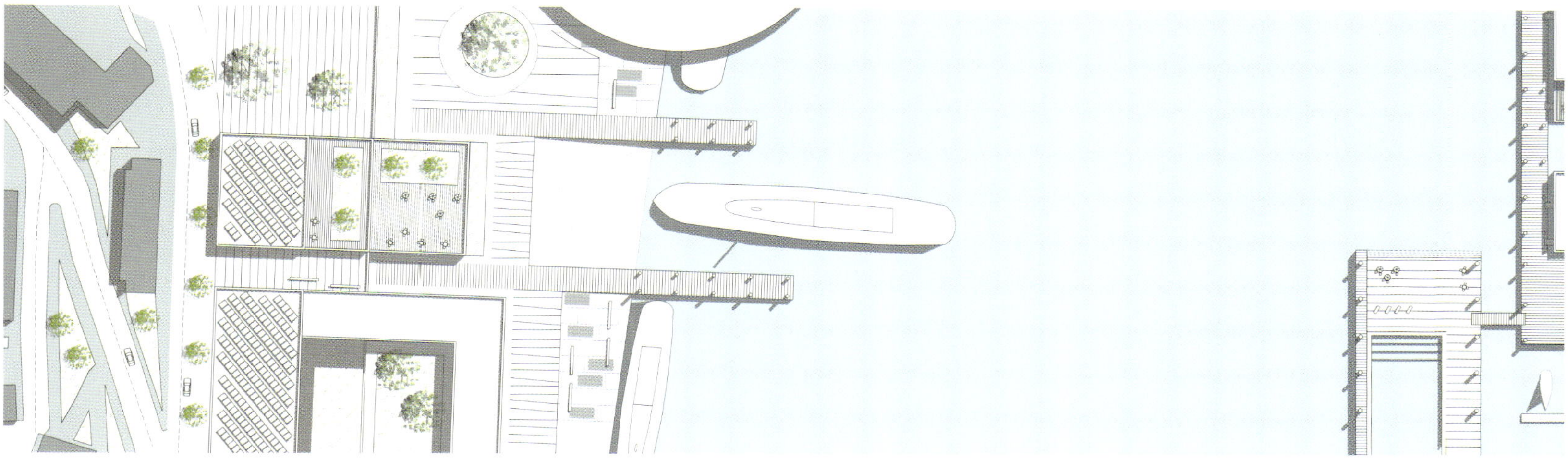

Herrera Park Balcony

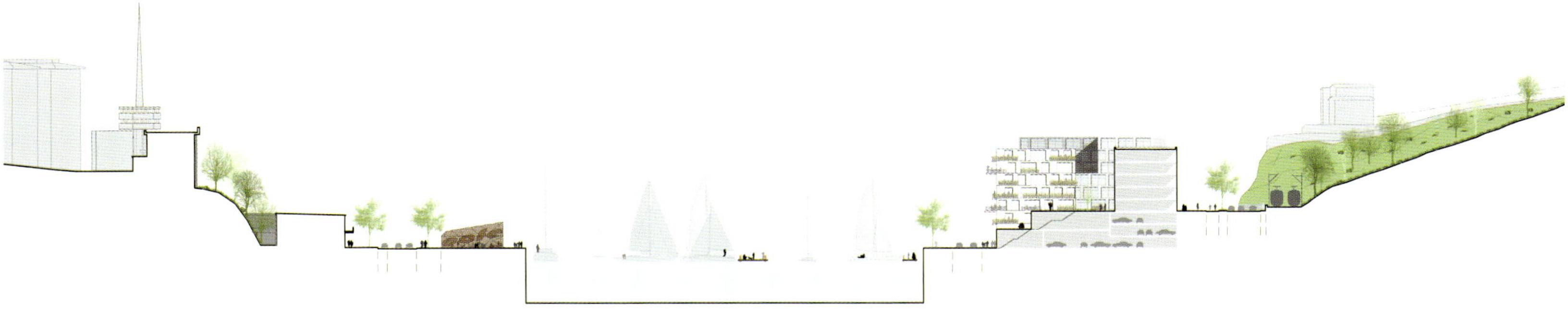

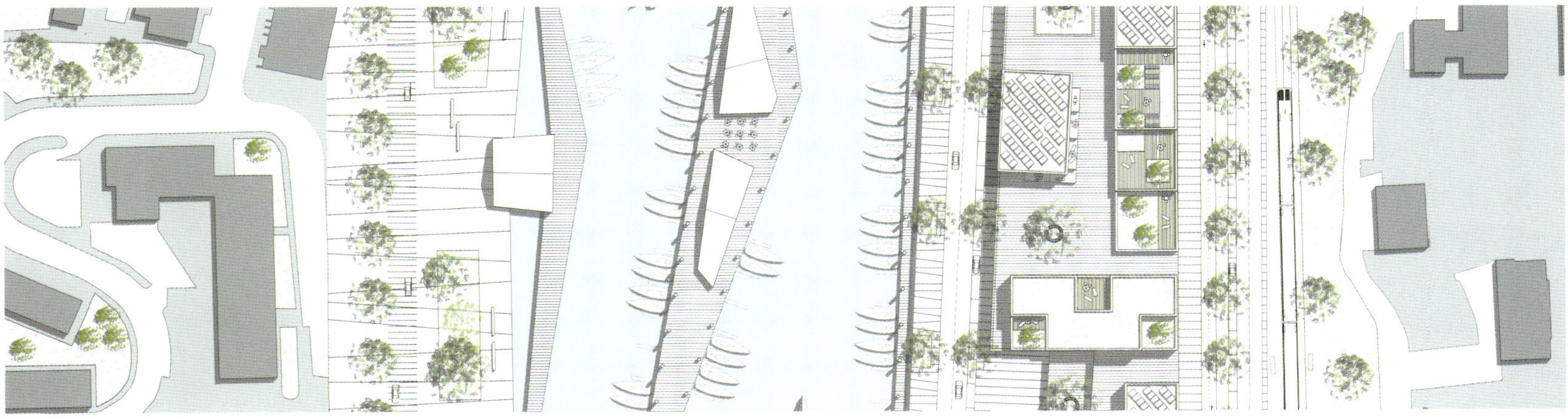

Lezo Village Waterfront

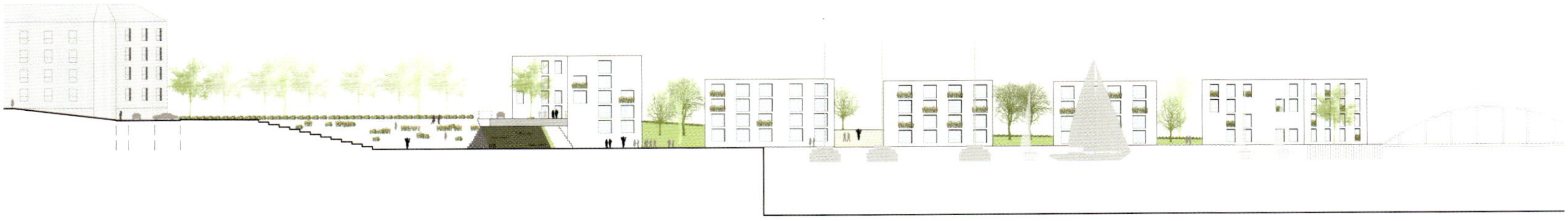

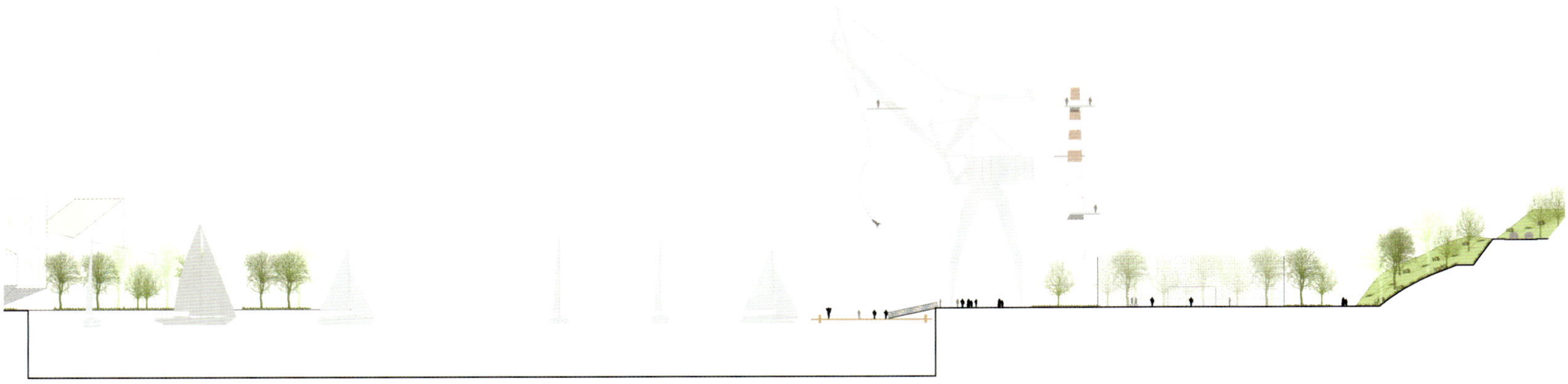

The 70 ha site at the 'Bahia de Pasaia' (Bay of Pasaia), San Sebastian is currently used for harbor activity and will gradually become available for redevelopment. The project is considered as a key opportunity for the surrounding communities which are in need of regeneration and will have an important impact on the entire region.

The masterplan defines an archipelago of varied waterfront developments, setting broad conditions for local involvement of the different communities. The urban plan provides the local communities with space to grow and possibilities to extend their identity and exploit their diversity. The growth is concentrated around public transport hubs. The different character zones are unified by a public space network around the docks and along green boulevards. Within this robust structure, a sensitive network of public spaces is formed, allowing current and new inhabitants to enjoy the water for leisure and recreation. Around the bay, new landmark buildings such as a cultural center and a maritime museum will become regional destinations. Clusters of new economic activities will generate new employment in a broad range of sectors.

School for Planning and Architecture

新德里规划与建筑学校

Hackenbroich Architekten

International Competition, 2009
2009年国际竞赛参赛方案

Credits
Client: School for Planning and Architecture
Location: New Delhi India
Area: 50,000 m²

Team: Hackenbroich Architekten: Wilfried Hackenbroich and Rainer Mühr with Deborah Rowe and Dalia Butvidaite Local Architekt: Sumit Diwan, Delhi India

四通八达的地点，位于新德里最优秀的自然生态场所之一，为把生动的人文经历体验与壮观的自然生态环境结合起来，提供了独一无二的机会。这种自然生态环境从宏观层面与微观层面影响着该地点。从较大的范围来讲，该地点与德里岭一起作为一个整体，提供了一处广阔的自然风景区，并且其本身延伸到实际地点。从较小的范围来讲，这种延续性为该地点提供了一个独特的条件，自然的"峡谷"。这条峡谷在该地点的布局中发挥重要作用，因为它在大自然中，提供了一处休闲娱乐和社区场所，创建了规划与建筑学院的标志——人文校园。

为了新校区能够达到这种浓重的人文感，建筑彼此之间就近放置，能够方便地访问所有设施和其他建筑。各种功能与活动并存，使得各院系保持着与公共设施以及宿舍之间舒适的距离。景观和建筑之间以相同的方式连锁，提供了住宅公园、人文广场和中央自然区——峡谷。

构建方式包含两条原则。一条适用于住宅类建筑，封闭的立面以及窗户和阳台不规则的排列形成一种开放的模式。第二条原则被发展成适用于院系部门、公共设施以及带有玻璃幕墙，或者是更加封闭的立面、垂直元素和垂直窗户条纹的康乐建筑。玻璃幕墙用于沿着人行街道的大多数城市区域，位于街道上方的遮日光屏保护建筑立面免受阳光直射。这种建筑的开放性使得院系部门与公共设施之间的关系更加亲密，并在他们之间创造一种流体空间。建筑立面的垂直元素被用来保护建筑免受阳光直射。

除了建筑立面以外，建筑底层楼面在校园的生动性中也发挥了重要作用。每一幢建筑拥有宽敞的底层楼面空间，也向行人发出了邀请。该底层楼面是人行街道的延续，并且提供了一种被相关计划激活的交互区。

作为最优秀的可视元素之一，遮日光屏是由两个2米长2米宽，白色的混凝土预制构件，连续拼装接合而成，从而形成实际的遮阳结构（它是本土的印度贾（Indian Jali）的现代翻版）。这块遮日光屏与建筑相连接，而且在需要的地方被白色的混凝土柱支撑起来。

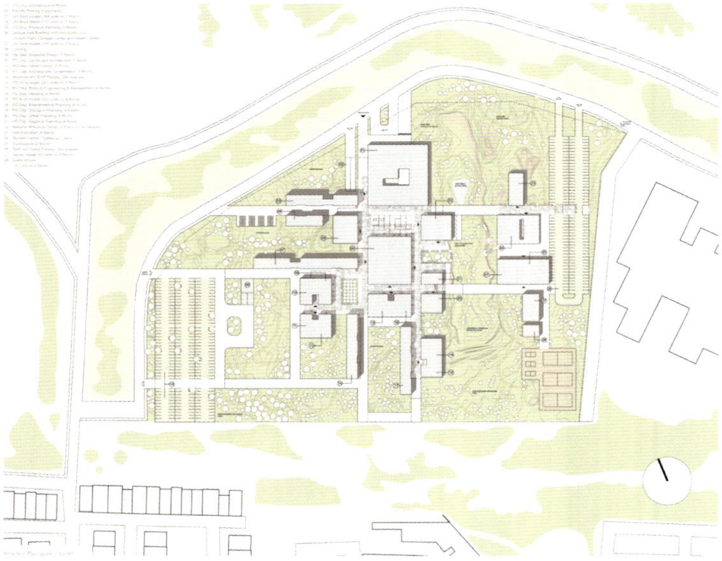

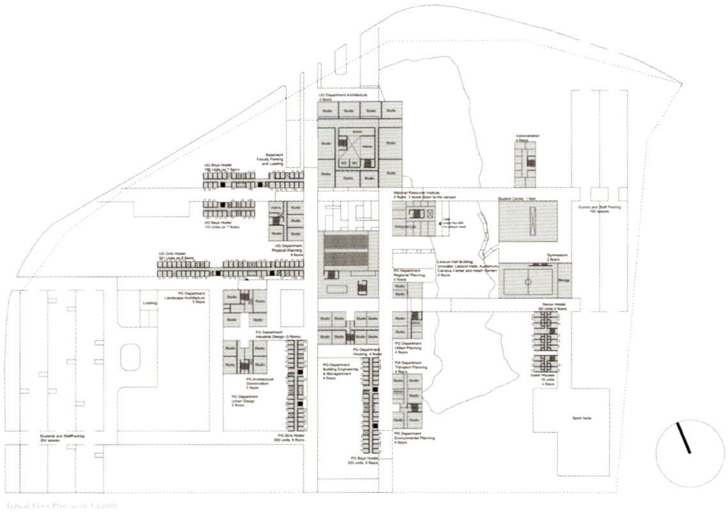

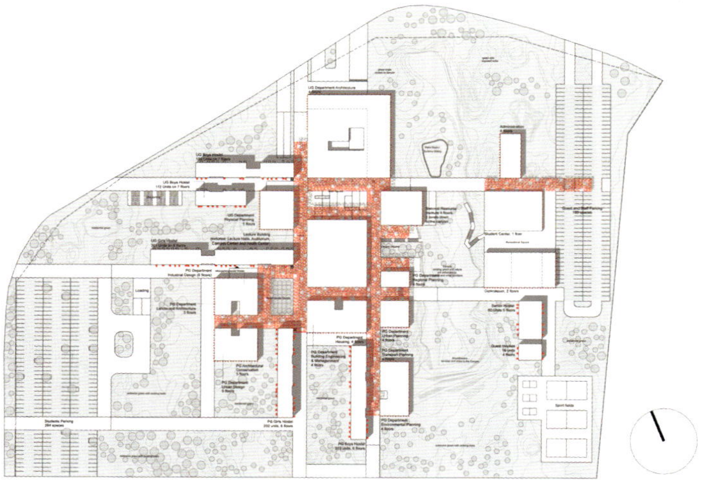

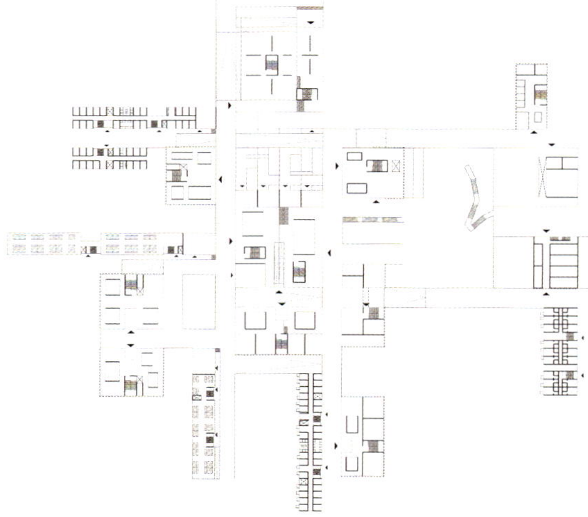

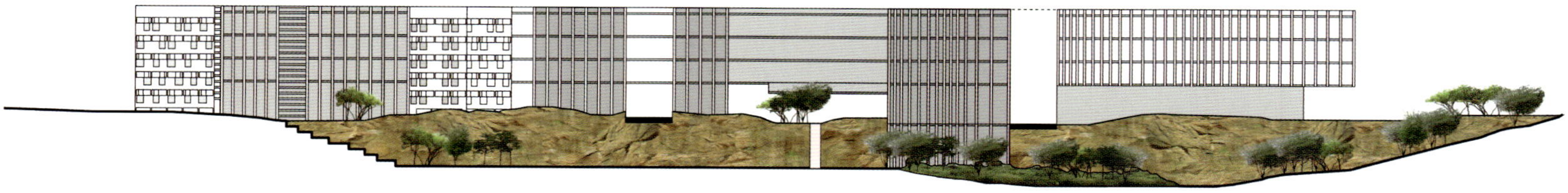

Canyon elevation

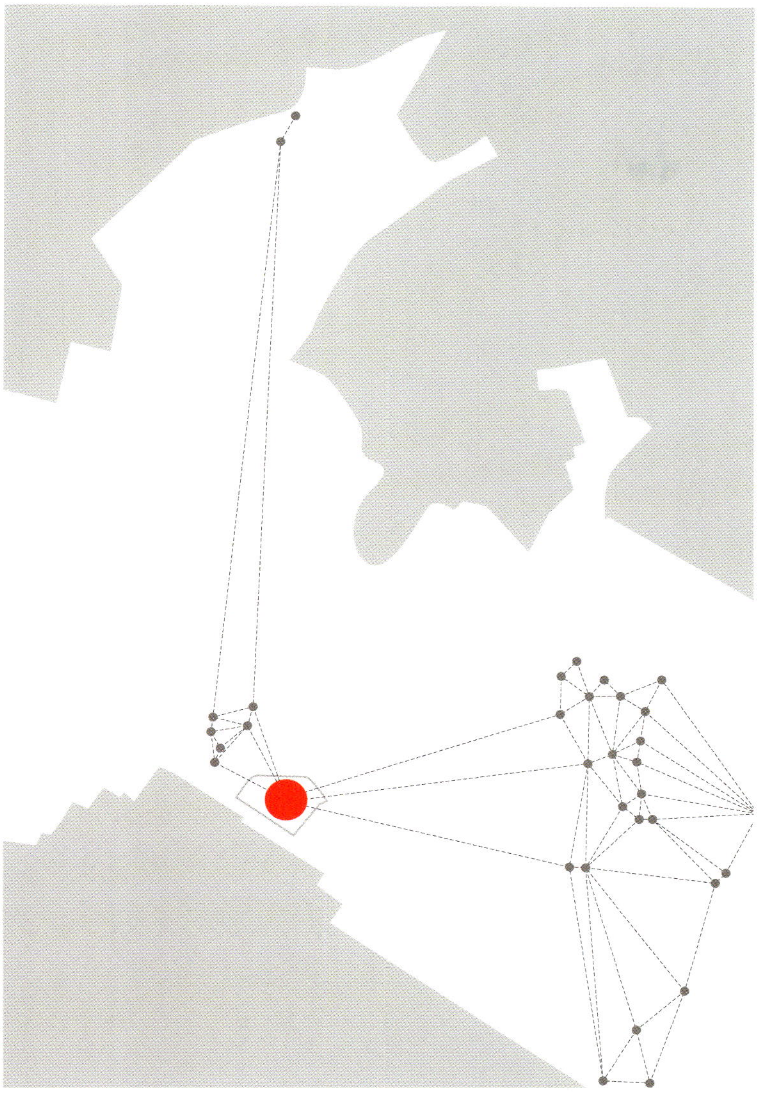

Network of education institutions

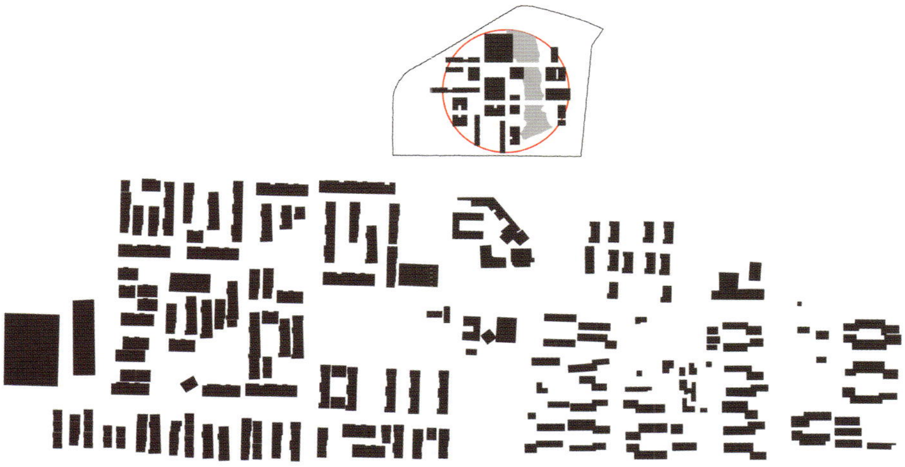

Context typologies II — buildings at the street

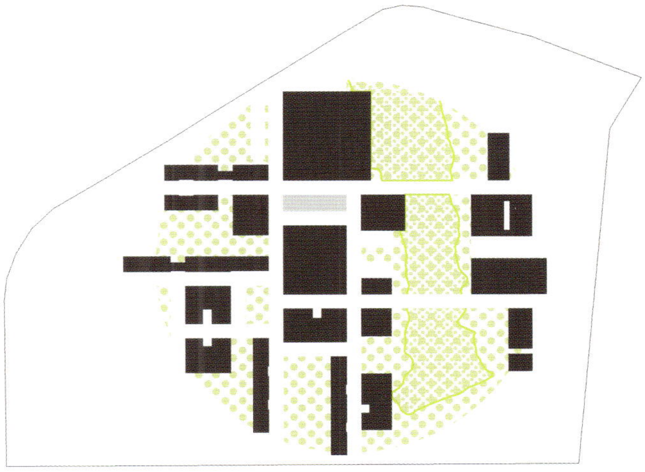

Urban Campus

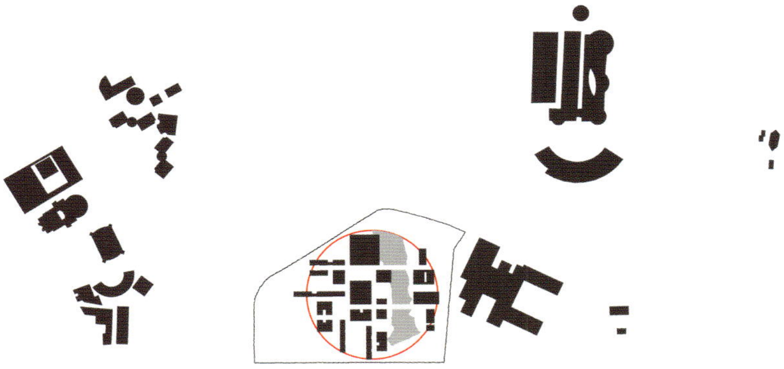

Context typologies I — buildings on the landscape

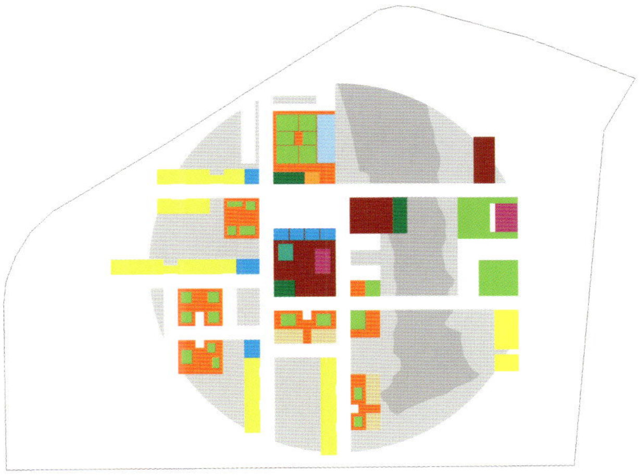

Interlocking functions

The well connected site, located in one of New Delhi's most outstanding natural settings, provides a unique opportunity to combine this vital urban experience with a spectacular natural environment. This natural environment influences the site on macroscopic as well as microscopic levels. On a broader scale, the site provides an extensive natural recreation area with the Delhi Ridge as a conglomerate which continues into the actual site itself. On a smaller scale, this continuation provides a unique condition on the site, the natural "canyon". This canyon plays a major role in the site layout as it provides a unique recreation and community area and creates the identity of the S.P.A. - the urban campus in nature. To achieve this urban intensity for the new campus, the buildings are placed proximal to each other, providing convenient access to all facilities and to each other. The various functions and activities are juxtaposed so that the different departments maintain a comfortable proximity to the common facilities as well as the housing. In the same way the landscape is interlocked with the buildings, providing residential parks, urban squares or the central nature area, the canyon. The build form consists of two principles, one for the residential buildings which have closed facades with windows and balconies arranged irregularly to form a pattern of opening. The second principle is developed for the departments, common facilities and the recreation buildings which have either a glass facade or a more closed facade with vertical elements and vertical window stripes. The glass facade is used in the most urban areas along the pedestrian boulevards where the facade is protected from direct sunlight by the sun-screen above the boulevards. This opening of the buildings allows for a more intimate relation between the departments and common facilities and provides a fluid space between them. The façade's vertical elements are used to protect the building from direct sunlight as well as enhancing the experience of being inside the campus and outside of it. Besides the facades of the buildings, the ground floors of the buildings play a major role for the vitality of the campus. Each building has a generous ground floor that provides an inviting space to the pedestrians. This ground floor space is the continuation of the pedestrian boulevards and provides an exchange zone which is activated by related programs. The sun-screen as one of the most outstanding visible elements is a construction of two precast white, concrete elements of 2 by 2 meters which are repeated and added to form the actual shading structure (It is a re incarnation of the indigenous Indian "Jali"). This sun-screen is connected to the buildings and supported where needed by white, concrete columns.

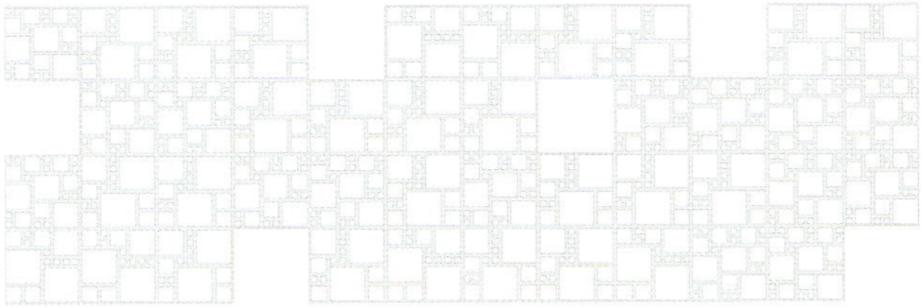

Sun-screen pattern

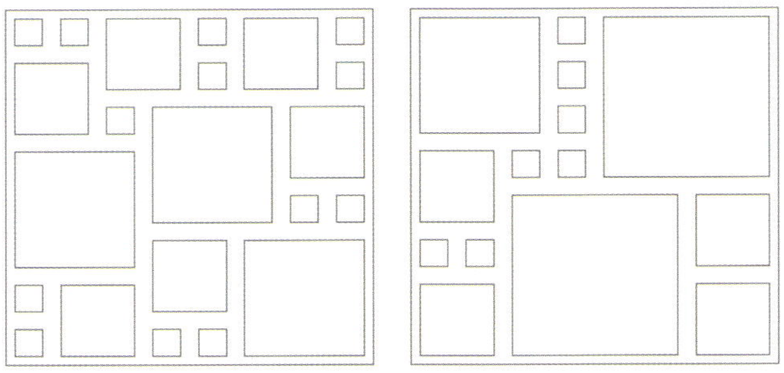

Sun-screen elements

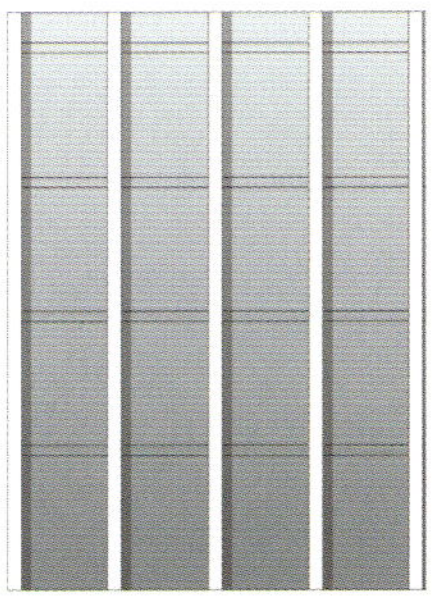

Vertical elements (2,5 m grid)

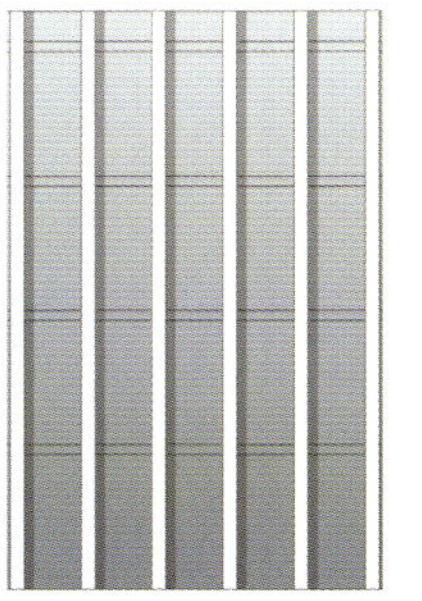

Vertical elements (1,5 m grid)

Vertical elements (1 m grid)

Facade of the Undergraduate Department of Architecture

Ujpest City Center

Ujpest 市中心规划

Studio 3LHD

Credits
Year: 2008
Program: retail, public, culture, business, leisure, sport, residential, urban planning
Address: Ujpest, District 4
City: Budapest
Country: Hungary
Type: International Competition

Geolocation: 47°33'49" N, 19°4'48" E
Site Area: 23,133 m²
Gross Floor Area: 120,860 m²
Volume: 473,181 m³
Footprint: 19,352 m²
Level: 109 m

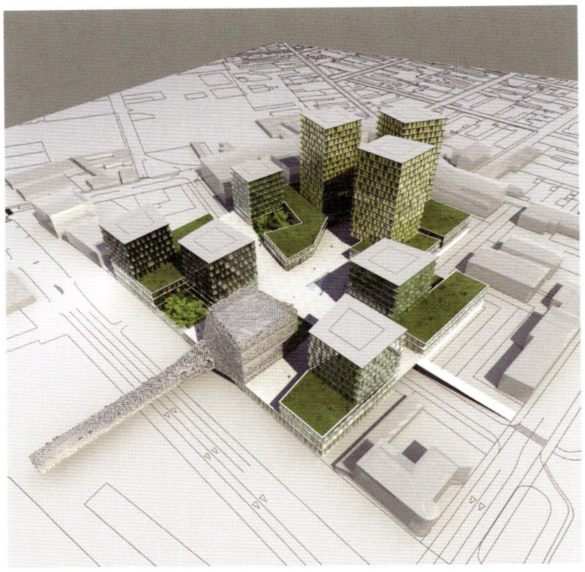

新开发的伊什特万镇中心屹立在多瑙河滨，沿北佩斯、Váci 路、Szabadaság公共花园和Újpest海湾的主干道而分布。

理念

通过对所有城市矩阵和现有(新老)建筑的仔细分析，设计师试图找到通向邻近地带所有可能的连接点。19和20世纪的城市矩阵是此次理念形成的基础。庞大的综合设施，所有关联的配备诸如交通、人行道、现有和既定的干道不仅使所有底层和地下室的活动连接为一体，而且也成为城市新风貌纵向元素展现的完美基地。这里的建筑群由九座不同高度和功能的塔楼组成。他们高耸于下沉广场、小广场和高架街道的上方，连接了所有邻近地点和建筑物。从城市不同角度观察的各式各样的轮廓尽量弱化大建筑群的整体效果，以应对不同的背景条件。

区域划分

九个组成部分——纵向建筑根据行人走廊和景观被精心地安排，以考虑现有建筑物的阴影。该地带中部是三座住房，高度是GF+16；两边边缘地带是高度在GF+7的商务楼和酒店，其中专门打造的第九空间，是公共集会、展览、餐馆和通向多瑙河长廊Ujpest湾公园的新人行天桥的所在。塔楼之下，海拔+109米处(李斯特胡大街)是公共设施，如餐馆、购物中心、画廊和咖啡馆。海拔+104米处（与瓦西路水平）是综合体育设施（健身和游泳池）的主要入口，也是酒店和会议中心的入口，同时还是地下室停车场的入口。一个两层车库位于酒店和体育中心下面（海拔+95米和+98米），还有一个位于零售中心下面（海拔+101米和+104米）。

使用这种方法，最大密度和宽广空间的要求都得到了满足。新的空间都有足够的间距以使每个空间都享有良好的光照和景色，从而使整体的空间得到了高质量的组织。对所有纵向建筑的规划都非常灵活，以便未来发展时在不改变城市矩阵的情况下，也能根据客户不同的需求转化成不同的功能。

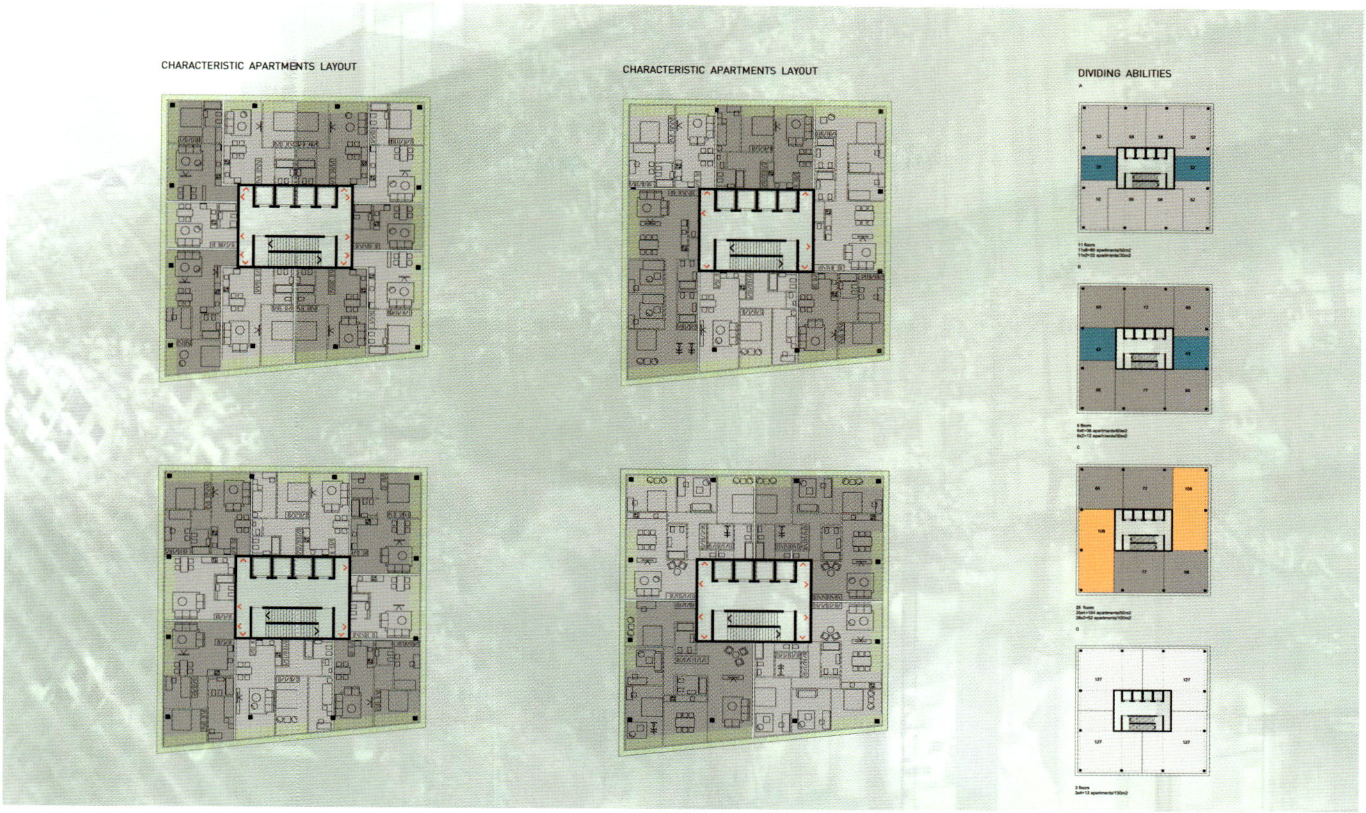
CHARACTERISTIC APARTMENTS LAYOUT
CHARACTERISTIC APARTMENTS LAYOUT
DIVIDING ABILITIES

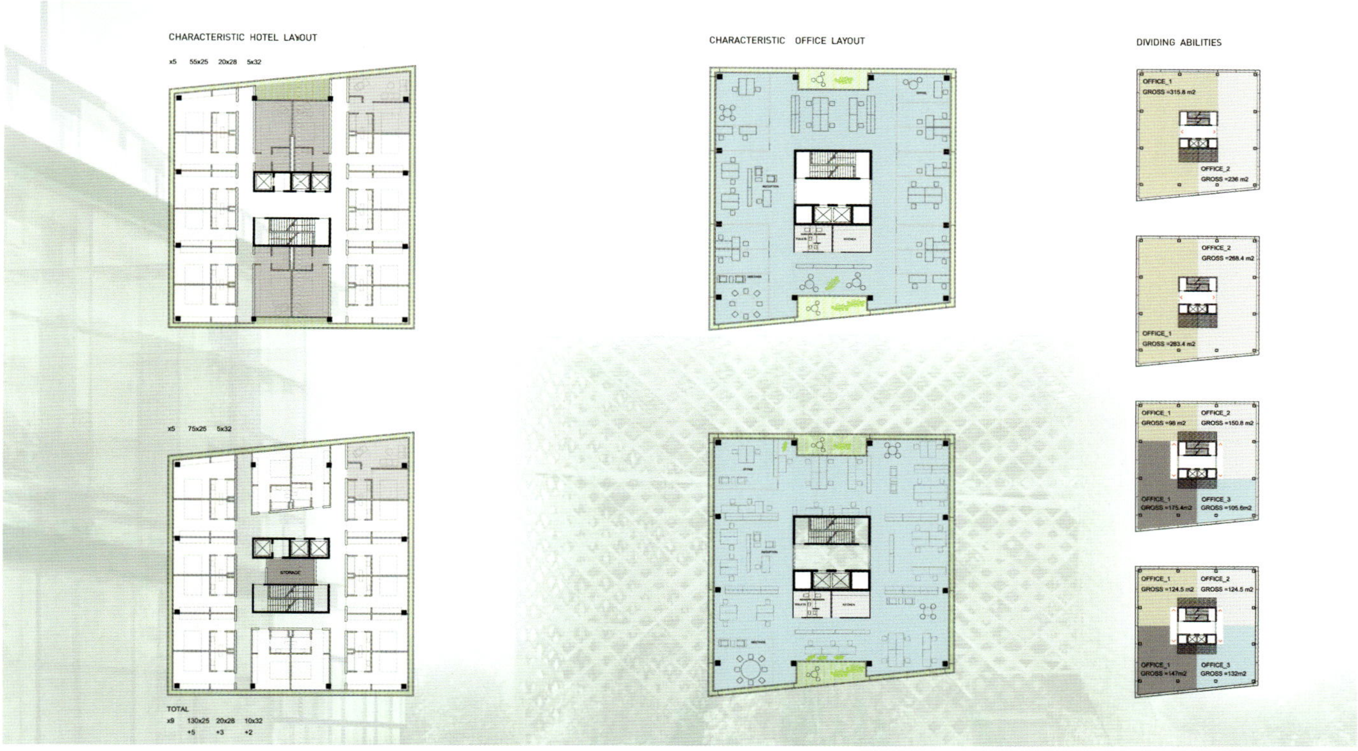
CHARACTERISTIC HOTEL LAYOUT
x5 55x25 20x28 5x32
x5 75x25 5x32
TOTAL
x9 130x25 20x28 10x32
+5 +3 +2
CHARACTERISTIC OFFICE LAYOUT
DIVIDING ABILITIES
OFFICE_1
GROSS =315.8 m2
OFFICE_2
GROSS =236 m2
OFFICE_2
GROSS =268.4 m2
OFFICE_1
GROSS =283.4 m2
OFFICE_1
GROSS =98 m2
OFFICE_2
GROSS =150.8 m2
OFFICE_1
GROSS =175.4m2
OFFICE_3
GROSS =105.6m2
OFFICE_1
GROSS =124.5 m2
OFFICE_2
GROSS =124.5 m2
OFFICE_1
GROSS =147m2
OFFICE_3
GROSS =132m2

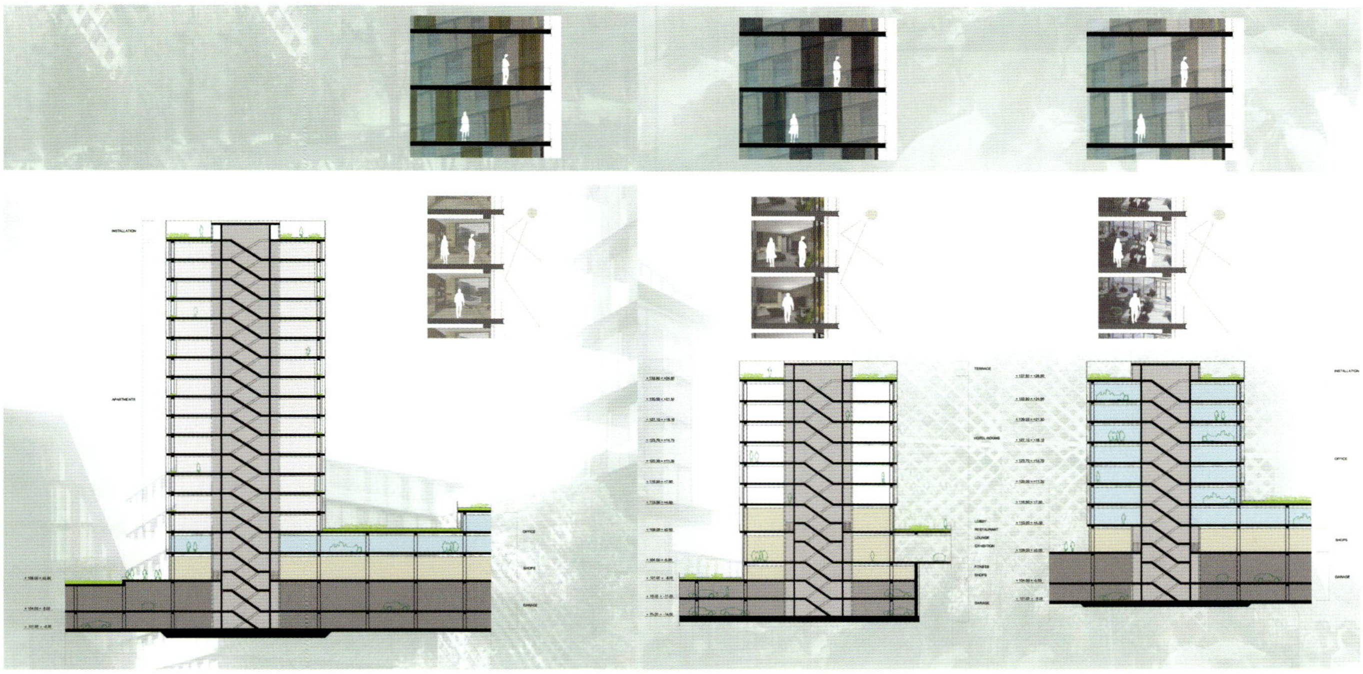

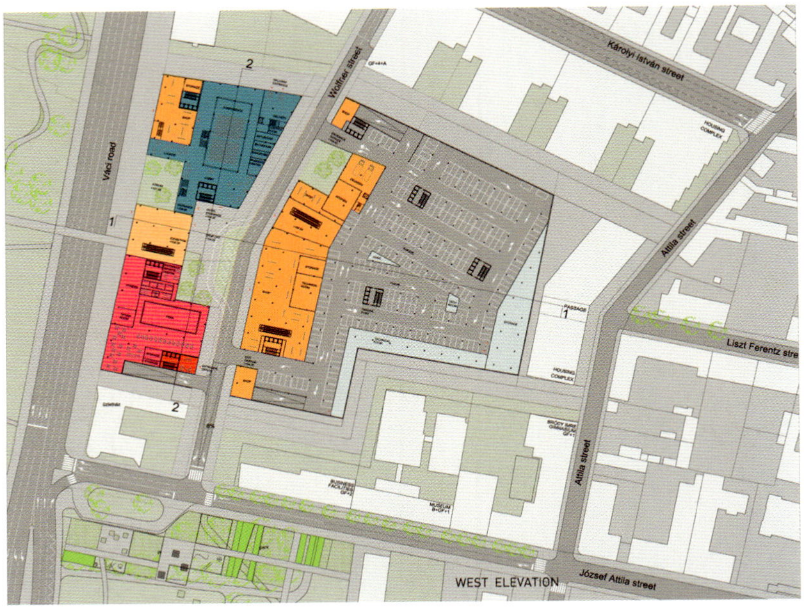

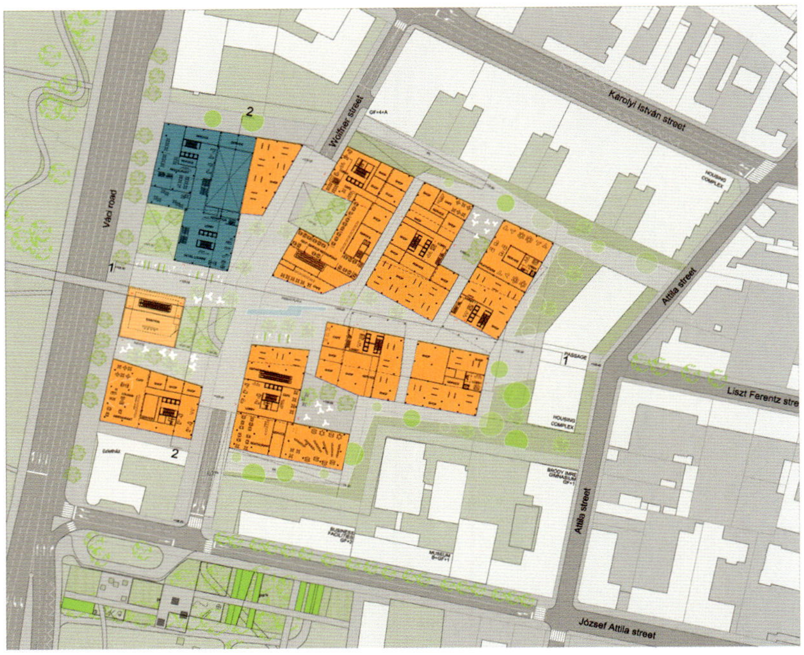

The new Károlyi István Town Center is positioned along the most important main road of the north Pest, Váci Road, Szabadaság public Park, and the Újpest Bay on Danube promenade.

Concept

Careful analysis of all urban matrices and existing (old and new) buildings, we try to find all possible connections to the neighboring sites. Urban matrices of the 19th and 20th century form the base for the development of the concept. The main sockle of the complex with all correlations; traffic, pedestrian routes, existing and given facts is not only main connector of all ground floor and basement activities but also perfect base for vertical elements of the site that will form New Townscape in the city. Buildings are compositions of nine towers with different heights and functions. They all rest on the complex basement with sunken plazas, small squares and elevated streets witch connect all neighboring sites and buildings. Various silhouettes seen from different points in the city try to reduce the overall effect of the big mass and to respond to the different contextual conditions.

Zoning

The nine elements - verticals are carefully positioned according to the pedestrian corridors and vistas, with taking care to the shadowing of the existing buildings. In the middle of the site, three verticals are housing facilities with height GF+16, and on the two edges verticals are business and hotel with height GF+7, with the special ninth volume that is dedicated to the public functions, exhibitions, restaurants and new pedestrian bridge that leads to the new Ujpest Bay park on the Danube promenade. Below the "towers" on level +109 m over the sea level (Liszt Ferenc street level) there are public facilities such as restaurants, shopping, galleries and cafes. On level +104 m over the sea level (Vaci road level) is main entrance in the complex with sport facilities (fitness and swimming pool), entrance to hotel and conference center, and entrances to the parking facilities in basement level – two floors garage below hotel and sport facilities (+95m and +98m over the sea level) and two floors garage below retail (+101m +104m over the sea level).

With this method requested maximum density and gross square meters is fulfilled. New volumes are positioned on the enough distance so that every volume has good light and vistas and they enable quality organization of space. Plans of all verticals are very flexible so in the further development, without changing urban matrices, they can be transformed into the different function, according to the client's needs.

SHOPPING

Ex Fonderie Riunite

工业建筑重建及其区域规划 Modostudio

Concorso internazionale di progettazione, winning project
国际建筑设计竞赛获奖项目

Credits
Location: Modena – Italy
Client: Comune di Modena
Design Team: MODOSTUDIO - Cibinel Laurenti Martocchia architetti associati + CCDP + Studio Cattinari
Areas: Faculty of design-DAST and services 23.000m², commercial and offices 2000 m², residential 8000 m²

Use: faculty of design + commercial + residential
Structure: mixed structure
Parking Plots: 1,050
Stories above the Ground: 8 floors
Stories under the Ground: 3 floors

该设计方案是对意大利Ex Fonderie Riunite地区的重建，这个地方是摩德纳城具有重要历史价值的地区。其基本的设计理念源于原有工业建筑楼重建及其所在区域的详细规划。

这个原有的工业综合体以其具有历史意义的办公楼为特色，设计保留了其作为意大利工人运动标志的地位。重建加入了新的功能服务区域与科技艺术设计中心（DAST）。

新的大楼将按照原有工业建筑的规划结构设计，并保留原有工厂设计的优越性。这些新的大楼将独立于其他建筑。同时，其北立面、南立面与屋顶外表面都将覆上一层金色穿孔金属板，使建筑物屋顶别具一格。

新建的金色房顶、金属材料的粗糙与金属板穿孔图案的温柔，都将激发人们回忆过去；充满历史回忆的金属楼，却又到处盛满现代的新特色。

金属的覆层在铁路上，甚至是连接城市两个分区的桥上就能看得见；很显然，它将成为摩德纳城市的象征。DAST的文化性目标，比如设计学院，是给建筑加入新的功能、使其作为研究创新的重要地方——这个综合体将加入商业与服务区域，使其立于城市中心重要位置。

原有建筑物外面设计了一个广场，可以作为商业、住宅与公园的综合体，最大化了此规划新功能的可利用性。

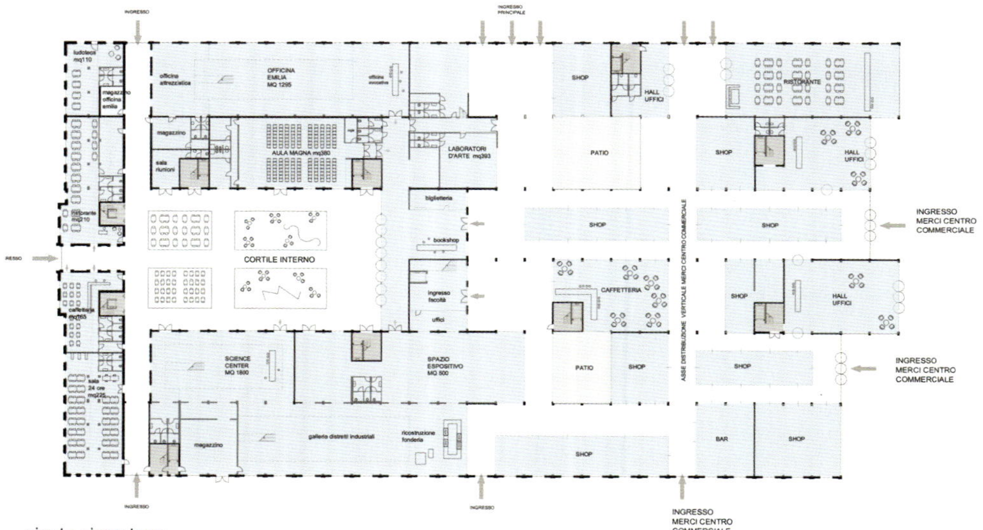

pianta piano terra

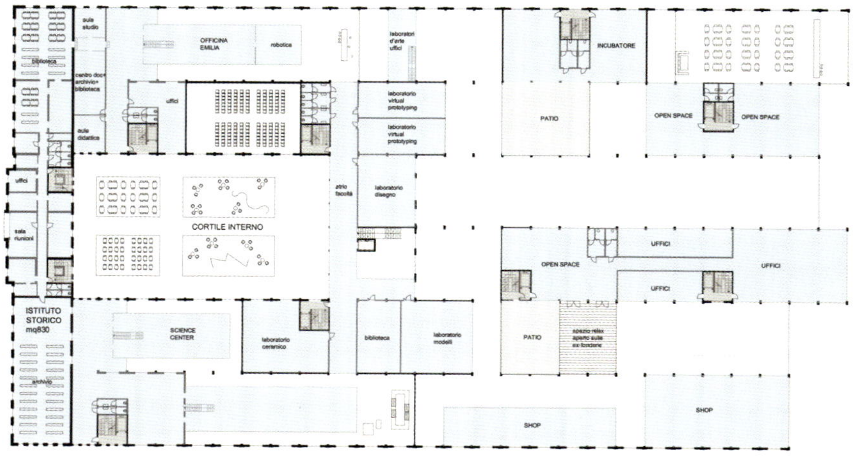

pianta piano primo

dast

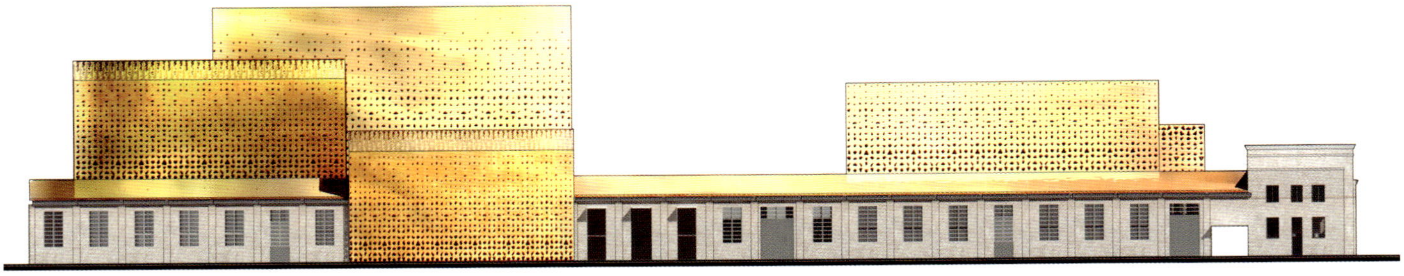

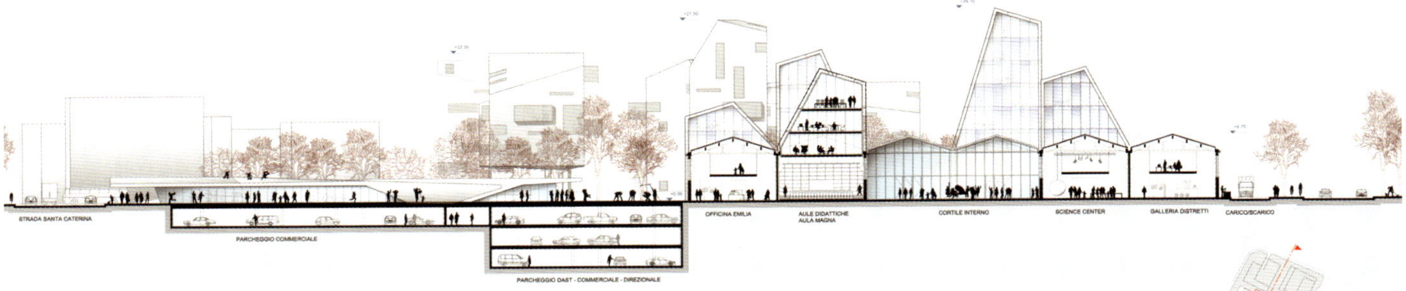
STRADA SANTA CATERINA
PARCHEGGIO COMMERCIALE
PARCHEGGIO DAST - COMMERCIALE - DIREZIONALE
OFFICINA EMILIA
AULE DIDATTICHE
AULA MAGNA
CORTILE INTERNO
SCIENCE CENTER
GALLERIA DISTRETTI
CARICO/SCARICO

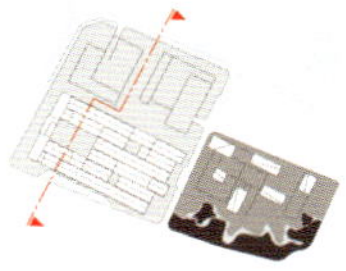

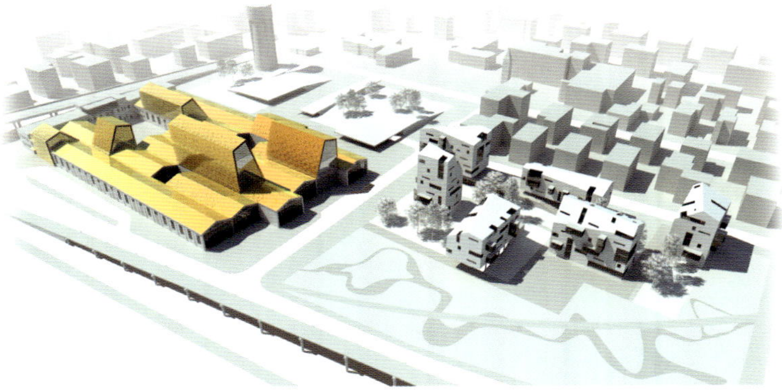

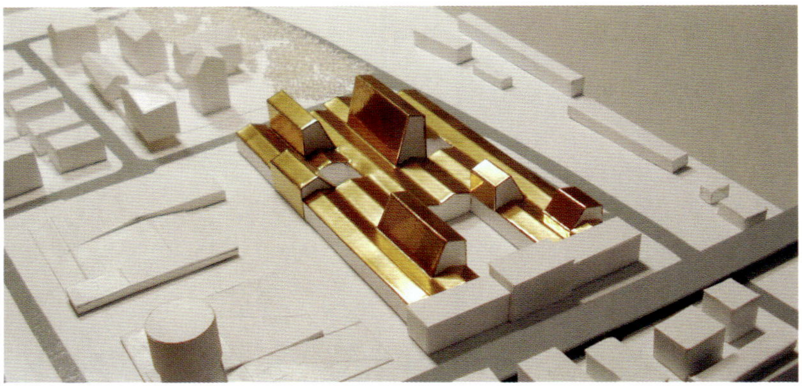

dast

schema delle funzioni

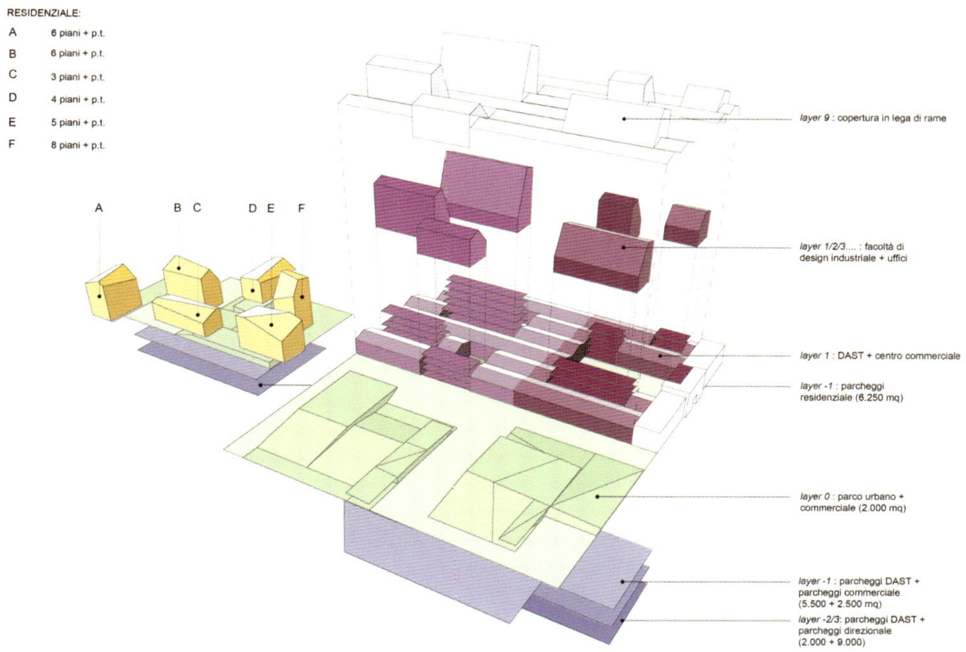

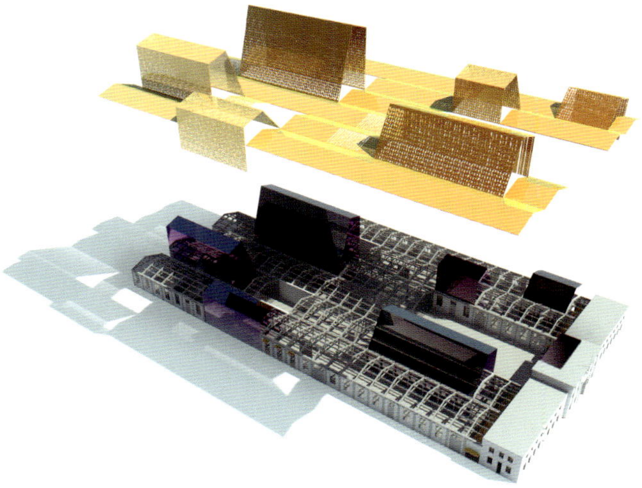

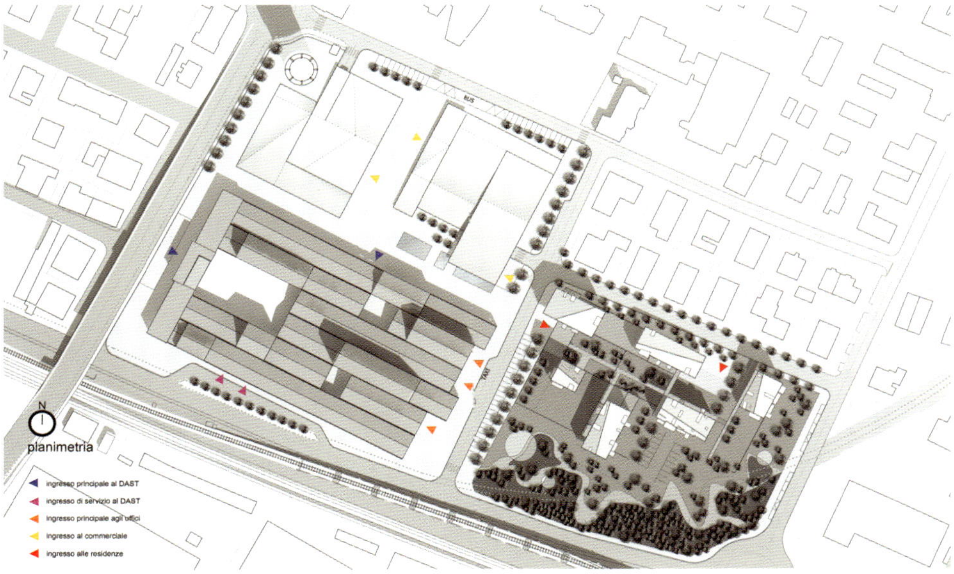

The competition brief was the regeneration of an urban area called Ex Fonderie Riunite, a strategic area for the city of Modena with important historical value. The basic idea was the recovery of a former industrial building and a new detailed urban plan in the other areas of the competition.

Our proposal for this existing industrial complex, characterized by historically protected office buildings, was to keep the identity as a symbol for the history of the Italian workers movement. The regeneration includes new uses and functions and the new spaces of the DAST faculty (Design, Art, Science and Technology).

We proposed to create buildings of different heights with plans that follow the structure of the industrial building and preserve the existing fascinating spaces of the old plant. These new volumes will be detached from the ground and they will be covered, on the north and south facades with a golden perforated metal sheet. This particular material is the same that will characterize the exterior skin of the roof throughout the rest of the building.

The golden colour of this new roof, the harshness of the metal material and at the same time, the softness of the perforated pattern will be able to evoke the past; a metal smelting complex building full of history, and at the same time to communicate unambiguously the birth of the new feature.

The cladding will be seen from the railroad and from an adjacent bridge that connects the two parts of the city and will become a symbol for the city of Modena. The cultural destination of DAST, such as the School of Design, will give a new feature to the building as a vital place pushing research and innovation. The complex will also contain other features such as commercial areas and services in support of the DAST, creating an important center for the city, available throughout the day.

The new urban design provides maximum accessibility for new functions with the creation of a square outside the former building, which will become a commercial, residential and public park.

Harbour Stones

石通港湾 C. F. Møller Architect

First Prize in Invited Competition in 2010
2010年获邀请赛一等奖

Credits
Client: Norra Älvstranden Utveckling AB, PEAB and Stadsbyggnad-skontoret in Gothenburg
Architect: C. F. Møller Architects
Landscape Architect: C. F. Møller Architects
Area: 70,000 m² (approximately 400 living units)
Location: Lindholmen, Gothenburg, Sweden

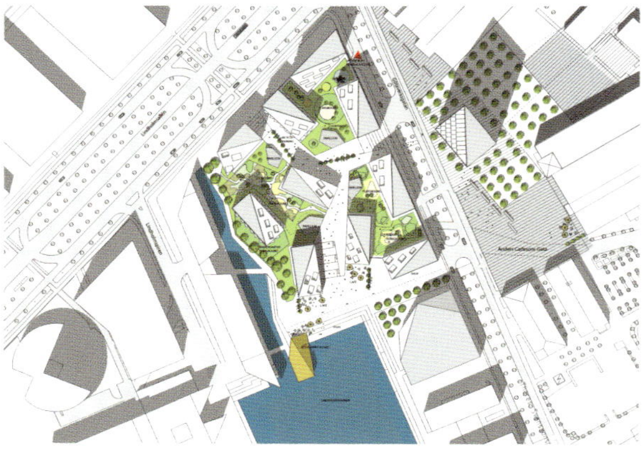

该项目是为位于哥德堡老工业港口的临德洪门区设计建造新居住区。临德洪门区曾经遍布的工业建筑和船坞，已经改造为大规模的商业区和开放广场。主要规划的提案，通过推敲大楼的方位、几何形状以及透明度，介绍了一种人性化的、不同的、关注于小规模城市空间的建筑。

方案包含10个大楼的雕塑空间，因而得名石头港湾，高度从4层楼到10层楼不等。方案同样包括一幢约22层楼高、将作为地标性建筑的大厦。设计使得大楼能最大程度利用太阳光照，提供了欣赏港口风景的视角，并且其循环用铝制的建筑立面体现出当地的造船历史。

对外部来讲，主要城市特点得到了延续，新区通过两条作为中央大道的公共阶梯和第三条通往瀑布的景观化阶梯，从内部敞开来。这创造了鸟瞰整个港口、和煦的、有屏障微气候的空间，同时为每一位居民开阔了视野。

C. F. Møller建筑事务所同另外三家实践事务所一起赢得了竞赛奖项，它们是Helen & Hard (NO)建筑事务所、Kjellander & Sjöberg (SE)建筑事务所以及KOD arkitekter (SE)建筑事务所。

DET GRÖNA STRÅKET
LÄGENHET
BUTIK

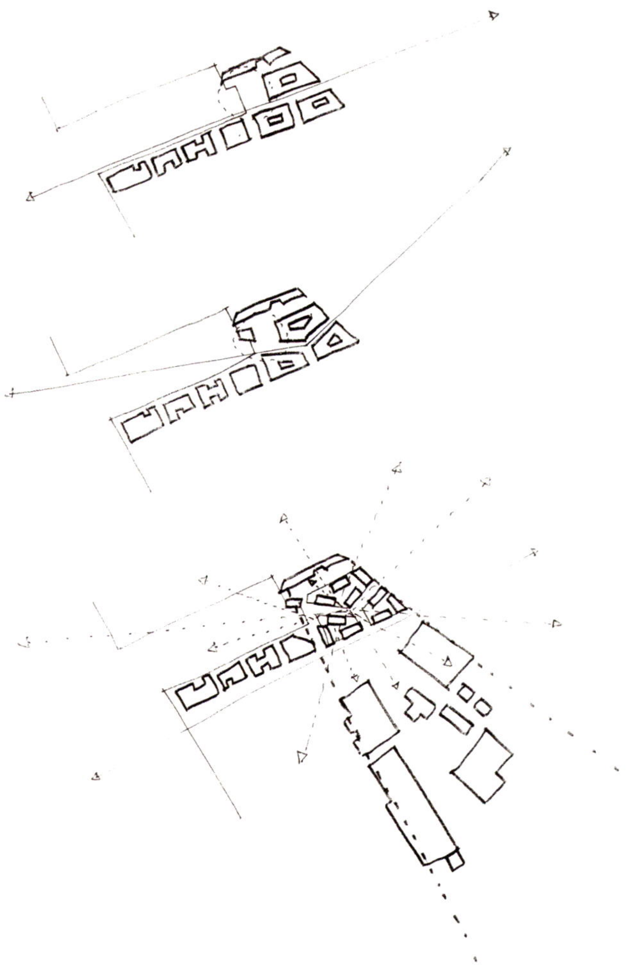

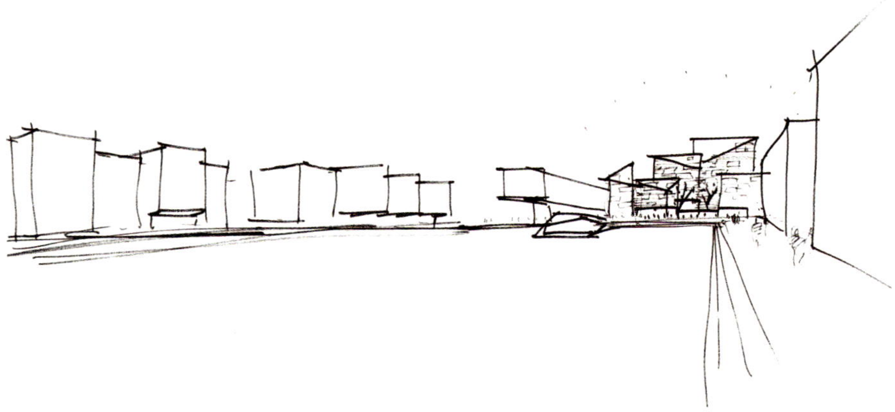

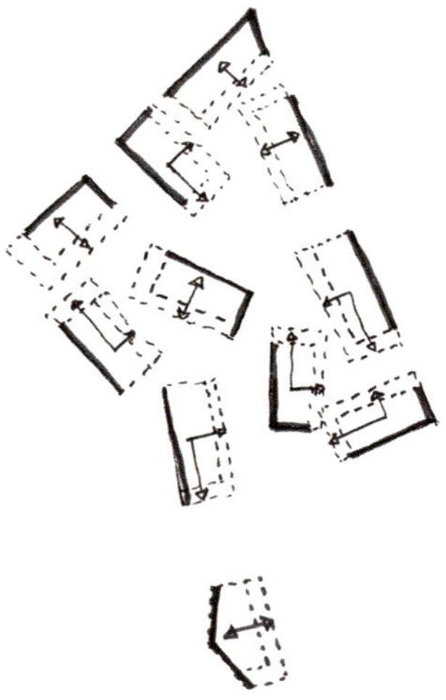

Building depths can be varied according to functional content without changing the overall structure

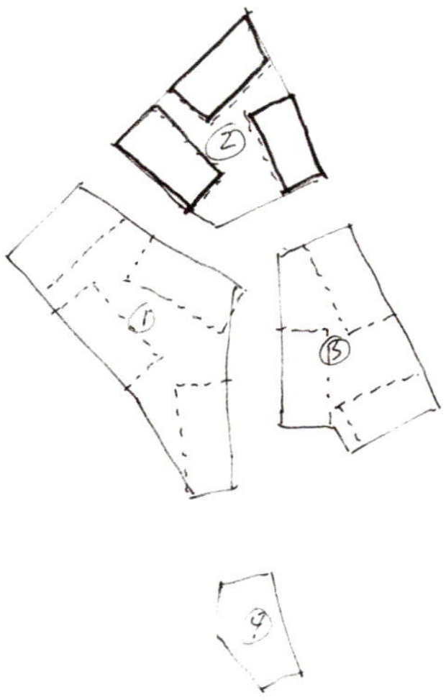

Phasing options

The task has been to design the master plan for a new housing district in the Lindholm area of Gothenburg's old industrial port.

The Lindholm area, dominated by former industrial buildings and shipyards, has been transformed into large-scale business districts and open plazas. The master plan proposal introduces a humane scale and a different architecture focusing on the small-scale urban spaces within the scheme by refining the positioning, geometry and transparency of the buildings.

The scheme consists of 10 sculptural building volumes, hence the name Harbour Stones, varying in height from four to ten storeys. The scheme also includes a tower of approximately 22 storeys which will rise as a landmark. The buildings are designed to maximise the potential for sunlight and views of the harbour, and their facades of re-cycled aluminium make a strong statement with reference to the ship-building history of the site.

Towards the exterior the major urban traits are continued, but the new area opens up inside by means of two public stairs acting as thoroughfares and a third landscaped stair to the waterfront. This creates a micro-climate with sunny, sheltered spaces directly overlooking the harbour, and also opens views out for every resident.

C. F. Møller Architects won the procedure in competition with three other practices: Helen & Hard (NO) Kjellander & Sjöberg (SE) and KOD arkitekter (SE).

A, 2 corner units + 4 single-aspect units

A, 2 corner units + 2 single-aspect units - balcony variation

A, 2 dual-aspect + 4 corner units

A, Ground floor living units

C, Long block (Housing units)

D, Narrow block (Housing units)

E, Office space

B, Student Housing units (narrow core)

B, Student Housing units (wide core)

Khedivial Cairo: An Evolved Metabolism

古都开罗的新城代谢

University of Arkansas Community Design Center

Urban Harmony: Opera and Ataba Square
城市和谐：歌剧与阿塔巴广场设计竞赛参赛方案

Credits
Location: Cairo, Egypt
Design Year: 2010
Architect: University of Arkansas Community Design Center

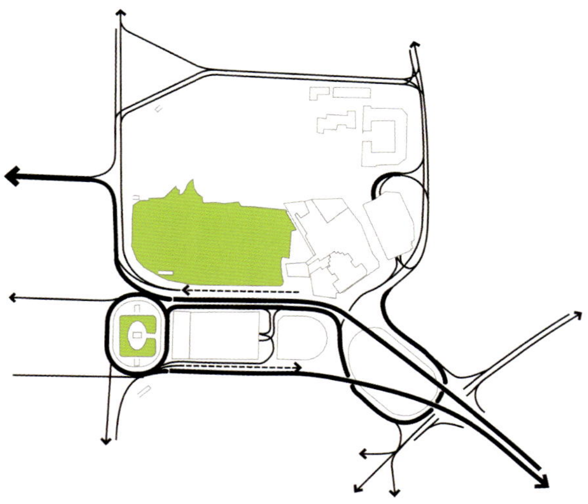

Current Traffic Solution

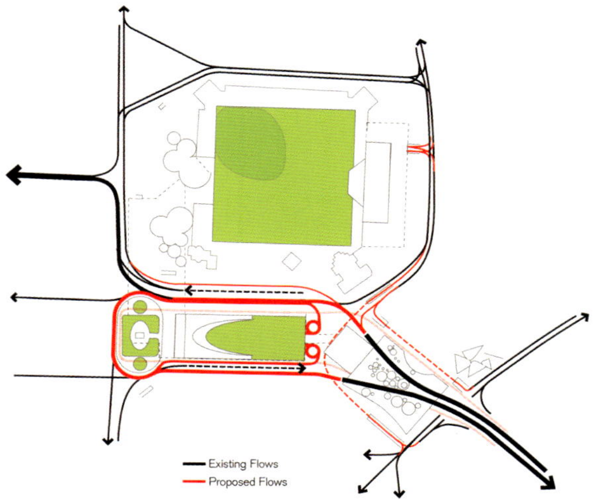

Proposed Traffic Solution

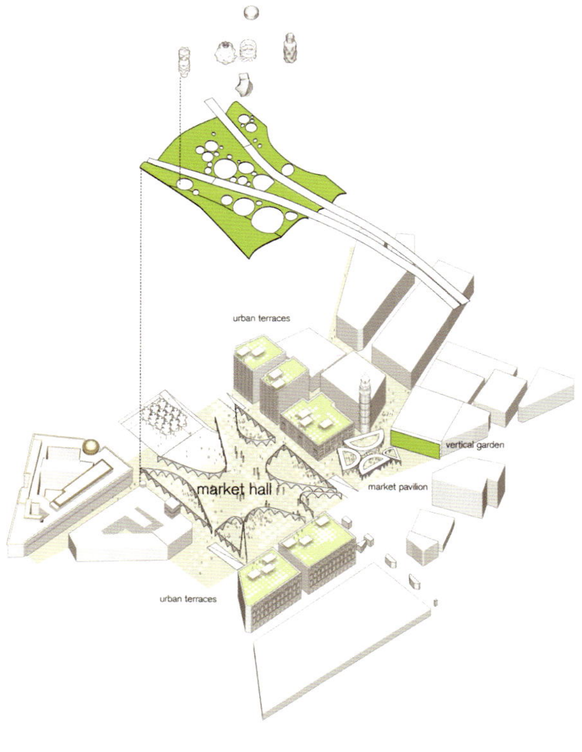

开罗如今有1 700万的人口，而目前的城市结构却是在其人口仅为3.5万时规划的。随着社会以及经济结构、交通、城市生态以及可意象性的发展，城市的新城代谢也随之进化。在19世纪，城市空间的规划注重的是适应性再利用，以应对现代城市的密集型发展。设计师提出的方案是设计一种生态型城市空间，实现这个19世纪城市系统与实用性城市新能源的和谐一致。这是一项艰巨的任务，高架公路、拥堵的公路、广阔的停车场以及民间市场被融为一体，成为一个新的开放式系统，使城市能源的利用达到最优化。19世纪的城市DNA与21世纪的竖向城市秩序相结合，实现了资源在四种新颖空间中的流动：1）扩展的歌剧广场；2）阿塔巴广场下的市集广场及立交桥；3）复合型林荫大道；4）Azbakiyya花园的大草坪。四种空间为这个可持续性的新开罗提供新的生态服务以及城市服务。

扩展的歌剧广场

歌剧广场新扩展了一片开放的草地，设有户外表演场地、入口地标、多式联运枢纽、地下停车场。入口处保留原有歌剧院的网屏设计，使广场东缘的城市属性失而复得。入口不仅形成草地上的表演舞台的背景幕布与工作台，同时也提供了通往下面的地铁站与停车场的通道。歌剧广场景观建立起一种新的城市后盾，与大开罗的新的社会能源相适应。

阿塔巴广场下的市集广场及立交桥

立交桥的建立对减缓交通压力来说是非常有必要的，于是阿塔巴广场表面被重新规划，下面形成一个带顶罩的人行区域，作为繁忙的市集活动区，上面的立交桥取代了之前的公路结构，显得更为文明。这样一种结构在开罗任何地方都可以见到。它包含了下面的气候友好型人行道景观以及上面的机动车景观，绿色屋顶及通风塔形成城市天窗。作为一种代表性的城市解决方案，这种结构多层化了城市空间，使同一空间中相冲突的结构功能得以分化。

复合型林荫大道

一个新的复合型林荫大道将供各类交通方式（公共汽车、地铁、汽车、步行）通行的道路综合在一起。人行桥和隧道相互交织，将歌剧广场与Azbakiyya花园联系在了一起。林荫大道弱化了热岛效应，同时通过公交站的迁移与新的电车服务的设立便利了多式交通服务。这样一种街道形式呈现的是资源丰富而高效的交通规划，与开罗的城市规划形成互补。

Azbakiyya花园的大草坪

在7月26日街道轴线的尽头是一个新的大草坪，不仅为人们提供了一个远离城市亲近自然的空间，更体现城市边缘文化项目的细致规划。草坪的西北象限上有一座大山丘，从上面可以俯视歌剧广场和阿塔巴广场，同时也使得花园外面的人能够欣赏到这片绿色景观。花园周围的景色充分展现了阿拉伯与西方景点规划传统。最令人称赞的阿拉伯花园中常见的水景，与相对的制冷设施互相配合运作。花园边缘实现了喧闹的街道与宁静的花园之间的完美过渡。边缘上设有地铁设施，西侧的林荫广场上还有一个图书市场。

绿地、草坪、复合型林荫大道、带遮盖的市集和立交桥组成了开罗这一地区标志性的城市结构，为开罗在21世纪的“新陈代谢”提供了解决方案。

Now located within a city of 17 million people, the urban structure of Khedivial Cairo was planned when the city's population consisted of only 350,000 people. The metabolism of Cairo has evolved, along with its structures of social and economic exchange, transportation, urban ecology, and its imagability—or comprehensibility. The 19th century structure of urban spaces requires adaptive reuse to address over development brought on by contemporary densities and movement systems. Our proposal suggests an ecological urbanism that reconciles the grand civic expressions of the 19th century urban system with the new energies of the functionalist city. This Gordian Knot of elevated highways, congested surface roads, extensive automobile parking and sprawling informal markets are integrated into a new open space system that optimizes the city's intensified energies. The DNA of 19th century urbanism is recombined with a 21st century verticalized urban order that manages an unprecedented magnitude of resource flows in four novel spaces: 1) an extended Opera Mall, 2) an Ataba Square Covered Market Plaza and Flyover Gateway, 3) the Hamdi Seif Al Nasr Multiway Boulevard, and 4) the new Great Lawn of Azbakiyya Garden. These four spaces deliver new ecological services combined with the expansion of urban services towards a new sustainable Cairo.

Opera Greensward

Opera Square is extended into a public greensward with new outdoor performance space, a gateway landmark, and an intermodal transit hub with underground parking. The proposed gateway structure, harboring a dot screen pattern of the former Opera Hall, reclaims the lost civic quality of the square's eastern edge. Housing a visitor overlook at its top, the gateway provides a new backdrop and staging for performance venues on the greensward and underground access to the metro and automobile parking underneath. Opera Greensward reestablishes a civic anchor recalibrated to the new social energies of a larger Cairo.

Ataba Square Covered Market Plaza and Flyover Gateway

Recognizing the necessity for flyover expressways to relieve surface traffic, a new "mat building" reorganizes the surface of Ataba Square as a covered pedestrian zone to accommodate spillover market activity. The Flyover Gateway atop the mat structure civilizes rogue highway engineering structures through the mat urban prototype that can be repeated throughout Cairo. This vertical urbanism shapes and shelters a climate-friendly pedestrian landscape underneath while proposing an automobile-scaled landscape above with green roofs and ventilation towers that function as urban lanterns. The mat prototype is an exemplary urban solution since it multiplies the urban surface (and shade), laminating functions of conflicting scales and speeds in the same space.

Hamdi Seif Al Nasr Multiway Boulevard

A new multiway boulevard integrates dedicated flow paths for various transportation modes involving bus, rail, auto, and pedestrian lanes. A cross grain of pedestrian bridges and tunnels establish new connections between Opera Greensward and Azbakiyya Garden. The vegetated multiway boulevard mitigates heat island effects while facilitating improved intermodal transportation services due to the relocation of the bus station and new tram service on the boulevard. This well-regarded street type introduces a resourceful and efficient traffic planning tool complementary to the urban planning vocabulary that defines Khedivial Cairo.

Great Lawn of Azbakiyya Garden

A new Great Lawn, terminating the axis of 26th of July Street, provides an urban refuge while organizing peripheral cultural programs at a more intimate scale. The northwest quadrant of the four-square lawn is punctuated by a large hill that overlooks Opera and Ataba Squares while making this green landscape visible to those outside of the gardens. The garden periphery is composed of fusion landscapes that celebrate both Arab and Western place-making traditions. Indeed, proposed water features, typical in the most celebrated Arab gardens, will function in tandem with a convective cooling infrastructure located beneath Azbakiyya Garden. The periphery gardens reestablish a well-defined and orderly transition between bustling streets and quiet gardens. The periphery absorbs the casual siting of the metro facilities and relocation of the book market to a shaded plaza at the western edge.

The greensward, lawn, multiway boulevard, covered market, and flyover gateway reconstitute a symbolic urban structure to this section of Cairo while solving for the new metabolism of a 21st century Cairo.

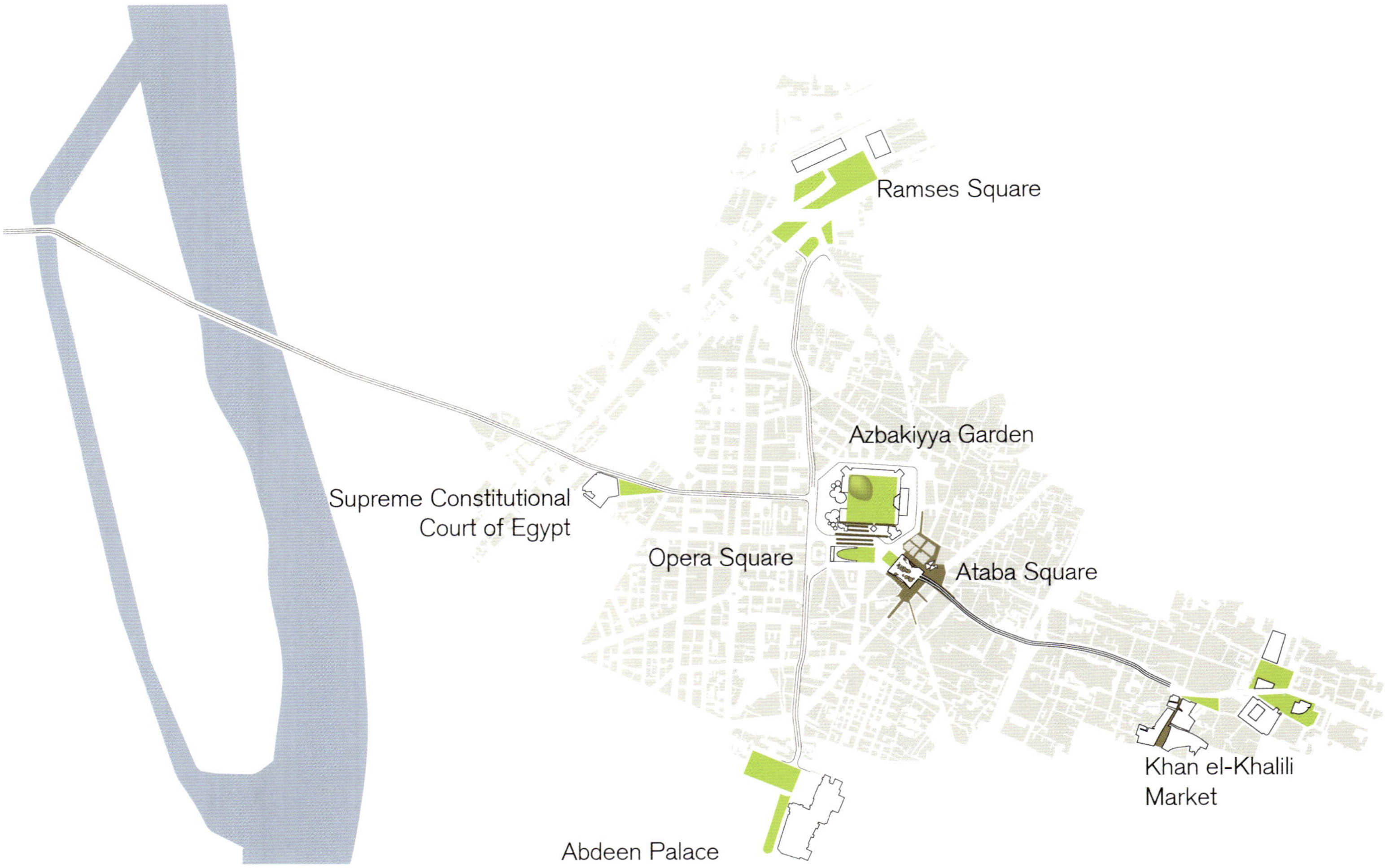

Ramses Square
Azbakiyya Garden
Supreme Constitutional Court of Egypt
Opera Square
Ataba Square
Khan el-Khalili Market
Abdeen Palace

tunnels and ventilation tower
Marker Pavilions
Ataba Square Covered Marker

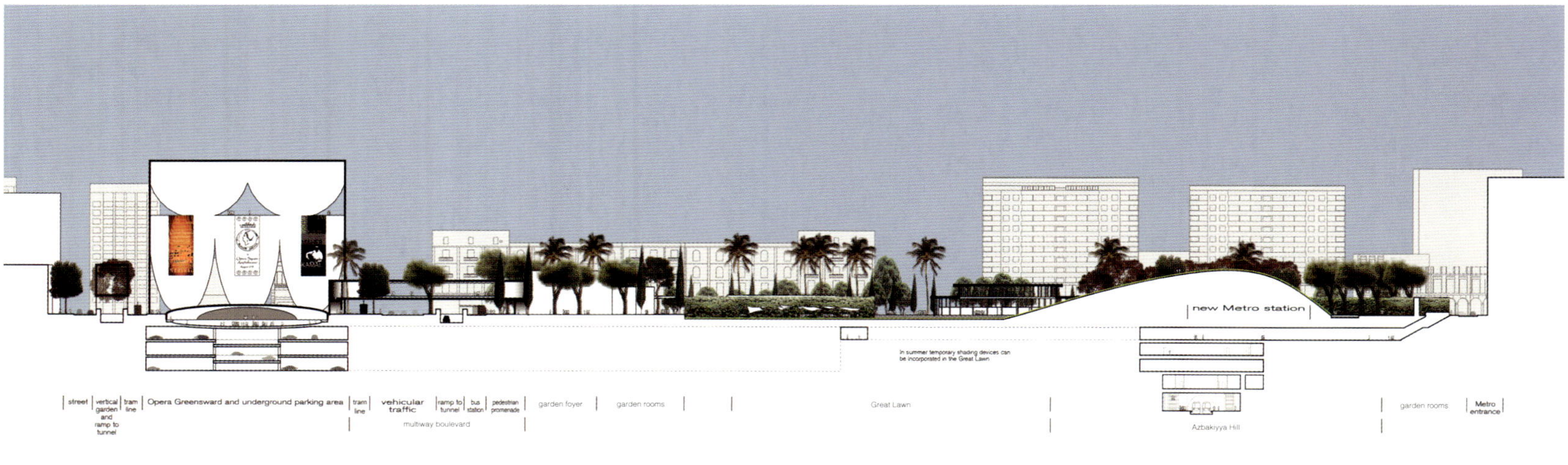

new Metro station
In summer temporary shading devices can be incorporated in the Great Lawn
street
vertical garden and ramp to tunnel
tram line
Opera Greensward and underground parking area
tram line
vehicular traffic
ramp to tunnel
bus station
pedestrian promenade
multiway boulevard
garden foyer
garden rooms
Great Lawn
Azbakiyya Hill
garden rooms
Metro entrance

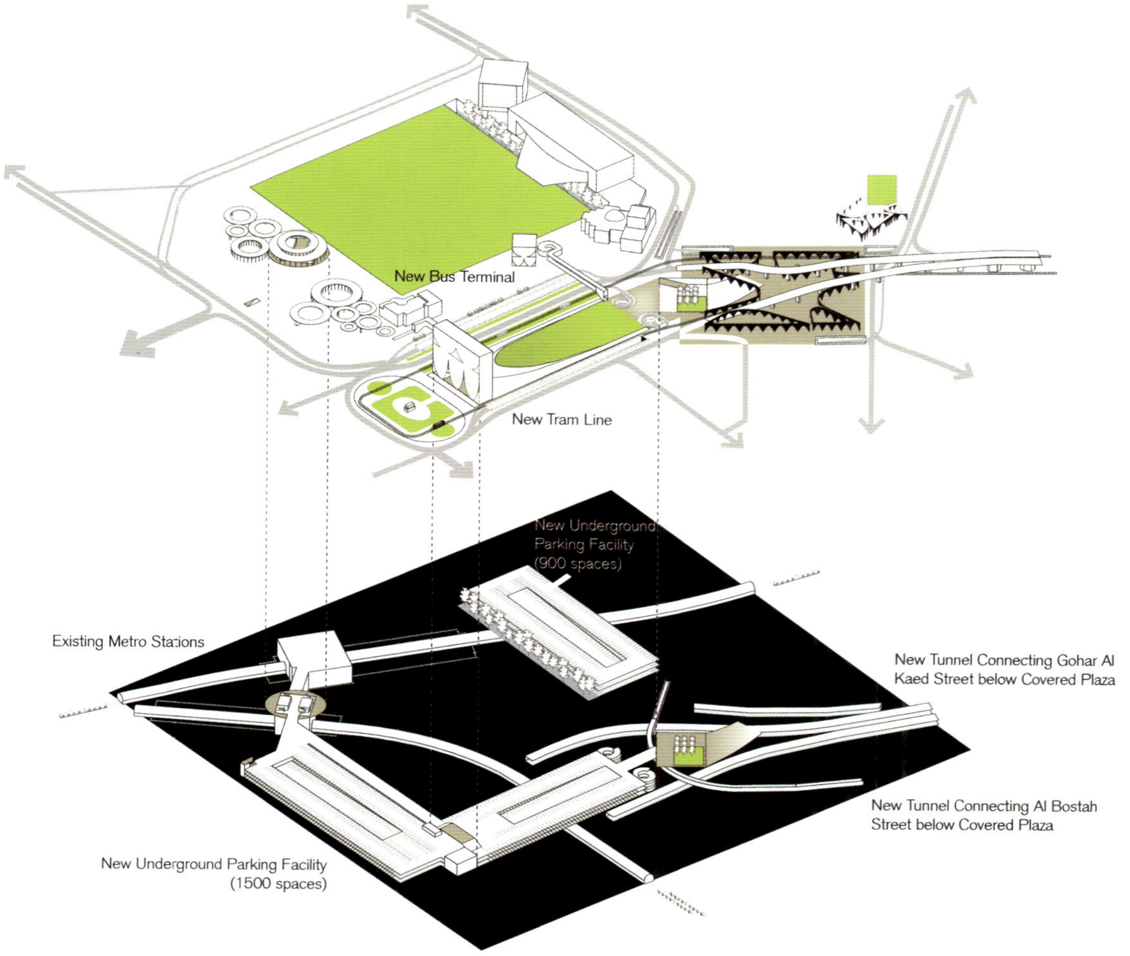
New Bus Terminal
New Tram Line
New Underground
Parking Facility
(900 spaces)
Existing Metro Stations
New Tunnel Connecting Gohar Al
Kaed Street below Covered Plaza
New Tunnel Connecting Al Bostah
Street below Covered Plaza
New Underground Parking Facility
(1500 spaces)

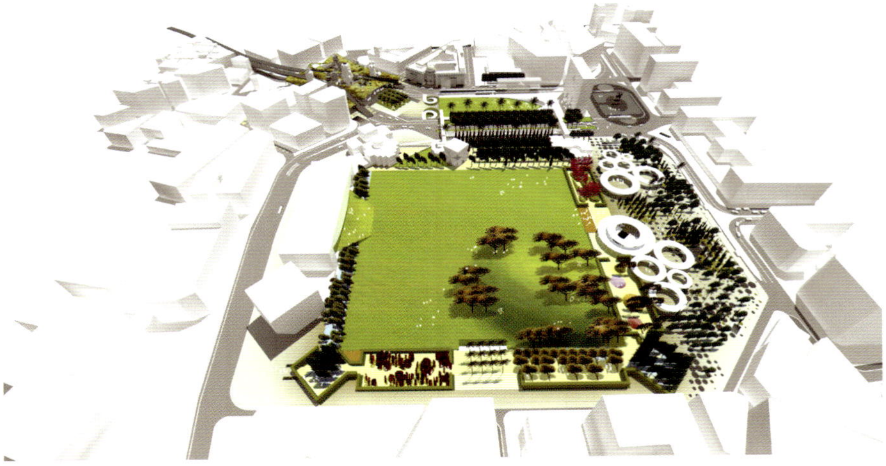

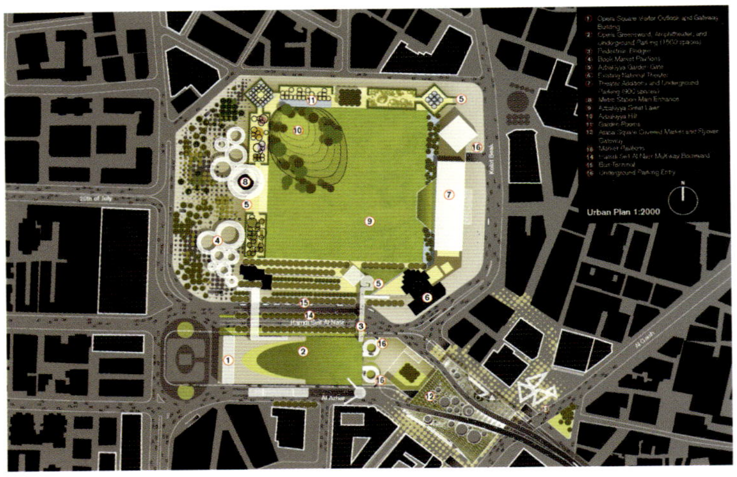
Urban Plan 1:2000

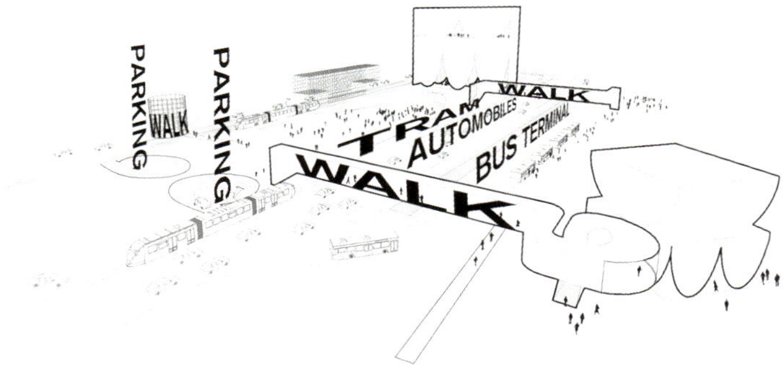
PARKING
PARKING
WALK
TRAM
AUTOMOBILES
BUS TERMINAL
WALK
WALK

ZRL Center Competition
ZRL 中心竞标方案 GAD

该地是位于商业区与住宅区之间的一个透明流畅的有机体——北面毗邻商业区，南面毗邻住宅区。

如此，设计师便把项目划分为商业或商业相关区、住宅或住宅相关区。设计师将办公楼与会议酒店设在商业与生活时尚酒店附近，老年住宅设在住宅区附近。将办公楼与会议酒店设在北面是合理的。同时，也是应ZRL要求，将办公楼设在邻近SH大楼的附近。在东北角处，也就是办公楼的附近，是会议酒店。办公楼南面，从高速公路E-5标志处可以清楚看到的位置，是生活时尚酒店的最佳位置——这个酒店将是这个综合体的优势建筑。老年住宅在该地块东侧，即生活时尚酒店东面。该地南面是欣赏博斯普鲁斯海峡的最佳位置——将占地35%的住宅区设在这里是最合适不过的。建筑群围绕环路排列，这样方便出行；如此一来，中央区域还可以建造花园或者购物广场。在地下沿施工线路建造一个服务型长廊，这样，方便每一个商店通往位于东南角的货物装卸台。将建筑群设在环路周围还有一个优势，那就是为街区与公寓楼的居民提供了露天式的购物娱乐中心与休闲吧。

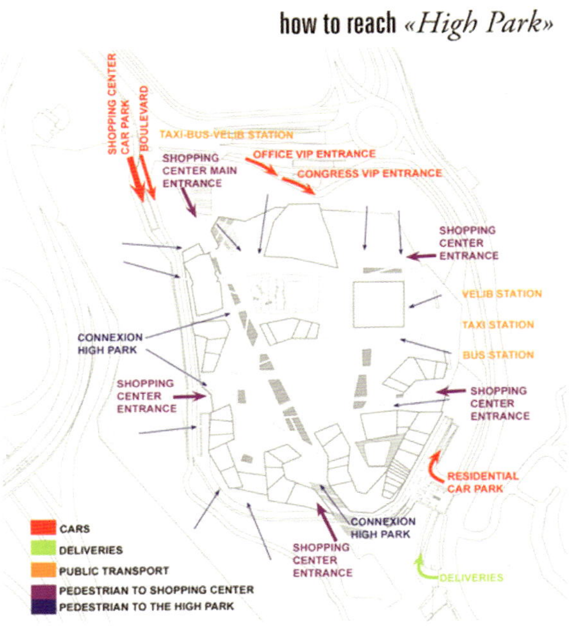

OVERALL PLAN

The site is located as a bridge between the business and housing district as transparent and fluent organism. North of the site is closer to the business district and south side is closer to residential district. In this case, we separate required program in two ways which are Business or business related and residential or residential related. As result, we have office building and conference hotel close to the business program and lifestyle hotel, elderly housing and residence close the residential program. It is a reasonable choice to place office and conference hotel to north of the site. Also it is required by ZRL, we positioned the Office building close to the SH building. Close to the Office building, and north east of the site, we placed conference hotel. South of the office building, most visible point from E-5 highway is most suitable place for lifestyle hotel. Lifestyle hotel will be one of the prestige buildings of complex. Elderly housing is sited the east of the lifestyle hotel, close to the east of the site. On the south of the site has the most view of Bosporus. It is a logical choice to put the residential units which has area of the 35% of the construction area. Buildings are placed to close the ring road for easy access. Locating building close on side of the ring road, center of site is suitable for garden and ground shopping. Creating a service corridor by around the permitted underground construction line, each store will have easy access to the loading dock which is placed in one location at the south east of the site. Another reason of why buildings are placed closed to the ring road is to create a wide open arena for shopping recreation and relax place by the blocks and apartments.

High Park

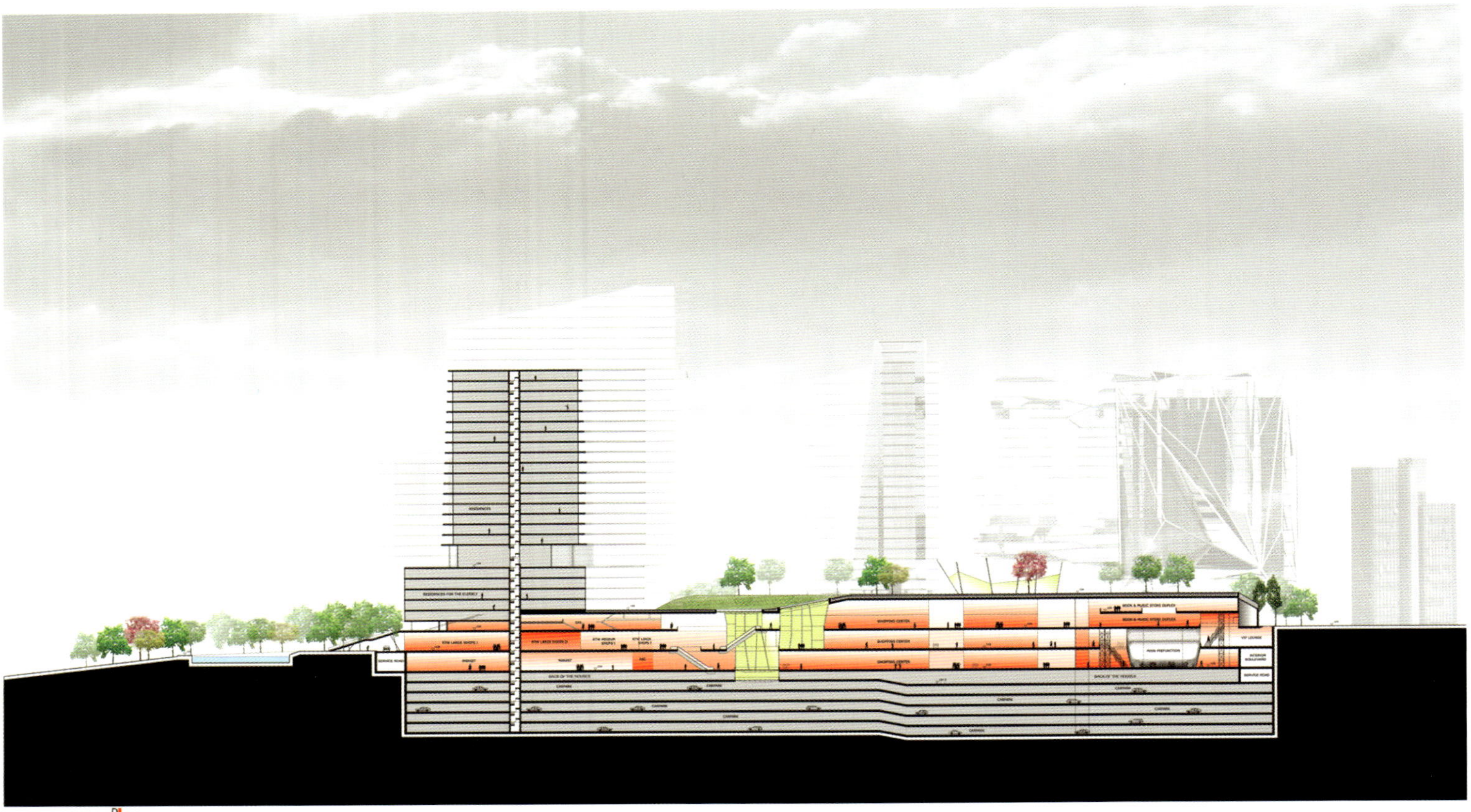

SITE SETION D/D

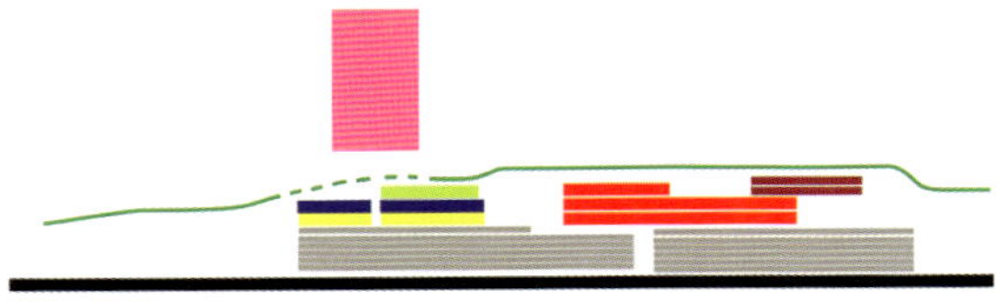

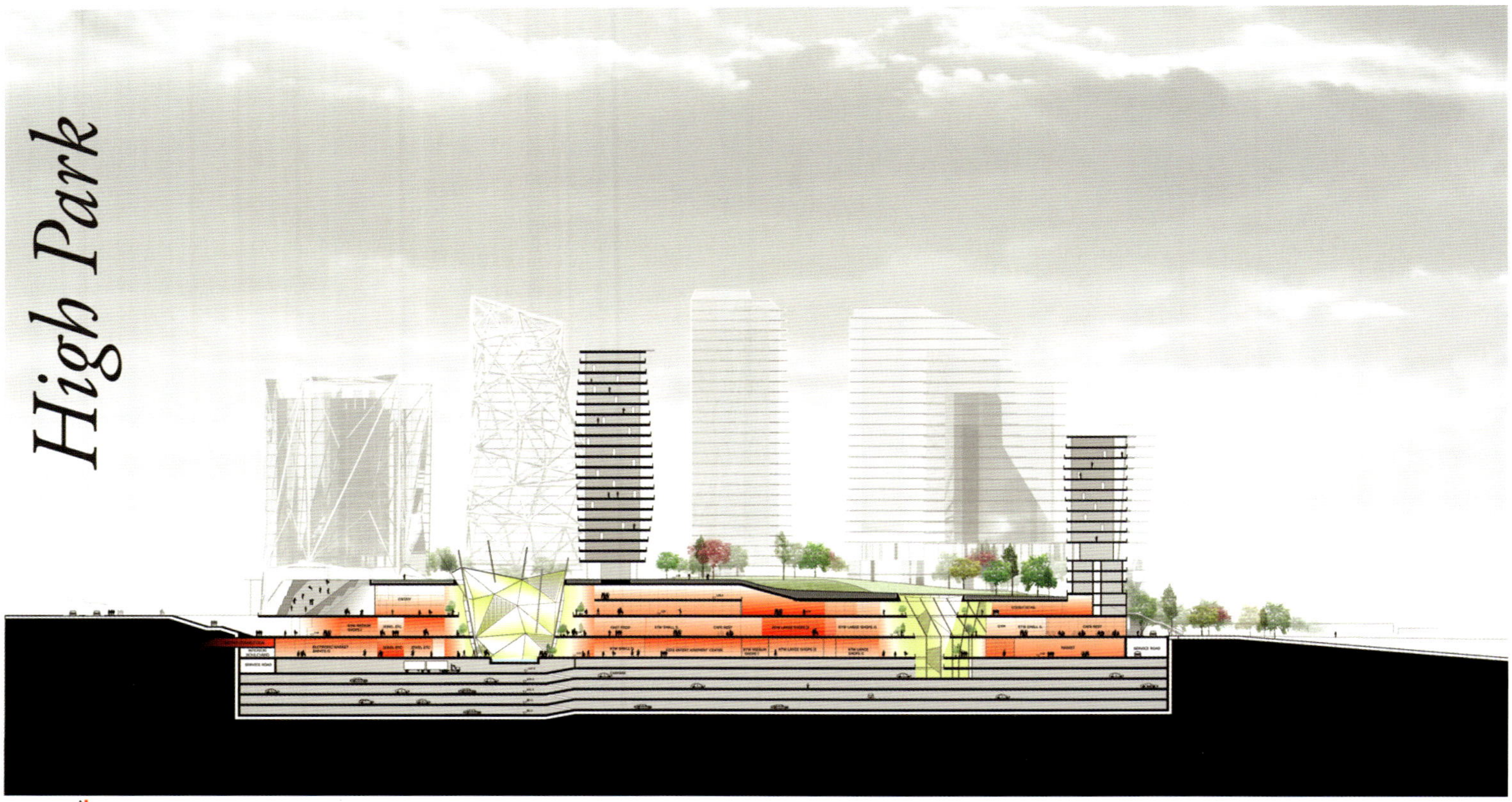

SITE SETION A/A

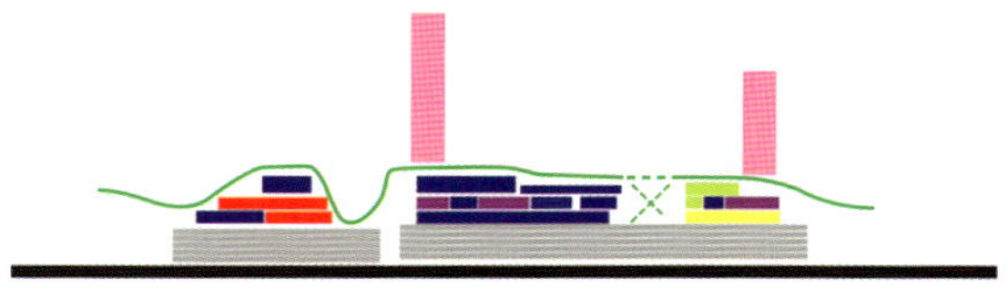

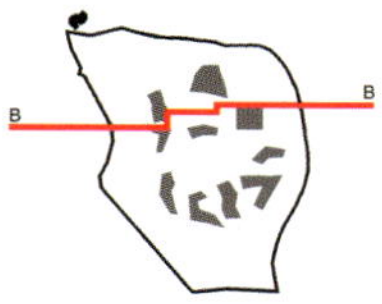

SITE SETION B/B

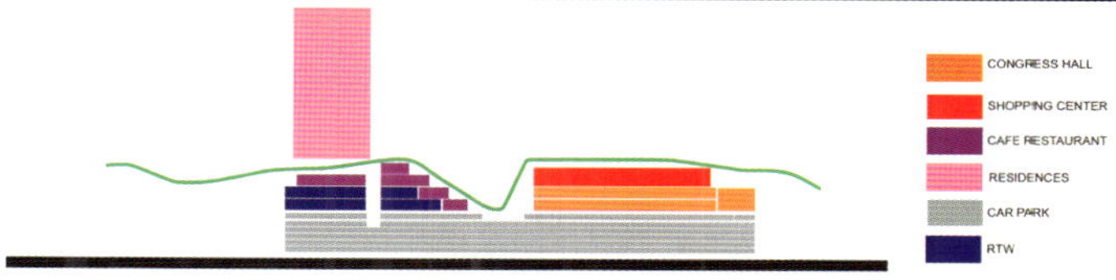

High Park

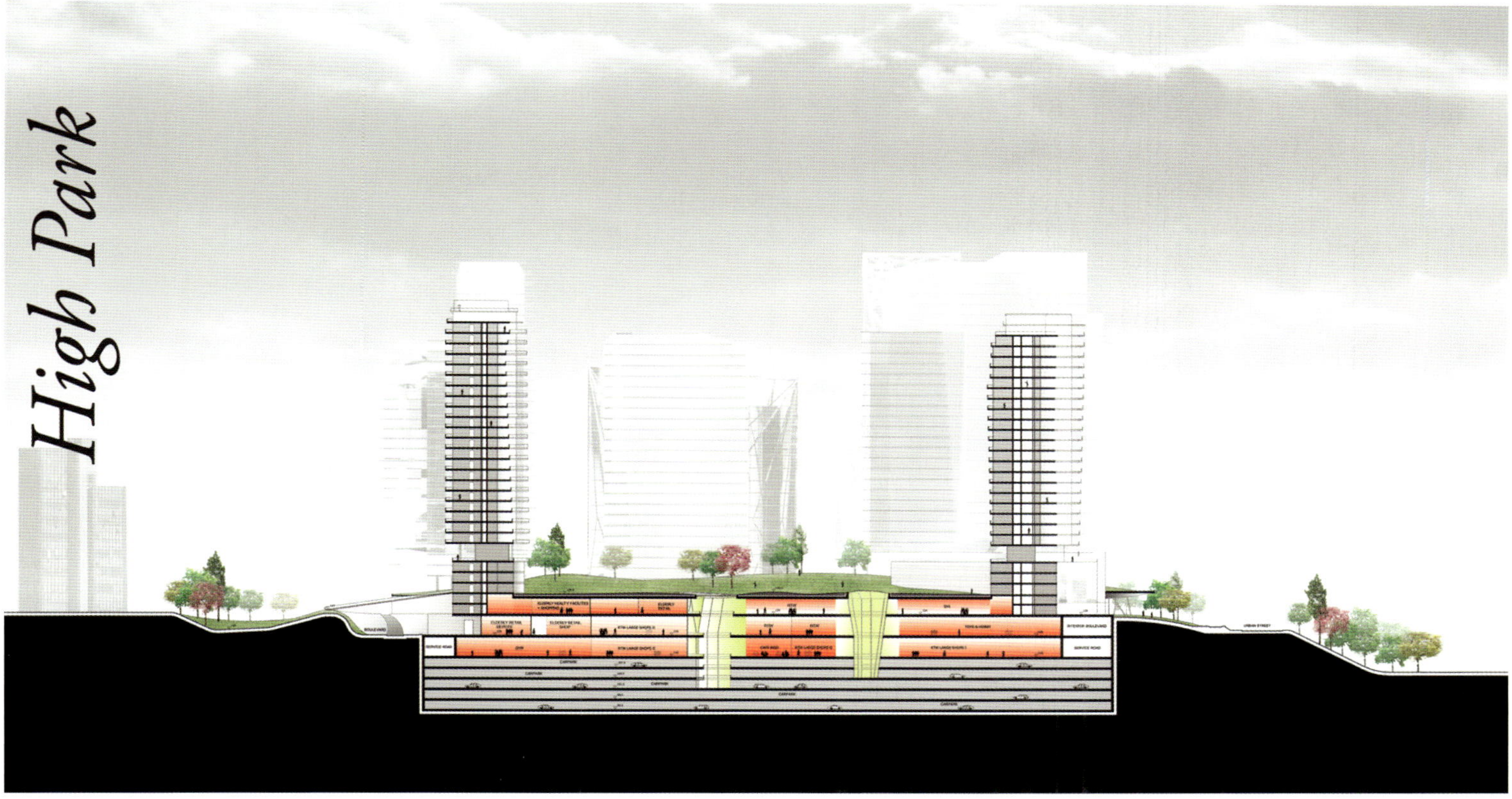

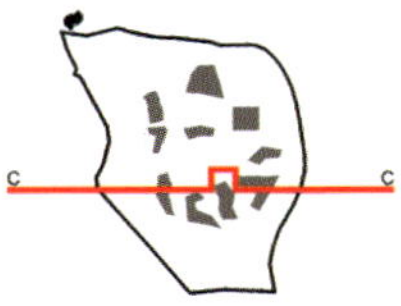

SITE SETION C/C

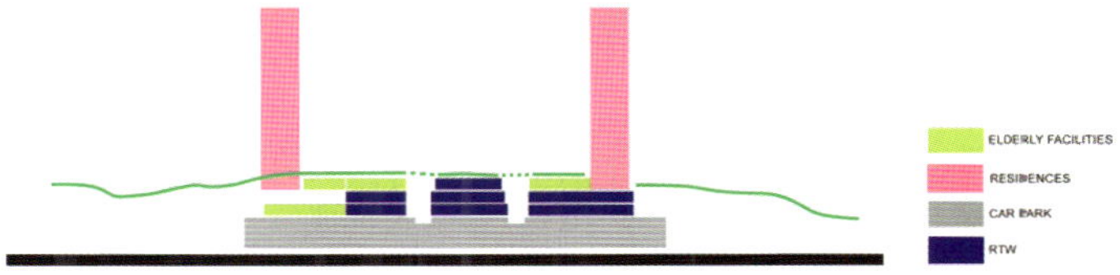

Ningbo Industrial Design & Creative Center Project

宁波工业设计创意街区竞赛方案

J. J. Pan and Partners, Architects and Planners

Credits

Location: Ningbo, Zhejiang Province, China
Design: 2007
Design Firm: J. J. Pan and Partners, Architects and Planners
Design Team: Joshua J. Pan, Jong-Yu Cheng, Pen Lee, Chia-Fang Chan, Yen-Chih Tseng, Chien-Yuan Liang, Chi-Wei Chen
Site Area: 78,000 m²
Total Floor Area: 147,770 m²(A+B1,2+C1,2+PARK)

宁波工业设计创意街区位于宁波甬江边上，孕育交织着宁波城市的过去未来、自然景观、人文历史之川流。

故事从过去工业发展的历史为起点，穿越历史轴线到江边的工业设计主题展场，往北转向纺织厂遗迹带出工业设计年表的景观轴线绿带，再向未来出发……动线回到宁波工业设计创意街区的三角制高点，象征的是追求卓越的决心及迈向国际舞台的愿景，在这个无形的历史回流里，从历史中看见的是宁波的未来，是一个高度的提升。园区南向的开放缓坡将人向上提升进入核心的国际设计展览中心，引领宁波工业设计产业与国际接轨。展览中心入口意象融合了历史的深度、滨江的景观及设计的精神。

黑色透明建筑体外冷内热，内部酝酿着充满无限生命力的半户外的中庭空间，提供的是一个交流的平台，同时也可以作为展览、休憩、设计的场所。各处办公空间里，可借由弹性的平面空间，空中花园绿意及中庭提供的双面采光，创造出具优质及创意的工作空间。

全区建筑群体的构架，以一个横向向上提升的设计，延续及呼应书城、渔轮的滨江天际线，带给宁波城市一个极具独特的都市景观价值。集中量体的开发，不仅提供给市民一个更大的绿带空间活化观光资源，亦提供一个未来发展的可能性。

宁波工业设计创意街区是一个与国际接轨的舞台，其创意及建筑设计不仅可丰富宁波的城市价值，更将这个极具江南之美的城市，推向国际舞台，展现独特的魅力及深度。

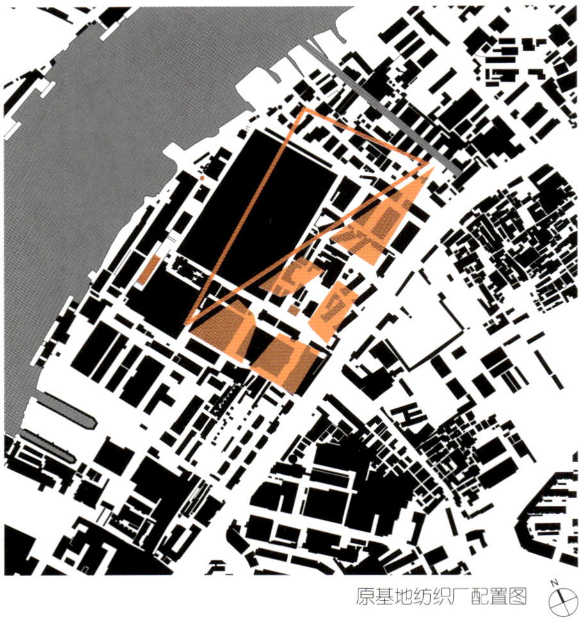

原基地纺织厂配置图

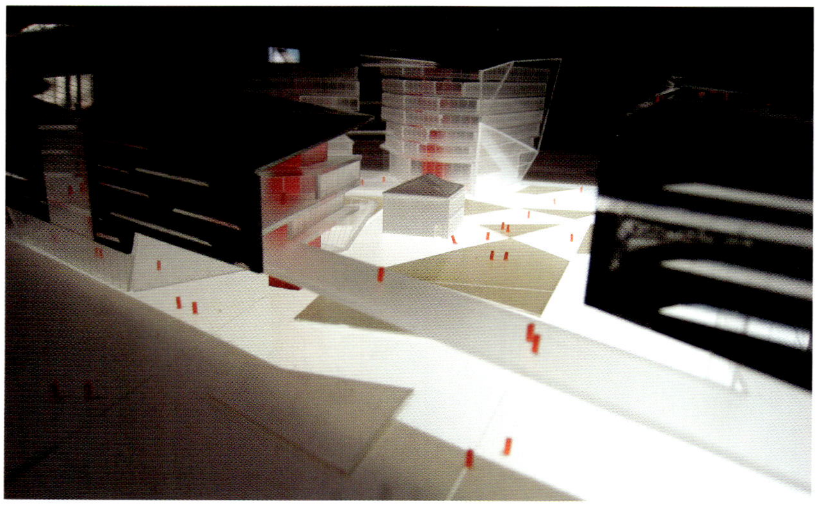

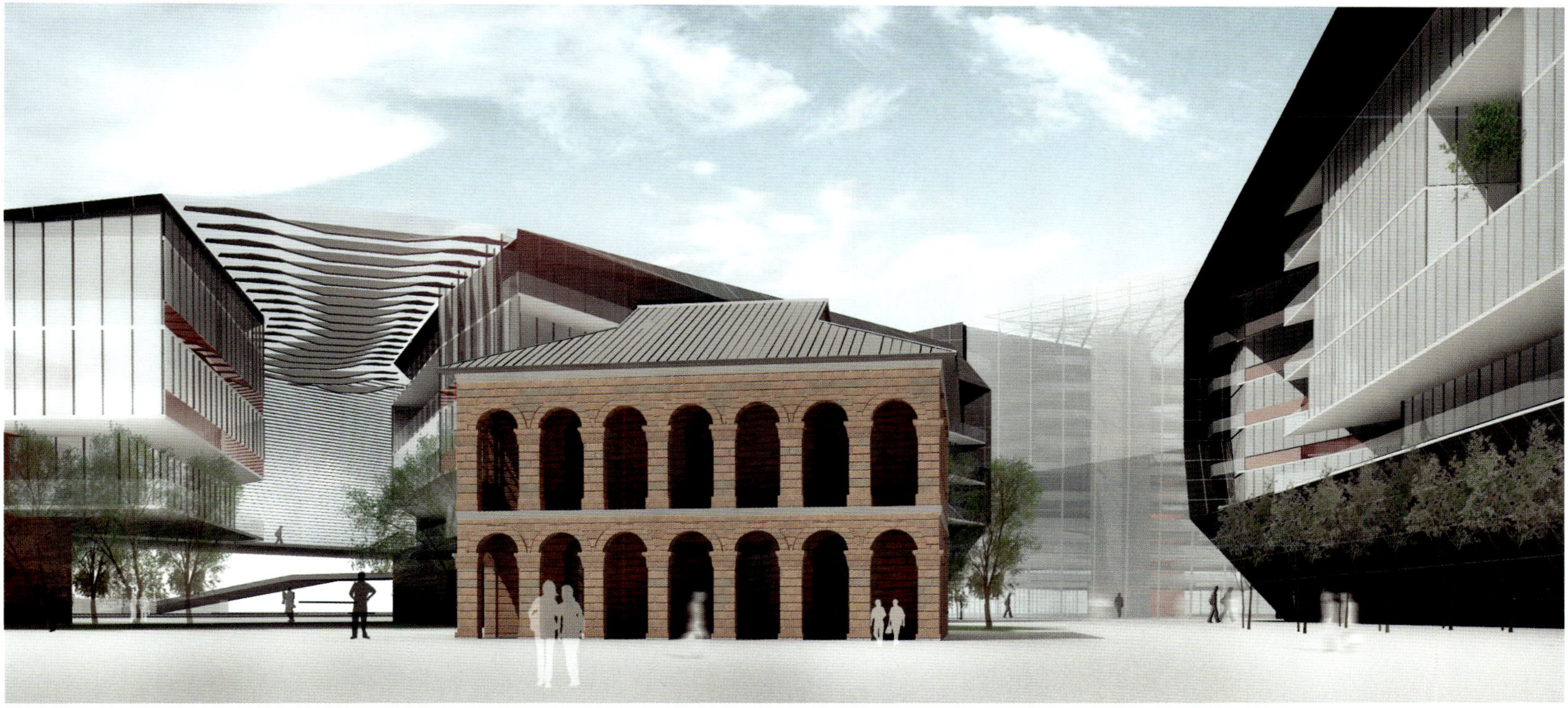

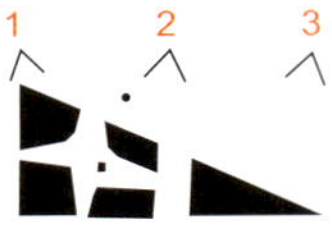
1
2
3

1

2

3

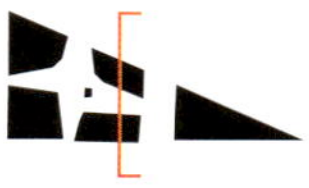

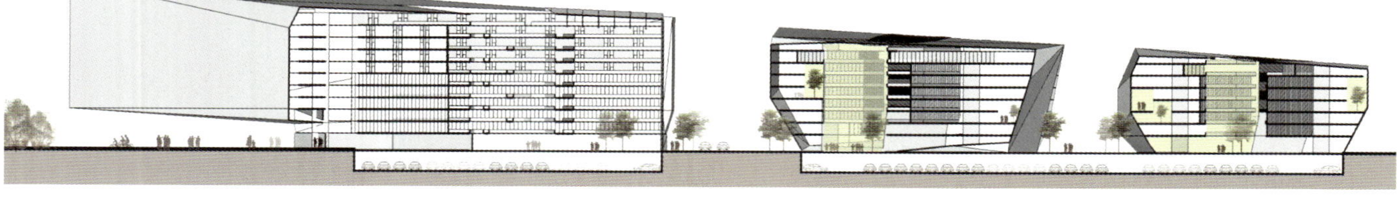

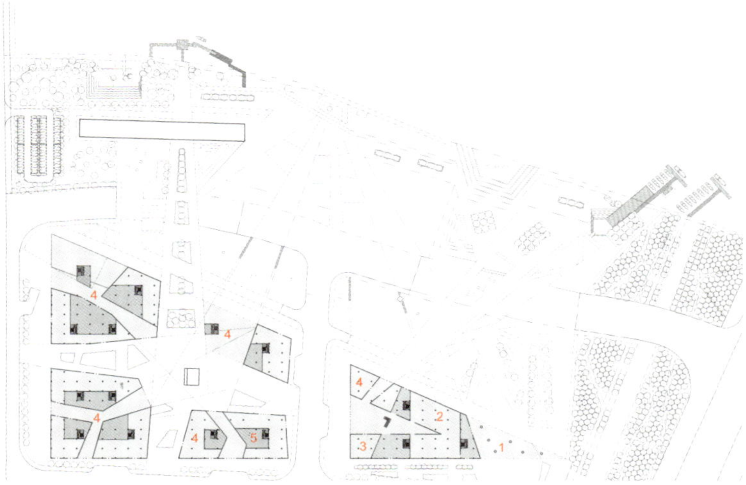

一层平面图

1 城市广场
2 展览空間
3 行政部门
4 零售商店
5 电梯大堂

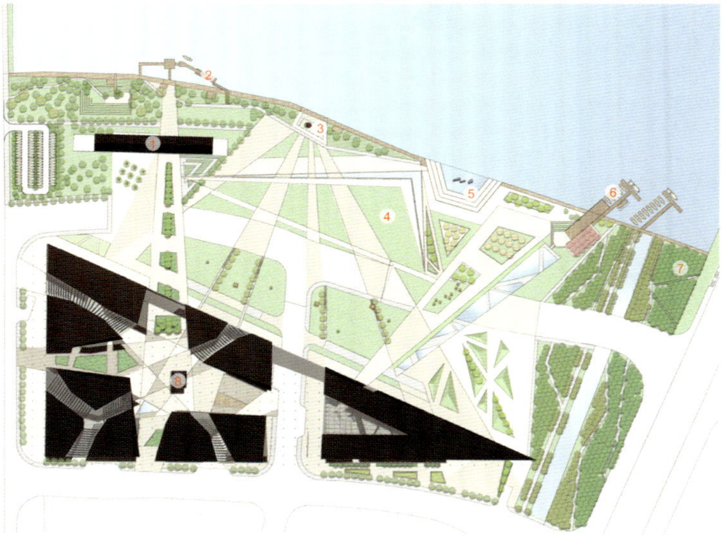

全区配置图

1 原有仓库
2 河岸步道
3 烟囱广场
4 户外展示区
5 亲子戏水区
6 码头
7 行人步道
8 原有纪念馆

主要展览中心

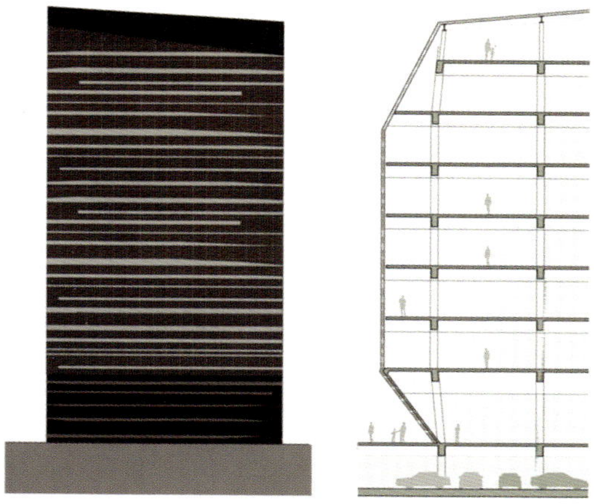

网点玻璃帷幕外观 剖面图

The Ningbo Creative Arts Park adopts one of the city's old textile plants along the Yong River. Prior to the acquisition of the land, preconceived decisions by the developer accompanied with massive demolition have left the site with almost no historical trace except for the residence of the previous owner, a landmark smokestack, and a row of abandoned warehouses along the waterfront. In late 2007, the city of Ningbo hosted a master planning competition in search for a solution to the revitalization of the precious property so as to reconnect the surrounding communities and to rejuvenate the urban riverfront spaces.

The proposed plan started with an idea to reintroduce buildings tracing the exact footprint of the historical factories on the site. Since an ordinary design approach would result in a massive construction, blocking visual and physical connections linking waterfront and the residential neighborhoods behind, the balance between the historical representation and the contemporary urbanity presents a major challenge to the design team. The design proposes a triangular building complex with segments of creative art studios, embracing a center courtyard built around the old mansion. The gesture of embracing historical context pays homage to the campus' past. Continuous ramps connect three separate studio buildings, leading to a crescendo at the triangle's tip, implying the pursuit of excellence. The building height steps down in gradual increments as the program gets closer to the river. The administrative office is strategically placed on the top floor of the triangular building to celebrate the project's pioneering status as a governmental institution.

The landscape design radiates out from the landmark smokestack, creating a secondary system stringing together the waterfront circulations, the river bank warehouses as well as restaurants and art shops in the park. The building's footprint symbolizes the major milestones of China's industrial development since the 19th Century, and the building's triangular tip leads to a focused façade, gesturing towards climactic grandeur.

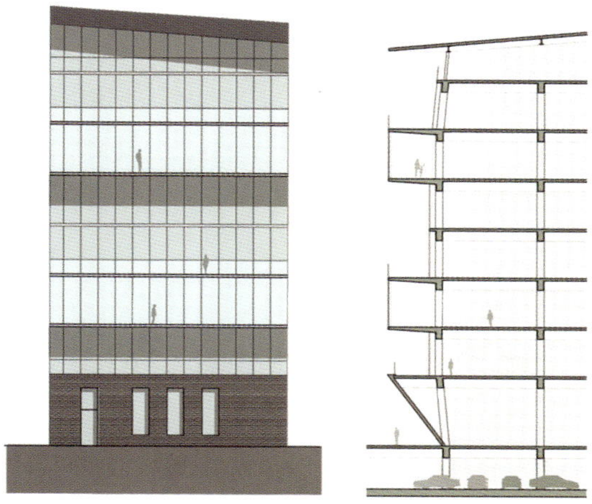

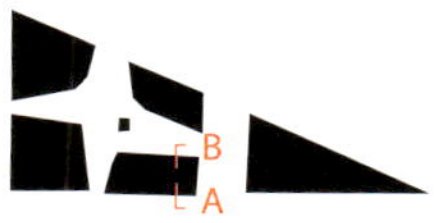

阳台外观/剖面图

图书在版编目（CIP）数据

建筑竞标方案 / 凤凰空间·上海编.
－南京 ：江苏人民出版社，2011.10
ISBN 978-7-214-07279-5

Ⅰ. ①建… Ⅱ. ①凤… Ⅲ. ①建筑设计－世界－现代－图集 Ⅳ. ①TU206

中国版本图书馆CIP数据核字(2011)第181000号

建筑竞标方案

凤凰空间·上海 主编

责任编辑：张晓华 王培娟 刘 焱
责任监印：彭李君
出 版：江苏人民出版社
发 行：天津凤凰空间文化传媒有限公司
销售电话：022-87893668
网 址：http://www.ifengspace.cn
集团地址：凤凰出版传媒集团（南京湖南路1号A楼 邮编：210009）
经 销：全国新华书店
印 刷：深圳当纳利印刷有限公司
开 本：1020毫米×1420毫米 1/16
印 张：26
字 数：208千字
版 次：2011年10月第1版
印 次：2011年10月第1次印刷
书 号：ISBN 978-7-214-07279-5
定 价：358.00元 (USD 65.00)
（本书若有印装质量问题，请向发行公司调换）